Spherical astronomy

ROBIN M. GREEN *Department of Astronomy, Glasgow University*

Spherical astronomy

CAMBRIDGE UNIVERSITY PRESS

Cambridge

London New York New Rochelle

Melbourne Sydney

Published by the Press Syndicate of the University of Cambridge
The Pitt Building, Trumpington Street, Cambridge CB2 1RP
32 East 57th Street, New York, NY 10022, USA
10 Stamford Road, Oakleigh, Melbourne 3166, Australia

First published 1985

Printed in Great Britain at the University Press, Cambridge

Library of Congress catalogue card number: 84-29208

British Library Cataloguing in Publication Data
Green, R.M.
Spherical astronomy.
1. Astronomy. Spherical and practical
I. Title
522′.7   QB145

ISBN 0-521-23988-5 hardcover
ISBN 0-521-31779-7 paperback

MP

CONTENTS

To Lynda, Nicola, Sally and Fern

PREFACE

For many years Smart's *Spherical Astronomy* has been regarded as a standard introductory text on positional astronomy. That book was first published in 1931, and, although there have been numerous revisions and updatings, the need for a new book at about the same level has been plain for some time. When I completed the most recent revision of Smart's text in 1977, the time was not appropriate for an entirely new book on the subject. Revisions were being planned in the *Astronomical Ephemeris* and in other almanacs prior to the introduction of a new system of astronomical constants. This transition, which was only completed in 1984, also involved the adoption of a new standard equinox and a new fundamental catalogue of the stars. Most important of all, relativity was to be recognized as fundamental to positional astronomy. Prior to this, it had been treated as a slightly esoteric addendum.

This book is intended as an undergraduate text; the methods used are mathematically simple. In this respect, it is similar to Smart's book, to which, naturally, I owe a considerable debt. Smart was also useful as a reference text for postgraduate students and research workers in other fields; it is hoped that this book will fulfil the same function.

I have been influenced by C. A. Murray's recently published book *Vectorial Astrometry*. While that book is intended primarily for the specialist and is considerably more advanced in a technical sense, its philosophy is similar to my own – general relativity must be introduced into astrometry at a fundamental level. Consequently, I have decided to treat the relationship of positional astronomy and general relativity at what is conceptually a fairly basic level. The methods of tensor calculus have been employed on occasion, and an Appendix has been included to explain the essentials of these methods to the reader who is not familiar with them. For the most part, however, the mathematical methods used involve spherical trigonometry and ordinary three-dimensional vector analysis.

As this book is primarily intended as an undergraduate text, it has been appropriate to include Problems at the end of the chapters. Some of these are of a numerical nature, and in such cases Answers have been included in an Appendix. Computation as such is not really part of the subject matter of the book; that has been admirably covered in *Practical Astronomy with Your Calculator* by P. Duffett-Smith. One word of warning, however; some of the numerical problems, particularly in the later chapters, may be more involved than they appear at first sight. It is assumed that at least a programmable calculator is available with which to undertake them. Such problems may be interpreted as the invitation to write a program employing a general procedure described in the text. In earlier chapters, where the procedures are more straightforward, worked examples have been included.

It is a pleasure to acknowledge the help and encouragement I have received from Professor Archie Roy. He read the entire manuscript and pointed out many errors and omissions. Those that remain are, of course, entirely my responsibility. At an earlier stage I received detailed help from Dr G. A. Wilkins, Superintendent of the Nautical Almanac Office, for which I am most grateful. I would also like to express my appreciation to Professor Peter Sweet and Dr Bernard Schutz with whom I have had helpful discussions, to Mrs L. Williamson for her competent, swift and considerate typing of the manuscript and to Gilbert Stewart for his timely help with the diagrams.

Robin M. Green

Glasgow 1984

ABBREVIATIONS

APFS	Apparent Places of the Fundamental Stars
AU	Astronomical unit
EHA	Ephemeris hour angle
EMS	Ephemeris mean sun
EST	Ephemeris sidereal time
ET	Ephemeris time
FK4	Fourth Fundamental Catalogue
FMS	Fictitious mean sun
GHA	Greenwich hour angle
GMST	Greenwich mean sidereal time
GST	Greenwich sidereal time
HA	Hour angle
H–R	Hertzsprung–Russell
IAU	International Astronomical Union
IF	Intermediate frequency
JD	Julian date
kpc	Kiloparsec
LSR	Local standard of rest
LST	Local sidereal time
MJD	Modified Julian date
Mpc	Megaparsec
NPD	North polar distance
pc	Parsec
PRF	Pulse repetition frequency
PZT	Photographic zenith tube
QSO	Quasistellar objects
RA	Right ascension
RF	Radio frequency
SBI	Short baseline interferometry
SI	Système International

TAI	International atomic time
TDB	Barycentric dynamical time
TDT	Terrestrial dynamical time
UMS	Universal mean sun
UT	Universal time
UTC	Coordinated universal time
VLBI	Very long baseline interferometry

1

Basic formulae

1.1 Introduction

Before the development of radar, there was no direct means of measuring the distance of any astronomical object. Even today the application of radar is severely limited. It is possible to measure the radar distances of some objects within the solar system, but this is not technically possible for more distant objects, like the stars, and is unlikely to be in the foreseeable future. The distances of the stars can only be inferred from slight periodic variations in their positions, variations that are caused by parallax. When the term position is used here, we mean the apparent direction in which the star is located. This is something that can be measured directly with great accuracy, and is neatly expressed as two angular coordinates. For positional purposes it is often convenient to ignore the distance of the star, since it is inaccessible to direct measurement, and to treat all the stars as if they were at the same distance, that is, as if they were situated on the surface of a sphere centred on the observer. This is what is meant by the 'celestial sphere'. Its radius is quite arbitrary, though presumably very large by any terrestrial standard. There is no loss of generality, however, if the radius of the celestial sphere is adopted as the unit of length in positional astronomy; indeed, this leads to considerable simplification.

1.2 Spherical geometry – great circle arcs

A sphere is defined as the surface whose points are all equidistant from a fixed point, the centre. The sphere is a two-dimensional surface which is finite but unbounded. Spherical geometry, meaning geometry performed on the *surface* of a sphere, is therefore, a two-dimensional geometry, but it differs significantly from the ordinary two-dimensional plane geometry of Euclid. In particular, there are no straight lines on the surface of a sphere. The equivalent curves are the arcs of great circles, which are defined as follows:

Definition. Any plane through the centre of the sphere intersects the sphere in a great circle. The *poles* of the great circle are the two extremities of the diameter of the sphere drawn perpendicular to that great circle.

It is immediately clear that a great circle is indeed a circle whose radius is the radius of the sphere, which we have taken as unity. A great circle AXB is shown in Fig. 1.1 and also the poles of this great circle, P and Q. These two points are said to be *diametrically opposite*. Another great circle is also drawn in Fig. 1.1, namely $PAQB$. It is easily seen, by rotating the plane of this great circle, that *any* great circle through P must also pass through the diametrically opposite point Q. This is a special property, however, of diametrically opposite points.

Consider, on the other hand, two general points of the sphere, such as A and X. The great circle on which they lie is uniquely determined. For the points A and X and the centre of the sphere O together define a unique plane, and this plane cuts the sphere in a great circle. The great circle arc AX is in fact the shortest curve that can be drawn on the sphere's surface to link these two points. This important property of great circle arcs means that

Figure 1.1

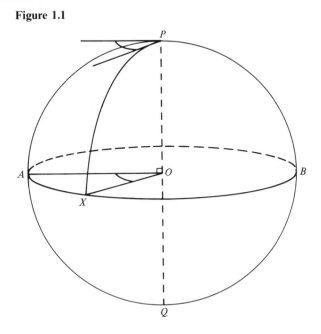

they are the geodesics of the sphere, analogous to straight lines in Euclidean geometry.

Since the sphere has unit radius, the following important conclusion is readily made: the length of a great circle arc is equal to the angle, in radians, that it subtends at the centre of the sphere. Although, strictly, radians should always be used, there is no ambiguity in practice in specifying the length of a great circle arc in degrees. For example, if we say that the great circle arc AX in Fig. 1.1 is $45°$, we mean that it subtends this angle at O and that its length is, therefore, $\frac{1}{4}\pi$. To be precise, there are two great circle arcs joining A and X, namely AX already discussed and the arc ABX measured in the opposite direction and of length $2\pi - AX$. When we speak of the great circle arc joining two points, however, we shall always intend the shorter of the two arcs, which is always less than π in length.

Now join PX (i.e. construct the great circle arc PX). Then PX and PA are two great circle arcs intersecting at P, and their intersection produces a spherical angle APX, which can be defined in several equivalent ways. For example:

Definition. The *spherical angle* between two intersecting great circle arcs is the angle between their planes.

Alternatively, the spherical angle may be defined as the angle between the tangents to the two intersecting great circle arcs at their point of intersection. With either definition, it is readily seen, from Fig. 1.1, that

Spherical angle $APX = A\hat{O}X$,

$$= \text{great circle arc } AX. \qquad (1.1)$$

Three points on a sphere which lie on the same great circle are analogous to three collinear points in plane geometry. Consider, however, three general points A, B, C which do not lie on the same great circle. Then, as shown in Fig. 1.2, these points may be joined in pairs by great circle arcs BC, CA, AB, each of which is less than π in length. The resulting figure is called a *spherical triangle*. Its parts consist of the three sides, which are the great circle arcs already mentioned, and the three included spherical angles. It will be convenient to denote each angle by the capital letter of its vertex and each side by the small letter corresponding to the opposite angle. Thus $BC = a$, $CA = b$, $AB = c$, as indicated in Fig. 1.2.

The three points A, B, C define a plane, and, since the three points do not lie on a single great circle, this plane does not pass through the centre of the sphere. We can construct a plane, however, passing through the sphere's centre parallel to the plane ABC. This latter plane divides the sphere into two hemispheres, and it is clear that the spherical triangle is confined to one of these hemispheres. It is then intuitively obvious that each of the angles of the spherical triangle is less than 180°.

Spherical triangles have certain properties in common with plane triangles. For example, any side is less than the sum of the other two. There are also important differences, however. In plane geometry the sum of the angles of a triangle is 180°. In a spherical triangle the sum of the angles is not fixed but always exceeds this value. A plane triangle may have one right angle, but only one. A spherical triangle may have one, two, or even three right angles. For example, referring back to Fig. 1.1, *PAX* is a spherical triangle in which angles *A* and *X* are both right angles. Trigonometrical formulae may be used to relate the parts of a spherical triangle. These will be developed in a later section, but it is useful to note one point at the outset. All parts of a spherical triangle lie between 0 and 180°; they are confined to the first two quadrants. The inverse cosine is single-valued in this range; the inverse sine is not. It is desirable, therefore, to use formulae that give the cosine rather than the sine of the part required, since, otherwise, an ambiguity inevitably exists.

Till now we have been concerned only with the geodesics of the sphere – great circles. Another curve of importance is defined as follows:

Definition. A plane that does not pass through the centre of the sphere intersects the sphere, if at all, in a *small circle*. The poles of the small circle

Figure 1.2

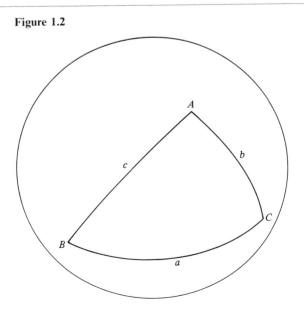

are the extremities of the diameter of the sphere that is perpendicular to the small circle's plane.

It is evident that a small circle is, indeed, a circle whose radius is less than that of the sphere, justifying the name. In Fig. 1.3, the small circle AB is shown, and also the parallel great circle CD. Both have the points P and Q as their poles. Let the radius of the small circle be r and let the great circle arc $AP = \theta$. Then from the plane triangle AOS, it is readily seen that

$$AS = AO \sin A\hat{O}S,$$

i.e.

$$r = \sin \theta. \tag{1.2}$$

Let E be any point on the small circle AB. Join the great circle arc PE and produce it to meet CD in F. Clearly $PE = \theta$. In fact, since all points on the small circle have the same separation on the sphere from the pole P, the small circle is the analogy of the circle in plane geometry. Let spherical angle $APE = \psi$. Then, equivalently, $C\hat{O}F = \psi$. Now, since AS and ES are, respectively, parallel to CO and FO, it follows that $A\hat{S}E = \psi$. The length of the small circle arc AE can now be derived as

$$\text{small circle arc } AE = rA\hat{S}E = \psi \sin \theta. \tag{1.3}$$

Figure 1.3

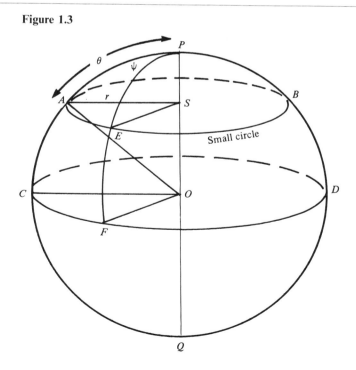

Notice that θ is the separation of the points on the small circle from their pole, and ψ is the spherical angle at the pole subtended by the arc in question.

The distance given in equation (1.3) is not, of course, the shortest distance on the sphere from A to E. The length of the *great* circle arc AE will be less than this. It is important to recognize that a small circle arc cannot be part of a spherical triangle. Certainly there is a spherical triangle APE, but it is not shown in Fig. 1.3. In particular, small circle arc AE must be replaced by the great circle arc, and, as a result, the angles at A and E will not be right angles as they appear in the diagram.

1.3   Spherical polar coordinates

A number of different coordinate systems can be set up on the celestial sphere. The systems used in practice are all basically similar and are just different forms of spherical polar coordinates.

Suppose a set of right-handed rectangular Cartesian axes $Oxyz$ are set up at the centre O of a unit sphere. Further, let the positive directions of these axes intersect the sphere in the points X, Y and Z, as shown in Fig. 1.4. Then the great circles XY and ZX represent the x–y and z–x planes respectively.

Figure 1.4

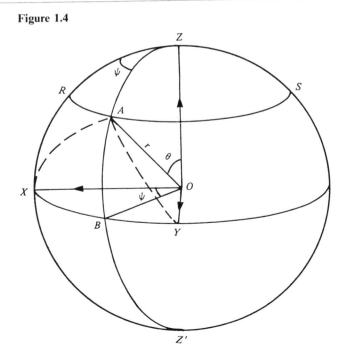

Let A be any point of the sphere with Cartesian coordinates (x, y, z). Then

$$x^2 + y^2 + z^2 = 1, \tag{1.4}$$

and so one of the coordinates is redundant. It is often more convenient, therefore, to employ spherical polar coordinates, (r, θ, ψ) say. With their usual definitions, the radial coordinate r of the point A is OA, the polar coordinate θ is $Z\hat{O}A$, and the azimuthal coordinate ψ is the angle between the plane ZOA and the z–x plane.

Since the point A is on the unit sphere, its radial coordinate $r = 1$. Moreover, its angular coordinates are simply related to the position of A on the sphere. From the definitions and results of the last section, it is immediately seen that the polar angle θ is the great circle arc length ZA, and the azimuthal angle ψ is the spherical angle XZA. To span the entire sphere, the coordinates θ and ψ must be in the ranges:

$$0 \leqslant \theta \leqslant \pi,$$
$$0 \leqslant \psi < 2\pi. \tag{1.5}$$

To set up a coordinate system on the celestial sphere, therefore, it is necessary to select a pole of the coordinate system Z, from which the polar angle θ is measured, and a reference great circle ZX from which the azimuthal angle ψ is measured. All the coordinate systems used in spherical astronomy are essentially of this form; their differences arise principally from a different choice of pole. Occasionally, it is true, a left-handed system may be used, and quite often the complement $(\frac{1}{2}\pi - \theta)$ is used rather than θ itself. Apart from these slight differences, however, the coordinate systems of spherical astronomy are all different examples of the spherical polar coordinates (θ, ψ) defined above. The coordinate grid is as follows: the curves $\theta = $ constant are small circles with pole Z, like RAS in Fig. 1.4, and the curves $\psi = $ constant are the semi-great circles like $ZABZ'$.

Despite the inherent redundancy, it can still be advantageous to use Cartesian coordinates. This often leads to a form of equations suitable for computation and allows the elegance of vector methods. Thus, if $\mathbf{i}$, $\mathbf{j}$ and $\mathbf{k}$ are unit vectors in the positive x-, y- and z-directions, then the location of A on the celestial sphere may be given by its position vector, $\mathbf{r}_A$, say, where

$$\mathbf{r}_A = x\mathbf{i} + y\mathbf{j} + z\mathbf{k}. \tag{1.6}$$

This is, of course, a unit vector, and (x, y, z) are the direction cosines of the line OA. So, in terms of great circle arcs,

$$x = \cos XA,$$
$$y = \cos YA, \tag{1.7}$$
$$z = \cos ZA.$$

Moreover, in terms of the spherical polar coordinates, we have the well-known transformations

$$x = \sin \theta \cos \psi,$$
$$y = \sin \theta \sin \psi, \tag{1.8}$$
$$z = \cos \theta.$$

Any problem in spherical astronomy may be treated by the methods of spherical trigonometry; alternatively, a three-dimensional vector approach may be adopted. The method chosen is largely a matter of personal preference. It has been assumed that the reader is already acquainted with basic vector methods, although, for reference purposes, essential vector formulae are summarized in Appendix A. The basic formulae of spherical trigonometry, on the other hand, are developed *ab initio* in the next section.

1.4   Spherical trigonometry – basic formulae

Let ABC be the spherical triangle shown in Fig. 1.5. If we adopt a spherical polar coordinate system (θ, ψ) with the point A as pole and the arc AB as the reference great circle, then the point B is given by $\theta = c, \psi = 0$, and the point C by $\theta = b, \psi = A$. Let $\mathbf{r}_B, \mathbf{r}_C$ be the position vectors of the points B and C respectively. Then, by equation (1.8), it follows that

$$\mathbf{r}_B = (\sin c, 0, \cos c), \tag{1.9}$$
$$\mathbf{r}_C = (\sin b \cos A, \sin b \sin A, \cos b). \tag{1.10}$$

Figure 1.5

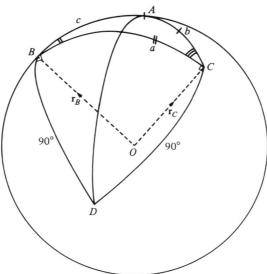

Now the angle between these two vectors is equal to the side BC of the spherical triangle. So, taking the scalar product, it follows that, since both are unit vectors, $\mathbf{r}_B \cdot \mathbf{r}_C = \cos a$. This scalar product can also be formed from equations (1.9) and (1.10) to yield the important result

$$\cos a = \cos b \cos c + \sin b \sin c \cos A. \tag{1.11}$$

This is the most fundamental formula of spherical trigonometry and will be referred to simply as the *cosine formula*. It is similar to the cosine formula for a plane triangle, and, like it, expresses one side of the triangle in terms of the other two sides and the included angle. There are two ready applications of equation (1.11). If the sides b and c and the included angle are known, it can be used as it stands to give the third side a. Alternatively, if all three sides are known, it may be used to derive the angle A. In addition to equation (1.11), companion formulae may be obtained by cyclical permutation of the symbols.

All the formulae of spherical trigonometry can be derived by successive applications of the cosine formula and subsequent trigonometric and algebraic manipulation. Two of these formulae can, however, be more directly obtained by considering the vector $\mathbf{r}_C \times \mathbf{r}_B$. Since the angle between these two unit vectors is equal to the arc BC, the vector product has a magnitude $\sin a$. Moreover, it will be directed towards a point on the sphere that is $90°$ from both C and B. This is the point D, shown in Fig. 1.5, which is the pole of the side BC of the original spherical triangle. If $\mathbf{r}_D$ is the position vector of the point D, then

$$\mathbf{r}_C \times \mathbf{r}_B = \sin a\, \mathbf{r}_D. \tag{1.12}$$

Now the left-hand side of this vector equation can be obtained from equations (1.9) and (1.10), giving the result

$$\mathbf{r}_C \times \mathbf{r}_B = (\sin b \cos c \sin A, \cos b \sin c$$
$$- \sin b \cos c \cos A, -\sin b \sin c \sin A). \tag{1.13}$$

Further, by equation (1.8), we may write the right-hand side of equation (1.12) as

$$\sin a\, \mathbf{r}_D = \sin a(\sin AD \cos BAD, \sin AD \sin BAD, \cos AD). \tag{1.14}$$

By equating the components of these last two equations, important new results are derived.

Consider spherical triangle BAD. Since D is the pole of BC, the arc $BD = 90°$ and it is perpendicular to BC. Therefore, the angle $ABD = 90° + B$. Applying the cosine formula to spherical triangle BAD then yields

$$\cos AD = \cos 90° \cos c + \sin 90° \sin c \cos (90° + B),$$

i.e.

$$\cos AD = -\sin c \sin B.$$

Substitute this result into the z-component of equation (1.14). Then equating z-components of equations (1.13) and (1.14) will yield

$$\sin b \sin c \sin A = \sin a \sin c \sin B,$$

or

$$\frac{\sin A}{\sin a} = \frac{\sin B}{\sin b}.$$

By symmetry, this last equation can be written in more complete form as

$$\frac{\sin A}{\sin a} = \frac{\sin B}{\sin b} = \frac{\sin C}{\sin c}. \qquad (1.15)$$

This result will be known as the *sine formula*. Its similarity to the sine formula in a plane triangle is evident and it suffers from the same inherent defect of ambiguity.

If the sine formula is now applied to the spherical triangle BAD, the following result is obtained

$$\sin AD \sin BAD = \sin BD \sin ABD = \cos B.$$

Now substitute this result into the y-component of equation (1.14), then equating y-components of (1.13) and (1.14) will give the important result

$$\sin a \cos B = \cos b \sin c - \sin b \cos c \cos A. \qquad (1.16)$$

This will be known as the *analogue formula*.

It will be found that the applications of the analogue formula are generally similar to those of the sine formula. It is, however, patently more complicated and perhaps slightly difficult to remember. On close inspection readers will probably discover for themselves some symmetry in the structure of the formula, but a visual aid to memory is given in Fig. 1.6. This

Figure 1.6

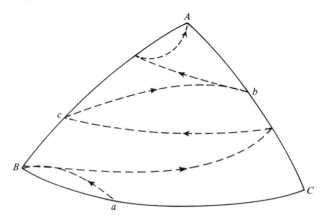

indicates the order in which the parts of the triangle occur in equation (1.16), namely $aBbcbcA$. If the roles of B and C (and, therefore, of b and c) are interchanged, a different version of the analogue formula is obtained,

$$\sin a \cos C = \cos c \sin b - \sin c \cos b \cos A. \qquad (1.17)$$

When all permutations are allowed for, triangle ABC will admit six cases of the analogue formula.

No further results can be derived from the vector equation (1.12). It is true that the x-component has not been considered, but this only leads to a restatement of the sine formula. The last of the important formulae that we wish to derive, the four-parts formula, may be deduced from the cosine and sine formulae as follows. Apply the cosine formula to triangle ABC (Fig. 1.5), obtaining

$$\cos b = \cos a \cos c + \sin a \sin c \cos B,$$

$$\cos c = \cos a \cos b + \sin a \sin b \cos C.$$

Eliminate $\cos c$ from the first equation, and replace $\sin c$ by using the sine formula. The result is that

$$\cos b = \cos a(\cos a \cos b + \sin a \sin b \cos C)$$

$$+ \sin a\left(\frac{\sin b \sin C}{\sin B}\right)\cos B,$$

or $\sin^2 a \cos b = \sin a \sin b(\cos a \cos C + \sin C \cot B)$. Finally divide throughout by $\sin a \sin b$, to obtain

$$\cos a \cos C = \sin a \cot b - \sin C \cot B. \qquad (1.18)$$

This is the *four-parts formula*. Referring to Fig. 1.5, it will be seen that the parts of the triangle involved in equation (1.18) are the four consecutive parts B, a, C, b. The side a and the angle C may be referred to as the 'inner side' and the 'inner angle' respectively. The four-parts formula may then be expressed in the general – and easily memorized – form as

cos (inner side) cos (inner angle)

= sin (inner side) cot (other side)

− sin (inner angle) cot (other angle).

When all the permutations are allowed for, it is found that there are six possible formulations of the four-parts formula in spherical triangle ABC. The four-parts formula is probably of more limited application than the other three, but is still a useful formula, particularly when right angles are involved. We conclude this section with a summary of the important results.

Summary. The basic formulae of a spherical triangle, used throughout the remainder of the book, are as follows:

(a) The cosine formula

$$\cos a = \cos b \cos c + \sin b \sin c \cos A. \tag{1.11}$$

(b) The sine formula

$$\frac{\sin A}{\sin a} = \frac{\sin B}{\sin b} = \frac{\sin C}{\sin c}. \tag{1.15}$$

(c) The analogue formula

$$\sin a \cos B = \cos b \sin c - \sin b \cos c \cos A. \tag{1.16}$$

(d) The four-parts formula

$$\cos a \cos C = \sin a \cot b - \sin C \cot B. \tag{1.18}$$

1.5   Terrestrial latitude and longitude

The concepts that have been introduced in previous sections are well illustrated with reference to the earth. As a first approximation, the earth may be regarded as a sphere spinning once per day about a fixed rotational axis. This axis intersects the earth's surface in the geographical north and south poles – the diametrically opposite points N and S in Fig. 1.7. The great circle which has the points N and S as its poles is termed the

Figure 1.7

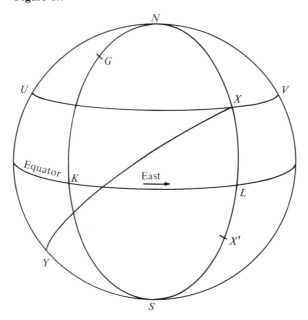

equator. Any semi-great circle perpendicular to the equator and terminating in the points N and S is called a *meridian of longitude* or simply a *meridian.*

The point N is a natural choice for the pole of a coordinate system on the surface of the earth. To specify the coordinate system completely, it is also necessary to adopt a reference great circle through N. This is the *prime meridian, NGKS* in Fig. 1.7. This selection, which is essentially arbitrary, was made by international agreement in the 19th century when the prime meridian was defined as that which passed through the principal positional telescope at the Royal Observatory, Greenwich. The position of any point X on the earth's surface is now determined by the great circle arc NX and the spherical angle GNX. The *latitude* ϕ and the *longitude* λ of this point are formally defined by the equations

$$\phi = 90° - NX \tag{1.19}$$

$$\lambda = GNX. \tag{1.20}$$

The arc NX is called the *colatitude* of the point X.

Extend the great circle arc NX to complete the meridian $NXLS$, intersecting the equator in L. All points on this meridian have the same longitude. Construct the small circle UXV through X which has N and S as its poles. All points on this small circle have the same latitude, and it is, therefore, termed a *parallel of latitude.* Clearly the parallels of latitude and the meridians of longitude provide a coordinate grid on the surface of the earth.

For a point on the equator, like L, the latitude is clearly zero. It follows from the definition (1.19) that if the point X were displaced south of the equator, to X', say, then ϕ would be negative. In practice, latitude and longitude are frequently written as positive with the additional specification of north or south and east or west. In that case, latitude varies from 0 to 90° N or S, longitude from 0 to 180° E or W. It is algebraically more satisfactory, however, to use equations (1.19) and (1.20) as they stand and to regard longitude as positive when measured in an anticlockwise direction as shown. Then ϕ and λ will cover the range

$$-90° \leqslant \phi \leqslant 90°$$

$$-180° < \lambda \leqslant 180°. \tag{1.21}$$

South latitude and west longitude are both then regarded as negative. For the point X in Fig. 1.7 both ϕ and λ are positive, while for Y they will both be negative.

We consider next the problem of calculating the distance between two points on the earth's surface – such as X and Y. As before, let the latitude and longitude of X be (ϕ, λ), and for Y let them be (ϕ', λ'). The shortest distance between these two points is the great circle arc XY which is a part of the spherical triangle NXY. In this triangle, $NX = 90° - \phi$, $NY = 90° - \phi'$. Moreover, $GNX = \lambda$ and, since λ' is negative in this case, $YNG = -\lambda'$. So $YNX = \lambda - \lambda'$. In fact, a little care is sometimes required to ensure that this spherical angle is between 0 and 180°. It does not affect the immediate purpose of calculating the separation XY. This is achieved by applying the cosine formula (1.11), to derive

$$\cos XY = \sin \phi \sin \phi' + \cos \phi \cos \phi' \cos (\lambda - \lambda'). \qquad (1.22)$$

The separation XY is thus calculated, but as an angle. To convert to kilometres, it is necessary to express this angle in circular measure and multiply by the radius of the earth which has until now been treated as unity. Alternatively, it is possible to express the separation in terms of the nautical mile, which is defined as the length of a great circle arc which subtends an angle of one arc minute at the earth's centre. The length of XY in nautical miles is, therefore, simply this arc XY expressed in arc minutes. The nautical mile is slightly in excess of the statute mile and is equal to 1.855 km.

1.6   Right ascension and declination

For the greater part of this book our concern will not be with the earth, but with the celestial sphere. The geometrical considerations, however, are very similar, and there is a coordinate system used on the celestial sphere which closely corresponds to terrestrial latitude and longitude. This is termed the *equatorial* system, and the two coordinates are known as *right ascension* and *declination*. It is by far the most important of all the coordinate systems in use in spherical astronomy, and just as a position on the earth is specified by its terrestrial latitude and longitude, so the position of a star on the celestial sphere is given by its right ascension and declination. These two coordinates are virtually fixed for each star; they are unaffected by the earth's diurnal rotation. Declination corresponds to terrestrial latitude, and right ascension to longitude east.

Imagine the celestial sphere with the earth situated at its centre, as in Fig. 1.8, and let us suppose that the radius of the celestial sphere exceeds the dimensions of the earth by a very large factor. Let N and S be the geographic north and south poles. Produce the earth's rotational axis NS to intersect the celestial sphere in P and Q. These two points are called the *north* and *south celestial poles*. The plane of the earth's equator will cut the celestial sphere in a great circle AB which is called the *celestial equator*. The coordinate grid of declination and right ascension which may be drawn on the celestial sphere is exactly similar to that of terrestrial latitude and longitude. In particular, we have *parallels of declination*, which are small circles parallel to the celestial equator, and *meridians of right ascension*, which are semi-great circles terminating in the celestial poles.

The right ascension α and the declination δ of a general point X on the celestial sphere are formally defined as

$$\delta = 90° - PX, \tag{1.23}$$

$$\alpha = \Upsilon PX, \tag{1.24}$$

where Υ is a fixed point on the celestial equator. This point, which defines the zero of right ascension, corresponds to the sun's position at the northern vernal equinox, around March 21, when the sun crosses the celestial equator from south to north. The definition of this point Υ, which is simply referred to as *the equinox*, will be made more fully in the next chapter. For the present it is sufficient to note that it is a fixed reference point on the celestial equator. Notice that right ascension increases in an easterly direction, as indicated by the arrow in Fig. 1.8. Moreover, on the celestial equator $\delta = 0$; while at the north and south celestial poles $\delta = +90°$ and $-90°$ respectively.

Although right ascension may be expressed in degrees and is then within the range $0 \leqslant \alpha < 360°$, it is normally expressed in time measure, equating

Figure 1.8

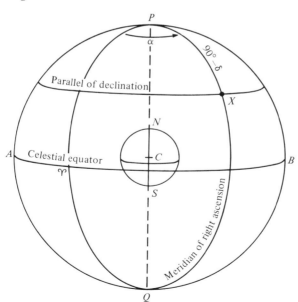

360° with 24 hours. The same convention is occasionally applied to longitude. The following relations apply; since

$$24^h = 360°,$$

we have

$$1^h = 15°, \quad 1^m = 15', \quad 1^s = 15''$$

and   (1.25)

$$1° = 4^m, \quad 1' = 4^s, \quad 1'' = \tfrac{1}{15}^s.$$

A system of *rectangular equatorial coordinates* may be introduced that is related to right ascension and declination. Let C be the centre of the celestial sphere. Then CP is chosen as the z-axis, $C\,\Upsilon$ as the x-axis, and the y-axis is chosen to complete a right-handed set. This means that it is in the direction $\alpha = 6^h, \delta = 0$. Equivalently to equation (1.8), the coordinates (x, y, z) of the point X are given by

$$x = \cos \delta \cos \alpha,$$

$$y = \cos \delta \sin \alpha,$$ (1.26)

$$z = \sin \delta,$$

where, as usual, the radius of the celestial sphere has been taken as unity. Alternatively, should some information be available about the distance of the celestial object, this can be incorporated in equation (1.26) by multiplying the right-hand sides by this distance.

1.7   The displacement of a star on the celestial sphere

Many problems in spherical astronomy involve the displacement of a star's position on the celestial sphere due to one of a number of causes. These displacements have a fairly standard form. For a particular case, the displacement will depend on the star's position but in such a way that the displacement is along the great circle joining the star to a definite fixed point on the celestial sphere. For example, due to annual parallax, the displacement is always directed towards the sun, while due to diurnal aberration, the displacement is towards the east point of the observer's horizon. The common features in the displacements allow them all to be treated as particular cases of the general result derived below.

Suppose that, due to some unspecified cause, the star $X\,(\alpha, \delta)$ is displaced by a *small* amount to a point X' and that the displacement is along the great circle joining X to the point O. Let O have equatorial coordinates (α_0, δ_0) and denote the arc OX by θ. Then XX' may be denoted by $d\theta$, and we are assuming that $d\theta$ is a small angle. The situation is illustrated in Fig. 1.9 – in which $d\theta$ is positive. Let us write the displacement as

$$d\theta = k \sin \theta.$$ (1.27)

In practice, it often transpires that k is either a positive or negative constant independent of the star chosen; but this is not necessary to the argument. Finally, let the equatorial coordinates of X' be $(\alpha + d\alpha, \delta + d\delta)$. As drawn in Fig. 1.9, both $d\alpha$ and $d\delta$ are positive.

Construct a small circle arc with pole P through X' to intersect PX in U. Now spherical angle $\Upsilon PX = \alpha$ and $\Upsilon PX' = \alpha + d\alpha$. So the angle subtended by the small circle arc at its pole, i.e. $UPX' = d\alpha$. Further, $PX' = PU = 90° - (\delta + d\delta)$. So, by equation (1.3),

$$UX' = d\alpha \cos (\delta + d\delta) = d\alpha \cos \delta,$$

to first-order accuracy. Moreover, since $PX = 90° - \delta$, it follows that

$$UX = d\delta.$$

Now, for convenience, denote the spherical angle OXP by χ. Then $UXX' = 180° - \chi$. The figure UXX' is not strictly a spherical triangle, but, since its dimensions are small compared with the celestial sphere, it may approximately be treated as a plane triangle, right-angled at U. Then,

$$UX = XX' \cos (180° - \chi) = - XX' \cos \chi,$$

and

$$UX' = XX' \sin (180° - \chi) = XX' \sin \chi.$$

Figure 1.9

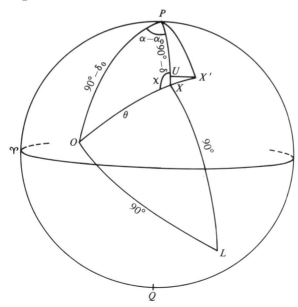

By our earlier two results and (1.27), it follows that

$$\cos \delta \, d\alpha = k \sin \theta \sin \chi,$$

$$d\delta = -k \sin \theta \cos \chi.$$

(1.28)

Consider now spherical triangle OPX. Since $\Upsilon PX = \alpha$ and $\Upsilon PO = \alpha_0$, it follows that $OPX = \alpha - \alpha_0$. Moreover, $PX = 90° - \delta$, $PO = 90° - \delta_0$, $OX = \theta$, and $OXP = \chi$. Now using the sine and analogue formulae, (1.15) and (1.16), it is found that

$$\sin \theta \sin \chi = \sin (90° - \delta_0) \sin (\alpha - \alpha_0),$$

$$\sin \theta \cos \chi = \cos (90° - \delta_0) \sin (90° - \delta)$$

$$- \sin (90° - \delta_0) \cos (90° - \delta) \cos (\alpha - \alpha_0).$$

Hence equations (1.28) give the final result

$$d\alpha = k \sec \delta \cos \delta_0 \sin (\alpha - \alpha_0),$$

$$d\delta = k(\sin \delta \cos \delta_0 \cos (\alpha - \alpha_0) - \cos \delta \sin \delta_0).$$

(1.29)

To make use of this general result in any particular case, it is only necessary to insert the value of k and the coordinates of the point O. It is important to recognize that these equations only apply for a small displacement. So, if (1.29) is expressed in radians, then k is a small quantity, and it must be remembered that in deriving these equations terms $O(k^2)$ were neglected. Where large displacements are involved, it is advisable to proceed by an *ad hoc* method.

The general results (1.29) can be expressed more neatly, but probably less usefully, in vector form. Let $\mathbf{s}$ be the position vector of X and $\mathbf{s}_0$ that of the point O. Both are unit vectors; so the vector $\mathbf{s} \times \mathbf{s}_0$ has magnitude $\sin \theta$ and is directed towards a point L on the celestial sphere $90°$ from both O and X – L is a pole of the great circle OX. Let X' have position vector $\mathbf{s} + d\mathbf{s}$. Since $\mathbf{s}$ is a unit vector, $\mathbf{s} \cdot \mathbf{s} = 1$, and, taking differentials, yields, for any small displacement,

$$\mathbf{s} \cdot d\mathbf{s} = 0.$$

(1.30)

The vector $d\mathbf{s}$ is, therefore, perpendicular to $\mathbf{s}$. Moreover, since the displacement is along OX, it is also perpendicular to $\mathbf{s} \times \mathbf{s}_0$. So, making allowance for the correct handedness, $d\mathbf{s}$ will be in the direction $\mathbf{s} \times (\mathbf{s} \times \mathbf{s}_0)$. Further, it is easily seen that the magnitude of this triple vector product is $\sin \theta$. Hence, by equation (1.27),

$$d\mathbf{s} = k\mathbf{s} \times (\mathbf{s} \times \mathbf{s}_0).$$

(1.31)

This form is more general than equations (1.29), for it applies in any coordinate system, not just for equatorial coordinates. In practice, it must be used in scalar form, and the scalar form in equatorial coordinates is exactly equations (1.29). The justification of this last statement is left as an exercise for the reader.

It is usually convenient to work in radians when carrying out analysis. In applying the results, however, more practical units are required, and many small angles will be expressed in arc seconds. Now, approximately,

$$1 \text{ radian} = 57°\,17'\,45'' = 206\,265''. \tag{1.32}$$

This last number is, therefore, an important conversion factor. One advantage of radians is that small-angle approximations may be used. For, if θ is a small angle expressed in circular measure, then

$$\sin\theta \approx \theta, \quad \cos\theta \approx 1, \quad \tan\theta \approx \theta. \tag{1.33}$$

Then it follows from equation (1.32) that

$$\sin 1'' = \tfrac{1}{206\,265}. \tag{1.34}$$

This provides a neat way of writing the conversion factor. For example, if θ'' is the number of arc seconds in the small angle θ, then, by equations (1.32) and (1.33),

$$\sin\theta = \theta'' \sin 1''. \tag{1.35}$$

Now suppose that in applying equations (1.29), arc seconds are to be used. To avoid confusion, let k be the value of that parameter in radians and k'' in arc seconds. Then, since the same conversion factor is to be applied to each side of the equations, $d\alpha$ and $d\delta$ are readily obtained in arc seconds by replacing k by k''. The equations are, however, accurate only to first order, and it is clear that there will be an error, ε say, where, in radians,

$$\varepsilon = O(k^2) = O((k'' \sin 1'')^2).$$

Hence, in arc seconds, the error is

$$\varepsilon'' = O(k''^2 \sin 1''). \tag{1.36}$$

This gives a useful guideline to accuracy. For example, for a displacement itself of order one arc second, we may expect an error in first-order formulae of order 5×10^{-6} arc seconds, which is totally negligible in all circumstances. On the other hand, for a displacement even as small as $15''$ or one time second, the error is of order $0''.001$, which does not quite match the accuracy of the most refined astrometric measurements, while for a displacement of one arc minute the error will be of order $0''.02$. Although the accuracy of some optical and radio observations is an improvement on this, many predictions, for example, in the *Astronomical Almanac*, do not claim a superior accuracy. The conclusions must be that, while first-order formulae are very useful, they must be used with an awareness of their limitation, and this awareness should be quantitative.

Finally, we express the general result (1.29) in practical units. If the parameter k is expressed in arc seconds, then $d\alpha$ in time seconds and $d\delta$ in arc seconds are given by

$$d\alpha^s = \tfrac{1}{15}k'' \sec \delta \cos \delta_0 \sin (\alpha - \alpha_0),$$
$$d\delta'' = k''(\sin \delta \cos \delta_0 \cos (\alpha - \alpha_0) - \cos \delta \sin \delta_0). \qquad (1.37)$$

Problems

1.1 Prove that the value of a spherical angle on the unit sphere is equal to the separation on the sphere's surface between the poles of the two great circle arcs forming the angle.

1.2 For any spherical triangle ABC, the *polar triangle $A'B'C'$* is defined as follows: A' is the pole of the side BC chosen so that $AA' < 90°$, and B' and C' are similarly defined. Prove that sides and angles in the two triangles are related by the formulae

$$A' = 180° - a, \quad B' = 180° - b, \quad C' = 180° - c.$$
$$a' = 180° - A, \quad b' = 180° - B, \quad c' = 180° - C.$$

1.3 Prove the *principle of duality* which states that, from any general formula involving the parts of a spherical triangle, a dual formula may be derived by replacing each side by the supplement of the opposite angle and each angle by the supplement of the opposite side. Show that the sine and four-parts formulae are self-dual.

1.4 Prove that the following formulae apply in any spherical triangle

$$\cos A = \sin B \sin C \cos a - \cos B \cos C,$$
$$\sin A \cos b = \cos B \sin C + \sin B \cos C \cos A.$$

1.5 A and B are two places on the earth's surface with the same latitude ϕ; the difference of longitude between A and B is $2l$. Prove that (i) the highest latitude reached by the great circle AB is $\tan^{-1}(\tan \phi \sec l)$, and (ii) the distance measured along the parallel of latitude between A and B exceeds the great circle distance AB by

$$2 \operatorname{cosec} 1' [l \cos \phi - \sin^{-1}(\sin l \cos \phi)] \text{ nautical miles.}$$

[Smart (1977)]

1.6 An aircraft leaves Lima (12° 10′ S, 77° 05′ W) and flies directly to Rome (41° 53′ N, 12° 33′ E). Calculate the distance travelled in nautical miles and the longitude at which the aircraft crosses the equator.

1.7 The most southerly latitude reached by the great circle joining a place A on the equator to a place B in south latitude ϕ is ϕ_1. Prove that the difference in longitude between A and B is $90° + \cos^{-1}(\tan \phi \cot \phi_1)$.

[Smart (1977)]

1.8 The spherical triangle ABC is subjected to a small distortion; show that the resulting increments in the parts of this triangle are related by the equations

$$da = \cos C \, db + \cos B \, dc + K \sin b \sin c \, dA,$$

$$dA = -\cos c \, dB - \cos b \, dC + K^{-1} \sin B \sin C \, da, \quad \text{etc.,}$$

where K is the ratio

$$K = \frac{\sin A}{\sin a} = \frac{\sin B}{\sin b} = \frac{\sin C}{\sin c}.$$

1.9 Prove that, in an equilateral spherical triangle, the sides and angles satisfy the condition that

$$\sec A - \sec a = 1.$$

1.10 Compute the length of the most direct air route from San Francisco (37° 40′ N, 122° 25′ W) to Tokyo (35° 48′ N, 139° 45′ E). Determine the direction in which the aircraft must depart from San Francisco and calculate the latitude and longitude of the most northerly point of the route.

2

The celestial sphere

2.1   Introduction

The celestial sphere has been introduced as a sphere of unit radius on which it is convenient to imagine all celestial objects are located. It was stated that the centre of the sphere was the observer – who will generally be situated on the surface of the earth. The positions of celestial objects will differ slightly from one observing site to another, and may vary periodically, at any one observing site, due to the earth's diurnal rotation and annual revolution round the sun. It will be convenient, therefore, to introduce standard viewpoints corresponding to hypothetical observers situated at the centre of the earth and at the centre of the solar system, effectively the sun. The celestial spheres corresponding to these two standard viewpoints will be referred to as the *geocentric* and the *heliocentric* celestial spheres respectively. On the other hand, when an actual observer is regarded as central, one speaks of the *topocentric* celestial sphere. The coordinates of a celestial object will be referred to as the topocentric, geocentric or heliocentric coordinates according to the centre of the celestial sphere that has been selected.

The differences between these coordinates arise from the effects of parallax and aberration and they will be discussed, in basic principle, in the next chapter, and, in computational detail, in chapters 4 and 8. For stars at least, the differences are small, and they will be largely ignored in the present chapter. The purpose of this chapter is to introduce the various coordinate systems that are used in positional astronomy and to explain the conventions connected with them. In addition to the equatorial system already introduced, we shall describe the alt-azimuth system and the systems of ecliptic and galactic coordinates. Each system is a form of spherical polar coordinates having the basic properties introduced in section 1.3. With each, an equivalent system of rectangular coordinates may be introduced, related to the spherical polars by equation (1.8). Moreover, in a particular system (be it equatorial, ecliptic or whatever) topocentric, geocentric and heliocentric coordinates can be defined as coordinates referred to different origins, but to *strictly parallel* coordinate axes.

The equatorial system of coordinates, right ascension and declination, was described in chapter 1. The definition given in section 1.6 (cf. Fig. 1.8), in fact, corresponds to geocentric right ascension and declination. Topocentric and heliocentric right ascension and declination would be defined similarly, merely by altering the origin to the appropriate point and maintaining the reference directions parallel to CP and $C\Upsilon$.

The distinction between topocentric, geocentric and heliocentric coordinates is important, and a precise meaning can only be assigned to a set of coordinates if it is understood which of the alternative viewpoints is intended. This distinction will not be emphasized in this chapter, but it is essential to recognize that, however a system of coordinates may be defined here, that system could equally well be established in its topocentric, geocentric, or heliocentric form.

2.2   The alt-azimuth system

Consider the celestial sphere centred on the observer, O, say. The most natural reference direction on the surface of the earth is the vertical, the direction defined by the earth's gravity. Suppose that the upward vertical intersects the celestial sphere in the point Z, Fig. 2.1. This point is called the *zenith*, and the diametrically opposite point the *nadir*. The great circle which has the zenith and the nadir as its poles will be termed the *celestial horizon*, or simply the horizon. This great circle divides the celestial sphere into two hemispheres – the upper hemisphere is the visible

Figure 2.1

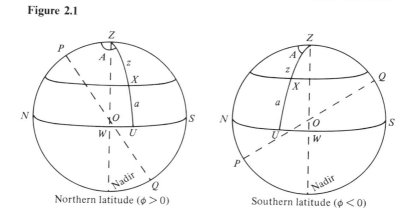

Northern latitude ($\phi > 0$)   Southern latitude ($\phi < 0$)

hemisphere, while the lower hemisphere is hidden from the observer by the earth itself.

The line drawn through O parallel to the earth's rotational axis intersects the celestial sphere in the two points P and Q, which are the north and south celestial poles respectively. Only one will be within the visible hemisphere for any particular observer. The situations for observers at northern and southern latitudes are exhibited separately in the two diagrams of Fig. 2.1. Although the reader will probably wish to concentrate on one of these diagrams, the argument that follows applies to them both.

Due to the earth's axial rotation, the observer will see the stars continually changing their positions in the sky. The diurnal rotation of the earth about its axis, from west to east, produces an apparent rotation of the entire celestial sphere about a parallel axis, namely POQ. As a result, a star situated at either of the celestial poles would be unaffected by this rotation and would appear fixed. The pole star, Polaris, is, in fact, a bright star within one degree of the north celestial pole and is easily visible to the naked eye, but there is no equivalent in the southern hemisphere.

Construct the great circle arc ZP and produce it, if necessary, to intersect the horizon in N. This is the north point of the horizon. The diametrically opposite point S is the south point of the horizon; the east and west points are at right angles to N and S. Notice that an observer facing north has W on his left and E on his right. The four points N, E, S, W are called the *cardinal points of the horizon.*

The alt–azimuth coordinate system is based on the point Z as pole. As reference great circle we select ZN, or equivalently ZP. Then, if X is any point on the celestial sphere, its *zenith distance* z and its *azimuth* A are defined as

$$z = ZX,$$
$$A = PZX. \qquad (2.1)$$

This definition of azimuth is strictly azimuth west (of north). It is one of a number of different conventions that are in use. With this definition, the azimuth increases in an anticlockwise direction, and the coordinate system is right-handed. Azimuth can also be measured east, but this is better regarded as a negative azimuth. We shall, therefore, use definition (2.1) within the range $-180° < A \leqslant 180°$.

Great circle arcs terminating in Z and at a point on the horizon are called *vertical circles.* The vertical circle through X meets the horizon at U, and it is seen that all points on this vertical circle have the same azimuth A which equals NU. The vertical circles through W and E are, a little confusingly, termed *prime verticals.*

The *altitude a* may be used as an alternative to zenith distance. For the point X, $a = XU$, and it is readily seen that

$$a = 90° - z. \tag{2.2}$$

Small circles with Z as pole are clearly curves of equal altitude – or zenith distance – and are termed *parallels of altitude*. The parallel of altitude through X is indicated in Fig. 2.1.

If the approximation is made that the earth is perfectly spherical, then OZ is a continuation of the outward radial direction from the centre of the earth. This makes an angle ϕ with the equator, where ϕ is the observer's geographical latitude. It follows that the arc PZ is equal to the observer's colatitude, yielding the important result

$$PZ = 90° - \phi. \tag{2.3}$$

Notice that, with the algebraic convention of latitude introduced in chapter 1, equation (2.3) holds for both northern and southern observers.

Zenith distance and azimuth provide a coordinate system that is easily realized in practice, as it is based on reference directions, Z and P, that can be established by direct observation. It has certain disadvantages, however. First of all, the zenith distance and azimuth of a star depend on the observing site. In no way can they be standardized for the whole earth. They are really intended as local coordinates, and a topocentric definition, as given here, is, therefore, the most natural. It is still possible to transform, for example, to a geocentric viewpoint, but this must retain, rather artificially, the zenith direction corresponding to the original observing site. Secondly, the alt–azimuth coordinates of a star will vary with the diurnal motion, and, what is more important, this variation is decidedly complicated. But it is only by considering the zenith distance that it can be established when a particular star will be visible from a particular observing site. For the star will be invisible whenever $z > 90°$. Indeed, the times of a star's rising and setting are computed as the times for which $z = 90°$.

2.3 Hour angle and declination

The alt–azimuth coordinate frame is fixed with respect to the observer, but not, of course, with respect to the stars. The complicated way in which the zenith distance and azimuth of a star change with time is due to the fact that this coordinate system, and, in particular, its pole Z, are unrelated to the cause of the star's diurnal motion, namely the earth's axial rotation. A coordinate system based on P as pole, however, overcomes this difficulty. If, in addition, PZ is selected as the reference great circle, the resulting coordinate system is still fixed with reference to the observer's immediate environment. The new coordinates are called *hour angle* and *declination*; the latter has already been introduced in a slightly different context.

The celestial sphere illustrated in Fig. 2.2 is centred on the observer O, and the zenith point, the celestial poles and the cardinal points of the horizon are shown exactly as before. For definiteness, an observer at a northern latitude is considered, but this is not essential to the main argument or to the following definitions. For any star on the celestial sphere, at X say, its declination δ and its hour angle H are defined as

$$\delta = 90° - PX$$

$$H = ZPX. \tag{2.4}$$

The arc PX itself is referred to as the *north polar distance* (NPD) of the star.

Semi-great circles terminating at the celestial poles, like PXQ are called *meridians of hour angle*. Along one of these curves the hour angle is clearly constant. The meridian of zero hour angle $PZSQ$ is known as the *observer's meridian*, and when a star crosses this meridian it is said to *transit*. It should be noted that hour angle is measured *westwards* from the observer's meridian.

Small circles with poles P and Q are termed *parallels of declination*. Now, since the diurnal motion of the stars is equivalent to a uniform rotation of the whole celestial sphere about an axis through P and Q, it follows that the diurnal motion of the star X is along its parallel of declination, shown in Fig. 2.2 as the arc $XDLRTX$. So, in the course of one day, the star

Figure 2.2

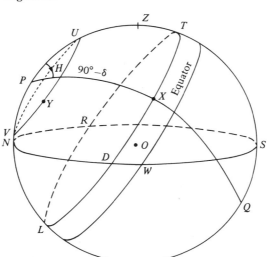

progresses westwards from the point X to D, where it sets; it then reaches its greatest distance below the horizon at L, when its hour angle is $180°$, and after that it approaches the horizon from below, rising at R; its altitude then increases reaching a maximum at transit T. Thereafter, its altitude decreases and it returns to the point X again after 24 hours have elapsed. In its diurnal movement, the star describes its parallel of declination and describes it at a uniform rate. The declination is constant, therefore, for any star, and the hour angle increases uniformly. So it is fairly straightforward to predict a star's hour angle and declination for any given time. To simplify matters further, hour angle is usually expressed, not in degrees, but in time measure like right ascension, equating 24^h with $360°$.

From Fig. 2.2, it will be seen that hour angle increases in a clockwise direction about P. The coordinate system is, therefore, left-handed, but this is unavoidable if hour angle is to increase with time. Another important conclusion may be drawn from the diagram by inspection: If a star's azimuth is west, its hour angle lies between 0^h and 12^h, while, if its azimuth is east, the hour angle is between 12^h and 24^h.

Declination zero corresponds to the celestial equator introduced in chapter 1. This is the great circle with poles P and Q. Consider the points E and W. They are $90°$ from zenith and from S. So they are the poles of the observer's meridian and are, therefore, $90°$ from P in particular. This implies that they must lie on the celestial equator. Putting this another way, the celestial equator intersects the horizon at its east and west points.

A star on the celestial equator will rise at E and set at W. It will, therefore, be above the horizon for exactly 12 hours. On the other hand, a star of negative declination rises south of east and sets south of west; it will be in the visible hemisphere for less than 12 hours. A star like X, however, which has positive declination, will be visible for more than 12 hours each day. The last two conclusions only apply, of course, to an observer at northern latitude. The reverse is true in the southern hemisphere. While the declination of a *star* is fixed, the sun's declination has a regular annual variation between limits of approximately $\pm 23°.5$. It is this variation that produces the seasonal differences in the duration of daylight hours that are pronounced at high latitudes.

Referring again to Fig. 2.2, it is seen that a star of sufficiently high declination does not rise or set at all; it is continuously visible. Such a star is said to be *circumpolar*. An example of a circumpolar star is shown by the point Y in Fig. 2.2 with its parallel of declination UYV, which represents the star's diurnal motion. The point V, where the star's altitude is least – its hour angle is then 12^h – is termed its *lower culmination* or *lower transit*. The point U, where it crosses the observer's meridian, has already been defined as its transit. To distinguish it from V, it is also referred to as the *upper transit* or *upper culmination*. The nomenclature is not entirely standard on this point, but the meaning is usually clear from the context.

The limiting case of a circumpolar star is when lower transit occurs at the north point of the horizon. Then $PN = 90° - \delta = \phi$, and the condition for a star to be circumpolar is

$$\delta > 90° - \phi. \tag{2.5}$$

In contrast, there is part of the celestial sphere that is never visible. By symmetry, it can be seen that the condition for this is

$$-\delta > 90° - \phi. \tag{2.6}$$

These two inequalities only hold for an observer in the northern hemisphere. The signs of δ and ϕ must both be changed for a southern observer, as the reader can readily verify.

This section will be concluded with a brief discussion of the transformation of coordinates from the alt–azimuth to the equatorial system and vice versa. While the trigonometric details of the method can be varied to some extent, the problem is essentially the solution of the spherical triangle PZX. This triangle is formed by the star itself and the poles of the two coordinate systems under consideration. For the sake of clarity, the triangle has been extracted from Fig. 2.2 and is shown in its bare essentials in Fig. 2.3. From the definitions of azimuth and zenith distance, equations (2.1), $PZX = A$, and $ZX = z$. Equally well, from the definitions of hour angle and declination, equations (2.4), $ZPX = H$ and $PX = 90° - \delta$. Moreover, $PZ = 90° - \phi$, where ϕ is the observer's latitude. In all, therefore, five parts may be labelled in the new diagram. This permits two applications of the cosine formula, which yield the following results

$$\sin \delta = \cos z \sin \phi + \sin z \cos \phi \cos A, \tag{2.7}$$

$$\cos z = \sin \delta \sin \phi + \cos \delta \cos \phi \cos H. \tag{2.8}$$

Figure 2.3

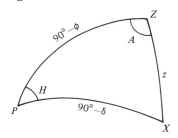

These two equations are sufficient to effect the desired transformation of coordinates, as shown by the two examples below.

Figure 2.3 applies when the star's azimuth is west. There are some slight modifications if the azimuth is east. The resulting equations are, however, still (2.7) and (2.8), which also apply for a southern observer. Finally, before proceeding with the examples of coordinate transformations, it is convenient to express one earlier conclusion in formal mathematical language:

$$0 < A < 180° \Leftrightarrow 0 < H < 12^h$$
$$0 > A > -180° \Leftrightarrow 12^h < H < 24^h. \tag{2.9}$$

Example. (a) A star has zenith distance $57°.57$ and azimuth $137°.60$ east at an observing site of north latitude $41°.36$. Calculate its hour angle and declination.

(b) For the same observing site, calculate the alt–azimuth coordinates of a star of declination $63° 43'$ nine hours after transit.

(a) The right-hand side of equation (2.7) may be calculated from the data, yielding

$$\sin \delta = -0.113\,47.$$

There is no ambiguity here; the principal value of the inverse sine is taken to give

$$\delta = -6° 31'.$$

To determine the hour angle, equation (2.8) is written in the form

$$\cos H = \cos z \sec \delta \sec \phi - \tan \delta \tan \phi.$$

Since the declination is now known, this yields

$$\cos H = 0.819\,67.$$

The azimuth A is negative, so we do not take the principal value (cf. equation (2.9)), but obtain

$$H = 325°.0513.$$

After dividing by 15, the result, in time measure, is found to be

$$H = 21^h 40^m.2.$$

Thus the equatorial coordinates (H, δ) have been obtained.

(b) This is the inverse problem of (a). It is again solved by using equations (2.7) and (2.8), but their roles are reversed. Everything is known on the right-hand side of (2.8), which accordingly yields

$$z = \cos^{-1} (0.357\,46) = 69°.06.$$

There is no ambiguity here – the principal value is taken.

The azimuth may be obtained by writing equation (2.7) in the form

$$\cos A = \sin \delta \, \text{cosec} \, z \, \sec \phi - \cot z \tan \phi,$$

since the right-hand side of this is now entirely known. It is then found that

$$A = \pm 19°.59.$$

The plus sign must be taken, because $H < 12^h$. So the star's azimuth is $19°.59$ west.

2.4   Sidereal time and right ascension

Consider the geocentric celestial sphere shown in Fig. 2.4, in which C represents the centre of the earth. Let p and q denote the earth's geographical poles. Then Cp and Cq, when produced, intersect the celestial sphere in the north and south celestial poles P and Q respectively. Let g represent the position of Greenwich, and o that of an observer at east longitude λ on the surface of the earth. Now let Cg and Co produced intersect the celestial sphere in the points G and O respectively. G is then the zenith point at Greenwich and so PGQ is the observer's meridian for the

Figure 2.4

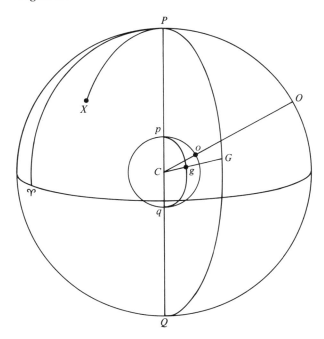

Greenwich observer. Similarly *POQ* is the observer's meridian for the observer at *o*, and spherical angle $GPO = \lambda$.

Suppose next that *X* is the position of any star on the celestial sphere. Then the hour angle of this star for the Greenwich observer is *GPX*. This is termed the *Greenwich hour angle* (GHA). On the other hand, the hour angle (HA) for the observer at longitude λ is angle *OPX*, and it follows that

$$HA = GHA + \lambda. \qquad (2.10)$$

In this equation λ is the *east* longitude of the observer. Since the two hour angles are presumably expressed in time measure, λ must also be expressed in these units.

Two equatorial systems of coordinates have been described, hour angle and declination in the previous section, and right ascension and declination in section 1.6. Each system has the same pole *P*, and they only differ in the choice of the reference great circle through *P*. It is the observer's meridian *PO* for hour angle, while for right ascension it is the great circle *P*♈. Further consideration will be given to the equinox ♈ in the next section; for the present, it is sufficient to recall that it is a point on the celestial equator fixed *with reference to the stars*. If α is the right ascension (RA) of the star at *X*, then $\alpha = $ ♈*PX*. Although the hour angle of the star will increase with time, the right ascension is fixed.

The equinox also provides a reference point for time measurement, but for time with reference to the stars, not the sun, as in civil timekeeping. It is termed *sidereal time*, therefore. The definition is straightforward: the local sidereal time (LST) is defined as

$$LST = HA \text{ ♈}. \qquad (2.11)$$

Similarly, for the observer at Greenwich, the Greenwich sidereal time (GST) will be given by

$$GST = GHA \text{ ♈}. \qquad (2.12)$$

It follows from equation (2.10) that the two times are related by the equation

$$LST = GST + \lambda. \qquad (2.13)$$

It will be recognized from the above that the sidereal time must increase by 24^h in the period of precisely one rotation of the earth about its axis. This period is called the *sidereal day*. It is not the same as the solar day; in fact, the sidereal day corresponds to about $23^h 56^m$ of solar time. The reason for this difference is that the reference point for sidereal time, namely ♈, is, unlike the sun, fixed with respect to the stellar background.

Sidereal time provides an important service as the link between hour angle and right ascension. Consider again Fig. 2.4. For the observer at *o*, the sidereal time is, by equation (2.11), the spherical angle $OP\Upsilon$. As already noted, the hour angle of X for this observer is OPX, and X has right ascension $\alpha = \Upsilon PX$. It follows that

$$\text{LST} = \text{HA } X + \text{RA } X. \tag{2.14}$$

This equation holds for any celestial object X and for any observer on the earth's surface. Moreover, it will be seen that (2.11) and (2.12) are special cases of this result. Equation (2.14) is all that is required to transform from the coordinate system of hour angle and declination to that of right ascension and declination.

2.5   The ecliptic and ecliptic coordinates

The earth's orbital motion round the sun provides the basis for an alternative coordinate system, and one that is particularly useful in solar system dynamics. The plane of the earth's orbit is known as the *plane of the ecliptic*, and the great circle in which this plane intersects the celestial sphere is itself termed the *ecliptic*. Throughout the annual motion, the earth's rotational axis maintains a fixed sidereal direction, making an angle of about $23\frac{1}{2}°$ with the normal to the plane of the ecliptic. This angle will be denoted by ε and is called the *obliquity of the ecliptic*.

The earth's orbital motion causes the apparent direction of the sun with reference to the fixed stars to change continuously. At any instant the sun will lie on the ecliptic, and it will describe the whole of this great circle in the course of one year. Its apparent sidereal direction, therefore, changes by nearly one degree per day. In Fig. 2.5, the two great circles shown are the equator and the ecliptic, and points P and K are their respective poles. P is, of course, the north celestial pole, and K is the *north pole of the ecliptic*, being the pole that is within $90°$ of P. By definition, the arc $KP = \varepsilon$, the obliquity of the ecliptic. Moreover, it is easily verified that the angle of intersection between the equator and the ecliptic is also ε, as shown in the diagram. The direction of the sun's apparent movement along the ecliptic is also indicated – it is anticlockwise when viewed from the north pole of the ecliptic. This sense of revolution or rotation in the solar system (or on the celestial sphere) is termed *direct*. The opposite sense is termed *retrograde*.

The equator and the ecliptic intersect in two points, one of which is shown in Fig. 2.5. This is the fundamental reference point Υ, the (northern vernal) equinox, the point at which the sun crosses the equator from south to north. When the term the *equinox* is used, without qualification, it is always this point Υ that is intended. The equinox provides a zero point for right ascension and also for ecliptic longitude, which is defined below.

In ecliptic coordinates, the point K is selected as pole of the coordinate system, and the great circle $K \Upsilon$ is chosen as the reference great circle. Then, if X is any point on the celestial sphere, its *ecliptic latitude* β and its *ecliptic longitude* λ are respectively defined as

$$\beta = 90° - KX$$
$$\lambda = \Upsilon KX. \tag{2.15}$$

Both β and λ are normally expressed in degrees, and it will be clear that β is confined to the range $-90° \leqslant \beta \leqslant 90°$, while λ may cover the entire range $0 \leqslant \lambda < 360°$. Ecliptic longitude increases in the sense that has previously been described as direct. In particular, the ecliptic latitude of the sun is zero, and its ecliptic longitude increases monotonically. For the planets, on the other hand, the motion again is normally direct, but, at certain configurations, the resultant of the earth's and the planet's orbital motion, each of which is direct, may produce an apparent retrograde motion.

The relationship between a celestial object's ecliptic and equatorial coordinates may be established by considering the spherical triangle PKX. Let X have right ascension and declination (α, δ). The sides of this triangle are $KP = \varepsilon$, $PX = 90° - \delta$, $KX = 90° - \beta$. Moreover, since ΥKP and ΥPK are both right angles, it follows that $PKX = 90° - \lambda$ and $KPX = 90° + \alpha$.

Figure 2.5

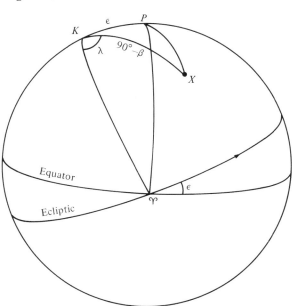

With five parts of the triangle known, applications of the standard trigonometric formulae allow the transformation from ecliptic coordinates (λ, β) to the equatorial coordinates (α, δ), and vice versa. An alternative, and perhaps more direct, procedure, however, is to work in terms of rectangular coordinates.

The system of rectangular equatorial coordinates, (x, y, z), was introduced in section 1.6, and the coordinates themselves are given by equations (1.26). The x-axis is directed towards Υ, the z-axis towards P, and the y-axis completes the right-handed set. A similar system of rectangular ecliptic coordinates, (ξ, η, ζ) say, may be introduced with the ζ-axis directed towards K, and the ξ-axis again directed towards Υ. The η-axis then points in the direction $\lambda = 90°$, $\beta = 0$. Further, the coordinates of X are given by

$$\xi = \cos \beta \cos \lambda,$$
$$\eta = \cos \beta \sin \lambda, \tag{2.16}$$
$$\zeta = \sin \beta.$$

Now the transformation from the one coordinate system to the other is equivalent to a rotation through an angle ε about the common x- and ξ-axis. We may, therefore, readily establish the transformation equations in the form

$$\xi = x, \qquad\qquad x = \xi,$$
$$\eta = y \cos \varepsilon + z \sin \varepsilon, \qquad y = \eta \cos \varepsilon - \zeta \sin \varepsilon, \tag{2.17}$$
$$\zeta = -y \sin \varepsilon + z \cos \varepsilon, \qquad z = \eta \sin \varepsilon + \zeta \cos \varepsilon.$$

These results may be expressed in terms of the spherical coordinates by using equations (1.26) and (2.16). The following two sets of results are derived:

$$\cos \beta \cos \lambda = \cos \delta \cos \alpha,$$
$$\cos \beta \sin \lambda = \sin \delta \sin \varepsilon + \cos \delta \cos \varepsilon \sin \alpha, \tag{2.18}$$
$$\sin \beta = \sin \delta \cos \varepsilon - \cos \delta \sin \varepsilon \sin \alpha.$$

$$\cos \delta \cos \alpha = \cos \beta \cos \lambda,$$
$$\cos \delta \sin \alpha = -\sin \beta \sin \varepsilon + \cos \beta \cos \varepsilon \sin \lambda, \tag{2.19}$$
$$\sin \delta = \sin \beta \cos \varepsilon + \cos \beta \sin \varepsilon \sin \lambda.$$

These two sets of equations are sufficient to transform between the equatorial coordinates (α, δ) and the ecliptic coordinates (λ, β). The reader may wish to verify that these equations can be established by applications of the sine, analogue and cosine formulae in the spherical triangle PKX. In many problems, however, the rectangular coordinates will be all that is required, and their transformation is clearly simpler and more direct.

2.6 Apparent and mean solar time

Sidereal time was defined, in section 2.4, as the hour angle of the equinox. This provides a time system of very high uniformity, and one that has many astronomical applications, but it is unsuitable for everyday purposes. Civil time regulation must depend rather on the hour angle of the sun. The time of the sun's passage across the observer's meridian is defined as *apparent solar noon*, and the interval between two successive passages defines the *apparent solar day*. It is convenient, however, to regard the solar day as commencing at midnight rather than at noon, so the following definition is made:

$$\text{Local apparent solar time} = 12^{\text{h}} + \text{HA} \odot, \tag{2.20}$$

where the astronomical symbol $\odot$ has been used to indicate the sun. To relate solar time to sidereal time, it is only necessary to apply equation (2.14), identifying the point X with the sun. Then,

$$\text{Local apparent solar time} = \text{LST} + 12^{\text{h}} - \text{RA} \odot. \tag{2.21}$$

In the course of one year, the sun's right ascension increases by 24^{h}, so this equation indicates that the number of sidereal days in the year is exactly one more than the number of solar days. Neglecting minor precessional effects, which will be considered in chapter 10, sidereal time depends only on the rotation of the earth, which is highly regular. Apparent solar time, on the other hand, also involves the sun's right ascension, which in turn depends on the earth's orbital motion.

The first basically correct description of planetary motion was given, in the 17th century, by Johannes Kepler in his famous three laws. These are

    (i) The orbit of a planet about the sun is an ellipse, with the sun situated at one focus.

  (ii) The planet moves with constant areal velocity, i.e. in such a way that the radius vector from the sun sweeps out equal areas in equal times.

 (iii) The cube of the semimajor axis of the planet's orbit is proportional to the square of its orbital period.

The first two of these laws will concern us immediately.

Figure 2.6 represents the earth's elliptical orbit. S denotes the position of the sun, and AB the major axis. The ellipticity of the earth's orbit has been greatly exaggerated – the orbit is, in fact, nearly circular, its eccentricity being only 0.016. The point A where the earth's distance from the sun is a minimum is called *perihelion*. The diametrically opposite point in the orbit, corresponding to the maximum distance, is called *aphelion*. The earth passes through perihelion in early January. The length of the semimajor axis of the earth's orbit is termed the *astronomical unit* (AU). This is a convenient unit of distance in solar system dynamics, equal to 1.496×10^8 km.

In Fig. 2.6, C represents the earth's position at the vernal equinox, when the sun appears in the sidereal direction of ♈. Let E represent the earth's position at some subsequent time; the sun will then appear in the sidereal direction R, as indicated. Now the angle ♈SR, or CSE, is the longitude of the sun, $\lambda_\odot$, say, corresponding to that time. The earth's angular velocity about the sun is not constant. This follows from Kepler's second law, which is equivalent, in fact, to the constancy of angular momentum, not velocity. Consequently, the longitude of the sun does not increase uniformly during the year, but its rate of change is greatest at perihelion and least at aphelion. This variation in the rate of increase of the sun's longitude is reflected in a corresponding variation in the rate of increase of its right ascension. This, in turn, produces irregularities in the apparent solar time, as may be seen immediately from equation (2.21).

There is, however, a further cause of irregularity in the sun's right ascension arising from the obliquity of the ecliptic. Let Fig. 2.7 represent the geocentric celestial sphere on which the ecliptic is shown as the great circle UA♈SV. The points V and U, at which the sun has its maximum and minimum declination of $\pm\varepsilon$, are called the (northern) *summer* and *winter solstices*. The point A represents the sun's position at the time of perihelion, and S is a general position of the sun. Let $(\alpha_\odot, \delta_\odot)$ be the right ascension and declination of S. Join the great circle arc PS and produce it to cut the

Figure 2.6. The earth's orbit viewed from the north ecliptic pole.

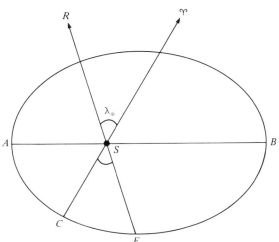

equator in T. Then, in spherical triangle ΥST, the following parts are known: $\Upsilon S = \lambda_\odot$, $\Upsilon T = \alpha_\odot$, $TS = \delta_\odot$, $\Upsilon TS = 90°$, and $S\Upsilon T = \varepsilon$. Spherical trigonometrical formulae may be applied to this triangle to relate the sun's right ascension and declination to its longitude. For example, by the four-parts formula, it follows that

$$\tan \alpha_\odot = \cos \varepsilon \tan \lambda_\odot. \tag{2.22}$$

This relationship implies that the sun's right ascension does not increase uniformly with longitude. In fact, the rate of increase is a minimum at the two equinoxes and a maximum at the two solstices.

The above discussion demonstrates that the right ascension of the sun varies non-uniformly with the sun's longitude, which itself does not increase uniformly with time. As a result, the true sun is an inadequate time-keeper, and an imaginary body, called the *fictitious mean sun*, is used instead. This may be defined geometrically as follows. Let τ be the time of perihelion. The sun's position on the celestial sphere is then A. Further, let n denote the mean angular velocity of the earth in its orbit, i.e. 360° per year. We consider an imaginary body moving round the ecliptic with angular velocity n in such a way that its position coincides with the true sun at perihelion – and also, by symmetry, at aphelion. This body is called the *dynamical mean sun*. Suppose that at time t the true sun is at S and the dynamical mean sun has reached the point D. Then $AD = n(t - \tau)$, provided the times are expressed in years.

Figure 2.7

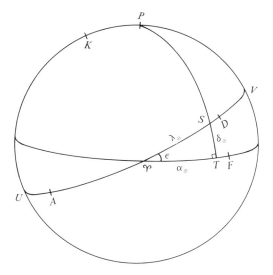

The device of the dynamical mean sun irons out the irregularities in the sun's longitude, but it does not remove the influence of the obliquity of the ecliptic. To achieve this, a second imaginary body, called the *fictitious mean sun*, is introduced. This is defined to move with constant angular velocity n round the *equator* and to coincide in position with the dynamical mean sun at the two equinoxes. If, at time t, the fictitious mean sun is situated at the point F, then, by our definitions, $\Upsilon F = \Upsilon D$. The fictitious mean sun is then a body whose right ascension increases uniformly and approximates to that of the true sun. It, therefore, provides a convenient reference point to define a system of solar time from which the gross irregularities have been eliminated. This is *mean solar time* which depends on the hour angle of the fictitious mean sun (FMS) in the same way as apparent solar time depends on the hour angle of the true sun, namely

$$\text{Local mean solar time} = 12^{\text{h}} + \text{HA FMS}. \tag{2.23}$$

Moreover, the relationship with local sidereal time is obtained from (2.14) as before, yielding

$$\text{Local mean solar time} = \text{LST} + 12^{\text{h}} - \text{RA FMS}. \tag{2.24}$$

The difference between apparent and mean solar time is known as the *equation of time*. It is equal to the arc TF in Fig. 2.7, as can be seen from equations (2.21) and (2.24), and is usually defined formally as

$$\text{Equation of time} = \text{RA FMS} - \text{RA} \odot. \tag{2.25}$$

This difference varies in a complicated way in the course of the year and may be as large as 15 minutes of time. The need for mean solar time is, therefore, evident.

The definitions of apparent and mean solar time given above are local definitions applicable to a particular observing site. The local solar time (either apparent or mean) depends on the observer's terrestrial longitude in exactly the same way as sidereal time, and standardization is achieved in a similar manner. The mean solar time on the Greenwich meridian is called *universal time* (UT), and it is easily established, by (2.10) and (2.23), that, for an observer at east longitude λ,

$$\text{Local mean solar time} = \text{UT} + \lambda. \tag{2.26}$$

As before, the longitude is, of course, expressed in time measure.

The local mean solar time is seldom required in practice. It is obviously essential that there should be some synchronization in civil time-keeping. In a small country, the mean solar time of a standard meridian is adopted as the civil time throughout. For example, in the United Kingdom or the Republic of Ireland the Greenwich meridian is chosen, so that civil time corresponds to UT – during the winter months at least; clocks are advanced one hour in summer for daylight-saving purposes, and the civil time then corresponds to the mean solar time on the meridian of longitude 15° east. For larger countries, like the United States, on the other hand, a single standard time is not practical, and these countries are divided into a number of time zones, each keeping the mean solar time of a standard meridian, usually spaced at intervals of 15° in longitude. In every case, the zone time is given by the following slight variant of equation (2.26),

$$\text{Zone time} = \text{UT} + \lambda_M \tag{2.27}$$

where λ_M is now the (east) longitude of the standard meridian of that time zone.

While the discussion of this section gives an outline of solar time-keeping, it does not deal with all the intricacies of the problem. These will be considered in chapter 10, and complete formal definitions will be given there. In particular, *ephemeris time* (ET) will be precisely defined. This may be loosely described as universal time corrected for irregularities in the earth's rate of axial rotation. For both universal and sidereal time are defined in terms of the hour angle of certain reference points on the celestial sphere, and, however accurately these points have been defined, the rate of change of their hour angles can only be as uniform as the earth's rate of rotation. Modern atomic clocks, and even the precise timing of some astronomical phenomena, indicate that there are many irregularities in the earth's angular velocity, in addition to a steady secular decrease. Moreover, these irregularities cannot be predicted in detail. Astronomical ephemerides have, therefore, been computed against a time-scale of ET in which such irregularities do not arise. The difference between ephemeris and universal time is denoted by ΔT, in the sense

$$\Delta T = \text{ET} - \text{UT}. \tag{2.28}$$

This difference ΔT cannot be tabulated in advance, but the relationship between universal and sidereal time can be; this tabulation is included in the *Astronomical Almanac*. The Greenwich sidereal time is given there for 0^h UT at daily intervals. The approximate relationship between the two time-scales may be stated as

$$1 \text{ mean solar day} = 24^h\,03^m\,56^s.6 \text{ of sidereal time}$$
$$1 \text{ sidereal day} = 23^h\,56^m\,04^s.1 \text{ of mean solar time.} \tag{2.29}$$

More detailed discussion of this and other relationships involving time is deferred until chapter 10, where more accurate formulae will be given.

2.7   Galactic coordinates

A system of galactic coordinates is also desirable as the plane of the galaxy is a natural reference plane for the positions of the stars and their motions. The two points on the celestial sphere which are 90° from the galactic plane are referred to as the *galactic poles*, and a coordinate system may be based on these points and the galactic plane in a conventional manner.

The determination of the galactic plane, however, depends on the statistical reduction of a large body of observational material. This was extended, after the second world war, with the addition of radio observations of the galaxy, and, in particular, by 21 cm line studies. As a result, the precision with which the galactic plane, and, therefore, the galactic poles, could be determined was considerably enhanced. A complete revision of the system of galactic coordinates became necessary. At the same time, a new convention was introduced identifying the zero of galactic longitude with the galactic centre – something that had not been attempted in the earlier system. Although the fact that two systems had been in use was, for a while, somewhat confusing – it was then necessary to state rather carefully which system was being employed – nonetheless the 'new' system has now been in almost exclusive use for over twenty years. It is this system that is described below.

In Fig. 2.8, *P* represents the north celestial pole and the great circle *UCNV* represents the galactic plane, intersecting the celestial equator in *N*.

Figure 2.8

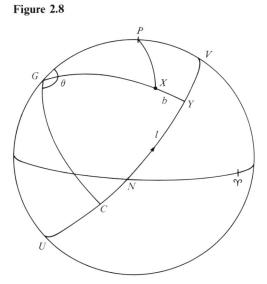

This great circle is termed the *galactic equator*. The point G is the *north galactic pole*, and C represents the direction of the centre of the galaxy. Now let X be the position of any star. Join the great circle arc GX and produce it, if necessary, to cut the galactic equator in Y. Then XY is the galactic latitude; it will be positive or negative according as X is north or south of the galactic equator. Notice, however, that by north of the galactic equator we mean that the star is on the same side of the galactic equator as G, it need not have positive declination. On the other hand, the galactic longitude of the star X is the arc CY measured in the direction indicated. Both galactic longitude and latitude are normally expressed in degrees.

More formally, the *galactic latitude* and *galactic longitude* of the star X are respectively defined as

$$b = 90° - GX,$$
$$l = CGX. \tag{2.30}$$

The galactic latitude may take values in the range $-90° \leqslant b \leqslant 90°$, while the galactic longitude may take values in all four quadrants, i.e. $0 \leqslant l < 360°$.

In order to relate the galactic coordinates of a star to its equatorial coordinates, it is necessary to know the positions of the two points G and C. Suppose that (α_G, δ_G) are the right ascension and declination of the north galactic pole. This specifies the position of G satisfactorily, and, since $GC = 90°$, the spherical angle PGC will then determine the point C. Let us call this angle θ. It is usually referred to as the *position angle of the galactic centre*, but it is also the galactic longitude of the north celestial pole. The values adopted for these three parameters, for the epoch 1950.0, are

$$\alpha_G = 12^h 49^m,$$
$$\delta_G = 27°.4, \tag{2.31}$$
$$\theta = 123°.$$

It is necessary to specify the epoch, since the north celestial pole has a precessional movement against the background of the stars – this is discussed in chapter 9. Galactic coordinates are, therefore, computed by the method described below only for this epoch, and it is necessary to ensure, before applying the method, that the star's equatorial coordinates are referred to the equator and equinox of 1950.0.

Consider spherical triangle GPX in Fig. 2.8. If the point X has equatorial coordinates (α, δ) and galactic coordinates (l, b), then the following identifications of parts in this triangle can be made:

$$PX = 90° - \delta, \quad GX = 90° - b, \quad GP = 90° - \delta_G,$$
$$GPX = \alpha - \alpha_G, \quad PGX = \theta - l.$$

Now apply the cosine formula to derive the side GX and, therefore, the latitude b. The result is

$$\sin b = \sin \delta_G \sin \delta + \cos \delta_G \cos \delta \cos (\alpha - \alpha_G). \tag{2.32}$$

This will determine the galactic latitude unambiguously. To do the same for galactic longitude, however, requires the use of both the sine and the analogue formulae, which give

$$\cos b \sin (\theta - l) = \cos \delta \sin (\alpha - \alpha_G),$$
$$\cos b \cos (\theta - l) = \cos \delta_G \sin \delta - \sin \delta_G \cos \delta \cos (\alpha - \alpha_G). \tag{2.33}$$

Since b is already determined, these two equations allow the determination of $\sin (\theta - l)$ and $\cos (\theta - l)$, and hence $(\theta - l)$ itself without ambiguity. The longitude l is then derived. The details of the above argument apply, strictly speaking, only when $\alpha - \alpha_G < 180°$, otherwise slight modifications are necessary to the diagram, but the resulting equations (2.32) and (2.33) are correct in all circumstances.

The reverse transformations, from galactic to equatorial coordinates, may be derived by a very similar argument. Again, from spherical triangle GPX, the relevant formulae are found to be

$$\sin \delta = \sin \delta_G \sin b + \cos \delta_G \cos b \cos (\theta - l),$$
$$\cos \delta \sin (\alpha - \alpha_G) = \cos b \sin (\theta - l), \tag{2.34}$$
$$\cos \delta \cos (\alpha - \alpha_G) = \cos \delta_G \sin b - \sin \delta_G \cos b \cos (\theta - l).$$

These equations will, of course, give the right ascension and declination referred to the equator and equinox of 1950.0.

As already stated, the above discussion refers to galactic coordinates in the 'new' system which is the only system in current use. The recommended convention is to denote these coordinates simply as (l, b), but, particularly in earlier literature when confusion with the 'old' system was likely, they have sometimes been written as (l^{II}, b^{II}), the superscript Roman numeral indicating that the new system is being used. The old system was then written as (l^{I}, b^{I}). It is not intended to discuss the old system in any detail here. It is defined in Problem 2.10 at the end of this chapter, where transformation from the old to the new system is also considered. Suffice it to say that, if Fig. 2.8 referred to the old system, the equatorial coordinates of G would be slightly altered from those given in equations (2.31) and longitude would be measured from the point N rather than from the point C. We conclude this section with the following example.

Example. The Crab Nebula has equatorial coordinates $(5^h 31^m.5, +21° 59')$ referred to the equator and equinox of 1950.0. Compute its galactic latitude and longitude.

This is a straightforward application of equations (2.32) and (2.33). From the data and (2.31), it is found that equation (2.32) yields

$$\sin b = -0.100\,85.$$

Hence

$$b = -5° 47'.$$

This value is now used in equations (2.33) to give the results

$$\sin(\theta - l) = -0.879\,26,$$

$$\cos(\theta - l) = 0.476\,34.$$

So $(\theta - l)$ is in the fourth quadrant and is $-61° 33'$. Substituting $123°$ for θ yields the value of l. The galactic coordinates of the Crab Nebula are thus found to be

$$l = 184° 33',$$

$$b = -5° 47'.$$

2.8　Setting a telescope

The problem of setting a ground-based telescope in a preassigned direction will now be considered. Suppose, for example, coordinates, normally right ascension and declination, are extracted from a star catalogue. The first step must be to convert the right ascension to an hour angle – although this is not always done explicitly – so that the coordinates are known in a frame that is fixed with respect to the observer's immediate environment. This requires a knowledge of the local sidereal time.

The observatory will normally be equipped with a clock that is keeping local sidereal time so that it can be read directly. Even so, the procedure for computing LST from the data in the *Astronomical Almanac* will be examined in the example given below. For this is essentially the same procedure as that required to set and maintain the local sidereal clock. The computation requires an accurate knowledge of the geographical longitude of the observing site.

Once the hour angle and declination of the star are known, further coordinate transformation may or may not be necessary depending on the mounting of the telescope. Most optical telescopes for general astronomical use have an *equatorial mount*. This means that the principal axis of the mounting is a polar axis directed towards the north celestial pole. It is, therefore, inclined to the vertical at an angle of $90° - \phi$, the colatitude of the observatory. Suppose now that the telescope has been set to point in a direction making an angle θ with the polar axis. The telescope is then pointed in a direction whose declination is $90° - \theta$. If now the whole equipment is rotated about the polar axis while maintaining the angle θ

fixed, then the telescope is sweeping out a parallel of declination on the sky. It is possible, therefore, to follow the diurnal movement of any celestial object by having the telescope driven to rotate about its polar axis at a rate of one revolution per sidereal day.

The advantages of an equatorial mount are, therefore, obvious. To direct such a telescope towards a particular object will require an initial setting in both declination and hour angle. The first of these will thereafter be kept constant, while the second will be adjusted automatically by the driving mechanism, which will keep the object within the field of view of the telescope. The setting in declination is straightforward, and that in hour angle may be simplified by having a scale with an adjustable zero point. This may be set for the appropriate right ascension, and the setting of the telescope then becomes a setting for LST rather than for hour angle, which need never be explicitly worked out.

An equatorial mount is normal for an optical telescope unless it has a specialized function. A number of instruments have been designed with special mounts in order to undertake absolute astrometric measurements. It is intended to discuss such instruments and their operation more fully in chapter 5, but we shall give brief consideration here to the most fundamental of these instruments, the meridian circle, because it provides direct measurements of right ascension and declination. The *meridian circle* or *transit instrument*, is usually a fairly small refractor that is mounted about a horizontal east–west axis. It is only free to rotate about this axis and, consequently, it is not fully steerable. As it is rotated about its axis, it sweeps out the observer's meridian, and observations are confined to stars at or near transit. In the focal plane of the instrument there is a cross-wire which represents the observer's meridian, and the observation consists of accurately timing the passage of the star across the cross-wire. Since the hour angle of a star is zero at transit, the observed local sidereal time of transit gives a direct measurement of the star's right ascension (cf. equation (2.14)).

A direct measurement of declination is also possible simply by reading off the inclination at which the telescope has been set. This will give the altitude of the star above the horizon, but at transit this is simply related to declination. If the zero point of the scale has been set to correspond with the celestial equator rather than with the horizon, a direct reading of declination is available. Thus the meridian circle used in conjunction with a local sidereal clock can give direct measurements of both right ascension and declination.

Large radio telescopes are often not fully steerable, and a transit-type arrangement is possible. Moreover, the capability of radio instruments may be extended by connecting them over long baselines as interferometers. The greatest positional accuracy in radio astrometry is obtained by this means. The analysis, however, becomes decidedly more complicated, and detailed consideration of radio astrometry is deferred until chapters 15 and 16. At present we shall concentrate on the problem of setting a single fully steerable radio antenna on a source of known right ascension and declination.

If the radio telescope has an equatorial mount, the problem is no different from that already considered. The basic requirements are still hour angle and declination. The mechanical disadvantages of an equatorial mount are considerable, however, since the principal axis is inclined to the vertical, and many large radio telescopes, therefore, use a simpler alt–azimuth mount for structural reasons. In this case, it is clearly necessary to compute both the altitude and the azimuth of the source in order to set the telescope correctly. A large radio telescope will be computer controlled, but this, of course, does not alter the spherical principles involved. The method of converting hour angle and declination to alt–azimuth coordinates has already been considered in section 2.3, and the following example employs the method developed there. To illustrate all the points in the present section, we have further included the detailed procedure required to determine local sidereal time, which is also necessary for equatorial mounts.

Example. An alt–azimuth radio telescope is sited at longitude $\lambda = 83° 31'$ W and latitude $\phi = 40° 15'$ N. It is intended to set the telescope on the radio source 3C 273, whose equatorial coordinates are $\alpha = 12^h 28^m.3$, $\delta = 2° 08'$. The date of the observation is 1985, January 7, $14^h 42^m$ UT. Compute the coordinates necessary to set the telescope.

First of all, the local sidereal time must be computed. From the *Astronomical Almanac* it is found that for 1985, January 7, 0^h UT

$$GST = 7^h 06^m 01^s.$$

To this must be added $14^h 42^m$ UT, but converted to sidereal time. This is best done by expressing this time interval in hours and applying the conversion factor which is 1.002 737 909 4. As a result, it is found that, for the time of observation

$$GST = 21^h.8406.$$

To derive LST, it is only necessary to *subtract* the west longitude, cf. equation (2.13), yielding

$$LST = 16^h 16^m 22^s.$$

The hour angle H of 3C 273 is now computed by (2.14) from the known value of its right ascension giving

$$H = 3^h 48^m 04^s = 57°.0174. \tag{2.35}$$

The formulae required to derive altitude and azimuth are equations (2.7) and (2.8). These are restated here in the partially modified form

$$\sin a = \sin \delta \sin \phi + \cos \delta \cos \phi \cos H, \tag{2.36}$$

$$\cos A = \sin \delta \sec a \sec \phi - \tan a \tan \phi. \tag{2.37}$$

The reader may find it helpful to refer to Fig. 2.3 which illustrates the spherical triangle from which the original formulae were derived. Equation (2.36) may now be used to compute the altitude as

$$a = 26°.0564.$$

Since the hour angle is, by (2.35), less than 12^h, the source's azimuth must be west. This removes any ambiguity in equation (2.37), which now yields

$$A = 111°.0779.$$

The required coordinates for setting the radio telescope are thus found to be

$$\text{altitude} = \quad 26° \ 03',$$

$$\text{azimuth} = 111° \ 04' \text{ west.}$$

Problems

(The following symbols are used below:

$\phi = $ observer's latitude

$\delta = $ star's declination

$\varepsilon = $ obliquity of the ecliptic $= 23° \ 26'.5$.)

2.1 Prove that the azimuth of a circumpolar star can have any value if $\delta < \phi$, but must be less than $\sin^{-1} (\cos \delta \sec \phi)$ when $\delta > \phi$.

2.2 Show that the angle ψ that a star's path makes with the horizon at rising or setting is given by

$$\psi = \cos^{-1} (\sin \phi \sec \delta).$$

2.3 If H is the hour angle of a star at rising, show that

$$\tan^2 \tfrac{1}{2}H = \frac{\cos (\phi - \delta)}{\cos (\phi + \delta)}.$$

[Smart (1977)]

2.4 Two stars A and B (declinations δ_1 and δ_2 respectively) are observed to rise at the same moment and A transits when B is setting. Prove that

$$\tan\phi\tan\delta_1 = 1 - 2\tan^2\phi\tan^2\delta_2.$$

2.5 Prove that, at the equator, all stars are visible for exactly 12^h and that their azimuths at rising and setting are $90° - \delta$.

2.6 Assuming that naked eye stars are distributed evenly over the celestial sphere, show that the ratio of such circumpolar to non-circumpolar stars visible at any observing site is $\frac{1}{2}(\sec\phi - 1)$.

2.7 Prove that the zenith distance z of the north pole of the ecliptic is given by

$$z = \cos^{-1}(\cos\varepsilon\sin\phi - \sin\varepsilon\cos\phi\sin T),$$

where T is the local sidereal time.

2.8 Show the obliquity of the ecliptic produces a six-monthly variation in the rate at which the sun's right ascension increases. Show further that, if the eccentricity of the earth's orbit is neglected, then this rate is a maximum at the solstices and a minimum at the equinoxes.

2.9 Using the data from equation (2.31), calculate the inclination of the galactic plane to the ecliptic. Show that the sun crosses the galactic plane within a day of each solstice and determine the galactic longitudes of these crossing points.

2.10 The galactic pole under the old system of galactic coordinates (l', b') is approximately $1°.5$ from the pole of the present system (l, b) along the meridian $l = 200°$. The longitude of galactic centre in the old system was $327°.7$, what was its galactic latitude in this system?

   Show that the galactic coordinates of any star in the two systems are approximately related by the equations

$$b' = b - 1°.5\cos(l - 20°)$$
$$l' = l - 32°.3 - 1°.5\tan b\sin(l - 20°).$$

3

The reference frame

3.1 Inertial frames of reference

Positional astronomy provides the observational basis for dynamical astronomy, which, for the solar system, in particular, is understood in terms of the overriding force of gravity. It is no coincidence that the law of gravity and the laws of dynamics were formulated at the same time by Sir Isaac Newton, for Newton's motivation was to explain planetary motions. The basic laws of dynamics are indeed well known as Newton's laws of motion. In view of their importance in what follows, we shall state them here, with some care, as

   (i) A body not subject to any external force moves with constant velocity, and, therefore, in a straight line.

  (ii) The rate of change of momentum of any body, normally its mass times its acceleration, is equal to the force applied externally to that body.

 (iii) Action and reaction are equal and opposite; this applies, for example, to the forces between two bodies.

The first law can be regarded as a special case of the second, but it is so important that it is usually stated separately.

In the case of solar system dynamics, the applied force is, of course, the force of gravity, primarily that of the sun. Now to apply these dynamical laws and the law of gravity, and to perform the subsequent analysis, it is necessary to have a coordinate system, or a frame of reference. But this must be an inertial frame. Indeed, we may define an inertial frame of reference as one in which Newton's three laws of motion apply. The identification of an inertial frame is, therefore, a matter of prime importance.

The system of right ascension and declination would appear at first sight to provide an inertial frame, or at least an approximation to it. It would certainly appear to be a better candidate than that of hour angle and declination which is rotating once every 24 hours with respect to the sidereal background. This raises a separate issue, however. For there appears to be no *logical* reason why an inertial frame should not be rotating with respect to the fixed stars. Certainly it would be astonishing if the frame of hour angle and declination proved to be inertial and that of right ascension and declination did not. In fact, it is quite easy to verify that hour angle and declination is not an inertial frame. The experiment that proves this is that of Foucault's pendulum, in which a pendulum is found to vary its plane of oscillation in the course of a day. But this does not prove that a frame fixed with reference to the sidereal background is necessarily inertial, it proves only that it is approximately so.

In the nineteenth century, the philosopher Ernst Mach formulated his principle that the inertia of a body is caused by the distribution of the rest of the material in the universe. If Mach's principle is correct, then an inertial frame cannot be rotating with respect to the rest of the universe as a whole. While the absolute correctness of Mach's principle is still a subject of considerable controversy, it is certainly a principle of great intuitive appeal and one that appears to be correct to a very high degree of approximation. For the most part, we shall adopt Mach's principle is this book, and, therefore, accept two alternative definitions of an inertial frame, namely, one in which Newton's laws of motion may be applied, and, secondly, one that is fixed with respect to the rest of the universe.

The principal coordinate system used in positional astronomy is, of course, that of right ascension and declination. It is intended that this system should be sufficiently refined so that ultimately it will provide an inertial frame of the highest possible accuracy. This places two requirements on the system. First of all, the coordinate axes that define the system should not be rotating, that is, not rotating with respect to the rest of the universe as a whole, and secondly, that the origin of this coordinate system should be unaccelerated with respect to the rest of the universe. In subsequent sections of this chapter these two requirements will be examined in some detail.

3.2   Precession, nutation and proper motion

Consider Fig. 3.1, in which the equator and the ecliptic are shown and their intersection at the equinox ♈. The equatorial coordinate system, right ascension and declination, is completely determined when these two planes are specified, or equivalently when the north celestial pole P and the north pole of the ecliptic K are given. The same is true of the ecliptic system of coordinates; again, it is determined by the two points P and K. To make these two dependencies explicit, consider a star X with equatorial coordinates (α, δ) and ecliptic coordinates (λ, β). Then the three sides in spherical triangle PKX are

$$KP = \varepsilon, \quad PX = 90° - \delta, \quad KX = 90° - \beta. \tag{3.1}$$

Moreover, since $KP\Upsilon$ and $PK\Upsilon$ are both right angles, two angles of this triangle are seen to be

$$KPX = 90° + \alpha, \quad PKX = 90° - \lambda. \tag{3.2}$$

Now let us insist that the centre C of the celestial sphere in Fig. 3.1 is the origin of *some* inertial frame of reference. This will place certain restrictions on the point C to be considered in the next section. For the moment no attempt is made to identify C, it is only required to be the origin of an inertial frame. What will be considered qualitatively here is any movement of the points P, K and X with respect to this inertial frame. It was tacitly assumed in chapter 2 that each of these points could be regarded as fixed. This is certainly a good first approximation, but no more. Each of the points is displaced due to a different cause. The displacement of P is much the greatest and is due to *luni–solar precession* and *nutation*, while the displacement of K is termed *planetary precession*, and that of the star X itself is termed *proper motion*. The three points and their displacements are considered separately below. The displacement of any one of them produces changes in the star's equatorial and ecliptic coordinates.

The line CP is, by definition, parallel to the earth's rotation axis. It, therefore, defines the direction of the earth's angular momentum vector.

Figure 3.1

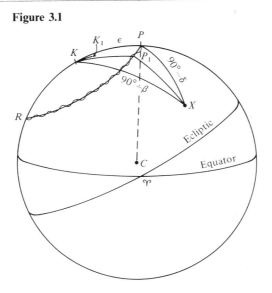

The forces being applied to the earth are the gravitational attractions of the sun, moon and planets. Now the gravitational attraction between two perfect spheres produces no torque. So, in the first approximation, the angular momentum remains constant, and CP is a fixed direction; the point P is fixed on the celestial sphere. Due principally to its axial rotation, however, the earth has developed a departure from spherical symmetry in the form of an equatorial oblateness. Its equatorial radius exceeds its polar radius by about 0.3 per cent. As a consequence, both the sun and the moon exert a torque which is slowly changing the direction of the earth's angular momentum vector, and thereby causing a displacement of the point P on the celestial sphere. The gravitational torque is directly proportional to the mass of the attracting body and inversely proportional to the cube (not the square) of its distance. Consequently the moon's effect is about twice as large as that of the sun, while even the strongest torques from the planets, those caused by Jupiter and Venus, are a factor of 10^5 weaker and are normally omitted from consideration.

The resultant torque exerted by the moon and the sun on the earth varies with the changing configurations and distances of the three bodies. The corresponding movement of the point P on the celestial sphere is, therefore, rather complicated. It is convenient to divide this movement into two parts. The long-term average or *secular* movement of the north celestial pole is called *luni–solar precession*, while short-term periodic oscillations about the mean position are termed *nutation*. The precessional movement is shown in Fig. 3.1 by the solid curve PP_1R, while the wavy curve represents the actual movement of the pole and includes nutation as well.

The effect of nutation is rather small, the displacement never being greater than 15″. It was discovered, in the 18th century, by the English astronomer Bradley, who detected periodic variations in his meridian measurements of the declinations of stars. The corresponding variations in right ascension were not detected at that time since clocks of sufficiently high precision did not exist. By contrast, the effect of luni–solar precession can build up over long periods to a very considerable displacement. Indeed, the effect was known to the ancient Greeks. In the second century BC, Hipparchus of Rhodes compared his own observations of the positions of stars with those recorded 150 years earlier. He deduced that, whereas the ecliptic latitudes of the stars appeared constant, there was a secular increase in their ecliptic longitudes. The increase amounts to about 50″ per year. This comparatively crude description of precession corresponds to a uniform movement of the north celestial pole in a small circle with pole K.

Let ψ be the *annual rate of luni–solar precession*. Then, if P was the initial position of the pole, and P_1 its position t years later, the spherical angle

$PKP_1 = \psi t$. Further, $KP_1 = KP$, so the obliquity of the ecliptic is unchanged. Now, if (λ_1, β_1) are the ecliptic coordinates of the star X at the later epoch, then, by comparison with equations (3.1) and (3.2),

$$P_1 K X = 90° - \lambda_1, \quad K X = 90° - \beta_1.$$

But, since $P_1 K X = PK X - PK P_1$, it follows that

$$\lambda_1 = \lambda + \psi t. \tag{3.3}$$

Finally, since KX is unchanged, the ecliptic latitude is fixed. The corresponding changes in right ascension and declination are more complicated and must be derived by considering spherical triangles KPX and $KP_1 X$.

The north celestial pole completes one circuit about the pole of the ecliptic in roughly 26 000 years. The motion is often compared with the motion of a rapidly spinning top. The axis of the top precesses while maintaining a fixed inclination to the vertical. In the case of luni–solar precession, the earth's rotational axis maintains a constant inclination to the normal to the ecliptic. There will be considerable changes in a star's equatorial coordinates over an entire precessional period. For example, suppose the initial epoch, referred to earlier, corresponds to the present, then the star Polaris will be close to the point P. After 13 000 years have elapsed the north celestial pole will be situated at R, and the north polar distance of Polaris will be $PR = 2\varepsilon$ (about 47°). Polaris will then be far from the north celestial pole, and, for latitudes less than 47°, it will not even be circumpolar.

The above description of precession was described as comparatively crude since it is based on two assumptions, neither of which is exact. These are that the obliquity of the ecliptic ε and the rate of luni–solar precession are both constant. Until now, we have not considered the motion of the point K, the pole of the ecliptic. According to Newtonian gravitational dynamics, the earth's motion round the sun is approximately a Keplerian orbit whose plane is fixed. This defines the pole of the ecliptic. This conclusion is, however, derived by considering the two-body problem (cf. chapter 6) which involves only the earth and the sun. The influences of the other planets are omitted. When these are included, they produce small perturbations of the earth's Keplerian orbit and, in particular, of the position of K on the celestial sphere.

Suppose that the north pole of the ecliptic is displaced from K to a point K_1. This displacement will be small; it amounts to about $0''.5$ per year. Let us consider the effect of the displacement KK_1 on the equatorial coordinates of the star X, regarding the north celestial pole as fixed at P. Since $PX = 90° - \delta$, a displacement of K can have no effect on the star's declination. On the other hand, the angle KPX in Fig. 3.1, which is equal to $90° + \alpha$, is reduced by the angle KPK_1. Consequently, the right ascension is reduced by this amount, which is independent of the position of the star. We conclude that the effect of the displacement of the point K due to planetary perturbations is the same for all stars: right ascensions are reduced each year by an amount usually denoted λ' and called the *rate of planetary precession*; declinations are unaffected. There is a concomitant reduction in the obliquity of the ecliptic.

The use of the term precession implies that the effect is regarded as secular. Two points should be noted, however. Firstly, as will be clear from the diagram, the rate of planetary precession is not an absolute constant but will have a slow variation as both P and K are displaced. Secondly, the omission of short period terms, similar to nutational terms, requires some justification. These terms are in fact very small, but they can, in any event, be omitted from positional measurements of the stars. For, as we shall see in chapter 5, meridian observations of the stars provide absolute values of declination, but only relative values of right ascension, and, as already demonstrated, any displacement of K produces no change in declination at all and only a uniform change in right ascension.

The secular change in the obliquity, caused by planetary perturbations, influences the rate of luni–solar precession. For this rate is determined by the average torque exerted by the moon and the sun on the earth, which, in turn, depends on the inclination of the earth's rotational axis to its orbital plane. This means that, like the rate of planetary precession, the rate of luni–solar precession is not an absolute constant but has a slow time variation.

The two rates of precession defined above have quite different dynamical origins, but, from a positional point of view, they are similar in that each produces a secular change in the equatorial coordinates of the stars. It is convenient, therefore, to combine them and to define a *rate of general precession*, p, given by

$$p = \psi - \lambda' \cos \varepsilon. \tag{3.4}$$

This is the general precession in longitude, and, since planetary precession is in right ascension, the factor of $\cos \varepsilon$ in the above equation is readily accounted for.

The total effect of general precession may be considered over short periods, say, less than one year, as the simple superposition of luni–solar precession and planetary precession. This was the procedure adopted above, where displacements of the points P and K were considered independently of each other. This procedure will not be sufficiently accurate over longer periods of time. Rigorous formulae must then be used which are very complicated. They will be discussed in chapter 9. The coefficients that occur in these formulae must, however, ultimately be calculated from the precessional rates defined above, which have the following values:

$$\text{Luni–solar precession } \psi = 50''.3878 + 0''.0049 \, T$$
$$\text{Planetary precession } \lambda' = 0''.1055 - 0''.0189 \, T \tag{3.5}$$
$$\text{General precession } p = 50''.2910 + 0''.0222 \, T.$$

Here T is the time in centuries from 2000.0.

We concentrated, in the above discussion, on the points P and K rather than on the equator and the ecliptic as such. This allowed us to fix ideas on a single spherical triangle. It is more normal practice, however, to refer to the reference planes themselves rather than to their poles. It has now been established that the ecliptic varies in a secular manner, and so it is necessary to specify the date to identify a unique ecliptic. The equator for this same date may be defined in two different ways, however, depending on whether nutation is included or not. If only luni–solar precession is allowed for, the equator varies in an entirely secular manner and is termed the *mean equator*. If the short-period nutational variation is also included, the resulting equator is called the *true equator*. Similar definitions may be made of the equinox: the *mean* (*true*) equinox is the intersection of the *mean* (*true*) equator with the ecliptic of that date. The catalogued positions of stars are usually referred to a mean equator and equinox; so, for comparison, meridian observations must first be corrected for nutation and, indeed, for some other effects.

If it is intended to compare a number of observations taken at different dates, it is desirable to refer them all to the *same* equator and equinox. The equatorial coordinates are then computed from each of the observations but referred to the mean equator and equinox of some standard epoch. This has most usually been the epoch of 1950, but 1900 and, more recently, 2000 have also been widely used. Indeed, any epoch, either in the past or future, could be used, provided a single epoch has been selected. This step is essential if the observations are to be the basis of some dynamical analysis. Suppose, for example, that a series of observations have been made of an asteroid and it is intended to compute the orbit from these observations. If the coordinates derived from each observation were referred to the equator and equinox corresponding to the date of that observation, then the coordinates would refer to a rotating frame of reference, and the subsequent dynamical analysis would be invalid. On the other hand, provided that the corrections for precession have been accurately made, referring observations to the mean equator and equinox of the standard epoch ensures that they are referred to an inertial frame.

The proviso just mentioned is, however, a very important one. If the stars could be regarded as fixed, then the changes in their coordinates found by meridian observations could be attributed entirely to precessional effects. Each star, however, has its own *proper motion*, and this means that the point X is itself being displaced in the celestial sphere represented in Fig. 3.1. The displacement may be in any direction.

The resulting changes in the star's coordinates are generally small compared with the precessional change. A very few stars have proper motions in excess of one arc second per year, but normally the annual proper motion is only a small fraction of this. A star's proper motion will depend on its motion relative to the centre of the celestial sphere, but also on its distance. Generally speaking, distant faint stars will have very small proper motions.

The working assumption can be made that the proper motions of stars will be random in direction. This would allow the systematic effects of precession in meridian observations to be disentangled from the proper motions. Both the precessional constants and the individual proper motions of the stars would then, in principle, be determined. It is important to recognize, however, that the accuracy of one determination limits the accuracy of the other. Consequently, although proper motions should be referred to an inertial frame, this can only be achieved from meridian observations to the precision to which the precessional constants are known.

Proper motions are derived by comparing observations taken many years apart. A long enough time interval must be allowed to let the star's displacement build up to a measurable amount. The components of annual proper motion in right ascension and declination are then obtained to a high degree of accuracy. The most usual method involves the precise measurements of a series of astrographic plates taken at a succession of dates. The measurements made on each photographic plate are reduced (see chapter 13) in order to derive right ascensions and declinations referred to a standard mean equator and equinox. Precessional effects are omitted altogether in this way. The measurement of a photographic plate can, however, only yield relative values of the stars' positions. For it is necessary to use certain stars on the plate as reference stars and to assume that their positions are known in advance for each epoch. This implies that the proper motions of the reference stars are also known in advance. Consequently, although the *relative* positions and proper motions of stars may be determined to a higher accuracy than might be achieved from absolute meridian observations, nevertheless, this advantage is largely dissipated by the ultimate dependence of the assumed positions of the reference stars on meridian observations. No increase in absolute positional accuracy is possible, and proper motions determined from astrographic plates may still suffer from systematic errors in the adopted reference frame. The latter difficulty can be minimized by using as reference stars objects at such great distances that their proper motions may be entirely neglected. Before discussing the practical realization of this, it is necessary to consider any systematic trends in the motions of the stars themselves.

The heliocentric celestial sphere is normally used when investigating proper motions of the stars. The cause of a star's proper motion is then its transverse velocity relative to the sun. Now the star and the sun each have their own particular motions within the galactic system, for the galaxy as a whole is in rotation. It is thought that the sun is at a distance of about 10 kiloparsecs from the galactic centre and, like other stars in its vicinity, is moving in a roughly circular orbit about that point. It is estimated that its velocity is between 200 and 250 km s^{-1}, and so one circuit is completed in a period of perhaps a quarter of a billion years. These conclusions follow from studies of galactic kinematics, which are based in part on proper motion surveys, but which are supplemented by radial velocity information. The latter has the advantage of being independent of distance, since it is derived from Doppler shift measurements in stellar spectral lines.

The average velocity of all the stars in the immediate solar neighbourhood defines what is known as the *local standard of rest* (LSR). Each star has its own peculiar velocity relative to the LSR, and this is also true of the sun. Since the proper motion of a star depends on its velocity *relative to the sun*, the effect of the sun's peculiar velocity, called the *local solar motion*, is present in the proper motion of every star. Assuming, therefore, that the peculiar velocities are randomly distributed, it is possible, from a statistical study of proper motions, to isolate the systematic effect of the local solar motion and to determine its magnitude and direction.

The LSR itself has a velocity with respect to the galactic centre. This is just the general rotational velocity of the galaxy and is in the direction $l = 90°$, $b = 0$. The galactic velocity of rotation varies, however, with distance from the galactic centre, and this produces small but systematic effects on the proper motions of the stars. The vast majority of stars that are observed and studied are in the same part of the galaxy as the sun and are at distances no greater than a few kiloparsecs. Consequently, the immediate concern is with differential galactic rotation, rather than with the rotation of the galaxy as a whole. It will be shown in chapter 14 that differential galactic rotation produces a proper motion μ for any star in the galactic plane, given by

$$\mu \propto A \cos 2l + B, \tag{3.6}$$

where *l* is the star's galactic longitude. This proper motion is in the direction of increasing galactic longitude and is independent of the distance of the star. The two constants *A* and *B* are known as *Oort's constants*. They are both of the order of 0″.01 per year, but *B*, in particular, is poorly determined.

The effect of the proper motion given by equation (3.6) is to introduce a virtually constant proper motion to all stars on a photographic plate. The assumption that faint stars, statistically at least, will be at great distances and will, therefore, have small proper motions is not entirely valid. They cannot, without amendment, provide a satisfactory reference frame – one that is completely inertial. The most satisfactory way of overcoming this difficulty is to use as reference points on the photographic plate objects which are far outside our galaxy, that is external galaxies. There is a difficulty here, however. Stars, being point-like objects, may have their positions accurately determined and make suitable reference points. Galaxies, on the other hand, have diffuse images on an astrographic plate, and it is more difficult to be precise in their coordinates. The discovery of quasars (QSOs – quasi-stellar objects) has, however, provided sources which are of a more suitable point-like nature. Moreover, many QSOs are radio objects whose positions may be ascertained with a very high precision by the methods of radio astrometry (chapter 16). They are generally thought to be galactic nuclei at enormous cosmological distances and are, therefore, unaffected by any galactic proper motions. Such extragalactic sources must ultimately determine the reference frame. At the moment, this is a hope that is becoming a reality.

This procedure does, however, assume that extragalactic objects, at sufficiently great distances, will exhibit no proper motions when referred to an inertial frame. This is akin to Mach's principle, requiring that the universe can have no overall rotation. This is accepted in most, but not all, cosmological theories, and entirely in those which adopt isotropic models of an expanding universe. The observational justification of such models is derived from the distant galaxies which seem to show an isotropic distribution over the whole sky and have radial velocities that increase systematically with distance. The clearest evidence of isotropy, however, is provided by the microwave background radiation, which is uniform over the celestial sphere to a very high degree. This radiation originated at a very early stage in the expansion of the universe and, therefore, demonstrates the isotropic nature of the universe at very great distances ($\sim 10^{10}$ light years).

In an isotropic model of the universe, any point may be treated as central. The observed radial velocities increase systematically with distance, approaching the velocity of light at the 'visible edge of the universe' – the horizon, as it is called. The transverse velocities, on the other hand, will only have a random distribution about zero, and this will be independent of distance. The proper motions of extragalactic objects tends to zero, therefore, with increasing distance. They are, in any event, expected to be very small for all extragalactic objects. An estimate of the random transverse velocities is provided by the sun's own velocity with respect to the universal background. Recent studies of the microwave background radiation suggest that this velocity is about 400 km s^{-1}, which would correspond to a negligible proper motion of 10^{-4} arc seconds per year at a distance of even one megaparsec. A reference frame based on the positions of extragalactic sources, therefore, fulfils all the requirements of an inertial frame of reference.

3.3   The origin of the reference frame

Attention must now be focussed on the origin of the celestial sphere, a point that was not investigated in the last section. Three different origins are in general use – the actual observing site, the centre of the earth, and the centre of the solar system, the sun. These correspond to the centres of the topocentric, geocentric and heliocentric celestial spheres, already mentioned in section 2.1. Each of these is important and useful in a different way. Observations are, of necessity, made with reference to the topocentric celestial sphere; while the ephemerides of planets and other bodies in the solar system are conveniently referred to the geocentric celestial sphere. This provides a natural standard with which observers at different terrestrial locations can compare their observations. Similarly, the heliocentric celestial sphere is a suitable standard for observations of the stellar universe. For in studying long-term effects like proper motions, annual variations present in the geocentric positions must be removed.

Consider first the topocentric origin. This is in motion due to the earth's diurnal rotation. The motion is variable (in direction) and consequently accelerated. This origin cannot, therefore, be regarded as the origin of an inertial coordinate system. The motion exhibits itself in two ways: firstly, it introduces a variable component in radial velocities; and, secondly, there are variable positional displacements due to parallax and aberration. The aberrational and radial velocity effects both depend on the observer's velocity expressed as a fraction of the velocity of light. The observer's velocity will be greatest on the equator where the rotational velocity amounts to 0.465 km s^{-1}. Parallax, on the other hand, depends on the displacement from one origin to another – in this case, it is effectively of the order of the earth's radius which is 6.378 × 10^3 km. The effect of parallax decreases with the distance of the source and is negligible for all bodies outside the solar system.

Similar considerations apply to the geocentric origin. Due to the earth's annual revolution about the sun, this origin also has an accelerated motion and cannot be regarded as the origin of an inertial coordinate system. The accelerated motion results in an annual variation in radial velocities, an annual parallax and an annual aberration. The earth's orbital velocity is nearly 30 km s^{-1}, corresponding to 10^{-4} times the velocity of light. The displacement responsible for annual parallax is of the order of an astronomical unit (AU). It is, of course, such parallax measurements that provide the key to stellar distances.

The situation is quite different when the origin of the celestial sphere is transferred to the sun. While it is true that the sun is in motion about the galactic centre, its velocity is essentially constant. Its acceleration a may be estimated as follows. Treat the sun as moving in a circle of radius $r = 10^4$ parsecs (1 pc $= 3.1 \times 10^{16}$ m) in a period $T = 2.5 \times 10^8$ years. Then, in SI units, its angular velocity

$$\omega = \frac{2\pi}{T} = 8.0 \times 10^{-16}$$

So

$$a = r\omega^2 = 2.0 \times 10^{-10} \text{ m s}^{-2}.$$

This very small acceleration is quite undetectable by any positional or radial velocity measurement. It amounts to a change in velocity of less than one metre per second in a century. In practice, therefore, the heliocentric origin may be treated as the origin of an inertial coordinate system. One fine point of distinction must, however, be made. It is the solar system as a whole that is in effectively uniform motion within the galaxy, not just the sun. The point that should be adopted as the origin of the inertial coordinate system is, therefore, the barycentre of the solar system. When the masses of the major planets are allowed for, it is found that the sun is displaced from the barycentre by a varying amount, which is of order 10^6 km. The distinction should, therefore, be made between the *barycentric* and the *heliocentric* celestial sphere. It is the barycentric celestial sphere, with a fixed equator and equinox, that must be used to provide an inertial frame of reference.

Consider now the positional effect associated with the transformation from some standard origin C to an observer, either actual or hypothetical, at O. Let the position vector of O (with respect to C) be **R**, and let **r** be the position vector of a source at S (Fig. 3.2). The apparent direction of the source for the observer at O is given by the vector **r**′, where

$$\mathbf{r}' = \mathbf{r} - \mathbf{R}. \tag{3.7}$$

The difference in the directions of the two vectors **r** and **r**′ is the parallax. Frequently, the magnitude of the displacement vector **R** is very small compared with that of the position vector of the source. It may then be convenient to use approximate first-order formulae. The exact equation (3.7) is so simple, however, that the approximate formulae will hardly appear as a simplification. The advantage, however, resides in the computational convenience rather than in the neatness of the analytic expressions.

Since our concern is primarily with directions, let $\mathbf{s}$, $\mathbf{s}'$ and $\mathbf{s}_0$ be unit vectors in the directions of $\mathbf{r}$, $\mathbf{r}'$ and $\mathbf{R}$ respectively, writing

$$\mathbf{r} = r\mathbf{s}, \quad \mathbf{r}' = r'\mathbf{s}', \quad \mathbf{R} = R\mathbf{s}_0. \tag{3.8}$$

Then equation (3.7) can be written as

$$r'\mathbf{s}' = r\mathbf{s} - R\mathbf{s}_0. \tag{3.9}$$

Take the vector product of this equation twice with the vector $\mathbf{s}$ to derive

$$\mathbf{s}' - (\mathbf{s} \cdot \mathbf{s}')\mathbf{s} = \frac{R}{r'}\mathbf{s} \times (\mathbf{s} \times \mathbf{s}_0).$$

Now, if $R \ll r$, the parallactic displacement is small. We may then write $r' = r$ *and* $\mathbf{s} \cdot \mathbf{s}' = 1$. The last equation then gives the approximate formula

$$d\mathbf{s} = \mathbf{s}' - \mathbf{s} = \frac{R}{r}\mathbf{s} \times (\mathbf{s} \times \mathbf{s}_0). \tag{3.10}$$

This is exactly the form of displacement considered in section 1.7; equation (3.10) is identical with equation (1.31), if k is identified with R/r. The formulae developed in that section may, therefore, be applied to any small

Figure 3.2

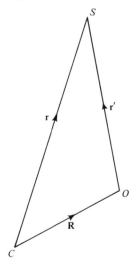

parallactic effect. However, if second- and higher-order effects in R/r cannot be ignored, it is best to use the original vector formula (3.7), which is, of course, exact.

In Fig. 3.2, the points S and O represent the source and the observer. Since both of these will, in general, be in motion relative to C, the points S and O in the diagram need to be defined with some care. In what follows, let us regard C as fixed, and let O have a velocity V with respect to this origin. Suppose that an observation of the direction of the source is made from O at time t. Let τ be the time that it takes the observed radiation to move from source to observer. It is then convenient to state that, in Fig. 3.2, O represents the observer's position at time t, and S the source's position at the earlier time $(t-\tau)$. With this refinement, equation (3.7) remains exact, but any effects of the source's motion are removed.

The line SO now represents the path of the observed photons, but in a coordinate system centred on C. The vector $\mathbf{s}'$ considered earlier, therefore, gives the direction of the source for a hypothetical *stationary* observer situated at O. The actual observer has a velocity V relative to this, and as a result the apparent direction of the source has an aberrational displacement. Let us consider this first from a classical point of view.

Denote the magnitude of the velocity of the photons in C's frame by c. Then the photons have a velocity vector $-c\mathbf{s}'$ in this frame. The frame itself has a velocity $-V$ relative to the observer, so the photons are observed to have a velocity vector

$$\mathbf{u} = -c\mathbf{s}' - \mathbf{V}. \tag{3.11}$$

Since this is a classical approach, the magnitude of $\mathbf{u}$ is not necessarily c, as it would be in a relativistic treatment. So put $\mathbf{u} = -c^*\mathbf{s}^*$, and $\mathbf{V} = V\mathbf{n}$, where $\mathbf{s}^*$ and $\mathbf{n}$ are unit vectors, the former giving the observed direction of the source. Then, by (3.11),

$$c^*\mathbf{s}^* = c\mathbf{s}' + V\mathbf{n}. \tag{3.12}$$

This equation is very similar to (3.9) and may be treated in the same way. Taking the vector product twice with $\mathbf{s}'$, and assuming that $V \ll c$, yields the approximate result

$$\mathbf{s}^* - \mathbf{s}' = -\frac{V}{c}\mathbf{s}' \times (\mathbf{s}' \times \mathbf{n}). \tag{3.13}$$

This is only accurate to order V/c, and once again it is to be compared with equation (1.31) so that the formulae of section 1.7 may be used.

The total positional effect of a transference of the origin from the point C to the point O is derived by adding together equations (3.10) and (3.13). This maintains only first-order accuracy in both R/r and V/c, but this is sufficient for many purposes. To this accuracy, s' may be replaced on the right-hand side of (3.13) by s, yielding the result

$$s^* - s = \frac{R}{r} s \times (s \times s_0) - \frac{V}{c} s \times (s \times n). \tag{3.14}$$

To summarize the notation, s^* is the direction observed from O at time t, while s is the geometric direction of the source from C at time $t - \tau$. The displacement of the origin is Rs_0, and the new origin has a velocity Vn relative to C.

Displacements of this kind are considered in transforming between topocentric and geocentric coordinates in chapter 4, and between geocentric and heliocentric coordinates in chapter 8. Where higher accuracy is required in the parallax displacement, this is easily obtained by using equation (3.7) directly. If higher accuracy is required in the aberrational displacement, however, the methods of special relativity must be used.

3.4   Selective review of special relativity

In treating aberration from a classical standpoint, we were forced to allow the velocity of light to vary from one frame of reference to another. This is in contradiction of the underlying principle of Special Relativity in which the velocity of light is a universal constant, independent of the observer. This principle requires such an interconnection of the space and time coordinates that time must be regarded as a fourth dimension, certainly having a different dimensionality but having an equivalent status to the three space dimensions. Whereas, in Newtonian physics, space can be regarded as relative but time is absolute, in the Special Theory of Relativity, both space and time are of a relative nature depending on the frame of reference, or equivalently the particular observer.

A four-dimensional manifold of 'points' is, therefore, adopted, each point being specified by three space coordinates and one time coordinate. A point represents an *event*, since its coordinates determine a unique point of space at a unique instant of time. For example, an observation is an event, whose time coordinate gives the time at which the observation took place, and whose space coordinates give the position of the observer at that time. The coordinate system may, of course, be changed, different coordinate systems being more naturally associated with different observers. When such a transformation is made, the time coordinate of an event is altered in formally much the same way as any of the space coordinates. In particular, the time difference between events is not conserved. What is conserved is the *interval* between the two events, which is defined as follows.

Consider two neighbouring events with coordinates (x, y, z, t) and

$(x+dx, y+dy, z+dz, t+dt)$. The element of interval between these two events will be denoted by ds and is given by the *metric*. In special relativity, which describes spacetime in the absence of gravity, the metric takes the form,

$$ds^2 = dt^2 - \frac{1}{c^2}(dx^2 + dy^2 + dz^2). \tag{3.15}$$

This very simple form of the metric occurs only if the coordinate system is inertial. In this form, x, y and z are rectangular Cartesian space coordinates, and t is the time as measured by any observer at rest in this coordinate system. If the coordinate system is not inertial, the metric will take a more complicated form. For example, suppose that (x, y, z, t) is an inertial coordinate system, while (x', y', z', t') is one that is rotating with constant angular velocity ω about the z-axis. The two systems are related by the transformation equations

$$\begin{aligned}
x &= x' \cos \omega t' - y' \sin \omega t', \\
y &= x' \sin \omega t' + y' \cos \omega t', \\
z &= z', \\
t &= t'.
\end{aligned} \tag{3.16}$$

Taking differentials and substituting into equation (3.15) yields the metric in the dashed coordinate system, after a little reduction, as

$$\begin{aligned}
ds^2 &= [1 - (x'^2 + y'^2)\omega^2/c^2]\, dt'^2 - c^{-2}(dx'^2 + dy'^2 + dz'^2) \\
&\quad + \frac{2\omega y'}{c^2}\, dx'\, dt' - \frac{2\omega x'}{c^2}\, dy'\, dt'.
\end{aligned} \tag{3.17}$$

The additional terms in this form of the metric are found to give rise to the so-called fictitious forces – centrifugal force and Coriolis force – that occur in a rotating frame of reference. In the Special Theory of Relativity, one normally works exclusively in inertial coordinate systems, but this is a matter of convenience, not of necessity.

The element of interval ds between two neighbouring events is defined by the metric and it is invariant under any coordinate transformation. This interval is said to be *time-like* or *space-like* according as ds is real or imaginary. If $ds^2 = 0$, the element of interval is said to be *null*.

The interval between two non-neighbouring events cannot be defined uniquely, at least without qualification. For suppose E_1 and E_2 are two such events; they are represented by two points in the four-dimensional spacetime. Let Γ be any open curve in the spacetime that joins E_1 and E_2. Then the interval s *along this curve* may be derived by integrating the element of interval ds along it, the latter being derived from the metric. Then

$$s = \int_{\Gamma E_1}^{E_2} \mathrm{d}s. \tag{3.18}$$

This interval is path dependent – its value depending on the choice of the curve Γ. In general, s will be a complex number, since parts of the curve may be space-like and parts time-like or even null. The interval between the points E_1 and E_2 can be defined uniquely, however, if Γ is chosen to be the *geodesic* joining the two points. This is the curve for which the interval s in equation (3.18) has a *stationary* value – it need not be, and in fact is not, a minimum. It can be shown, however, that any geodesic has the property that it is either time-like, space-like, or null along its entire length. So any two events are joined by a geodesic, which is a uniquely determined curve, and the interval between the two events may be unambiguously described as either space-like, time-like or null according as the interval along the geodesic is imaginary, real or zero.

A curve in four dimensions, whether a geodesic or not, is a one parameter family of points, which may be represented as

$$x = x(p), \quad y = y(p), \quad z = z(p), \quad t = t(p). \tag{3.19}$$

The curve parameter p can be chosen in many ways. In particular, if the curve is time-like, the interval s, measured along the curve from some fixed initial point, may be chosen as the curve parameter. This choice is possible, but inconvenient, for a space-like curve, since s is then imaginary; the choice is not possible, if the curve is null. Now the curve represents a continuous sequence of events. If the events are exactly those appertaining to a particular body, the curve is said to be that body's *world line*. Suppose that the body has velocity $\mathbf{v} = (v_x, v_y, v_z)$. Then, since $v_x = \mathrm{d}x/\mathrm{d}t$, etc., we may write

$$\frac{\mathrm{d}x}{\mathrm{d}s} = \gamma v_x, \quad \frac{\mathrm{d}y}{\mathrm{d}s} = \gamma v_y, \quad \frac{\mathrm{d}z}{\mathrm{d}s} = \gamma v_z,$$

where

$$\gamma = \frac{\mathrm{d}t}{\mathrm{d}s} = \left(1 - \frac{v^2}{c^2}\right)^{-1/2}. \tag{3.20}$$

The last result follows from the metric (3.15).

For a real material body, $v < c$, and so γ is real – in fact, $\gamma > 1$. Since the time coordinate is real, it follows that $\mathrm{d}s$ is real. The body's world line is time-like, and s may certainly be adopted as the curve parameter. Suppose next that the body is moving freely and is not subject to any external force. Then, by Newton's first law, v_x, v_y, v_z, and therefore γ, are all constant throughout the motion. The world line then has an algebraically linear form in four dimensions. For, if (x_0, y_0, z_0, t_0) is any event on the world line, and if *s* is the interval measured from this event, then integrating equations (3.20) yields

$$x = x_0 + \gamma v_x s,$$
$$y = y_0 + \gamma v_y s,$$
$$z = z_0 + \gamma v_z s,$$
$$t = t_0 + \gamma s.$$

(3.21)

The world line of a free particle may be described as a straight line in four dimensions. More precisely, it is a geodesic. The general form of the geodesics' differential equations is given in equation (3.36). For the simple form of the metric given by (3.15), however, these differential equations reduce to

$$\frac{d^2x}{ds^2} = \frac{d^2y}{ds^2} = \frac{d^2z}{ds^2} = \frac{d^2t}{ds^2} = 0.$$

(3.22)

Two integrations of (3.22) lead to the parametric solution (3.21), already established. The simple linear form of the geodesics is a property of the inertial coordinate system. For a non-inertial coordinate system, the geodesics will have a more complicated parametric form. Since geodesics are invariant curves, however, the geodesic, even in a non-inertial coordinate system, still gives the world line of the free particle. This illustrates in part the *geodesic principle*, which may be stated as follows: The world lines of free particles are time-like geodesics. Further, as an extension, the world lines of photons are null geodesics. This principle is retained in the General Theory of Relativity.

The geodesic principle can be regarded as a generalization of Newton's first law of motion. Whenever the metric has the simple form of (3.15), the geodesics have the corresponding simple form (3.21). In such a coordinate system, a free particle moves in a straight line with constant velocity. The coordinate system is inertial. So, in a transformation from one inertial coordinate system to another, the form of the metric must be unchanged. The transformation with this property is termed the *generalized Lorentz transformation*. It is a combination of three simpler transformations, namely (i) a translation of the origin, (ii) a rotation of the coordinate space axes, and (iii) a boost. Let us consider these in turn.

(i) A translation of the origin of the spacetime to a new point, say, the event (x_0, y_0, z_0, t_0), will clearly not affect any of the differentials. The metric is unchanged.

(ii) A simple rotation of axes does not affect the time coordinate at all. The space part of the metric is that of three-dimensional Euclidean geometry; this is unaltered by a rotation of axes. So the metric as a whole is unchanged.

(iii) A boost maintains the origin of the spacetime, but the *space* origin in the new coordinate system is moving with a constant velocity relative to the original coordinate system. For a boost in the *x*-direction with velocity V, the coordinate transformation equations are

$$\tilde{x} = \beta(x - Vt),$$
$$\tilde{y} = y, \quad \tilde{z} = z,$$
$$\tilde{t} = \beta(t - Vx/c^2),$$
(3.23)

where

$$\beta = (1 - V^2/c^2)^{-1/2}$$

It is easily verified that, as a result of this transformation,

$$d\tilde{t}^2 - \frac{1}{c^2}(d\tilde{x}^2 + d\tilde{y}^2 + d\tilde{z}^2) = dt^2 - \frac{1}{c^2}(dx^2 + dy^2 + dz^2), \quad (3.24)$$

as required. The inverse transformation is obtained by writing $-V$ for V, i.e. it is easily shown to be

$$x = \beta(\tilde{x} + V\tilde{t}),$$
$$y = \tilde{y}, \quad z = \tilde{z},$$
$$t = \beta(\tilde{t} + V\tilde{x}/c^2).$$
(3.25)

The transformations (3.23) and (3.25) are often referred to simply as Lorentz transformations. For our present purpose, however, they are better referred to as boosts in the *x*-direction. For the generalized Lorentz transformation, as already mentioned, is a combination of the transformations (i), (ii) and (iii) listed above.

As an example, let us consider a translation to a new origin (X, Y, Z, T) followed by a boost with velocity $\mathbf{V} = (V_x, V_y, V_z)$ in a direction which is not along one of the coordinate axes. First, the axes must be rotated to bring the *x*-axis into the direction of $\mathbf{V}$, then the transformation (3.23) is applied, and finally the axes must be rotated back through the same angle. The resulting transformation equations are rather complicated, but equation (3.24) will still hold. In fact, after considerable reduction (cf. Problems 3.7 and 3.8 at the end of this chapter) these transformation equations turn out to be

$$x^* = \left(1 + \frac{\beta^2 V_x^2}{(\beta+1)c^2}\right)(x - X) + \frac{\beta^2 V_x V_y}{(\beta+1)c^2}(y - Y) + \frac{\beta^2 V_x V_z}{(\beta+1)c^2}(z - Z) - \beta V_x(t - T),$$

$$y^* = \frac{\beta^2 V_y V_x}{(\beta+1)c^2}(x - X) + \left(1 + \frac{\beta^2 V_y^2}{(\beta+1)c^2}\right)(y - Y) + \frac{\beta^2 V_y V_z}{(\beta+1)c^2}(z - Z) - \beta V_y(t - T),$$

$$z^* = \frac{\beta^2 V_z V_x}{(\beta+1)c^2}(x-X) + \frac{\beta^2 V_z V_y}{(\beta+1)c^2}(y-Y) + \left(1 + \frac{\beta^2 V_z^2}{(\beta+1)c^2}\right)(z-Z) - \beta V_z(t-T),$$

$$t^* = -\frac{\beta V_x}{c^2}(x-X) - \frac{\beta V_y}{c^2}(y-Y) - \frac{\beta V_z}{c^2}(z-Z) + \beta(t-T). \qquad (3.26)$$

Transformations of this type are important in positional astronomy in determining the aberrational displacement resulting from a change of origin of the type considered in the last section.

Suppose, as before, that C is the space-origin of an inertial coordinate system (x, y, z, t). To fix ideas, we might regard it as the barycentric origin. An observer O makes an observation (Fig. 3.3) at the instant $t = T$, when he is situated at the point with position vector $\mathbf{R} = (X, Y, Z)$. Then the observation is the event (X, Y, Z, T). Let the observer's velocity at the instant of observation be $\mathbf{V} = (V_x, V_y, V_z)$. Then the coordinate system (x^*, y^*, z^*, t^*) defined by equations (3.26) is that of an inertial observer coincident with O at time T and moving with the same instantaneous velocity. We may refer to it as O's coordinate system. Moreover, the observation itself is now the origin of the spacetime in the new coordinate system.

Let the observed photons have velocity $\mathbf{v} = (v_x, v_y, v_z)$ in the original

Figure 3.3

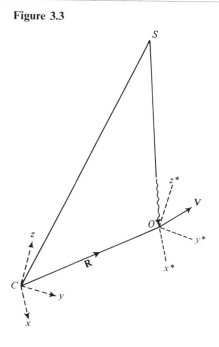

coordinate system. Once corrected for parallax, the vector **v** gives the direction of the source in C's system. The parallax correction follows from equation (3.7) and is not subject to any special relativistic effect. The classical treatment is exact. Let the photons' velocity in the starred coordinate system be $\mathbf{u} = (u_x, u_y, u_z)$. This corresponds to the apparent direction of the source, and the difference between the two velocity vectors is essentially the aberrational displacement.

Now, using the time coordinate t as the curve parameter, the photons' world line is

$$
\begin{aligned}
x &= X + v_x(t - T), \\
y &= Y + v_y(t - T), \\
z &= Z + v_z(t - T).
\end{aligned}
\tag{3.27}
$$

This is in C's coordinate system and, of course,

$$
\mathbf{v} = \left(\frac{\mathrm{d}x}{\mathrm{d}t}, \frac{\mathrm{d}y}{\mathrm{d}t}, \frac{\mathrm{d}z}{\mathrm{d}t} \right).
\tag{3.28}
$$

Since the world line is a null geodesic, it follows from the metric that $v_x^2 + v_y^2 + v_z^2 = c^2$. In O's coordinate system the world line becomes

$$
\begin{aligned}
x^* &= u_x t^*, \\
y^* &= u_y t^*, \\
z^* &= u_z t^*,
\end{aligned}
\tag{3.29}
$$

and

$$
\mathbf{u} = \left(\frac{\mathrm{d}x^*}{\mathrm{d}t^*}, \frac{\mathrm{d}y^*}{\mathrm{d}t^*}, \frac{\mathrm{d}z^*}{\mathrm{d}t^*} \right).
\tag{3.30}
$$

Then the unchanged form of the metric ensures that the photons' velocity is unchanged – cf. equation (3.24). The individual components of this velocity are, however, altered and these must be determined by using the coordinate transformation.

This can be achieved as follows: substitute for x, y and z from (3.27) into (3.26), to derive

$$
\begin{aligned}
x^* &= \left[v_x + \frac{\beta^2 V_x}{(\beta + 1)c^2} (v_x V_x + v_y V_y + v_z V_z) - \beta V_x \right](t - T), \\
y^* &= \left[v_y + \frac{\beta^2 V_y}{(\beta + 1)c^2} (v_x V_x + v_y V_y + v_z V_z) - \beta V_y \right](t - T), \\
z^* &= \left[v_z + \frac{\beta^2 V_z}{(\beta + 1)c^2} (v_x V_x + v_y V_y + v_z V_z) - \beta V_z \right](t - T), \\
t^* &= \beta \left[1 - \frac{(v_x V_x + v_y V_y + v_z V_z)}{c^2} \right](t - T).
\end{aligned}
\tag{3.31}
$$

The last of these equations may be used to eliminate $(t - T)$ from the other three. Comparing the result with (3.29) then yields (u_x, u_y, u_z). In vector form this result is

$$\mathbf{u} = \left[\frac{\mathbf{v}}{\beta} - \mathbf{V} + \frac{\beta \mathbf{v} \cdot \mathbf{V}}{(\beta + 1)c^2} \mathbf{V} \right] \bigg/ \left(1 - \frac{\mathbf{v} \cdot \mathbf{V}}{c^2} \right). \tag{3.32}$$

This formula indicates how the velocity transforms under a generalized Lorentz transformation. It can be applied to the velocity of a material particle as well as that of a photon. As a check, the reader may like to verify, by squaring equation (3.32), that when $\mathbf{v}$ has magnitude c, $\mathbf{u}$ also has this magnitude.

Let us write equation (3.32) in terms of unit vectors, so that the orders of magnitude become clear. Therefore, put $\mathbf{v} = -c\mathbf{s}'$, $\mathbf{u} = -c\mathbf{s}^*$, and $\mathbf{V} = V\mathbf{n}$, where $\mathbf{s}'$, $\mathbf{s}^*$ and $\mathbf{n}$ are all unit vectors. Then (3.32) can be written as

$$\mathbf{s}^* = \left(1 + \frac{V}{c} \mathbf{s}' \cdot \mathbf{n} \right)^{-1} \left[\frac{\mathbf{s}'}{\beta} + \frac{V}{c} \mathbf{n} + \frac{\beta V^2 (\mathbf{s}' \cdot \mathbf{n})}{(\beta + 1)c^2} \mathbf{n} \right]. \tag{3.33}$$

This formula gives the exact aberrational effect according to special relativity. The vector $\mathbf{s}^*$ gives the observed direction of the source, and $\mathbf{s}'$ the direction of the source for an observer instantaneously coincident with O but at rest in C's coordinate system. It is not the direction of the source as seen by C, since parallax is not included. The notation corresponds to that in the classical treatment given earlier (cf. equation (3.12)). To first order in (V/c), the classical and special relativistic treatments give the same result. Indeed, the first-order formula (3.13) can be re-established by taking the vector product of (3.33) twice with $\mathbf{s}'$ and neglecting higher-order terms, to obtain

$$\mathbf{s}^* - \mathbf{s}' = -\frac{V}{c} \mathbf{s}' \times (\mathbf{s}' \times \mathbf{n}). \tag{3.13}$$

As previously mentioned, this last formula is sufficiently accurate for most purposes, but not all. In no case, however, need terms of an order higher than the second be included. It is convenient, therefore, to expand (3.33) by the binomial theorem, to derive

$$\mathbf{s}^* = \left[1 - \frac{V}{c} (\mathbf{s}' \cdot \mathbf{n}) + \frac{V^2}{c^2} (\mathbf{s}' \cdot \mathbf{n})^2 \ldots \right] \left[\frac{\mathbf{s}'}{\beta} + \frac{V}{c} \mathbf{n} + \left(\frac{\beta}{\beta + 1} \right) \frac{V^2}{c^2} (\mathbf{s}' \cdot \mathbf{n}) \mathbf{n} \right].$$

The quantity β must also be expressed as a power series in (V/c), giving

$$\beta = 1 + \tfrac{1}{2} V^2 / c^2 \ldots$$

After a little reduction, the second-order formula is found to be

$$\mathbf{s} = \mathbf{s}' - \frac{V}{c} \mathbf{s}' \times (\mathbf{s}' \times \mathbf{n}) + \frac{1}{2} \frac{V^2}{c^2} \{ [2(\mathbf{s}' \cdot \mathbf{n})^2 - 1] \mathbf{s}' - (\mathbf{s}' \cdot \mathbf{n}) \mathbf{n} \}. \tag{3.34}$$

This degree of accuracy is only required for the reduction of the most precise measurements in positional astronomy. When this is the case, however, there is another effect of the same order of magnitude to be considered which arises in the General Theory of Relativity. This is the gravitational deflection of the light itself. The deflection would imply that the figure *CSO* in the diagram (Fig. 3.3) can no longer be regarded as a plane triangle, for space itself is non-Euclidean. The use of three-dimensional vectors is then suspect, and they should be regarded as no more than a shorthand notation. The basic argument must be formulated in terms of four-dimensional vectors, as explained in the next section.

3.5   General relativity

The spacetime in the Special Theory of Relativity is said to be *flat*. Its geometrical features, which are entirely determined by the metric, are an extension to four dimensions of ordinary Euclidean geometry. The flatness of the spacetime reveals itself in the geodesics, which can be regarded as straight lines in four dimensions. This is a consequence of the simple form of the metric (3.15), in which the coefficients of the squares of the differentials are all constant. The above is only true in an inertial frame of reference, and it is this that accounts for the importance and usefulness of such reference frames.

The effects of gravity are included in the General Theory of Relativity. Consequently, the spacetime has a curvature, and the metric cannot be expressed in the simple form (3.15). This means that there is no coordinate system that is inertial; none has this special status. Einstein, therefore, formulated the principle of covariance which states that the laws of physics may be expressed in the same form in *all* coordinate systems. This is only possible if the techniques of the tensor calculus are used. The reader who is not familiar with the tensor calculus is advised, before proceeding further with this section, to consult Appendix A, where the basic features of the tensor calculus are discussed and explained. It is beyond the scope of this book to give a full exposition of the General Theory of Relativity. We are only concerned with its relevance in the interpretation of positional observations. There is no need to consider in detail how the geometry of the spacetime is related to the distribution of gravitating material. This is expressed through Einstein's field equations which, in turn, determine the metric. We shall omit all consideration of these equations and take the metric itself as our starting point.

As before, spacetime is represented by a four-dimensional manifold of points. Each point represents an event, and it is specified by four real coordinates x^{λ} $(\lambda = 1, \ldots, 4)$. Henceforth the single appearance of a suffix will mean that the suffix may take any of these values. The coordinate system may be quite general. The individual coordinates might be angular coordinates, or have the dimensions of length or time, or even a combination of these. The coordinates of a point should be regarded as no more than labels identifying the particular event. The geometry of the spacetime is determined by the metric, which has the general Riemannian form

$$ds^2 = g_{\mu\nu}\, dx^\mu\, dx^\nu. \tag{3.35}$$

Here the dummy suffix notation is being used, the repeated suffices μ and ν indicating a double sum with these suffices each taking values $1, \ldots, 4$ independently. Now $g_{\mu\nu}$ is a covariant tensor of rank 2, called the metrical tensor. Since the differentials dx^μ are components of a contravariant vector, equation (3.35) indicates that ds^2 is a scalar quantity independent of the coordinate system in value.

The geodesics are curves $x^\lambda = x^\lambda(p)$ which satisfy the geodesic differential equations, namely

$$\frac{d}{dp}\left(g_{\lambda\mu}\frac{dx^\mu}{dp} \right) - \frac{1}{2}\frac{\partial g_{\mu\nu}}{\partial x^\lambda}\frac{dx^\mu}{dp}\frac{dx^\nu}{dp} = 0. \tag{3.36}$$

These equations allow the geodesics to be derived once the metric is known. As before, the geodesics may be time-like, space-like or null. Moreover, the geodesic principle may again be applied, that is (i) the world lines of free particles are time-like geodesics, and (ii) the world lines of photons are null geodesics. The term 'free particle' is now to be understood, however, as a particle not subject to any external force, but free to move under gravity. For gravitation is now a property of the spacetime and is not to be regarded as an externally applied force. The curve parameter p occurring in equation (3.36) is not arbitrary but must be an 'affine parameter'. For a time-like geodesic this may be the interval s measured from an arbitrary event on the geodesic.

Consider now a time-like curve, possibly, but not necessarily, a geodesic, expressed in the form $x^\lambda = x^\lambda(s)$. Let q^λ be the contravariant vector defined by

$$q^\lambda = \frac{dx^\lambda}{ds} \tag{3.37}$$

at every point of the curve. This is the *unit tangent vector* to the curve. It is a unit vector since

$$q_\mu q^\mu = g_{\mu\nu}q^\nu q^\mu = g_{\mu\nu}\frac{dx^\mu}{ds}\frac{dx^\nu}{ds} = 1. \tag{3.38}$$

On the other hand, if $x^\lambda = x^\lambda(p)$ is a null curve, the tangent vector, u^λ, say, may be defined similarly, but it is not a unit vector. For writing

$$u^{\lambda} = \frac{\mathrm{d}x^{\lambda}}{\mathrm{d}p}, \tag{3.39}$$

it follows that

$$u_{\mu}u^{\mu} = g_{\mu\nu}\frac{\mathrm{d}x^{\mu}}{\mathrm{d}p}\frac{\mathrm{d}x^{\nu}}{\mathrm{d}p} = \left(\frac{\mathrm{d}s}{\mathrm{d}p}\right)^{2} = 0. \tag{3.40}$$

A null vector, like u^{λ}, is in fact orthogonal to itself.

Suppose now that a number of positional observations are taken at different times by the one observer. Each observation is an event situated on the observer's world line, which is normally a time-like geodesic – it is certainly a time-like curve. Since coordinates are no more than labels, all observational features must be expressible in an invariant form. Consider first the time interval between two observations as measured by the observer. This may be identified with the interval s between these two events on the observer's world line. This provides a satisfactory invariant definition of time.

The definition of the observed direction of a source in general relativity is rather more subtle. It is conveniently specified by three direction cosines, that is by the angles that the photons' velocity makes with the three coordinate axes. The photon itself is represented by a null geodesic, but the identification of the coordinate axes is not immediately obvious. A comparison with special relativity is helpful here. Consider the coordinate system of an inertial observer (x, y, z, t). The observer's world line is the t-axis, and its tangent vector has components $(0, 0, 0, 1)$. It is seen then that vectors in the directions of the coordinate axes are orthogonal to the observer's world line. This fact is used in making an invariant definition of the observed direction of a source in general relativity.

Consider now an observation which corresponds to an event O. Let q^{λ} be the unit tangent vector to the observer's world line at this point, and let u^{λ} be the tangent vector, again at O, to the null geodesic that represents the observed photons. The observer's coordinate axes are represented by space-like unit vectors at $O, i^{\lambda}, j^{\lambda}, k^{\lambda}$, say. These must be mutually orthogonal and each must be orthogonal to q^{λ}; that is, formally,

$$i^{\lambda}i_{\lambda} = j^{\lambda}j_{\lambda} = k^{\lambda}k_{\lambda} = -1,$$
$$i^{\lambda}j_{\lambda} = j^{\lambda}k_{\lambda} = k^{\lambda}i_{\lambda} = 0, \tag{3.41}$$

and

$$i^{\lambda}q_{\lambda} = j^{\lambda}q_{\lambda} = k^{\lambda}q_{\lambda} = 0. \tag{3.42}$$

The null vector u^{λ} may be resolved into a component parallel to q^{λ} and a component orthogonal to this vector. This means that

$$u^{\lambda} = \alpha q^{\lambda} + \beta e^{\lambda}, \tag{3.43}$$

where e^λ is a unit space-like vector, and $e^\lambda q_\lambda = 0$. Since u^λ is null, it follows from (3.40) that

$$\alpha = \beta = u^\lambda q_\lambda. \tag{3.44}$$

Now the unit vector e^λ determines the direction of the source as seen by the observer. In fact, the source's direction cosines (l, m, n) are given by the inner products of e^λ with each of the unit vectors i^λ, j^λ and k^λ. It is then deduced from equations (3.43), (3.42) and (3.44) that

$$l = (u^\lambda i_\lambda)/(u^\lambda q_\lambda),$$
$$m = (u^\lambda j_\lambda)/(u^\lambda q_\lambda), \tag{3.45}$$
$$n = (u^\lambda k_\lambda)/(u^\lambda q_\lambda).$$

The conditions (3.41) and (3.42) are sufficient to ensure that the vectors $i^\lambda, j^\lambda, k^\lambda$ represent a consistent set of mutually orthogonal axes for the observer, but not that these axes are oriented in the desired directions. The solution of this problem is best considered specifically in terms of the spacetime that represents the sun's gravitational field.

The *Schwarzschild spacetime* is spherically symmetric and has a single point-singularity. As stated below, it describes the gravitational field about a particle of mass M, or the field external to a spherically symmetric body of this mass. It is convenient to put $x^1 = r$, $x^2 = \theta$, $x^3 = \phi$, and $x^4 = t$. Then the simplest form of the metric of Schwarzschild spacetime is

$$ds^2 = (1 - 2m/r) \, dt^2 - c^{-2}[(1 - 2m/r)^{-1} \, dr^2 + r^2 \, d\theta^2 + r^2 \sin^2 \theta \, d\phi^2],$$
$$\tag{3.46}$$

where

$$m = GM/c^2, \tag{3.47}$$

G being the Newtonian gravitational constant. The quantity m has the dimensions of a length, and, if the mass of the sun is inserted for M in (3.47), it is found that $m \approx 1.5$ km.

This metric may be compared with the metric of special relativity expressed in terms of spherical polar coordinates. Applying the transformation

$$x = r \sin \theta \cos \phi,$$
$$y = r \sin \theta \sin \phi, \tag{3.48}$$
$$z = r \cos \theta,$$

to the metric (3.15) yields

$$ds^2 = dt^2 - c^{-2}(dr^2 + r^2 \, d\theta^2 + r^2 \sin^2 \theta \, d\phi^2). \tag{3.49}$$

The similarity between this and equation (3.46) is evident. Moreover, the Schwarzschild metric is to be applied at values of the radial coordinate of the order of 1 AU, so that $r \gg m$. This means that the differences between the two metrics are very small. Further, most problems in solar system gravitational dynamics may be treated adequately using only Newtonian theory. This is true, in particular, of planetary perturbations – the only significant effects of purely general relativistic origin arise from the sun's influence. The Schwarzschild metric (3.46), which includes only the gravitational effect of the central body in the solar system, is, in these circumstances, satisfactory.

The planets are bodies that are moving freely in the sun's gravitational field. Newtonian gravitation predicts Keplerian elliptic orbits that are fixed in space. The planets' world lines in general relativity will be time-like geodesics. These may be derived from equation (3.36), using the components of the metrical tensor that are supplied by the metric itself (3.46). This procedure will be followed in detail in chapter 6. It is found that the planetary orbits are best described as ellipses that are very slowly rotating in their orbital planes. The effect is greatest for the planet Mercury for which it amounts to nearly $0''.5$ per year. Positional observations of Mercury had detected this effect, over 50 years before Einstein formulated the General Theory of Relativity and provided the explanation for it.

Another effect of general relativistic origin that is important in positional astronomy is the gravitational deflection of light. This can be studied by investigating the null geodesics of the metric (3.46). These curves are no longer straight lines but are bent under the sun's gravitational influence. They are slightly concave towards the sun. The total gravitational deflection, when expressed in radians, is of the order of m/r, where r is the closest distance that the light path approaches to the centre of the sun. The effect, predicted by Einstein in 1915, was first detected observationally four years later during a solar eclipse. For, at the time of totality, stars can be seen in the vicinity of the sun itself, and the ray-paths of light from these stars pass within a few solar radii of the sun's centre. The quantity m/r is then of the order of an arc second. The effect has been amply verified by positional observations of radio sources that lie close to the ecliptic. At the appropriate time of year, such radio sources will be close to the sun in the sky, and, of course, daytime observations of radio sources present no insurmountable problems.

The gravitational deflection was regarded until recently as an effect that could be detected by positional observations of sources in close proximity to the sun and, as such, was an important test of the theory of general relativity. The effect was disregarded under more normal circumstances. The very high astrometric accuracy that can now be achieved makes this disregard unjustifiable. For any earth-based observation at all, the minimum distance of the ray-path from the sun cannot exceed 1 AU, and the ratio m/r for this distance is equivalent to an angle of $0''.002$. This is of the same order of magnitude as the second-order terms due to annual aberration. What it more, this is no coincidence. For it follows from Newtonian gravitational theory (cf. chapter 6) that the velocity of a body moving in a circular orbit of radius a about the sun is given by $V^2 = GM/a$. It follows that for any earth-based observation

$$m/r \gtrsim V^2/c^2. \tag{3.50}$$

The gravitational deflection of light is, therefore, expected to be at least as large as the second-order aberrational effect.

3.6 General relativistic treatment of parallax, aberration and light deflection

The three positional effects of annual parallax, annual aberration and light deflection are usually treated separately. This is certainly justifiable if they are all sufficiently small, which is usually a reasonable supposition. Parallax is then treated from a classical standpoint, aberration from the standpoint of special relativity, and only the light deflection requires a general relativistic formulation. While this may be a practical procedure, its theoretical justification depends on the existence of a satisfactory synthesis of the three effects within the General Theory of Relativity.

The problem is essentially to relate the heliocentric position of a source to the position observed by a hypothetical geocentric observer. Although this problem may be formulated in any coordinate system, its formulation is greatly facilitated by using a particular system that relates readily to Cartesian coordinates and to direction cosines. The Schwarzschild metric (3.46) is transformed to a new coordinate system (r', θ', ϕ', t') where

$$r = r'(1 + \tfrac{1}{2}m/r')^2,$$
$$\theta = \theta', \quad \phi = \phi', \quad t = t'. \tag{3.51}$$

When this transformation is made, we obtain what is known as the isotropic form of the metric, which, dropping the dashes, is given by

$$ds^2 = \frac{\left(1 - \dfrac{m}{2r}\right)^2}{\left(1 + \dfrac{m}{2r}\right)^2} dt^2 - \frac{1}{c^2}\left(1 + \frac{m}{2r}\right)^4 (dr^2 + r^2 d\theta^2 + r^2 \sin^2 \theta \, d\phi^2). \tag{3.52}$$

The final bracket on the right-hand side of this equation is just the metric of ordinary three-dimensional Euclidean geometry expressed in spherical polar coordinates. It is possible, then, to make the formal transformation to

Cartesian coordinates, namely equation (3.48), and to write the metric in the quasi-Cartesian form as

$$ds^2 = \frac{\left(1 - \dfrac{m}{2r}\right)^2}{\left(1 + \dfrac{m}{2r}\right)^2} dt^2 - \frac{1}{c^2}\left(1 + \frac{m}{2r}\right)^4 (dx^2 + dy^2 + dz^2), \qquad (3.53)$$

where

$$r^2 = x^2 + y^2 + z^2. \qquad (3.54)$$

This is the form of the metric that will be used. It should always be remembered that coordinates in general relativity are basically only labels for the points. We may speak of t as the time coordinate and x, y, z as the space coordinates, but they should not rashly be identified with actual times and distances. These must be defined in an invariant manner. In particular, r is not strictly the distance from the sun, although it is, of course, approximately this.

Suppose that an observation is made at space coordinates (x_0, y_0, z_0) and at coordinate time t_0. This observation is then the event, O say, with coordinates (x_0, y_0, z_0, t_0). Suppose, further, that at the time of emission of the observed photons, the source had space coordinates (x_e, y_e, z_e). The emission of these photons is also an event, E say, with coordinates (x_e, y_e, z_e, t_e), where $t_e = t_0 - \tau$, τ being the coordinate travel time of the light between the two events. But the two events E and O are linked by a null geodesic. It is, therefore, possible to determine τ by solving the geodesic differential equations (3.36), the necessary data being the space coordinates of the two events. Now the heliocentric direction of the source can be defined as the unit (three) vector $\mathbf{s}$, where

$$\mathbf{s} = (x_e/r_e, \ y_e/r_e, \ z_e/r_e), \qquad (3.55)$$

and, of course,

$$r_e^2 = x_e^2 + y_e^2 + z_e^2. \qquad (3.56)$$

Let us denote the geocentric direction of the source by the unit vector $\mathbf{s^*}$. Then from the last section,

$$\mathbf{s^*} = (l, m, n), \qquad (3.57)$$

where these direction cosines were defined in equation (3.45), namely

$$\begin{aligned} l &= (u^\lambda i_\lambda)/(u^\lambda q_\lambda), \\ m &= (u^\lambda j_\lambda)/(u^\lambda q_\lambda), \\ n &= (u^\lambda k_\lambda)/(u^\lambda q_\lambda). \end{aligned} \qquad (3.45)$$

Here u^λ is the tangent vector at O to the null geodesic. Since this geodesic has already been derived, u^λ will be known. Again q^λ is the unit tangent vector at O to the observer's world line. The three unit vectors $i^\lambda, j^\lambda, k^\lambda$, however, which represent the observer's coordinate axes, have not been precisely specified, and this should be done in a way that makes them effectively parallel to the coordinate axes of the heliocentric observer.

The first step in achieving this objective is to define the mutually orthogonal vectors $I^\lambda, J^\lambda, K^\lambda, Q^\lambda$ as follows

$$I^\lambda = (c(1 + \tfrac{1}{2}m/r_0)^2, 0, 0, 0),$$
$$J^\lambda = (0, c(1 + \tfrac{1}{2}m/r_0)^2, 0, 0),$$
$$K^\lambda = (0, 0, c(1 + \tfrac{1}{2}m/r_0)^2, 0),$$
$$Q^\lambda = (0, 0, 0, (1 + \tfrac{1}{2}m/r_0)(1 - \tfrac{1}{2}m/r_0)^{-1}).$$

(3.58)

where

$$r_0{}^2 = x_0{}^2 + y_0{}^2 + z_0{}^2.$$

(3.59)

It is easily verified, using the metric (3.53), that these are unit vectors. Suppose S is an observer who is at rest in the coordinate system at the point (x_0, y_0, z_0). Then Q^λ is the unit tangent vector to S's world line. Moreover, I^λ, J^λ, and K^λ define locally Cartesian coordinate axes for S in the x-, y- and z-directions. It is our intention to construct similar axes for the actual observer. They will be represented by the mutually orthogonal unit vectors $i^\lambda, j^\lambda, k^\lambda$, each of which is orthogonal to q^λ. We shall make use of the fact that locally, that is, within the neighbourhood of the event O, the spacetime may be matched with that of special relativity.

The equivalent vectors to $I^\lambda, J^\lambda, K^\lambda$ and Q^λ in the special theory are obtained by putting $m = 0$ in equations (3.58). Moreover, if $\mathbf{V} = (V_x, V_y, V_z)$ is the observer's velocity, then the vectors equivalent to $i^\lambda, j^\lambda, k^\lambda, q^\lambda$ are the unit vectors in the coordinate directions after a Lorentz transformation with velocity $\mathbf{V}$ has been applied, essentially equation (3.26). It is convenient to consider the inverse transformation, which is

$$x - x_0 = \left[1 + \frac{\beta^2}{(\beta+1)c^2}V_x{}^2\right]x^* + \frac{\beta^2}{(\beta+1)c^2}V_xV_yy^* + \frac{\beta^2}{(\beta+1)}\frac{V_xV_z}{c^2}z^* + \beta V_x t^*,$$

$$y - y_0 = \frac{\beta^2}{(\beta+1)}\frac{V_yV_x}{c^2}x^* + \left[1 + \frac{\beta^2}{(\beta+1)c^2}V_y{}^2\right]y^* + \frac{\beta^2}{(\beta+1)}\frac{V_yV_z}{c^2}z^* + \beta V_y t^*,$$

$$z - z_0 = \frac{\beta^2}{(\beta+1)}\frac{V_zV_x}{c^2}x^* + \frac{\beta^2}{(\beta+1)}\frac{V_zV_y}{c^2}y^* + \left[1 + \frac{\beta^2}{(\beta+1)}\frac{V_z{}^2}{c^2}\right]z^* + \beta V_z t^*,$$

$$t - t_0 = \frac{\beta V_x}{c^2}x^* + \frac{\beta V_y}{c^2}y^* + \frac{\beta V_z}{c^2}z^* + \beta t^*.$$

(3.60)

We may then read off from these equations the components of the vectors i^λ, j^λ, k^λ and q^λ; for example, i^λ is the direction $y^* = z^* = t^* = 0$. The results, after normalization, are

$$
i^\lambda = c\left[1+\frac{\beta^2}{(\beta+1)}\frac{V_x^2}{c^2},\ \frac{\beta^2}{(\beta+1)}\frac{V_xV_y}{c^2},\ \frac{\beta^2}{(\beta+1)}\frac{V_xV_z}{c^2},\ \frac{\beta V_x}{c^2}\right],
$$

$$
j^\lambda = c\left[\frac{\beta^2}{(\beta+1)}\frac{V_yV_x}{c^2},\ 1+\frac{\beta^2}{(\beta+1)}\frac{V_y^2}{c^2},\ \frac{\beta^2}{(\beta+1)}\frac{V_yV_z}{c^2},\ \frac{\beta V_y}{c^2}\right],
$$

$$
k^\lambda = c\left[\frac{\beta^2}{(\beta+1)}\frac{V_zV_x}{c^2},\ \frac{\beta^2}{(\beta+1)}\frac{V_zV_y}{c^2},\ 1+\frac{\beta^2}{(\beta+1)}\frac{V_z^2}{c^2},\ \frac{\beta V_z}{c^2}\right],
$$

$$
q^\lambda = (\beta V_x,\ \beta V_y,\ \beta V_z,\ \beta). \tag{3.61}
$$

Since it is intended to use these results in the general theory, it is necessary to express them in vector form. First of all, using equations (3.58), with $m=0$, yields

$$
q^\lambda = \beta[Q^\lambda + (V_x/c)I^\lambda + (V_y/c)J^\lambda + (V_z/c)K^\lambda]. \tag{3.62}
$$

Then (3.61) gives the three space-like unit vectors as

$$
i^\lambda = I^\lambda + \frac{\beta}{(\beta+1)}\frac{V_x}{c}(q^\lambda + Q^\lambda),
$$

$$
j^\lambda = J^\lambda + \frac{\beta}{(\beta+1)}\frac{V_y}{c}(q^\lambda + Q^\lambda), \tag{3.63}
$$

$$
k^\lambda = K^\lambda + \frac{\beta}{(\beta+1)}\frac{V_z}{c}(q^\lambda + Q^\lambda).
$$

The components of the velocity **V**, and indeed β, may now be eliminated. For, taking successive inner products of (3.62) it is found that

$$
\beta V_x = -cq^\lambda I_\lambda, \quad \beta V_y = -cq^\lambda J_\lambda, \quad \beta V_z = -cq^\lambda K_\lambda, \quad \beta = q^\lambda Q_\lambda. \tag{3.64}
$$

Substituting these results into (3.63) yields the final expressions,

$$
i^\lambda = I^\lambda - \frac{(q_\mu I^\mu)}{(1+q_\nu Q^\nu)}(q^\lambda + Q^\lambda),
$$

$$
j^\lambda = J^\lambda - \frac{(q_\mu J^\mu)}{(1+q_\nu Q^\nu)}(q^\lambda + Q^\lambda), \tag{3.65}
$$

$$
k^\lambda = K^\lambda - \frac{(q_\mu K^\mu)}{(1+q_\nu Q^\nu)}(q^\lambda + Q^\lambda).
$$

The three vectors i^λ, j^λ, k^λ given by equations (3.65) are mutually orthogonal unit vectors, each of which is orthogonal to q^λ – points that may be verified by the reader. The final expressions for these vectors are tensor equations, involving only quantities that are already defined in the General Theory of Relativity. While the special theory was used as a means of identifying these vectors in a way that makes physical sense, the end result is tensor equations that may be adopted as definitions in the general theory.

The procedure for determining the observed direction of the source in general relativity may be summarized as follows:

(i) Derive from (3.36) the null geodesic that links the event of observation O (x_0, y_0, z_0, t_0) with the event of emission E (x_e, y_e, z_e, t_e). The tangent vector u^λ to this geodesic at O is then known.

(ii) Derive the unit vectors I^λ, J^λ, K^λ, Q^λ from equations (3.58).

(iii) Combine these with q^λ, the unit tangent vector to the observer's world line at O, to derive i^λ, j^λ, k^λ from equations (3.65).

(iv) The observed direction cosines (l, m, n) may now be derived from (3.45). The resulting unit 3-vector s^*, equation (3.57), gives the observed direction of the source.

(v) This may now be compared with the heliocentric direction of the source s given by equation (3.55). The difference between the two vectors s and s^* includes the effects of what may separately be called parallax, aberration and light deflection.

The full relativistic treatment described above is only necessary when a positional accuracy of the order of a milli-arc second can be achieved. Even then, considerable simplification is possible, by using approximate, but sufficiently accurate, methods. This simplification is essential, if the procedure just summarized is to be made explicit, and, in particular, if the difference $s^* - s$ is to be expressed algebraically. This will be undertaken in chapter 8. The application of general relativity to orbital dynamics is considered in chapter 6.

At the beginning of the present chapter, it was stated that the identification of an inertial frame of reference is a prime concern of positional astronomy. There is an apparent contradiction, therefore, in introducing general relativity into positional astronomy, since, in general relativity, inertial frames do not exist. General relativistic effects, however, are only significant in the vicinity of the sun, at least from a positional point of view. At large distances from the sun they may be entirely neglected, and the coordinate system used in equation (3.53) tends to that of special relativity, as r tends to infinity. The earlier considerations of the reference frame, for example, in relation to proper motion, are, therefore, in no way vitiated by the inclusion of general relativity. On the contrary, its inclusion is necessary to identify a reference frame that is inertial at large distances from the sun.

Problems

(ε = obliquity of the ecliptic = $23°.5$.)

3.1 Show that the total variation in the declination of a star due to precession is either 2ε or $180° - 2|\beta|$ depending on its ecliptic latitude β. Show that in all circumstances the maximum and minimum declinations are the principal values of

$$\sin^{-1}[\sin(\beta + \varepsilon)] \quad \text{and} \quad \sin^{-1}[\sin(\beta - \varepsilon)].$$

3.2 Show that precession produces a monotonic increase in the right ascension of stars of low ecliptic latitude, but that, for stars near the ecliptic poles, the right ascension is confined to a range of magnitude $2\sin^{-1}(\cos\beta \operatorname{cosec}\varepsilon)$.

3.3 Show that a southern star of declination $-\delta$ and right ascension α must at some epoch have been circumpolar at a station of north latitude ϕ, provided that

$$\sin\delta\cos\varepsilon + \cos\delta\sin\varepsilon\sin\alpha < -\cos(\phi - \varepsilon).$$

3.4 Make a rough estimate of the declination of the present Pole Star in the year 44 BC.

3.5 Two events, E_1 and E_2, are linked by a time-like geodesic in special relativity. Show that there exists a generalized Lorentz transformation which will transform the geodesic E_1E_2 into the time-axis. Make the equivalent statement for two events joined by a space-like geodesic and justify it.

3.6 Prove that the interval measured along a time-like geodesic in special relativity is a maximum in the sense that it is greater than the interval along any other time-like curve linking the end-points of the geodesic.

3.7 The coordinates (x, y, z) refer to a standard set of rectangular equatorial axes. The x-axis is redirected to the point on the celestial sphere with coordinates (α, δ) by applying a rotation through an angle α about the original z-axis followed by a rotation through an angle $-\delta$ about the resulting y-axis. Show that the new coordinates (x', y', z') are related to the old by the equations

$$x' = x\cos\alpha\cos\delta + y\sin\alpha\cos\delta + z\sin\delta$$
$$y' = -x\sin\alpha + y\cos\alpha$$
$$z' = -x\cos\alpha\sin\delta - y\sin\alpha\sin\delta + z\cos\delta.$$

Show further that the inverse transformation is

$$x = x'\cos\alpha\cos\delta - y'\sin\alpha - z'\cos\alpha\sin\delta$$
$$y = x'\sin\alpha\cos\delta + y'\cos\alpha - z'\sin\alpha\sin\delta$$
$$z = x'\sin\delta + z'\cos\delta.$$

3.8 If β denotes the Lorentz factor, namely

$$\beta = (1 - V^2/c^2)^{-1/2},$$

prove that

$$V^2 = \frac{c^2(\beta^2 - 1)}{\beta^2}.$$

Use this result and the results established in the previous question to derive the transformation equations (3.26).

3.9 The four vectors $I^\lambda, J^\lambda, K^\lambda$ and Q^λ are defined at a point in spacetime and are mutually orthogonal. The first three are unit space-like vectors, while the fourth is a unit time-like vector. Prove that, for any time unit vector q^λ defined at the same point, equations (3.65) define a set of mutually orthogonal space-like unit vectors, each of which is orthogonal to q^λ.

4

Geocentric coordinates

4.1   Introduction

Considerable attention was paid in the last chapter to the theoretical aspects of a change of origin from one point to another. This chapter will be more specific. It is concerned with the practical details involved in the particular case of the transfer from a topocentric to the geocentric origin. In later sections of the chapter, the exact nature of the origin's displacement is considered and the resulting positional effects of aberration and parallax. The motion of the observer relative to the geocentric origin, which is the cause of the aberration, is simply the diurnal rotation of the earth. The aberrational effect is, therefore, termed *diurnal aberration*. Likewise, the accompanying parallax effect is called *diurnal parallax*, although the term *geocentric parallax* is also used.

There is one other effect, however, that must be included for any ground-based observer, namely the effect of atmospheric refraction. For, as the rays of light from a star pass through the earth's atmosphere, they encounter layers of progressively higher air density and of correspondingly greater refractive index. The path of the rays is not, therefore, a straight line, but the rays are gradually refracted towards the vertical. This is a very difficult effect to treat accurately, since the amount of refraction depends on the atmospheric conditions which vary from day to day. It is, therefore, essential to define the topocentric coordinates to mean the coordinates referred to the topocentric origin, but with the effects of atmospheric refraction completely eliminated.

4.2   Approximate refraction formulae

The approximate formulae developed in this section are based on the assumption that the atmosphere may be treated as if it were stratified in plane layers, as shown in Fig. 4.1. This, of course, neglects the curvature of the earth, but still leads to formulae that are useful approximations for stars close to the zenith. At the zenith point itself, the effect of refraction vanishes by symmetry. The justification of this assumption of plane stratification resides in the fact that the significant refraction is taking place in the densest part of the atmosphere, the troposphere, which only extends a few kilometres above the earth's surface. The radius of curvature of the troposphere, therefore, greatly exceeds its vertical extent, and for tropospheric refraction the approximation of plane stratification is reasonable. For optical sources, only refraction in the troposphere is important, and the refraction in the upper reaches of the atmosphere can generally be neglected. This is not true, however, at radio wavelengths at which ionospheric refraction may be important, particularly at long wavelengths. Since the ionosphere extends over 1000 km above the earth's surface, the approximation of plane stratification is clearly not justified in that case. It should be borne in mind, therefore, that the approximation on which the present section is based is of restricted validity.

Imagine the atmosphere as consisting of N horizontal strata with refractive indices $n_0, n_1, \ldots, n_{N-1}$, as shown in Fig. 4.1. The refractive index decreases with height, n_0 is its value at ground level, and above the uppermost layer there is the virtual vacuum of outer space with unit refractive index. Consider a ray which enters the atmosphere at an angle z to the vertical. Let z_i be its inclination to the vertical in the layer with refractive index n_i. Then z_0 is the observed zenith distance of the source. The laws of refraction may be applied at each interface of the layers. In general, applying Snell's law to the interface of the ith and $(i+1)$th level yields

$$n_i \sin z_i = n_{i-1} \sin z_{i-1}. \tag{4.1}$$

Figure 4.1. Refraction in a plane-parallel atmosphere.

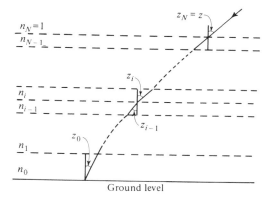

It is seen, therefore, that the quantity $n_i \sin z_i$ is conserved at each level, and we may deduce that

$$n_0 \sin z_0 = n_N \sin z_N = \sin z. \tag{4.2}$$

This result is independent of the way in which the atmosphere has been divided up into discrete layers and, therefore, still holds as we proceed to the limit of a continuous variation of refractive index with height. The angle z gives the direction in which the source would appear in the absence of any atmospheric refraction. This means that z is the *topocentric zenith distance* of the source. Since $n_0 > 1$, equation (4.2) indicates that the observed zenith distance is less than the topocentric value. Moreover, since the refraction process clearly occurs in a vertical plane, the azimuth of the source is unchanged.

Denote the angle of refraction by R, that is, put

$$R = z - z_0. \tag{4.3}$$

Then equation (4.2) yields

$$\sin z_0 \cos R + \cos z_0 \sin R = n_0 \sin z_0.$$

Since the angle of refraction is a small angle, the usual small-angle approximations may be made, yielding

$$R = (n_0 - 1) \tan z_0. \tag{4.4}$$

This result is, of course, expressed in radians; converting to arc seconds yields

$$R = K \tan z_0, \tag{4.5}$$

where

$$K = 206\,265(n_0 - 1). \tag{4.6}$$

The formula (4.2) is exact within the limits of the assumption of plane stratification of the atmosphere. Equations (4.4) and (4.5) are accurate only to first order in $(n_0 - 1)$ and are useful for small zenith distances. There is no point, however, in deriving higher-order terms in $(n_0 - 1)$. For, if the zenith distance is sufficiently large for such terms to be significant, then the value of K required to maintain the same accuracy must be modified to include curvature effects.

An attractive feature of the formulae just derived is that the angle of refraction depends only on the observed zenith distance of the source and the value of the refractive index at ground level. The details of how the refractive index varies with height are irrelevant. All that is necessary to correct an observed zenith distance is a knowledge of the local atmospheric conditions, so that the local refractive index may be determined. Unfortunately, this advantageous feature does not survive the inclusion of curvature effects.

The quantity n_0 depends significantly on the local atmospheric conditions. Standard conditions are adopted as atmospheric pressure 760 mm Hg and atmospheric temperature 0 °C. The refractive index under these conditions is given by

$$n_0 = 1.000\ 292\ 7, \tag{4.7}$$

leading to

$$K = 60''.4. \tag{4.8}$$

This value of K is sometimes called the *constant of refraction*.

The value of K for non-standard conditions may be derived from the Dale–Gladstone law, which states that $(n_0 - 1)$ is proportional to the air density. Then, if P is the atmospheric pressure in mm Hg, and t is the temperature in degrees Centigrade, it follows from the gas law that

$$n_0 - 1 \propto \frac{P}{(273 + t)}. \tag{4.9}$$

Using the values already given for standard conditions, the refraction formula (4.5) may be expressed as

$$R = 60''.4 \, \frac{(P/760)}{(1 + t/273)} \tan z_0. \tag{4.10}$$

No further refinement of the refraction formula is usually attempted without including the effect of the earth's curvature. Two further points are, however, worth noting. The refractive index at ground level depends not only on the local pressure and temperature, as considered above, but also on the composition of the air and on the wavelength of the light involved. The atmospheric composition may be regarded as constant except for the variation in the water vapour content. The effect of this is sufficiently small to warrant its neglect in the approximate formulae being considered here. The situation is only slightly more serious when we consider the wavelength dependence of the refractive index. The value of n_0 quoted in equation (4.7) corresponds to the centre of the V-band used in defining the visual magnitude of a star. Over the entire visible spectrum, the quantity $(n_0 - 1)$ varies by about 2 per cent, with a corresponding variation in the value of K. As a result, the point-like image of a star is dispersed into a minute spectrum along a vertical circle with the violet end closest to zenith. This can introduce systematic effects in the measurements of the positions of stars of different colours, and the inclusion of the wavelength variation in equation (4.10) is marginally justified despite the equation's approximate nature.

The wavelength variation of the refractive index of air can be expressed as

$$n_0 - 1 = 2.871 \times 10^{-4} (1 + 0.005\ 67/\lambda^2), \tag{4.11}$$

where λ is the wavelength in microns (Allen, 1976). This variation can be incorporated in equation (4.10) to give the angle of refraction as

$$R = 21''.3 \frac{P(1 + 0.0057/\lambda^2)}{(273 + t)} \tan z_0. \tag{4.12}$$

This section on approximate refraction formulae will be concluded by deriving the resulting effects on the right ascension and declination of a star. Consider the celestial sphere shown in Fig. 4.2, in which X represents the topocentric position of a star and X' the observed position. The points P and Z are respectively the north celestial pole and the zenith point. The displacement of the star X is along the great circle XZ.

It is possible to use the general result for a small displacement of a star on the celestial sphere that was developed in section 1.7. The resulting changes $(d\alpha, d\delta)$ are given in equations (1.29). In applying this general result, we must identify the arc θ with the zenith distance z, and (α_0, δ_0) are the equatorial coordinates of the zenith point. Then

$$\alpha_0 = LST,$$

$$\delta_0 = \phi,$$

ϕ being the observer's latitude. By equation (2.14) it follows that $\alpha_0 - \alpha = H$, the star's hour angle. The star's displacement is given by (1.27) as $d\theta = k \sin \theta$. Comparing this with equation (4.5) indicates that, in this applica-

Figure 4.2.

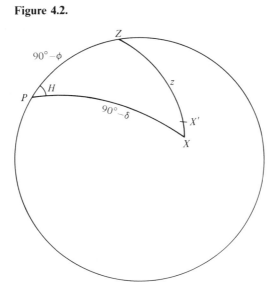

tion, the quantity k is not a constant but is given by

$$k = -K \sec z.$$

It is convenient, therefore, to eliminate z by using the cosine formula in spherical triangle PZX, which gives

$$\cos z = \sin \delta \sin \phi + \cos \delta \cos \phi \cos H.$$

Then, after a little simplification, equations (1.29) will yield the final results, as

$$d\alpha = \frac{K \sec^2 \delta \sin H}{(\cos H + \tan \phi \tan \delta)},$$

$$d\delta = K \frac{(\tan \phi - \tan \delta \cos H)}{(\cos H + \tan \phi \tan \delta)}.$$

$$(4.13)$$

4.3 Refraction in a radially symmetric atmosphere

The approximation of treating the earth's atmosphere as plane stratified leads to a simple refraction formula, but one that becomes increasingly inaccurate far from the zenith. A considerable improvement can be achieved by treating the atmosphere as radially symmetric. Then the air density, and, therefore, the refractive index, is a function only of distance from the earth's centre, the point C in Fig. 4.3. Suppose further that O represents the observer at a geocentric distance r_0, and P is a general point on the refracted ray-path OPS. Let $CP = r$ and angle $OCP = \theta$. Then (r, θ) are polar coordinates of the point P. Solving for θ as a function of r will determine the ray-path.

Produce the line CO to Z. Then OZ is the observer's zenith direction. Construct PZ' parallel to OZ, and produce CP to a point Q. Then $Z'PQ = \theta$. Now let us denote by z the angle that the ray makes at the point P with the direction of the observer's zenith. Let z_0 be the value of z at O, which is the observed zenith distance. Further, denote by ψ the angle that the ray makes with the radius vector CPQ. Then it follows that

$$z = \theta + \psi.$$

$$(4.14)$$

Notice that the angle ψ is just the angle between the radial direction and the tangent to the curve $\theta = \theta(r)$, and so

$$\tan \psi = r \frac{d\theta}{dr}.$$

$$(4.15)$$

Finally, construct the asymptote to the ray-path, and let it intersect OZ in the point A. We shall denote the height OA by h_0.

To proceed further, we must derive the refraction law. Since the refractive index $n = n(r)$, imagine the atmosphere as consisting of a finite number of concentric layers, similar to the treatment of the plane-parallel case. Figure 4.4 shows an enlarged view of a typical layer of refractive index n_i, and the angles ψ_i and ψ_{i+1} are shown. Let the angle $RPC = \chi$. Then, by the refraction law,

$$n_{i+1} \sin \psi_{i+1} = n_i \sin \chi.$$

The angle χ may be eliminated by applying the sine formula to plane triangle RPC. This gives

$$r_{i+1} \sin \chi = r_i \sin \psi_i.$$

Hence

$$r_{i+1} n_{i+1} \sin \psi_{i+1} = r_i n_i \sin \psi_i.$$

This implies that the quantity $rn \sin \psi$ is conserved at each level as we proceed to the limit of a continuous variation of the refractive index.

Figure 4.3. Refraction in a radially symmetric atmosphere.

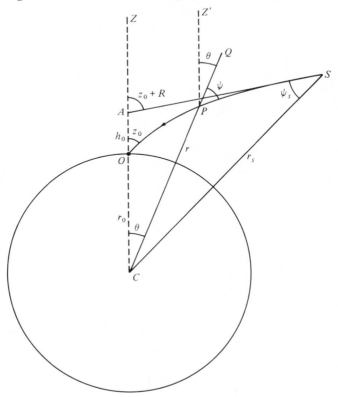

Equating this with its value at ground level then yields

$$rn \sin \psi = r_0 n_0 \sin z_0, \tag{4.16}$$

since, at the point O, $\theta = 0$, and so $\psi = z_0$.

The refraction of the ray is given, however, by the variation in z with height, not by the variation in ψ. It is found from equations (4.14) and (4.15) that

$$dz = d\psi + dr \tan \psi / r.$$

However, the differential form of the refraction law (4.16) may be written in the form

$$rn \cos \psi (d\psi + dr \tan \psi / r) = - dn \, r \sin \psi,$$

to yield the result that

$$dz = - dn \tan \psi / n. \tag{4.17}$$

In view of the form of this equation, it is convenient to regard n rather than r as the independent variable. Moreover, $\tan \psi$ may be derived from (4.16), which gives

$$\tan \psi = \frac{r_0 n_0 \sin z_0}{(r^2 n^2 - r_0^2 n_0^2 \sin^2 z_0)^{1/2}}. \tag{4.18}$$

Finally, substituting this into equation (4.17) and integrating yields the angle of refraction, as

$$R = r_0 n_0 \sin z_0 \int_1^{n_0} \frac{dn}{n(r^2 n^2 - r_0^2 n_0^2 \sin^2 z_0)^{1/2}}. \tag{4.19}$$

This formula is exact. If the variation of the refractive index with height is postulated, that is the function $r = r(n)$, equation (4.19) may be integrated numerically to derive the angle of refraction. Further, the actual ray-path

---

Figure 4.4

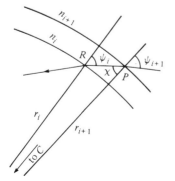

may also be computed for a given radial distribution of atmospheric density and, therefore, of refractive index. For equation (4.15) may be integrated, and, using (4.18), the resulting integral is

$$\theta = r_0 n_0 \sin z_0 \int_{r_0}^{r} \frac{\mathrm{d}r'}{r'(r'^2 n^2 - r_0^2 n_0^2 \sin^2 z_0)^{1/2}}. \tag{4.20}$$

Once the ray-path is known, its asymptote will be determined, and so the height h_0 (OA in Fig. 4.3) may be computed. A simple formula may be obtained for h_0, however, by considering the triangle ACS in Fig. 4.3. Here, the point S is to be regarded as a point on the ray-path at a great height, effectively above the atmosphere where $n = 1$. Let r_s and ψ_s be the value of r and ψ corresponding to this point. Then, since $ZAS = z_0 + R$, applying the sine formula to the triangle ACS yields

$$r_s \sin \psi_s = (r_0 + h_0) \sin (z_0 + R).$$

Now, if the refraction law (4.16) is applied to the left-hand side of this equation, the following result is obtained

$$h_0 = r_0 \left[\frac{n_0 \sin z_0}{\sin (z_0 + R)} - 1 \right]. \tag{4.21}$$

The height h_0 is very small, completely negligible at small zenith distances. Even for moderate zenith distances, it amounts only to a few metres. It does increase rapidly near the horizon, however, and is about 1.5 km for a source on the horizon itself. The significance of the height h_0 is this: when the correction for refraction is applied, the topocentric coordinates that are derived correspond to the position, not of the observing site itself, but rather for a height h_0 above the observer. It may be necessary to recognize this fine point of distinction if a subsequent correction is to be made for geocentric parallax. Due to the smallness of h_0, however, the correction need only be applied to observations of the moon, and then only at very low altitudes, as it is negligible for all other celestial objects. On the other hand, the effect is important for artificial satellites.

The general refraction formula in its exact form (4.19) requires a detailed knowledge of the refractive index, and consequently of the atmospheric conditions, for the whole of the earth's atmosphere. Table 4.1 gives the refractive index for mean (not standard) conditions as a function of height above the earth's surface. The final column in this table gives the parameter K as defined in equation (4.6). This gives approximately the amount of refraction taking place above the level in question, for a source at zenith distance $45°$. Inspection of Table 4.1 indicates an approximately exponential decrease in $(n - 1)$ with a scale height, H_0 say, that is less than 10 km. The justification of the earlier approximation, involving the

Table 4.1. *Variation of the refractive index and the angle of refraction with height above the earth's surface*

Height h (km)	$\log{(n-1)}$	K (arc seconds)
0	-3.55	57.8
10	-4.03	19.2
20	-4.69	4.2
30	-5.38	0.85
40	-6.04	0.19
50	-6.63	0.05
100	-9.96	2×10^{-5}

Based on data taken from C. W. Allen, 1976

complete neglect of curvature effects, is apparent. The normal procedure is to take this approximation one stage further by including only first-order effects of the earth's curvature. In developing this approximation, we shall use a series expansion of equation (4.19). We must recognize that there are two independent small quantities, namely $(n_0 - 1)$ and H_0/r_0. A precise definition of the scale height H_0 will be given in the course of the analysis.

Now consider equation (4.19), and write

$$r = r_0 + h \tag{4.22}$$

in that integral. Formally h tends to infinity as $n \to 1$, but it is clear from Table 4.1 that the integration may be curtailed at a lower limit differing insignificantly from unity, but corresponding to a value of h which is still very much less than r_0. The radius vector r may, therefore, be expanded in a power series in (h/r_0), and, if only zero and first-order terms are retained, equation (4.19) may be written as

$$R = R_1 - R_2 + O(h^2/r_0^2), \tag{4.23}$$

where

$$R_1 = n_0 \sin z_0 \int_1^{n_0} \frac{dn}{n(n^2 - n_0^2 \sin^2 z_0)^{1/2}}, \tag{4.24}$$

and

$$R_2 = \frac{n_0 \sin z_0}{r_0} \int_1^{n_0} \frac{hn \, dn}{(n^2 - n_0^2 \sin^2 z_0)^{3/2}}. \tag{4.25}$$

For this expansion to be valid it is necessary that $(n^2 - n_0^2 \sin^2 z_0)^{1/2}$ should not itself be a small quantity of the same order as h/r_0. This requires that z_0 shall be considerably less than $\sin^{-1}(1/n_0)$ – the source must not be close to the horizon.

The integral R_1 may be derived exactly, giving

$$R_1 = \left[-\sin^{-1}\left(\frac{n_0 \sin z_0}{n}\right)\right]_1^{n_0} = \sin^{-1}(n_0 \sin z_0) - z_0.$$

Alternatively, this result may be written as

$$\sin(z_0 + R_1) = n_0 \sin z_0,$$

in precise agreement with the result derived for a plane-parallel atmosphere, equation (4.2). It is convenient to expand R_1 in a power series in $(n_0 - 1)$, but it will now be necessary to include the term of order $(n_0 - 1)^2$, which was not previously considered. Correct to this order,

$$R_1 = (n_0 - 1)\tan z_0 + \tfrac{1}{2}(n_0 - 1)^2 \tan^3 z_0. \tag{4.26}$$

In evaluating the integral R_2, it is usual to work in terms of the air density ρ rather than in terms of the refractive index. By the Dale–Gladstone law, we may express n as

$$n = 1 + (n_0 - 1)\frac{\rho}{\rho_0}, \tag{4.27}$$

where ρ_0 is the air density at ground level. Moreover, since $\rho \leqslant \rho_0$, this facilitates an expansion in powers of $(n_0 - 1)$, leading to the result

$$R_2 = \frac{\tan z_0 \sec^2 z_0}{r_0}\frac{(n_0 - 1)}{\rho_0}\int_0^{\rho_0} h\,d\rho + \cdots,$$

where only the leading term has been retained. Now, integrating by parts, it is seen that

$$\int_0^{\rho_0} h\,d\rho = [h\rho]_0^{\rho_0} + \int_0^{\infty} \rho\,dh.$$

The first term on the right-hand side of this equation vanishes, since $h = 0$ when $\rho = \rho_0$, and $h\rho \to 0$ as $\rho \to 0$, due to the approximately exponential decrease of density with height. Further, let us define H_0 as the *height of an equivalent homogeneous atmosphere*, writing

$$H_0 = \frac{1}{\rho_0}\int_0^{\infty} \rho\,dh. \tag{4.28}$$

This is equivalent to the scale height that the atmosphere would have if its density distribution were exactly exponential, but the definition (4.28) does not require this assumption. Then, applying this definition, yields the result

$$R_2 = (n_0 - 1)\frac{H_0}{r_0}\tan z_0 \sec^2 z_0. \tag{4.29}$$

The two expressions in equations (4.26) and (4.29) may now be combined to give the total refraction. It is of the form

$$R = A \tan z_0 + B \tan^3 z_0 + \cdots, \tag{4.30}$$

where, in terms of the physical parameters,

$$A = (n_0 - 1)(1 - H_0/r_0),$$
$$B = -(n_0 - 1)(H_0/r_0 - \tfrac{1}{2}(n_0 - 1)). \tag{4.31}$$

The most usual procedure, however, is to employ equation (4.30) as the refraction law, but to regard A and B as empirically determined constants, since both the physical parameters vary from a number of causes.

4.4   The refraction constants and refraction tables

The following is one of the many observational methods that may be used to determine the two constants A and B, which were formally defined in the last section. A circumpolar star of declination δ is observed at both upper and lower transit, and the zenith distance is measured on these two occasions. The points X and Y in Fig. 4.5 represent the star's topocentric positions at upper and lower transit respectively, while X_1 and Y_1 represent the observed positions. Then $PX = PY = 90° - \delta$, and $PZ = 90° - \phi$, where ϕ is the latitude of the observer. The topocentric zenith distance at upper transit $ZX = PX - PZ = \phi - \delta$. If z_0 denotes the observed zenith distance at this time, then, by equation (4.30), it is seen that

$$\phi - \delta = z_0 + A \tan z_0 + B \tan^3 z_0.$$

A similar argument may be applied to the lower transit, yielding $ZY = PZ + PY = 180° - \phi - \delta$. Then if z_0' is the observed zenith distance, it follows that

$$180° - \phi - \delta = z_0' + A \tan z_0' + B \tan^3 z_0'.$$

The declination δ may be eliminated from these two equations to yield

$$180° - 2\phi = z_0' - z_0 + A(\tan z_0' - \tan z_0) + B(\tan^3 z_0' - \tan^3 z_0). \tag{4.32}$$

Figure 4.5

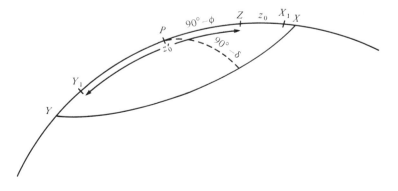

The quantities z_0 and z_0' appearing in this equation are known directly from observation – instruments capable of performing such measurements are considered in the next chapter. The unknown quantities in equation (4.32) are, therefore, the constants A and B, and possibly ϕ. Observations of three distinct stars will provide three equations similar to (4.32) which will be sufficient, in principle, to determine the refraction constants A and B and also the latitude of the observer. In practice, however, in order to minimize inevitable observational errors, the zenith distances of each star will be measured many times, and a large number of stars will be observed in this way. The quantities A, B and ϕ may then be determined statistically by the method of least squares.

The refraction constants A and B are expected, in view of their physical definitions, to depend on the meteorological conditions. Observationally determined values exhibit the expected variations. For standard conditions of pressure and temperature, however, the refraction formula (4.30) becomes

$$R = 60''.29 \tan z_0 - 0''.066\,88 \tan^3 z_0. \tag{4.33}$$

Some of the causes of variation in the refraction constants were considered in the previous section, namely atmospheric pressure, temperature, humidity and a variation with wavelength. Each of these will affect the value of n_0 occurring in (4.31). The value of the scale height H_0, however, also depends on the local gravity, and, since this depends principally on the observer's distance from the centre of the earth, a correction is required for the latitude and elevation above sea level of the observing site.

Since the causes of variation in refraction are so diverse and complicated, it has long been the practice to use specialized refraction tables, instead of applying an explicit formula which includes all the effects. Several major observatories have compiled comprehensive tables of refraction which include all the effects mentioned above, e.g. the Pulkova Refraction Tables (Orlov, 1956). All that even the most detailed refraction tables can accomplish, however, is to express the refraction as a function of local conditions. Now, the general refraction formula (4.19) shows that the total refraction depends not just on conditions at ground level, but on the entire variation of refractive index with height. It is only if the series expansion is restricted to its first two terms, as in (4.30), that the coefficients involved may be satisfactorily expressed entirely in terms of local conditions. The formula (4.30), however, becomes increasingly inaccurate at large zenith distances, and ceases to provide satisfactory results if the zenith distance exceeds $75°$. Higher-order terms depend on the detailed structure of the atmosphere, and, for $z_0 > 80°$, this dependence is critical. At normal zenith distances, refraction tables are based on a combination of theory and observation. Near the horizon, on the other hand, the tabulated values of refraction are based almost entirely on observation. Indeed, the angle of refraction at the horizon is very large, variable and rather uncertain, so that refraction tables can give only approximate values, and precise positional measurements become impossible.

When a celestial object is observed at a zenith distance of exactly 90°, the angle of refraction is termed the *horizontal refraction*. Although this angle is uncertain, it is sufficient to adopt a standard value of 34' for the purpose of computing times of apparent rising and setting, since a high accuracy is not normally then required. The times of sunrise and sunset are computed in this way in the *Astronomical Almanac*. A further adjustment is included, however, to allow for the size of the solar disc. For the time of sunrise or sunset is to be regarded as the time when the sun's upper limb is observed to cross the horizon. Taking the sun's angular semidiameter as 16' means that the sun's centre is then, in reality, 50' below the horizon. The tabulated times of sunrise and sunset correspond, therefore, to the times at which the topocentric zenith distance of the sun is 90° 50'. Similar considerations apply to the times of moonrise and moonset, which are also included in the *Astronomical Almanac*. In this case, however, a correction is included for geocentric parallax, a correction that is neglected for the sun.

It will be clear from the discussion in this, and in the previous two sections, that atmospheric refraction alters the coordinates of a star in a way that is not entirely predictable, and that the uncertainty increases rapidly with zenith distance. There are definite advantages, therefore, if positional observations are restricted to small zenith distances, particularly when high accuracy must be maintained. An instrument which, for this reason, concentrates on zenith observations is described in the next chapter.

The present discussion of refraction is relevant to refraction in the lower atmosphere, and we have concentrated on the optical case. Most of the considerations still apply to radio refraction, but the values of the constants are slightly different. Moreover, radio refraction, unlike optical, is strongly dependent on the humidity of the atmosphere. In addition, at long radio wavelengths, there is significant refraction taking place in the ionosphere. For this, the approximations of the present chapter are inadequate. A fuller discussion of radio refraction is given in chapter 16.

Positional observations in the optical band made with satellite-borne telescopes remove the problem of atmospheric refraction in its entirety. As the final entry in Table 4.1 indicates, no refraction correction of any kind is then required. In this case the topocentric coordinates are directly observed, apart from purely instrumental corrections.

4.5   Geocentric coordinates of the observer

The remainder of this chapter is concerned with the reduction of a star's coordinates from its topocentric to its geocentric place. This involves the corrections for geocentric parallax and diurnal aberration, which are considered in turn. The size of the parallax effect is dependent on the distance of the celestial object involved, and is negligible for bodies outside the solar system. In the earth's immediate vicinity, on the other hand, the parallax is considerable; it is nearly one degree for the moon, and may be far larger than this for artificial satellites. An exact treatment is essential in these circumstances, and a correspondingly accurate knowledge is required of the observer's position with respect to the centre of the earth. A simple spherical model is then an inadequate representation of the figure of the earth, which must be considered in more detail.

The mean ocean level is, of necessity, in gravitational equilibrium and is, therefore, constrained to follow an equipotential surface of apparent gravity. The potential function involved naturally includes a term representing the centrifugal effects arising from the earth's rotation. This equipotential surface covering the ocean level and continued under the land masses is termed the *geoid* and is what is normally meant by sea level. Notice that the direction of local gravity is by definition, everywhere normal to the geoid. This surface possesses certain small irregularities but it can be very closely fitted to an ellipsoid of revolution, that is, a spheroid, the axis of revolution coinciding with the earth's rotational axis. A *standard spheroid* is, therefore, adopted for reference purposes and is the basis for geodetic measurements.

Figure 4.6 represents a meridianal section of the standard spheroid which is an ellipse of semimajor axis a and semiminor axis b. The standard spheroid is usually specified by the equatorial radius a and the polar flattening f, where

$$b = a(1 - f).   \tag{4.34}$$

The equation of the ellipse may then be written as

$$\frac{x^2}{a^2} + \frac{y^2}{a^2(1-f)^2} = 1.   \tag{4.35}$$

Consider an observer O situated at a height h above the standard spheroid. Apart from a small *geodetic correction*, normally a few metres, this will be his height above sea level. Draw the normal through O to the spheroid, meeting it at the point O', and produce this line to intersect the x-axis in Q. Then $OO' = h$. No less than three independent definitions can be made of the zenith direction for the observer O. The actual direction of a plumb line defines *astronomical zenith*; this, however, is not shown in Fig.

4.6. The direction QO defines *geodetic zenith*. This would be identical with astronomical zenith if the geoid coincided precisely with the standard spheroid – they differ only due to local gravitational anomalies. The angle between these two zenith directions is called the *deflection of the vertical*. It is seldom more than a few arc seconds. The third zenith direction, *geocentric zenith*, is defined by the direction CO produced. The angle between this and geodetic zenith, v, say, is called the *angle of the vertical*.

With each definition of zenith there is an associated definition of latitude for the observer. It is simply the angle that the zenith direction makes with the equatorial plane. The *geodetic latitude* ϕ and the *geocentric latitude* ϕ' for the observer O are shown in Fig. 4.6. The *astronomical latitude* is not indicated, however, since the astronomical zenith direction is not, in general, in the plane of the diagram. Alternatively, latitudes, and indeed longitudes, may be defined in terms of the celestial sphere. For, if Z is any one of the zenith points and P is the north celestial pole, then the arc PZ is the colatitude. Moreover, the great circle PZ defines the observer's meridian, and the angle this makes with the Greenwich meridian is the observer's longitude. It is easily seen that the geocentric and the geodetic longitudes are identical. Let us denote them simply by λ.

Let ρ be the geocentric distance of the observer, i.e. $\rho = CO$. Then ρ,

Figure 4.6. Meridianal section of the standard spheroid.

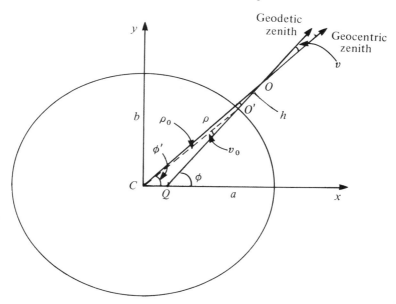

together with the observer's geocentric latitude and longitude (ϕ', λ) completely specify his position with respect to the centre of the earth. It is desirable that they should be expressed in terms of the observer's geodetic latitude and longitude and the height h, all of which may be presumed known. Let (x_0, y_0) be the Cartesian coordinates of the observer and (x, y) those of the point O'. Then it is easily seen that

$$\rho \cos \phi' = x_0 = x + h \cos \phi,$$
$$\rho \sin \phi' = y_0 = y + h \sin \phi. \tag{4.36}$$

Now the point (x, y) is on the ellipse given by (4.35), and $\tan \phi$ is the slope to the normal to this curve at that point, which means that $\tan \phi = -dx/dy$. Differentiating equation (4.35), therefore, yields

$$y = x(1-f)^2 \tan \phi. \tag{4.37}$$

This result may now be substituted back into (4.35) to give

$$x^2(1 + (1-f)^2 \tan^2 \phi) = a^2.$$

From this and (4.37), both x and y may be expressed in terms of the parameter ϕ. The results are usually written in the form

$$x = aC \cos \phi,$$
$$y = aS \sin \phi, \tag{4.38}$$

where C and S are the functions of ϕ defined by

$$C = [\cos^2 \phi + (1-f)^2 \sin^2 \phi]^{-1/2},$$
$$S = (1-f)^2 C. \tag{4.39}$$

These functions are tabulated in the *Explanatory Supplement*, but for a slightly different value of f to that adopted below. It is probably best, therefore, to compute them directly from (4.39) when they are needed. Combining equations (4.38) and (4.36) gives the desired expression for the observer's own geocentric coordinates, namely

$$\rho \cos \phi' = a \cos \phi(C + h/a),$$
$$\rho \sin \phi' = a \sin \phi(S + h/a). \tag{4.40}$$

These equations are exact.

It is sometimes convenient to have more definitive expressions for the observer's geocentric distance ρ and the angle of the vertical v $(\equiv \phi - \phi')$. They may be expressed as power series in f. To simplify matters, we shall, in the first instance, set $h = 0$ in equations (4.40), and let ρ_0 and v_0 be the resulting geocentric distance and angle of the vertical. These refer rigorously to the point O' in Fig. 4.6 rather than to the actual observer O, i.e. $\rho_0 = CO'$ and $CO'Q = v_0$. It will then be possible to avoid series expansions in both f and h – the quantities ρ_0 and v_0 are first derived and later related to ρ and v.

The geocentric latitude of O' is $\phi_0' = O'CQ$, and, by equations (4.40) and (4.39),

$$\tan \phi_0' = \frac{S}{C} \tan \phi = (1-f)^2 \tan \phi.$$

Then, since $v_0 = \phi - \phi_0'$,

$$\tan v_0 = \frac{\tan \phi - \tan \phi_0'}{1 + \tan \phi \tan \phi_0'} = \frac{(2f-f^2) \tan \phi}{1 + (1-f)^2 \tan^2 \phi}.$$

Multiplying above and below by $\cos^2 \phi$ allows this equation to be written, after some reduction, as

$$\tan v_0 = \frac{q \sin 2\phi}{1 + q \cos 2\phi}, \tag{4.41}$$

where

$$q = \frac{2f-f^2}{1 + (1-f)^2}. \tag{4.42}$$

If all the trigonometric functions in (4.41) are expressed in exponential form, it is found that

$$e^{2iv_0} = (1 + q\, e^{2i\phi})/(1 + q\, e^{-2i\phi}),$$

or, in logarithmic form,

$$2iv_0 = \log(1 + q\, e^{2i\phi}) - \log(1 + q\, e^{-2i\phi}).$$

Since q is a small quantity, the logarithmic expansions may be employed, leading to the result

$$v_0 = q \sin 2\phi - \tfrac{1}{2}q^2 \sin 4\phi + \tfrac{1}{3}q^3 \sin 6\phi \ldots. \tag{4.43}$$

Now the flattening f has a magnitude of about one part in 300, and q is of the same order of magnitude. Consequently, higher-order terms than the third may certainly be neglected in the series expansion.

A similar series expression may be derived for ρ_0 but not so neatly. Squaring and adding equations (4.40), with $h = 0$, yields

$$\rho_0^2 = a^2 C^2 (\cos^2 \phi + (1-f)^4 \sin^2 \phi).$$

If now a series expression for C is derived, then ρ_0 is expressible as a power series, which, after considerable labour, is found to be

$$\rho_0 = a[1 - \tfrac{1}{2}f + \tfrac{5}{16}f^2 + \tfrac{5}{32}f^3 + (\tfrac{1}{2}f - \tfrac{13}{64}f^3) \cos 2\phi$$
$$- (\tfrac{5}{16}f^2 + \tfrac{5}{32}f^3) \cos 4\phi + \tfrac{13}{64}f^3 \cos 6\phi \ldots], \tag{4.44}$$

where again terms of order f^4 have been neglected.

The expressions for ρ_0 and v_0, which refer rigorously to the point O', are useful approximations to ρ and v which may be improved by considering triangle OCO' in Fig. 4.6. An exaggerated view of this triangle is presented in Fig. 4.7. An application of the cosine formula to this triangle yields

$$\rho^2 = \rho_0{}^2 + h^2 + 2\rho_0 h \cos v_0.$$

Since v_0 is a small angle, we may expand $\cos v_0$ as $1 - \frac{1}{2}v_0{}^2 \ldots$, to derive

$$\rho = (\rho_0 + h)[1 - \frac{1}{2}\rho_0 h v_0{}^2 (\rho_0 + h)^{-2} \ldots].$$

The second term in the right-hand bracket may in fact be neglected. It is of order $v_0{}^2(h/a)$. Since v_0 is $O(f)$, the term is $O(f^2 h/a)$. Third-order terms in f are of marginal importance, and, since (h/a) is considerably less than f, this term may normally be omitted. To sufficient accuracy, therefore, we may write

$$\rho = \rho_0 + h. \tag{4.45}$$

Again, by applying the sine formula to triangle OPO', it is found that

$$\rho \sin v = \rho_0 \sin v_0.$$

Expanding the sines of the two small angles, and applying similar arguments, yields the result

$$v = (1 - h/a)v_0. \tag{4.46}$$

The quantities ρ and v may, therefore, be obtained by combining equations (4.43) to (4.46). The results will be correct to second order in both f and h, but will also include the terms in f^3, although not the other third-order terms. An accuracy of $0''.001$ is then expected in v, and of about 1 cm in ρ.

The dimensions of the standard spheroid that were adopted by the International Astronomical Union (IAU) in 1976 are as follows:

Equatorial radius $a = 6378.140$ km,

Flattening $f = 0.003\,352\,81 = 1/298.257.$ $\tag{4.47}$

Subsidiary quantities may then be derived such as

Polar radius $b = a(1 - f) = 6356.755\,308$ km,

Figure 4.7

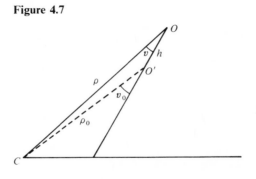

and

$$q = \frac{1-(1-f)^2}{1+(1-f)^2} = 0.003\ 358\ 431.$$

Then equations (4.43) and (4.44), expressed in numerical form, are

$$v_0 = 692''.7260 \sin 2\phi - 1''.1632 \sin 4\phi + 0''.0026 \sin 6\phi,$$

$$\rho_0 = 6367.470\ 098 + 10.692\ 297 \cos 2\phi \qquad (4.48)$$

$$-\ 0.022\ 443 \cos 4\phi + 0.000\ 049 \cos 6\phi \ \text{km}.$$

4.6 Geocentric parallax

In Fig. 4.8, O represents the observer, C the centre of the earth and S some nearby celestial object. The line CO, produced to Z', gives the direction of the observer's geocentric zenith. The plane of the diagram is the plane defined by the three points C, O and S, that is, the plane of the vertical circle containing S. So the profile of the earth indicated will not, in general, be a meridianal section. Denote the angle $Z'OS$ by z'. This is the observed zenith distance of the source referred to geocentric zenith. The directions of geodetic and astronomical zenith are not, in general, in the plane of the

Figure 4.8

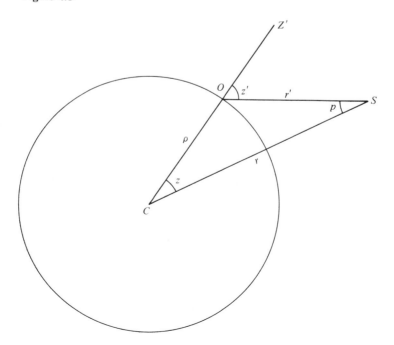

diagram. Let r' and r be the topocentric and geocentric distances of the source and ρ the geocentric distance of the observer – all as indicated in the diagram. The *geocentric parallax* p is the angle OSC. Then

$$z' = z + p, \tag{4.49}$$

where z is the angle OCS, which is formally the zenith distance as it would appear from the centre of the earth. The effect of geocentric parallax is, therefore, to increase the *geocentric* zenith distance by p and, since the displacement is in the plane OCS, to leave the (geocentric) azimuth unaltered.

The sine formula applied to triangle OCS yields

$$\sin p = \frac{\rho}{r} \sin z' = \frac{\rho}{r'} \sin z. \tag{4.50}$$

Clearly, the greater the distance of the celestial object the smaller will be the parallax. But a particular value of the parallax will depend on the zenith distance at the time and also on ρ. Some standardization is useful. The *equatorial horizontal parallax* P is defined as the parallax for an equatorial observer ($\rho = a$) when the source is on the horizon ($z' = 90°$). Then, by (4.50),

$$\sin P = \frac{a}{r}. \tag{4.51}$$

The equatorial horizontal parallax, often abbreviated to the *horizontal parallax*, is an inverse measure of the source's geocentric distance. The distance of the moon is tabulated in this way in the *Astronomical Almanac*. The parallax for a particular observer is given by

$$\sin p = \frac{\rho}{a} \sin P \sin z'. \tag{4.52}$$

Geocentric parallax measurements are relevant to bodies within the solar system. Due to the orbital motions of these bodies, and of the earth itself, the values of their horizontal parallaxes vary. For example, owing to the eccentricity of the moon's orbit about the earth, its horizontal parallax has a fairly wide variation between $54'$ and $61'$. Following Murray (1983), the mean value of the moon's horizontal parallax, consistent with the recommendations of the IAU, is P_0, given by

$$\sin P_0 = 3422.485 \sin 1'',$$

or equivalently

$$P_0 = 57'\,02''.6050. \tag{4.53}$$

The horizontal parallax of a planet is much less than this. For Saturn, it is about $1''$, and for the nearest planet, Venus, it varies between the very wide limits of $34''$ and $5''$. The horizontal parallax of a body at a distance of 1 AU is termed the *solar parallax*, since it corresponds, apart from some slight refinements, to the mean parallax of the sun itself.

Measurements of geocentric parallax provide a means of determining the distances of bodies within the solar system. This method has now been superseded by direct measurements by means of radar, but the parallax method merits some discussion in view of its historical importance. Moreover, the principles involved in interpreting distance measurements, whether made by parallax or radar, are the same.

Positional measurements of the planets must be interpreted in the light of gravitational dynamics. If we ignore the masses of the planets compared with the mass of the sun and, therefore, omit the considerable and complex refinements of celestial mechanics, the process of interpretation may be made plain. All bodies in heliocentric orbits may then be considered as moving only under the influence of the sun, and the orbits will be exact Keplerian ellipses. Kepler's third law of planetary motion may be applied stating that for each planet the cube of the semimajor axis (a_p) is proportional to the square of the orbital period (T_p), that is

$$k^2 a_p{}^3 = T_p{}^2, \tag{4.54}$$

where k^2 is a constant. Now positional observations taken over a long period of time allow a very accurate determination to be made of the orbital period of each planet. It follows that the relative sizes of the planetary orbits are known. All that is lacking for their absolute determination is the value of the single constant k^2 in equation (4.54), for this constant depends on the unknown mass of the sun. In effect, a scale model of the entire solar system may be constructed, but to determine the scale of the model in kilometres requires a measurement of distance. In principle, a single measurement will suffice.

Distance determinations based on geocentric parallax have a low inherent accuracy due to the small value of the parallax that is being measured. If only a single measurement is required, it is clearly desirable to select the body which will have the largest possible parallax. This will maximize the percentage accuracy of the distance measurement itself and, therefore, the derived scale of the solar system.

Table 4.2 gives the orbital elements of the earth and the two nearest planets, Venus and Mars. Further, these orbits are represented schematically in Fig. 4.9, the fact that the orbits are in slightly different planes being ignored. The points E_1, V_1, M_1 represent the positions of the three planets on a certain date when Mars, the earth and the sun are aligned in that order. Mars is then said to be at *opposition*. A few months later the planets have moved to positions E_2, V_2, M_2 and Venus is then directly between the earth

Table 4.2. *Orbital elements of the bodies represented in Fig. 4.9*

	Semimajor axis (a (AU))	Eccentricity (e)	Longitude of perihelion (ϖ)	Orbital inclination (i)
Venus	0.7233	0.0068	131°.5	3°.4
Earth	1	0.0167	102°.5	0
Mars	1.5236	0.0939	335°.7	1°.9
Eros	1.4581	0.2227	178°.5	10°.8

and the sun. This configuration is described as *inferior conjunction*. (The other possible alignment, *superior conjunction*, where the planet is on the far side of the sun from the earth, will not concern us here.) It is clear that the closest approaches of Venus and Mars to the earth correspond to inferior conjunction and opposition respectively. Mars has an orbital eccentricity of nearly 0.1, and so its distance from the earth at opposition varies significantly, depending on the ecliptic longitude (or time of year) at which the opposition occurs. On the other hand, the orbits of the earth and Venus are close to exact circles. There is, therefore, little variation in Venus' distance from the earth at successive inferior conjunctions.

Venus is, of course, not normally observable at an inferior conjunction due to its close proximity to the sun at that time – unless the very special circumstances occur when Venus actually transits the disc of the sun. In principle, parallax measurements of Venus should then be possible by

Figure 4.9. The orbits of Venus, Earth, Mars and Eros.

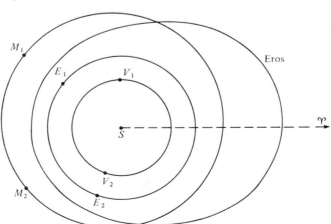

making accurate timings, from different points of the earth's surface, of its passage over the limb of the sun. The intractable influence of Venus' dense atmosphere, however, vitiated these measurements when they were attempted at the last transits of Venus which occurred in the late nineteenth century. Observational effort was, therefore, transferred to favourable oppositions first of Mars and then of the minor planet Eros, which had been discovered at about that time.

As Fig. 4.9 and Table 4.2 indicate, Eros moves in an eccentric orbit which can bring it within 0.16 AU of the earth. Major programmes of parallax measurement were, therefore, undertaken in 1901 and 1931 when favourable oppositions occurred. As a result, significant new determinations were made of the solar parallax and revisions of the scale of the planetary system. The solar parallax determination based on observations of the 1931 opposition was the generally accepted standard until the early 1960s.

Radar astronomy has provided a more direct method of making the necessary distance measurement. The signal to noise ratio in the radar echo is very sensitive to range, being inversely proportional to the fourth power of the target's distance. Venus is, therefore, the planet that affords the strongest radar echo, and radar contact was first achieved at the inferior conjunction of 1959. Consistent and definitive results were obtained at the next inferior conjunction in 1961. These and subsequent measurements have provided the basis of a new determination of the solar parallax and the astronomical unit. The values of these quantities adopted in the most recent IAU system of constants (IAU, 1977) are given by

$$1 \text{ AU}, \quad A = 1.495\,978\,70 \times 10^{11} \text{ m},$$

$$\text{Solar parallax}, \quad P_0 = 8''.794\,148. \tag{4.55}$$

These two constants are strictly related to each other. From equation (4.51)

$$\sin P_0 = a/A. \tag{4.56}$$

The astronomical unit is no longer defined as exactly the semimajor axis of the earth's orbit, which is, in any case, variable due to planetary perturbations. Instead, it is defined in terms of gravitational theory as follows. The constant k in equation (4.54) is known as the *Gaussian gravitational constant*. It is a defining constant in the IAU system of constants, and its value is chosen as

$$k = 0.017\,202\,098\,95. \tag{4.57}$$

This value has been maintained in recent revisions of the system of constants, and it is not anticipated that it will be altered in any future changes. It is approximately 2π divided by the number of days in a year. Now, the astronomical unit is defined as the unit of distance in which a_p

must be expressed in (4.54) when the above value of k is used and T_p is expressed in days. The value quoted in (4.55) corresponds to this definition of the astronomical unit and differs slightly from the earth's mean distance from the sun.

The value of the solar parallax now derived from radar observations is determined to a precision of one micro-arc second. This is far beyond the scope of any positional observations. The most recent favourable opposition of Eros in 1975 was not, therefore, the occasion of renewed parallax measurements, but instead afforded an opportunity of studying the physical nature of a minor planet at comparatively close quarters. In particular, radar studies of Eros were made to investigate its topographical features. For, while such investigations are possible for most of the major planets, they are generally precluded for minor planets due to their small target areas.

Parallax measurements of the moon can provide distance determinations that are decidedly more accurate than those of the planets using the same method. The reason for this is obvious – the moon's parallax is several hundred times greater. The parallax method of determining the moon's distance has also been superseded, however, first by radar and later by laser ranging. The reflectors necessary for the laser measurements were placed in position by the Apollo astronauts. In no case, therefore, are parallax measurements any longer the primary means of distance determination for bodies within the solar system. Certainly today, geocentric parallax has a less fundamental role; it is still, however, an essential positional effect that must be allowed for in a celestial object's coordinates.

4.7   The effect of geocentric parallax on right ascension and declination

It was established in the last section that the effect of geocentric parallax is to increase an object's geocentric zenith distance but to leave its azimuth unaltered. If the parallax is sufficiently small, equation (4.50) may be written as

$$dz = \frac{\rho}{r} \sin z. \tag{4.58}$$

The displacement on the celestial sphere is of the form considered in section 1.7, and the general formulae (1.29) may be used. It is only necessary to set k equal to ρ/r and to identify the point (α_0, δ_0) with the geocentric zenith point. Then $\delta_0 = \phi'$, the geocentric latitude of the observer, and α_0 is the local sidereal time. This implies that $\alpha_0 - \alpha = H$, the hour angle of the source.

Equations (1.29) then yield

$$d\alpha = -(\rho/r) \cos \phi' \sin H \sec \delta,$$
$$d\delta = (\rho/r)(\cos \phi' \cos H \sin \delta - \sin \phi' \cos \delta). \tag{4.59}$$

Since these formulae are only correct to first order in (ρ/r), they should not be used for the parallax of the moon. They are generally adequate, however, for the parallax of bodies in heliocentric orbits and not in the earth's immediate vicinity, when the angle of parallax will only be a few arc seconds. Further, if the parallax is very small, as in the case of an outer planet, it may not be necessary to distinguish between the geocentric and geodetic latitudes, and it may be sufficient to set ρ equal to the earth's equatorial radius. As always, first-order formulae should be used with circumspection; the rigorous formulae are easily derived in this case.

Selecting the centre of the earth as origin, let $\mathbf{r}$ and $\mathbf{R}$ be the position vectors of the source S and the observer O. Then, as already shown in equation (3.7), the vector OS, from observer to source, is given by

$$\mathbf{r}' = \mathbf{r} - \mathbf{R}. \tag{4.60}$$

This notation is compatible with that in our earlier diagram, Fig. 4.8. It may be assumed that the position vector $\mathbf{R}$ is accurately known. It is a vector of magnitude ρ in the direction of geocentric zenith which, in equatorial coordinates, is

$$\mathbf{R} = \rho(\cos \phi' \cos T, \cos \phi' \sin T, \sin \phi'), \tag{4.61}$$

where T is the local sidereal time of the observation. Equation (4.60) may be used to derive either $\mathbf{r}$ or $\mathbf{r}'$, assuming that the other one is known.

Suppose, for example, that the moon's geocentric right ascension and declination (α, δ) and its horizontal parallax P are all derived from the *Astronomical Almanac*. Then, by (4.51),

$$\mathbf{r} = a \operatorname{cosec} P(\cos \delta \cos \alpha, \cos \delta \sin \alpha, \sin \delta), \tag{4.62}$$

and each component of this vector may be computed. If now we denote the components of the vector $\mathbf{r}'$ by (x', y', z') and the observed right ascension and declination by (α', δ'), equation (4.60) in scalar form is

$$x' = r' \cos \delta' \cos \alpha' = a \operatorname{cosec} P \cos \delta \cos \alpha - \rho \cos \phi' \cos T,$$
$$y' = r' \cos \delta' \sin \alpha' = a \operatorname{cosec} P \cos \delta \sin \alpha - \rho \cos \phi' \sin T,$$
$$z' = r' \sin \delta' \qquad = a \operatorname{cosec} P \sin \delta - \rho \sin \phi' \tag{4.63}$$

Once the components (x', y', z') are computed from this equation, (α', δ') may be derived as

$$\alpha' = \tan^{-1} (y'/x'),$$
$$\delta' = \tan^{-1} [z'/(x'^2 + y'^2)^{1/2}]. \tag{4.64}$$

A little care is required in using these formulae to remove a possible ambiguity in right ascension. The principal value of the inverse tangent is taken for δ' in any event, but 12^h must be added to the principal value for α' if $x' < 0$.

We conclude this section by illustrating, with a numerical example, the reverse process of computing the geocentric place from the observed. The computation of the observer's own geocentric coordinates that was described in section 4.5 is also illustrated.

Example. Combined radio and optical observations are made of an artificial satellite from a station at geodetic latitude $39° 42' 48'' \text{N}$. The height of the station is 456 m above sea level. From these observations the range and equatorial coordinates of the satellite are found to be

$$r' = 1735.87 \text{ km},$$
$$\alpha' = 7^h 12^m 19^s,$$
$$\delta' = -21° 42' 21'',$$

and the observation is timed at $9^h 17^m 34^s$ local sidereal time. Compute the geocentric place and distance of the satellite.

We shall use the notation of the present section and section 4.5 without further explanation. First the position vector **R** of the observer must be computed. The standard spheroid is specified by

$$a = 6378.14 \text{ km},$$
$$f = 3.35281 \times 10^{-3}.$$

For this particular observing site, we have

$$\phi = 39°.7133,$$

and

$$h/a = 7.15 \times 10^{-5}.$$

Further,

$$T = 139°.3917.$$

Applying equations (4.39) and (4.40) yields successively

$$C(\phi) = 1.001\ 369\ 3,$$
$$S(\phi) = 0.994\ 665\ 8,$$
$$\rho \cos \phi' = 4913.459 \text{ km},$$
$$\rho \sin \phi' = 4053.845 \text{ km}.$$

Then, from equation (4.61), we compute

$$\mathbf{R} = (-3730.183,\ 3198.095,\ 4053.845) \text{ km}.$$

The vector $\mathbf{r}'$ is then computed from the observational data, namely,

$r' = 1735.87$ km,

$\alpha' = 108°.0792$,

$\delta' = -21°.7058$,

yielding

$\mathbf{r}' = (-500.498, 1533.162, -641.997)$ km.

So the geocentric vector $\mathbf{r}$ may be computed as

$\mathbf{r} = \mathbf{r}' + \mathbf{R} = (-4230.681, 4731.257, 3411.849)$ km.

Finally, converting from Cartesian to spherical coordinates yields

$r = 7205.843$ km,

$\alpha = 131°.8030 = 8^{\mathrm{h}} 47^{\mathrm{m}} 13^{\mathrm{s}}$,

$\delta = +28°.2607 = +28° 15' 38''$.

4.8 Diurnal aberration

Geocentric parallax is an important correction in the transformation from topocentric to geocentric coordinates, but only for bodies within the solar system. By contrast, the aberrational correction – diurnal aberration – is independent of the distance of the source and must be allowed for in the position of every celestial object. It is not a large effect, however. The earth's equatorial velocity of rotation is only 1.6×10^{-6} times the velocity of light, and so the displacement due to diurnal aberration can never exceed $\frac{1}{3}$ of an arc second. Consequently, the first-order classical treatment is all that is needed, and relativistic effects may be ignored.

Recall, therefore, equation (3.13) which gives the aberrational displacement for an observer moving with a velocity $V\mathbf{n}$, $\mathbf{n}$ being a unit vector. Simplifying the notation slightly, the displacement for a source in direction $\mathbf{s}$ is given by

$$\mathrm{d}\mathbf{s} = -(V/c)\mathbf{s} \times (\mathbf{s} \times \mathbf{n}). \tag{4.65}$$

This displacement is identical in form with that in equation (1.31), so the discussion of section 1.7 is again relevant. In particular, the changes in right ascension and declination are derived from equation (1.29) as

$$\mathrm{d}\alpha = (V/c) \sec \delta \cos \delta_0 \sin (\alpha_0 - \alpha)$$
$$\mathrm{d}\delta = (V/c)[\cos \delta \sin \delta_0 - \sin \delta \cos \delta_0 \cos (\alpha_0 - \alpha)], \tag{4.66}$$

where (α_0, δ_0) are the equatorial coordinates determined by the unit vector $\mathbf{n}$.

Consider now an observer at geocentric latitude ϕ', whose geocentric distance is ρ. If ω represents the earth's angular velocity, the observer's velocity with respect to the centre of the earth will be

$$V = \rho\omega \cos \phi'. \tag{4.67}$$

This velocity is in an easterly direction. This means that the vector **n** is directed towards the east point of the observer's horizon and consequently that

$$\alpha_0 = \text{LST} + 6^h$$
$$\delta_0 = 0. \tag{4.68}$$

Inserting these results into equations (4.66), and introducing the source's hour angle H, then leads to the results

$$d\alpha = (\rho\omega \cos \phi'/c) \sec \delta \cos H$$
$$d\delta = (\rho\omega \cos \phi'/c) \sin \delta \sin H. \tag{4.69}$$

These formulae are sufficiently accurate for all purposes. In fact, the effect is so small that some further approximations are often made. The distinction between geocentric and astronomical latitudes may usually be dropped, and it is normally sufficient to equate ρ to the earth's equatorial radius. Now substitute the value of ω as 2π radians per sidereal day, or in SI units

$$\omega = 7.292 \times 10^{-5}.$$

Then, with the simplifications mentioned above, equations (4.69) are expressed in practical units as

$$d\alpha = 0^s.0213 \cos \phi \sec \delta \cos H$$
$$d\delta = 0''.320 \cos \phi \sin \delta \sin H. \tag{4.70}$$

Strictly speaking, these formulae give the differences between the topocentric coordinates and those that would be obtained at the observing site by an observer stationary with respect to the earth's centre. To complete the transformation to geocentric coordinates, it is also necessary to make the parallax correction considered earlier. When both the effects are small, the order in which they are made is immaterial. If the parallax correction is large, however, it is necessary to remove the aberrational effect before applying the rigorous vector correction for parallax.

It was stated that relativistic effects can be ignored due to the smallness of (V/c). This refers to the special relativistic effects in aberration. The neglect of general relativistic light deflection in the earth's gravitational field requires separate justification. In fact, this deflection, $\delta\psi$ say, cannot exceed $2m/\rho$ radians, where m is one-half the Schwarzschild radius of the earth, i.e.

$$m = GM_\oplus/c^2 = 4.4 \text{ mm}. \tag{4.71}$$

It is then found that

$$\delta\psi < 0''.0003,$$ (4.72)

justifying its neglect.

Problems

4.1 If the declination of a star is unaffected by refraction at a given moment, prove that the azimuth is then a maximum.

[Smart (1977)]

4.2 Estimate for how long near the summer solstice some part of the solar disc is continuously over the horizon at a site on the Arctic Circle.

4.3 Theoretical sunset is defined as the instant when the topocentric zenith distance of the centre of the solar disc is 90°. Observed sunset means the final disappearance of the sun's upper limb. Show that the interval between observed and theoretical sunset at latitude ϕ is approximately given by

$$\Delta T = 3\tfrac{1}{3}^m (\cos^2 \phi - \sin^2 \delta)^{-1/2},$$

where δ is the sun's declination.

4.4 Astronomical twilight is defined to begin or end when the topocentric zenith distance of the sun is 108°. Show that there is a period of about three months during the summer when it is never astronomically dark at a latitude of 56°.

4.5 If the refraction formula had the *exact* form $R = K \tan z_0$, how would the refractive index vary with height? Determine K and the extent of the atmosphere in terms of the refractive index at ground level.

4.6 Assuming a refraction formula $R = K \tan z_0$, show that, due to refraction, the sun's disc appears as an ellipse of flattening f, given by

$$f = \frac{K \sec^2 z}{1 - K},$$

where z is the topocentric zenith distance of the sun's centre.

4.7 If a and b are the equatorial and polar radii of the earth (assumed spheroidal), show that the maximum value of the angle of the vertical is attained at geodetic latitude $\tan^{-1}(a/b)$.

4.8 Calculate the geocentric distance, the geocentric latitude and the angle of the vertical for an observer at sea level at latitude $55° 52'$.

4.9 Using the results of the previous question, calculate the maximum geodetic altitude that can be reached, at that observing site, by an earth satellite moving in a circular orbit of radius 8798 km inclined at an angle of $18° 36'$ to the equator.

4.10 Show that, if the horizontal parallax P of a body is small so that $\sin^2 P$ may be neglected, the apparent daily path of the body, as seen from a place in latitude ϕ, is a small circle of radius $90° - \delta + P \sin \phi \cos \delta$, described about a point depressed $P \cos \phi \sin \delta$ below the pole.

[Smart (1977)]

4.11 Define and distinguish between the astronomical, geodetic and geocentric zenith points. Explain what is meant by the angle of the vertical and show that, at geodetic latitude ϕ, it is given by

$$\tan v = \frac{e^2 \sin 2\phi}{2(1 - e^2 \sin^2 \phi)},$$

where e is the eccentricity of the standard spheroid.

Both atmospheric refraction and geocentric parallax change the zenith distance of a celestial object. To which zenith points are the displacements related?

5

Direct measurements of right ascension and declination

5.1   Introduction

The meridian circle, already briefly described in section 2.8, is the classical instrument of positional astronomy. The detailed theory of the instrument is rather specialized and is beyond the scope of this book. Some basic instrumental corrections will, however, be considered below. For the instrumental errors are of two basic types, namely, those arising from imperfections in the instrument itself, and those arising from inevitable slight misalignments in the construction and mounting of the instrument. The latter are of a purely geometrical character and are treated by the methods of spherical astronomy.

At the same time as the instrumental corrections are applied, it is the normal practice to make the corrections for refraction and diurnal aberration, and, if necessary, for geocentric parallax. In this way, the geocentric coordinates of the stars are obtained more or less directly from the observations. These are also referred to as the *apparent coordinates*, which are formally defined as follows.

Definition. The apparent coordinates of a star, at a particular date, are its coordinates on the geocentric celestial sphere referred to the true equator and equinox of that same date.

Being geocentric, apparent coordinates are independent of the particular observer, but they do vary with date, and fairly rapidly, principally due to annual aberration and precession. An annual almanac is published by the Astronomisches Rechen-Institut in Heidelberg entitled *The Apparent Places of the Fundamental Stars* (APFS). It is intended for use by meridian observers and gives apparent right ascensions and declinations of 1535 stars at 10-day intervals.

5.2   Errors of the meridian circle

A typical view through the eye-piece of a transit instrument is illustrated in Fig. 5.1. Although considerable variations in design of the instrument are possible, the observations will basically consist of timing the passage of the star across a series of vertical cross-wires. The mean of these times is then adopted as the observed time of transit of the star. This will correspond to the transit across a *mean thread* which will coincide closely, but not exactly, with the position of the central cross-wire *AB*. Let *T* be the time of transit over the mean thread as determined by an observatory clock keeping local sidereal time. Then, if ΔT is the *clock error*, the recorded right ascension of the star will be

$$\alpha' = T + \Delta T. \tag{5.1}$$

A fundamental determination of the clock error requires the location of the equinox in the sky to be known, since the local sidereal time is the hour angle of the equinox. A provisional value of ΔT may be obtained, however, by using radio time signals. These will give the universal time which may be converted to Greenwich sidereal time from tables in the *Astronomical Almanac*. Then the local sidereal time follows from

$$\text{LST} = \text{GST} + \lambda, \tag{5.2}$$

where λ is the longitude of the instrument. Observe that the longitude itself would need to be accurately known *a priori*. This cannot, in fact, be

Figure 5.1. Field of view of a transit telescope.

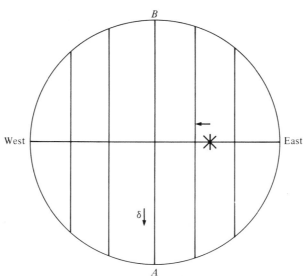

assumed, for it is found that both the longitude and latitude of the instrument are subject to slight variations due to polar motion. This is considered in section 5.5.

While the star is traversing the field of view shown in Fig. 5.1, the pointing elevation of the telescope must be adjusted so that the horizontal cross-wire bisects the image of the star. This allows an accurate measurement to be made of the star's declination, for the star's altitude at transit is just its declination plus the colatitude $(90° − \phi)$ of the instrument. With a suitable mounting of the vertical graduated circle that is attached to the rotation axis of the instrument, a direct reading D may be made of the star's declination. Then the recorded declination of the star can be defined as

$$\delta' = D + d. \tag{5.3}$$

Here d is a small constant which must be included, since the mounting mentioned above will not be exact.

When the transit telescope is rotated about its axis, each of the vertical cross-wires (more precisely, each point of intersection with the horizontal cross-wire) describes a curve on the celestial sphere. These curves will be a set of parallel small circles whose planes are perpendicular to the rotation axis of the instrument. The parallel great circle defines the *collimation plane*. If the cross-wires were ideally positioned, this would coincide with the mean thread. In practice, the mean thread is displaced from the collimation plane by a small angle c, which is known as the *collimation constant*. This will be regarded as positive if the mean thread is east of the collimation plane.

The direction of the instrument's rotation axis defines two diametrically opposite points on the celestial sphere, E' and W', say. For a perfectly aligned axis these points would coincide with the east and west points of the horizon. Two further parameters are required, however, to allow for any slight misalignment. The displacement of the axis in azimuth (south of west) is called the *azimuth constant a*, and the elevation of the west-end of the rotation axis is termed the *level constant b*. As shown in Fig. 5.2, both a and b are positive. They are defined as

$$a = WZW'$$
$$b = 90° − ZW'. \tag{5.4}$$

The five errors that have been defined are sufficient for the ensuing discussion of meridian observations. We expect all five errors will be small for a well-adjusted instrument!

It is often more convenient to express the errors in alignment of the rotation axis in equatorial, rather than alt–azimuth, form. The quantities m and n are defined as the departures of the point W' from the west point of the horizon in right ascension and declination respectively, i.e.

$$m = WPW'$$
$$n = 90° - PW'. \tag{5.5}$$

The two sets of quantities (m, n) and (a, b) may be related by solving spherical triangle PZW'. The sides in this triangle are:

$$PW' = 90° - n, \quad W'Z = 90° - b, \quad PZ = 90° - \phi;$$

while
$$W'PZ = 90° - m \quad \text{and} \quad W'ZP = 90° + a.$$

The four-parts formula, therefore, yields

$$-\sin \phi \sin a = \cos \phi \tan b - \cos a \tan m,$$

reducing to

$$\tan m = \tan a \sin \phi + \tan b \sec a \cos \phi. \tag{5.6}$$

The quantity n is derived from the cosine formula as

$$\sin n = \sin b \sin \phi - \sin a \cos b \cos \phi. \tag{5.7}$$

Since a, b, m and n are all small quantities, it is seldom necessary to use the expressions for m and n in their exact form. With the usual small-angle approximations, one can readily derive the following expressions:

$$m = a \sin \phi + b \cos \phi$$
$$n = b \sin \phi - a \cos \phi. \tag{5.8}$$

These are generally of sufficient accuracy.

Figure 5.2

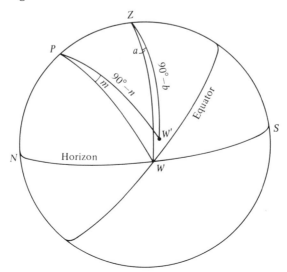

This last statement may seem a little surprising in view of earlier caveats about the shortcomings of first-order formulae. Equations (5.8) are, however, more accurate than is immediately apparent, since there are, in fact, no second-order terms. The constraints on the smallness of the instrumental errors are, therefore, less stringent than they might appear at first sight. The absence of second-order terms is a consequence of geometrical symmetry about the meridian. Whenever a first-order correction is necessary, the corresponding second-order correction automatically vanishes, and it is only when the first-order correction is itself zero that there are second-order terms to be considered. If such a circumstance does arise, this will be indicated in the text, but otherwise only first-order precision will be maintained in the subsequent discussion.

The instrumental constants are normally expressed in time measure, since they are primarily required to reduce the observations of right ascension. The only exception is the declination constant d which is more conveniently expressed in arc seconds.

5.3 The removal of instrumental effects

The coordinates (α', δ') discussed in the last section were described as the recorded right ascension and declination; they have been corrected only for the errors ΔT and d, which can be applied directly to the clock and circle readings. Let (α, δ) be the 'true' coordinates of the transiting star, that is what would be observed by an ideal perfectly adjusted instrument. But for the absence of a correction for refraction they would be the topocentric coordinates. Suppose X is the position of the star on the celestial sphere at the instant of recorded transit. As Fig. 5.3 indicates, this occurs earlier than 'true' transit by a time interval τ, where

$$\tau = \alpha - \alpha'. \tag{5.9}$$

Consider the spherical triangle PXW'. From the definitions of the instrumental constants, $PW' = 90° - n$, $W'X = 90° + c$, and $W'PX = 90° - m + \tau$. Further, $PX = 90° - \delta$. So applying the cosine formula to this triangle will yield

$$-\sin c = \sin n \sin \delta + \cos n \cos \delta \sin (m - \tau).$$

After making the usual small-angle approximations, this reduces to

$$\tau = \alpha - \alpha' = m + n \tan \delta + c \sec \delta. \tag{5.10}$$

This is known as Bessel's formula. It will enable the true right ascension to be determined.

The spherical angle $PW'X$ is simply related to the circle reading. In fact, with a suitable definition of the zero point, we may write

$$PW'X = 90° - \delta'. \tag{5.11}$$

Then the true declination is given by the cosine formula as

$$\sin \delta = -\sin n \sin c + \cos n \cos c \sin \delta'. \tag{5.12}$$

This equation indicates that δ differs from δ' only in the second order. It is immaterial, therefore, which declination is used on the right-hand side of equation (5.10). It may be necessary, however, to include the second-order terms for the accurate determination of the declination itself. So, proceeding to this order, equation (5.12) yields the approximate result

$$\delta' - \delta = \tfrac{1}{2}(n^2 + c^2) \tan \delta' + nc \sec \delta'.$$

Radians, however, have been used here, and the result expressed in practical units becomes

$$\delta' - \delta = 225'' \sin 1'' [\tfrac{1}{2}(n^2 + c^2) \tan \delta' + nc \sec \delta']. \tag{5.13}$$

An alternative to Bessel's formula is obtained by replacing (m, n) with (a, b). Using equations (5.8), the new result, known as Mayer's formula, is found to be

$$\tau = [a \sin (\phi - \delta) + b \cos (\phi - \delta) + c] \sec \delta. \tag{5.14}$$

Bessel's formula is preferred for the routine reduction of transit observations, but Mayer's formula is still useful.

All these results have been derived for the normal transit of a star.

Figure 5.3

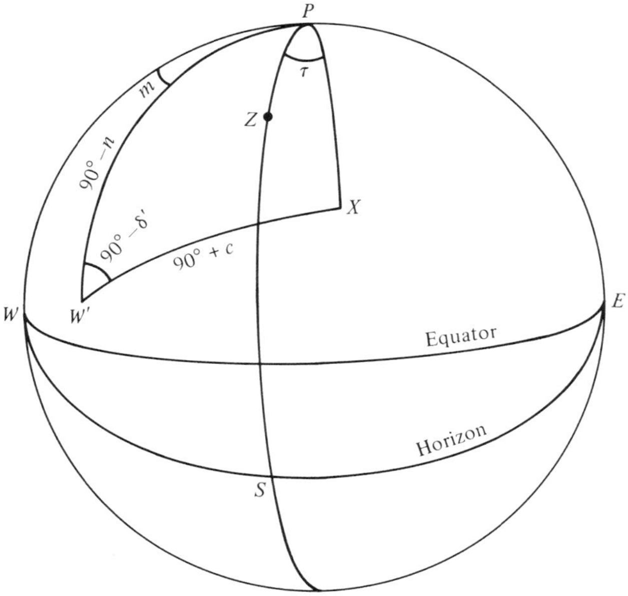

Equally well, they are valid for the upper transit of a circumpolar star, but for the lower transit slight differences are encountered. The analysis required is very similar to that already given, so the details will be omitted and only the results quoted. Bessel's formula for a lower transit has the form

$$\tau = \alpha - \alpha' = m - n \tan \delta - c \sec \delta, \tag{5.15}$$

where, once again, τ means the time of true transit minus the time of recorded transit. The quantity α' is now to be understood as the recorded local sidereal time of transit plus 12 hours. Further, Mayer's formula will now take the form

$$\tau = [a \sin (\phi + \delta) + b \cos (\phi + \delta) - c] \sec \delta. \tag{5.16}$$

The theory embodied in equations (5.10) to (5.16) is sufficient to allow observations of right ascension and declination to be corrected for the main instrumental errors, provided, of course, that these instrumental constants are themselves already known. The determination of some of these quantities involves star observation. Moreover, since it is desirable to monitor the instrumental constants continuously, their determination is, in practice, incorporated as an integral part of the programme of star observation. A detailed description of such a programme for the establishment of a fundamental system of star coordinates will be found in chapter 17 of Woolard and Clemence (1966). The procedure described below is artificial. It is only intended to illustrate the principles involved in determining each instrumental constant and to show that their determination will not compromise the independence of subsequent star observation. Moreover, for simplicity, the admittedly important effect of refraction is left out.

The transit instrument will be equipped with some means of making measurements in its focal plane. For example, one might include, in addition to the cross-wires shown in Fig. 5.1, another vertical wire which can be moved across the field of view by adjusting a micrometer to which it is attached. Now it is possible to devise a series of measurements with this micrometer that will yield both the collimation and level constants. Moreover, these measurements do not rely on the observations of any stars.

To determine the collimation constant c, two collimators are set up to provide reference marks near the north and south points of the horizon. They will not coincide exactly with the cardinal points, but they can be adjusted so that they are in precisely diametrically opposite directions. The transit telescope is then pointed in turn at each of these two collimators, and the position of its fixed system of vertical cross-wires is measured with reference to each of the defined directions. The reduction of these measurements leads to a value of the collimation constant c in micrometer units.

In order to measure the level constant b, it is necessary to have the transit telescope pointed in a vertical direction. It is usually directed towards the nadir, rather than towards the zenith, so that a trough of mercury may provide a horizontal reflecting surface. Comparison of the positions of the fixed wires with their reflections in the mercury surface enables b to be determined, again in micrometer units. One method of ensuring that the pointing of the telescope is accurately vertical is to adjust its altitude until the horizontal cross-wire coincides with its reflected image. This adjustment is not critical for the measurement of b described above, but it does provide an accurate means of determining the latitude. For the declination of the nadir is $-\phi$. So a reading of the declination scale is taken at this setting. This value of the latitude still requires correction for instrumental effects, mainly the declination constant d. This can be provided subsequently when that constant has itself been determined by the method considered below.

Suppose next that a circumpolar star is observed, both at upper and lower transit. The declination readings taken on the two occasions will differ, apart from any refraction effects which we are ignoring, since the declination constant d operates in opposite senses at the two transits. The mean of the two readings gives the star's declination – strictly δ'. The star may be timed across a measured distance in the field of view, that is, a distance measured in micrometer units. The angle it has traversed is simply the time interval multiplied by $\cos \delta$. So the micrometer scale is thereby calibrated in time seconds, and the earlier measurements of b and c may be converted into these units. Moreover, the declination constant d is determined as one-half the difference between the declination readings at the two transits.

Next let T_U and T_L be clock times that may be noted for the upper and lower transit of the circumpolar star. Equations (5.10) and (5.15) then yield for their respective transits

$$\alpha - T_U - \Delta T = m + n \tan \delta + c \sec \delta$$
$$\alpha - T_L - \Delta T - 12^{\mathrm{h}} = m - n \tan \delta - c \sec \delta. \tag{5.17}$$

The crude assumption has been made that the clock error will be unchanged over the 12-hour interval between the two transits. To overcome this difficulty, the change in ΔT must be monitored by comparison with time signals – cf. also Problem 5.3. Now eliminating the unknown right ascension and clock error, equations (5.17) provide a value for n as

$$n = [6^h - \tfrac{1}{2}(T_U - T_L)] \cot \delta - c \operatorname{cosec} \delta. \tag{5.18}$$

Finally the azimuth constant a and the quantity m may be derived from equations (5.8) using the values of the level constant and the latitude previously determined.

A fundamental determination of the clock error ΔT requires the position of the equinox among the stars to be known. Equivalently, an object must be observed whose right ascension is known *a priori*. The sun is an obvious candidate. Although a detailed dynamical theory of the earth's orbital motion is needed to deduce the sun's right ascension, the underlying principles can be demonstrated if it is assumed that the sun's ecliptic latitude is exactly zero. In Fig. 5.4, S represents the position of the sun, P the north celestial pole and Υ the equinox. Produce great circle arc PS to intersect the equator in U. Then the right ascension and declination of the sun $(\alpha_\odot, \delta_\odot)$ are given by

$$\alpha_\odot = \Upsilon U$$
$$\delta_\odot = SU. \tag{5.19}$$

The four-parts formula applied to spherical triangle $S\Upsilon U$, in which $S\Upsilon U = \varepsilon$, gives

$$\cos\alpha_\odot \cos\tfrac{1}{2}\pi = \sin\alpha_\odot \cot\delta_\odot - \sin\tfrac{1}{2}\pi \cot\varepsilon.$$

Hence

$$\alpha_\odot = \sin^{-1}(\cot\varepsilon \tan\delta_\odot). \tag{5.20}$$

Figure 5.4

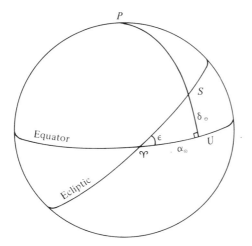

Now the sun's declination may be measured at the time of transit, and so equation (5.20) allows the calculation of $\alpha_\odot$ provided that the obliquity of the ecliptic is known. This, too, can be determined, in principle, from transit observations, since ε is the maximum value of the sun's declination at the summer solstice.

Now let T be the clock time of the sun's apparent transit. Then, by equations (5.1) and (5.9), the clock error is given by

$$\Delta T = \alpha_\odot - T - \tau. \tag{5.21}$$

The determination of the clock error means that the local sidereal time is now accurately known. It may be compared with Greenwich sidereal time by using radio time signals as explained in section 5.2. An accurate value of the observer's longitude may then be determined by inverting equation (5.2).

It will be necessary to consider the observer's geographical coordinates (λ, ϕ) in more detail, for both are variable due to *polar motion*. The effect is a very small one, however, and its discussion is deferred until section 5.5. The main neglect in the foregoing discussion is not polar motion but atmospheric refraction. This effect is so significant that its inclusion is essential, even in the most approximate reduction of declination measurements. The omission of refraction up till now, which was purely to simplify the discussion, will be speedily rectified in the next section.

5.4   Reduction to the apparent place

The procedure described in the last section is artificial and idealized. It does, however, demonstrate the possibility of determining each of the instrumental constants without previous knowledge of the positions of any of the stars. Subsequent measurements with the meridian circle may then be corrected for these instrumental errors to yield fundamental values of both right ascension and declination through equations (5.10) and (5.13). It is normal practice, however, to reduce the observations further to yield the apparent place. This involves making corrections for refraction, diurnal aberration and geocentric parallax, all of which have been fully discussed in chapter 4. In fact, considerable simplification is possible, since the corrections are to be applied to measurements made close to the observer's meridian.

For example, setting the hour angle equal to zero in equations (4.70) establishes the influence of diurnal aberration as

$$d\alpha = 0^s.0213 \cos \phi \sec \delta$$
$$d\delta = 0. \tag{5.22}$$

Declination measurements are, therefore, unaffected, while the correction for right ascension is similar to the collimation correction in that each is proportional to sec δ. This means that diurnal aberration may be entirely allowed for by using a modified collimation constant in Bessel's or Mayer's formula. We simply replace c in these formulae by c^*, where

$$c^* = c - 0^s.0213 \cos \phi. \tag{5.23}$$

The effect of atmospheric refraction is more serious, and this is usually the largest correction. Refraction reduces the zenith distance of a star, but leaves the azimuth unaltered. Since the observer's meridian is itself a vertical circle, refraction clearly causes a change in the observed declination, but the time of transit is unaffected. The angle of refraction R may be extracted from refraction tables, and this must be subtracted from the measured value of the declination – unless the star transits between zenith and the pole, when the sense of the correction is reversed.

The corrections for refraction and diurnal aberration may now be incorporated in the observational equations of the meridian circle. Geocentric parallax is negligible for a star or for any object beyond the bounds of the solar system, and no further correction will be required. Suppose that (α', δ') are again the star's recorded right ascension and declination, as defined in equations (5.1) and (5.3). Now, however, let (α, δ) represent the apparent place of the star. Then, to first order,

$$\alpha = \alpha' + m + n \tan \delta' + c \sec \delta', \tag{5.24}$$
$$\delta = \delta' - R. \tag{5.25}$$

Normally the instrumental constants are expressed in seconds of time, while the angle of refraction will be in arc seconds.

These last two equations may be improved by including second-order terms. Those for the declination equation have already been given in (5.13). Refraction also introduces a second-order correction for right ascension (cf. Problem 5.7). This may be important at large zenith distances.

For a celestial object that is located within the solar system, the correction for geocentric parallax must also be incorporated. This will usually be a small correction (a few arc seconds) which can be derived from equation (4.59) by setting $H = 0$. This indicates that the right ascension is unaffected, while, in the notation of section 4.7, the recorded declination must be increased by $(\rho/r) \sin (\phi' - \delta)$. Once again the sign of this correction is reversed between zenith and the celestial pole.

This treatment is inadequate, however, for the moon or artificial satellites, when the angle of parallax may not be treated as a small quantity. The best procedure then is first to apply the corrections for refraction and diurnal aberration, as described above. This will yield the geometric direction of the source for the topocentric observer. The transformation to the geocentric place is then completed by following the vector method described and illustrated in section 4.7.

5.5   Polar motion

The term *polar motion* is used to describe the slow and very slight movement of the geographic poles over the surface of the earth. Notice that the motion is with respect to the earth, not with respect to the stars. The instantaneous direction of the earth's rotational axis determines the *pole of rotation* on the surface of the earth and the true celestial pole on the celestial sphere. Certainly the latter becomes displaced with reference to the stars due to luni–solar precession and nutation (cf. chapter 9), but such effects are not considered in this section. For the present we shall treat the earth's rotational axis as a fixed sidereal direction. Our concern is with the apparent movement of this fixed direction with reference to the earth, but, in an absolute sense, it is the body of the earth that is in motion rather than the pole of rotation.

Since polar motion does not affect the sidereal direction of the pole, it produces no changes in the right ascension or declination of a star. It does, however, affect the reduction of measurements of these coordinates with a transit instrument. For the mounting of the instrument is fixed with reference to the earth's surface, and so the celestial pole, and indeed the whole sidereal background, is displaced with respect to the instrument. In effect, the instrumental constants are altered by polar motion, as we shall see below.

Figure 5.5 is intended to represent a 'bird's eye view' of the geographic north pole. The point P is the instantaneous pole of rotation, while P_0 is a mean position, which may be loosely termed the *pole of figure*. Unlike the pole of rotation, the pole of figure is a fixed point on the earth's surface. The displacement γ of P from P_0 is typically about $0''.3$, corresponding to a separation of some 10 m on the earth's surface. The east longitude Γ of the displacement is generally increasing so that the pole of rotation moves in a predominantly anticlockwise direction around P_0. The motion is complicated, however. It cannot be predicted in detail, but principal periods of revolution are recognizable, which are 428 days and one year in length.

The causes of the polar motion are almost entirely geophysical. If the earth were a freely spinning rigid body, the departure of the pole of rotation from the axis of symmetry would cause a uniform circular motion about the pole of figure with a period of 305 days. This period is considerably increased, however, by the yielding nature of the earth, and the 428-day period represents the modified free motion. The annual component is a forced motion imposed by continually changing geophysical conditions.

The displacement of P from P_0 is usually expressed as rectangular coordinates (x, y). These are respectively the displacements along the meridians of longitude zero and 270° (east), i.e.

$$x = \gamma \cos \Gamma,$$
$$y = -\gamma \sin \Gamma. \tag{5.26}$$

Values of x and y are published by the *International Polar Motion Service* and the *Bureau International de l'Heure*.

Polar motion produces changes in the latitude and longitude of any observing site. Suppose (λ_0, ϕ_0) are the geographic coordinates of an observer referred to the pole of figure, while (λ, ϕ) are those referred to the instantaneous pole of rotation. The definition of latitude is straightforward, but a little care is required in defining the meridian of zero longitude through P. As shown in Fig. 5.5, this is chosen as the meridian which is parallel to the prime meridian of figure. In fact, the two prime meridians will intersect at the equator, and the prime meridian of rotation will not normally pass through the fiducial point at Greenwich. This definition also clearly implies that the meridian of east longitude Γ is common to the two systems.

Figure 5.6 shows the celestial sphere for an observer O. The zenith point

Figure 5.5. Bird's eye view of the earth's north pole.

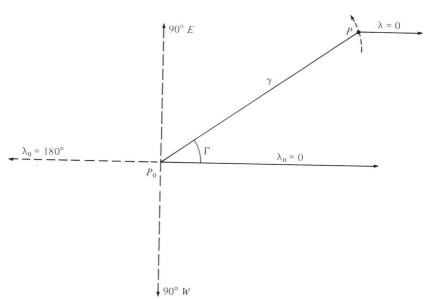

Z will be regarded as fixed. This is rigorously true if it is geocentric zenith that is intended. The instrumental constants are, however, determined with reference to astronomical zenith, and this is subject to small variations. For the direction of astronomical zenith is determined by apparent local gravity, and luni–solar tidal forces make a very small variable contribution. Moreover, these forces induce geophysical changes which further alter the direction of apparent gravity. The effect is purely geometric, however, and it will not corrupt the reduction of star observations, provided that the instrumental constants have been consistently determined with respect to a fixed zenith direction. The variation in the astronomical zenith point will, therefore, be ignored in the subsequent argument, and only the changes in astronomical latitude and longitude due to polar motion will be considered.

The points P_0 and P in Fig. 5.6 indicate the celestial poles corresponding to the pole of figure and of rotation respectively. Then $P_0 Z = 90° - \phi_0$, while $PZ = 90° - \phi$. Moreover, some cardinal points of the horizon have been indicated, N_0, W_0 and S_0 being determined by P_0, and N and W by the pole P. Construct a small circle, with pole Z, through the point P to cut $P_0 Z$ in Q. Then $P_0 Q = d\phi = \phi - \phi_0$. Further, $P_0 PQ$ may be treated as a plane

Figure 5.6

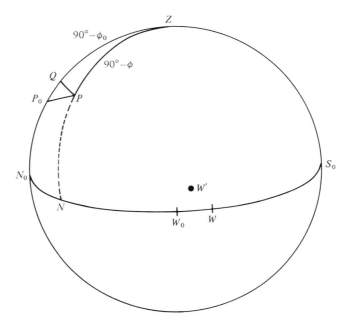

triangle right-angled at Q, yielding

$$\mathrm{d}\phi = P_0 P \cos\ QP_0 P = P_0 P \cos ZP_0 P.$$

Now $P_0 P = \gamma$. Moreover, $P_0 Z$ is the meridian of longitude λ_0, while $P_0 P$ is the meridian of longitude Γ, both based on P_0 as pole. So $ZP_0 P = \lambda_0 - \Gamma$, and

$$\mathrm{d}\phi = \gamma \cos (\lambda_0 - \Gamma). \tag{5.27}$$

Now, PZ represents the meridian of longitude λ, and PP_0 that of longitude $(180° + \Gamma)$, each now referred to P. So the spherical angle $P_0 PZ = 180° - (\lambda - \Gamma)$. The four-parts formula may, therefore, be applied to spherical triangle $P_0 ZP$ to yield

$$-\cos \gamma \cos (\lambda - \Gamma) = \sin \gamma \tan \phi - \sin (\lambda - \Gamma) \cot (\lambda_0 - \Gamma).$$

Since γ is a small angle, this gives approximately

$$\gamma \tan \phi \sin (\lambda_0 - \Gamma) = \sin (\lambda - \Gamma) \cos (\lambda_0 - \Gamma) - \cos (\lambda - \Gamma) \sin (\lambda_0 - \Gamma).$$

The result is, therefore, established to sufficient accuracy that

$$\mathrm{d}\lambda = \lambda - \lambda_0 = \gamma \tan \phi \sin (\lambda_0 - \Gamma). \tag{5.28}$$

Normally these changes are expressed in terms of the rectangular coordinates (x, y). Expand the right-hand sides of equations (5.27) and (5.28); then substituting from (5.26) yields

$$\phi = \phi_0 + x \cos \lambda_0 - y \sin \lambda_0,$$
$$\lambda = \lambda_0 + (x \sin \lambda_0 + y \cos \lambda_0) \tan \phi_0. \tag{5.29}$$

The section is concluded with a discussion of how polar motion will affect the instrumental constants of the meridian circle. The point W' located by the west end of the instrument's rotation axis is unaltered, and clearly there will be no change in the collimation constant. Moreover, since the zenith point is fixed, the level constant, too, is unchanged. The displacement of the pole will, however, reduce all west azimuths, including the azimuth constant, by the arc $N_0 N$. So

$$\mathrm{d}a = -P_0 ZP.$$

This small angle is readily derived from the sine formula in spherical triangle $P_0 ZP$. In fact,

$$\mathrm{d}a = -\gamma \sin (\lambda_0 - \Gamma) \sec \phi. \tag{5.30}$$

From (5.28), this result can also be expressed as

$$\mathrm{d}a = -\mathrm{d}\lambda \operatorname{cosec} \phi. \tag{5.31}$$

Finally it must be noted that the change in longitude will introduce an additional clock error of the same amount. Formally

$$\mathrm{d}(\Delta T) = \mathrm{d}\lambda. \tag{5.32}$$

The effect that polar motion will have on right ascension measurements may be determined from Mayer's equation, namely

$$\alpha - T = \Delta T + a \sin (\phi - \delta) \sec \delta + b \cos (\phi - \delta) \sec \delta + c \sec \delta.$$

$$(5.14)$$

Taking differentials of this equation gives, with sufficient accuracy,

$$d(\alpha - T) = d\lambda + da \sin (\phi - \delta) \sec \delta.$$

Remembering that the 'true' right ascension is unaffected and using equation (5.31) leads to the result

$$dT = -d\lambda \cot \phi \tan \delta. \qquad (5.33)$$

This gives the change in the clock time of transit due to polar motion.

The effect of polar motion on declination measurements is much simpler. An increase in the observer's latitude produces a corresponding reduction in the declination reading.

5.6   The Danjon astrolabe and the photographic zenith tube

The meridian circle was described as the classical instrument of positional astronomy. It has considerable advantages. It is very versatile in that it is capable of measuring the right ascension and declination of any star visible to the observer, whereas other instruments have some restriction on the declinations for which they can be used. Moreover, in principle at least, the observations are straightforward direct measurements of both right ascension and declination. Until the early years of this century, the meridian circle provided the most accurate means of determining the absolute coordinates of stars. Observations with these instruments became the basis of fundamental star catalogues. More recently, new instruments have been developed which surpass the meridian circle in accuracy, but not in versatility, and generally the reduction of the observations is more complicated.

Observation need not, of course, be confined to the observer's meridian. Let us consider a slight variant of the meridian circle, namely an identical instrument but mounted instead on a horizontal north–south axis. As this instrument is rotated, it will sweep out the prime vertical. As shown in Fig. 5.7, a *prime vertical transit instrument* can only be used to observe stars in the range $0 < \delta < \phi$. All such stars will make two transits, and the local sidereal time of each transit is recorded. The mean of the two times will provide the star's right ascension. The graduated circle reading will now give the star's zenith distance z at transit over the prime vertical; the hour angle of transit will also be known, since it is one-half of the interval between the east and west transit.

Let X (Fig. 5.7) be the position of a star at west transit. Then the parts of spherical triangle PZX are as indicated in the diagram, and the sine formula yields

$$\cos \delta = \sin z \operatorname{cosec} H. \tag{5.34}$$

So the star's declination may be computed. The latitude may also be derived from observations with this instrument, for the four-parts formula gives

$$\cos \phi = \tan z \cot H. \tag{5.35}$$

The instrumental errors of the prime vertical transit instrument are the same as those for the meridian circle. Clearly, however, the correction for these instrumental effects will be decidedly more complicated. The only advantage of the prime vertical instrument is that it allows a series of independent measurements of the observer's latitude.

The versatility of the meridian circle is a virtue that brings its own problems. These ultimately limit the reliability of observations with this instrument. For the transit circle is going to be used over a wide range of orientations, and the geometrical idealizations that were introduced to account for instrumental effects cannot be expected to apply exactly over the large angles involved. Problems arise from the flexure of the instrument

Figure 5.7

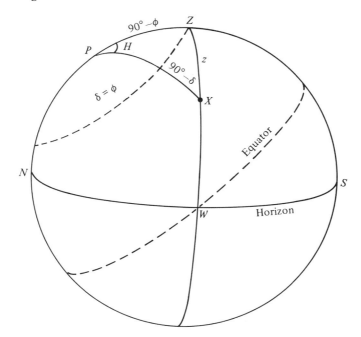

under its own weight. This can introduce systematic variations in the instrumental constants. Moreover, these systematic effects will be undetectable. For example, how can we know if the adopted value of the level constant is correct for modest zenith distances, when this constant was determined by measurements made with the telescope pointed to the nadir? While vigorous attempts continue to be made to control these systematic uncertainties and thus improve the performance of the transit instrument, it is generally agreed that the most accurate and consistent results are achieved with other instruments. The two most important are the photo zenith tube (PZT) and the impersonal astrolabe of Danjon. With the PZT the versatility of the meridian circle is entirely abandoned, and the observations are restricted to the zenith. On the other hand, the Danjon astrolabe may be rotated in azimuth, but all the observations are made at the same zenith distance, namely 30°, so that inconsistencies arising from flexure should not be important.

The basic principle of the astrolabe is shown in Fig. 5.8. An equilateral glass prism is placed, with one of its faces vertical, in front of a horizontal telescope. Two images of a star are formed from light passing through the prism – direct rays internally reflected off the lower face produce one, and rays reflected from a horizontal mercury surface and then internally from the upper face produce the other. These two images will coincide when the zenith distance of the star is exactly 30°. The simple instrument represented in the diagram has an inherent optical defect in that the separation of the images is influenced by a focussing adjustment of the telescope eye-piece. The instrument was, therefore, redesigned by Danjon to remove this defect, and additional features make the modern instrument essentially

Figure 5.8. Basic features of an astrolabe.

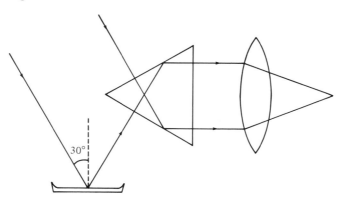

impersonal (Danjon, 1960). The time corresponding to a zenith distance of 30° is recorded automatically.

A star can be observed with the astrolabe provided it crosses the parallel of altitude 60° some time during its diurnal motion. This normally restricts observable stars to the range of declination $\phi + 30° > \delta > \phi - 30°$. Each such star will make two transits over the parallel of altitude, one east and one west of the observer's meridian. The average of the two recorded times of transit gives the star's right ascension; while one half of the time difference will give the hour angle H that corresponds to its zenith distance being 30°.

The point X in Fig. 5.9 represents a star's position on the celestial sphere when it is observed with a Danjon astrolabe. From the cosine formula, it follows that

$$\cos z = \sqrt{3}/2 = \sin \phi \sin \delta + \cos \phi \cos \delta \cos H. \tag{5.36}$$

This equation may be used to derive either the declination of the star or the latitude, assuming that the other is known. If the latitude is known, we proceed as follows. Introduce a subsidiary angle F, defined so that

$$\tan F = \cot \phi \cos H. \tag{5.37}$$

This angle may be computed. Then equation (5.36) may be expressed as

$$\sqrt{3}/2 = \sin \phi \sec F \sin (\delta + F).$$

So the declination is derived as

$$\delta = \sin^{-1} \left(\tfrac{1}{2}\sqrt{3} \cos F \, \mathrm{cosec} \, \phi \right) - F. \tag{5.38}$$

This equation has two solutions, due to the ambiguity in the inverse sine.

Figure 5.9

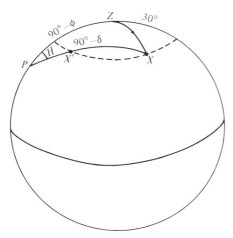

Table 5.1. *The variation of the two declinations with hour angle for a Danjon astrolabe at latitude 50°*

H	F	δ_1	δ_2
0	40°	80°	20°
10°	39°.57	79°.80	21°.06
20°	38°.26	79°.15	24°.34
30°	36°.00	77°.85	30°.14
40°	32°.73	75°.28	39°.26
45°	30°.68	72°.85	45°.79
50°	28°.34	67°.38	55°.93
51°	27°.84	63°.58	60°.74

These correspond to the points X and X' in Fig. 5.9. Normally, there is no problem in distinguishing the correct solution. However, near the value of δ which is the repeated root of (5.38) – this corresponds to the maximum hour angle for which z can be 30° – the above method of determining the declination becomes unacceptably inaccurate. The difficulty is illustrated in Table 5.1, which gives the solutions of equation (5.38) for latitude 50°. Clearly, declinations cannot, in practice, be determined over the whole of the expected range – in this case $20° < \delta < 80°$. In particular, declinations between roughly 50° and 65° will be virtually impossible to measure.

Despite this restriction, the astrolabe can determine declinations over a significant range, and, as already mentioned, it gives superior results to the meridian circle.

The other modern instrument that we wish to describe briefly is the photographic zenith tube (PZT). Its main features are sketched in Fig. 5.10. It consists of a horizontal objective lens with a plate carriage just below it. Star light is imaged on to the photographic plate after reflection from a mercury surface positioned below the apparatus.

An essential feature in the design of this instrument is that a nodal point of the lens should lie on the photographic emulsion (Markowitz, 1960). The distance of the mercury surface is then adjusted so that focussed images are obtained on the plate in this preassigned position. This effectively eliminates errors of level and collimation in setting up the instrument.

The plate carriage is driven across the field of view at the appropriate diurnal rate. This provides brighter images and allows faint stars to be satisfactorily photographed. The exposure is normally about 10 to 20 seconds. The time of mid-exposure is automatically recorded with very high accuracy.

In fact, a complete observation involves taking four exposures. Between each exposure the lens and the plate carriage are rotated, as a unit, through 180°. The four images of a star will form an approximate parallelogram, as shown in Fig. 5.11. Let $t_1, \ldots, t_4$ be the recorded times of the four exposures $X_1, \ldots, X_4$ – in the correct temporal order. If there were no reversals between exposures, the four images would, neglecting slight curvature effects, have been on a straight line. The lines $X_1 X_3$ and $X_2 X_4$, therefore, represent the diurnal motion for the plate in its two orientations. These lines will be parallel if the rotation is exactly 180°. Moreover, the plate scale may be determined, since $X_1 X_3$, for example, represents diurnal motion over the time interval $(t_3 - t_1)$.

When the rotation takes place the image of the zenith point on the plate remains fixed. It is not, of course, marked in any way, but we can be certain that it is equidistant from the lines $X_1 X_3$ and $X_2 X_4$. Let us set up a system of coordinates (x, y) on the plate with an arbitrary origin, but with the x-

Figure 5.10. Basic features of a photographic zenith tube.

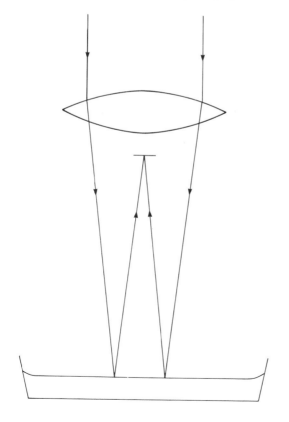

axis parallel to the two lines. Let (x_i, y_i) be the coordinates of $X_i, i = 1, \ldots, 4$, which can, of course, be measured. These coordinates may be readily converted into angular units, since the plate scale is known. Note that $y_1 = y_3, y_2 = y_4$, by construction. Let (x_0, y_0) be the coordinates of the zenith point. Then x_0 is unknown, but

$$y_0 = \tfrac{1}{2}(y_1 + y_4) = \tfrac{1}{2}(y_2 + y_3).$$

More important, the separation between the two lines will be twice the zenith distance of the star at transit. Hence

$$\phi - \delta = \tfrac{1}{2}(y_1 - y_4) = \tfrac{1}{2}(y_3 - y_2). \tag{5.39}$$

The declination of the star is, therefore, determined, provided the latitude is known.

The hour angle of the star at time t_1 is $(x_0 - x_1)$, while the hour angle at time t_4 is $(x_4 - x_0)$, since the plate is then reversed. Taking the mean of these two times, it follows that at time $\tfrac{1}{2}(t_1 + t_4)$ the hour angle is $\tfrac{1}{2}(x_4 - x_1)$. We could have used the second and third exposures instead, since there is some redundancy in the problem. Using all four images, it follows that, at time

$$t_0 = \tfrac{1}{4}(t_1 + t_2 + t_3 + t_4), \tag{5.40}$$

the star has hour angle

$$H_0 = \tfrac{1}{4}(x_4 - x_3 + x_2 - x_1). \tag{5.41}$$

Then the right ascension of the star is determined as

$$\alpha = t_0 - H_0. \tag{5.42}$$

Remember curvature effects have been neglected in this treatment. They are considered later in chapter 13.

Figure 5.11. Star images from a complete PZT observation.

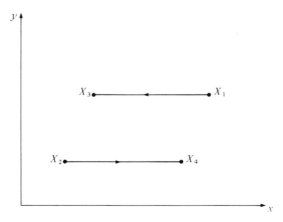

One great advantage of the PZT is that, since observations are made very close to the zenith, the uncertainties in refraction are reduced to a minimum. The disadvantage is that only a very small range of declinations, normally less than $1°$, can be covered.

Although superior in accuracy to the meridian circle, neither the PZT nor the Danjon astrolabe can determine right ascensions and declinations in as fundamental a manner. They can, however, accurately relate positions widely separated on the celestial sphere, and can reveal errors in fundamental catalogues based on meridian observations. Moreover, they are well suited to determining irregular changes in latitude and time, which arise, not only from polar motion, but also from variations in the earth's rotation rate; a subject to be considered in chapter 10.

Problems

5.1 Prove that the error in the time of transit of a star when observed with a meridian circle is given by Mayer's formula, namely:
$$\tau = [a \sin (\phi - \delta) + b \cos (\phi - \delta) + c] \sec \delta.$$

5.2 Derive Bessel's and Mayer's form of the observational equation for a transit instrument appropriate to the lower transit of a star, i.e. equations (5.15) and (5.16).

5.3 A star of declination $79° 35'$ is observed with a transit instrument at latitude $51° 16' 38''$. Subsequent to correction for the instrument's level and collimation errors, the universal times of lower and upper transit are found to be $18^h 03^m 18^s.4$ and $6^h 00^m 45^s.1$. Calculate the azimuth error of the instrument.

5.4 Define the azimuth, level and collimation constants of a meridian circle and also the two constants (m, n) used in Bessel's formula. Show that the error in the time of transit may be written in alternative forms as
$$\tau = b \sec \phi + n(\tan \delta - \tan \phi) + c \sec \delta$$
$$\tau = a \csc \phi + n(\tan \delta + \cot \phi) + c \sec \delta.$$
The first of these is known as *Hansen's formula*.

5.5 Prove that the true declination δ of a star is related to the value δ_m measured with a meridian circle by the equation
$$\delta = \delta_m + d + \tfrac{225}{2}[c^2 - (\tau - m)^2] \cot \delta \sin 1'',$$
where the symbols have their usual meanings and are expressed in conventional practical units.

5.6 A transit telescope is pointed vertically downwards towards a horizontal reflecting surface. Show that the angle between the centre cross-wire and its image by reflection is $2(c-b)$. Show further that this angle should be regarded as positive when the image is west of the cross-wire.

5.7 Prove that atmospheric refraction produces a first-order effect on declination measurements made with a meridian circle, but only a second-order effect in right ascension. Show that if the refraction formula is taken $R = K \tan z$, then Bessel's formula should be modified to

$$\alpha = \alpha' + (m + n \tan \delta' + c \sec \delta') \left(1 + \frac{K \sec^2 \delta'}{1 + \tan \phi \tan \delta'} \right)$$

where (α', δ') are the recorded right ascension and declination of a star.

5.8 Explain the principle of the astrolabe and show that it can be used to make positional measurements of stars within the range of declinations $\phi - 30° < \delta < \phi + 30°$. Prove that the precision of declination measurements is greatest at the ends of the range and goes to zero at $\delta = \sin^{-1} (2 \sin \phi / \sqrt{3})$, while the precision in right ascension is greatest at $\delta = \sin^{-1} (\sqrt{3} \sin \phi / 2)$.

5.9 A star of known declination δ is observed with an astrolabe at hour angle H. Prove that the latitude of the instrument may be determined as

$$\phi = \sin^{-1} \left(\frac{\sqrt{3}}{2} \cos G \operatorname{cosec} \phi \right) - G,$$

where G is the angle defined as

$$G = \tan^{-1} (\cot \delta \cos H).$$

5.10 A meridian circle, whose instrumental constants are unknown, has been set up at a site of only approximately known latitude and longitude. Devise a method for the determination of the following quantities $a, b, c, \Delta T, m, n, \phi$ and λ using observations of stars of known position.

6

Two-body orbital motion

6.1   Basic equations

The gravitational two-body problem is fundamental to positional astronomy. An obvious example is planetary motion about the sun, but other important applications are satellite motion and double stars. To fix ideas, however, the discussion will concentrate on the particular system comprising the sun S and a planet P, as shown in Fig. 6.1. Even so, the analysis that follows is quite general and may be applied to any two-body system.

The discussion is restricted to two bodies – an exact solution can then be found. Unfortunately no such analytic solution exists for systems containing more than two bodies. The treatment of the planet's motion is, therefore, being idealized, since the gravitational attractions of the other bodies in the solar system, principally the other major planets, are not included. Since the sun is by far the most massive body in the solar system, its attraction is the dominant force, and the two-body treatment provides a very useful first approximation. One other point of idealization is that the two bodies at S and P are each treated as if they were point masses.

Let m_1 and m_2 be the masses at the points S and P respectively, and let the point C be their centre of mass. This point may be chosen as the origin, and then the position vectors of the two masses $\mathbf{r}_1$ and $\mathbf{r}_2$ will satisfy the equation

$$m_1\mathbf{r}_1 + m_2\mathbf{r}_2 = 0. \tag{6.1}$$

Now it is convenient to refer the planet's position to that of the sun rather than to the centre of mass. We shall concentrate, therefore, on the vector $\mathbf{r} = \overrightarrow{SP}$, i.e.

$$\mathbf{r} = \mathbf{r}_2 - \mathbf{r}_1. \tag{6.2}$$

It immediately follows from the two above equations that

$$\mathbf{r}_1 = -\frac{m_2}{(m_1 + m_2)}\,\mathbf{r}$$

$$\mathbf{r}_2 = \frac{m_1}{(m_1 + m_2)}\,\mathbf{r}. \tag{6.3}$$

All the analysis will be directed to deriving the relative orbit, that is, of P about S; the parameters subsequently deduced will refer to this orbit. For a planet moving about the sun, when $m_1 \gg m_2$, this will approximate to its actual orbit in space. But this approximation is in no way vital; the treatment is valid for any ratio of m_1 to m_2. For the actual orbits of the two bodies in space may be recovered from the relative orbit through equations (6.3). The orbits of the two bodies about the fixed centre of mass are similar to the relative orbit and are in phase with it. The three orbits differ only in their sizes which are in the ratios

$$SC:PC:SP = m_2:m_1:m_1 + m_2. \tag{6.4}$$

The equations of motion for the two bodies are

$$m_1 \ddot{\mathbf{r}}_1 = -\mathbf{F}$$

$$m_2 \ddot{\mathbf{r}}_2 = \mathbf{F}, \tag{6.5}$$

where $\mathbf{F}$ is the gravitational force exerted on the body m_2. From the Newtonian law of gravity it follows that

$$\mathbf{F} = -\frac{Gm_1 m_2}{r^3}\,\mathbf{r}. \tag{6.6}$$

So the acceleration of each body may be deduced from (6.5), and the relative acceleration is then found to be

$$\ddot{\mathbf{r}} = \ddot{\mathbf{r}}_2 - \ddot{\mathbf{r}}_1 = -\frac{G(m_1 + m_2)}{r^3}\,\mathbf{r}.$$

For brevity introduce the quantity μ defined as

$$\mu = G(m_1 + m_2). \tag{6.7}$$

Then the relative motion of the two bodies is governed by the simple

---

Figure 6.1

equation

$$\ddot{\mathbf{r}} = -\frac{\mu}{r^3}\,\mathbf{r}. \tag{6.8}$$

All the important conclusions about two-body motion follow from this second-order differential equation. It readily admits two integrals. Taking the vector product with $\mathbf{r}$ yields

$$\mathbf{r} \times \ddot{\mathbf{r}} = 0,$$

which will integrate to give

$$\mathbf{r} \times \dot{\mathbf{r}} = \mathbf{h}. \tag{6.9}$$

Here $\mathbf{h}$ is a constant vector – which is proportional to the total angular momentum of the system.

On the other hand, consider the scalar product of (6.8) with $\dot{\mathbf{r}}$, namely

$$2\dot{\mathbf{r}} \cdot \ddot{\mathbf{r}} = -2(\mu/r^3)\mathbf{r} \cdot \dot{\mathbf{r}}.$$

Since $\mathbf{r} \cdot \dot{\mathbf{r}} = r\dot{r}$, this, too, integrates to yield

$$\dot{r}^2 = \frac{2\mu}{r} + C, \tag{6.10}$$

C being the constant of integration. This equation is equivalent to the conservation of energy.

The two conservation equations may be expressed in a neat scalar form by means of an appropriate choice of coordinate axes. Cylindrical polar coordinates (r, θ, z) are adopted, and the z-axis is chosen in the direction of the vector $\mathbf{h}$. Since $\mathbf{h} \cdot \mathbf{r} = 0$, the orbit is confined to the x–y plane, and the Cartesian components of the vectors $\mathbf{r}$ and $\mathbf{h}$ are given by

$$\mathbf{h} = h(0, 0, 1)$$
$$\mathbf{r} = (r\cos\theta, r\sin\theta, 0). \tag{6.11}$$

Equation (6.9) then reduces to the single scalar equation

$$r^2\dot{\theta} = h, \tag{6.12}$$

and equation (6.10) will take the form

$$\dot{r}^2 + r^2\dot{\theta}^2 = 2\mu/r + C.$$

In view of the result just proved, the latter may also be written as

$$\dot{r}^2 + h^2/r^2 = 2\mu/r + C. \tag{6.13}$$

6.2 Kepler's equation for a bound orbit

The constant C in (6.13) is proportional to the total energy of the system. Positive values of C correspond to unbound orbits, in which the total energy is sufficient to enable the two bodies to escape from each other's gravitational influence. The relative orbit is then a hyperbola.

Suppose, however, that C is negative. There is no loss of generality in writing

$$C = -\mu/a, \tag{6.14}$$

where a is a characteristic length of the system. Its geometrical significance (as the semimajor axis of the elliptic orbit) is established below. It is now seen from (6.12) and (6.13) that the relative velocity V of the two bodies is given by

$$V^2 = \dot{r}^2 + h^2/r^2 = \mu(2/r - 1/a). \tag{6.15}$$

The right-hand side of this equation cannot be negative. We therefore deduce that $r < 2a$, indicating that the orbit is bound. Clearly this is the type of orbit that is relevant to the study of planetary motion.

Even as it stands, equation (6.15) is a very important formula. It enables the magnitude of a planet's velocity to be computed provided that the planet's position is known for that instant of time. Moreover, the direction of the planet's velocity may be derived from (6.12), which gives the transverse component of its velocity V_T as

$$V_T = r\dot{\theta} = h/r. \tag{6.16}$$

A more difficult problem, however, is to derive the planet's position at a particular instant of time. We shall make the mathematical solution of this problem our first priority rather than deriving, for example, the overall shape of the orbit. This and the other Kepler laws will be established as by-products of the main analysis. The treatment that follows is, therefore, somewhat unconventional, and the reader may wish to consult a more standard treatment as well, e.g. Roy (1982).

A significant point in the planet's orbit is its closest approach to the sun, its perihelion. Let τ be the time corresponding to perihelion, when the planet is at the point A shown in Fig. 6.2. Let P be a general position of the planet at a subsequent time t. Its radius vector is then $r = SP$. The angle $v = PSA$ is called the *true anomaly*. It differs only from the polar angle θ in that it is measured from the direction of perihelion rather than from the x-axis. If we write

$$v = \theta - \omega, \tag{6.17}$$

the constant angle $\omega = xSA$ is called the *argument of perihelion*.

To determine the planet's position in its orbit we must relate the coordinates (r, v) to the time t. This involves integrating equation (6.15), which will be written as

$$\dot{r}^2 = -\frac{\mu}{ar^2} F(r). \tag{6.18}$$

Here, $F(r)$ is simply a shorthand for the quadratic form given by

$$F(r) = r^2 - 2ar + h^2 a/\mu. \tag{6.19}$$

Equation (6.18) implies that $F(r) \leqslant 0$. So we may conclude that the radius vector r is restricted to the range

$$r_1 \leqslant r \leqslant r_2, \tag{6.20}$$

where r_1 and r_2 are the two roots of $F(r) = 0$. In fact, they must correspond to the perihelion and aphelion distances.

It should be noted, from equation (6.19), that $r_1 + r_2 = 2a$. Moreover, it is convenient to introduce the parameter e – later to be identified with the orbital eccentricity – as

$$e = \frac{r_2 - r_1}{r_2 + r_1}. \tag{6.21}$$

Then, working in terms of a and e, it is seen that

$$\begin{aligned} r_1 &= a(1 - e) \\ r_2 &= a(1 + e). \end{aligned} \tag{6.22}$$

Further, it is readily seen from (6.19) that

$$r_1 r_2 = h^2 a/\mu.$$

This leads to the important relationship that

$$h^2 = \mu a(1 - e^2). \tag{6.23}$$

In view of the restrictions on r, a suitable parametric representation of the

---

Figure 6.2

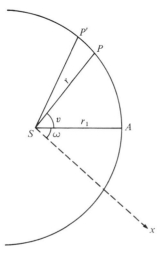

radius vector is

$$r = a(1 - e \cos E). \tag{6.24}$$

The parameter E is known as the *eccentric anomaly*. It is zero at perihelion and will increase by 2π in the course of one orbit. Now this change of variable, together with equation (6.23), allows the quadratic form $F(r)$ to be expressed as

$$F(r) = -a^2 e^2 \sin^2 E.$$

This facilitates the integration of (6.18) which now reduces to

$$(1 - e \cos E)\, \mathrm{d}E = (\mu/a^3)^{1/2}\, \mathrm{d}t.$$

The integration of this is straightforward, but first define the *mean motion n* so that

$$n^2 a^3 = \mu. \tag{6.25}$$

Then, since $E = 0$ at perihelion, it follows that

$$M \equiv n(t - \tau) = E - e \sin E. \tag{6.26}$$

This very important result is known as *Kepler's equation*. The quantity $n(t - \tau)$ is usually denoted by M and is called the *mean anomaly*. Like the other two anomalies, it is an angle which is zero at time τ and which increases by 2π in the course of one orbital period. Unlike v and E, however, M increases uniformly with time; its rate of increase is the mean motion n, which justifies that choice of name.

The eccentric anomaly is a very useful mathematical parameter. The radius vector and the time have both now been expressed in terms of it, and the same is possible for the angular coordinate θ, or equivalently the true anomaly. The condition that $r^2 \dot{v} = h$ may, by (6.23) to (6.26), be written as

$$\frac{\mathrm{d}v}{\mathrm{d}E} = \frac{(1 - e^2)^{1/2}}{(1 - e \cos E)}.$$

Now, by changing the variable to $\tan \frac{1}{2}E$, it may be verified that

$$\int \frac{(1 - e^2)^{1/2}\, \mathrm{d}E}{(1 - e \cos E)} = 2 \tan^{-1}\left[\left(\frac{1 + e}{1 - e}\right)^{1/2} \tan \frac{1}{2}E\right]. \tag{6.27}$$

Consequently, since both v and E are zero at perihelion, we derive the result that

$$\tan \frac{1}{2}v = \left(\frac{1 + e}{1 - e}\right)^{1/2} \tan \frac{1}{2}E. \tag{6.28}$$

The position of the planet (r, v) is now completely expressed in terms of the eccentric anomaly by equations (6.24) and (6.28). These, together with Kepler's equation (6.26), provide a complete parametric solution of the problem.

6.3 Kepler's three laws

Kepler's three laws of planetary motion can be regarded as a geometrical statement of the analytical results which were formally derived in the last section. They require only one slight modification to be in precise agreement with the exact solution of the two-body problem. Let us consider the three laws in turn.

(i) Consider the integration process that led to the parametric solution for v, equation (6.28). We can write this as follows:

$$\frac{dv}{dE} = \frac{(1-e^2)^{1/2}}{(1-e\cos E)} \Leftrightarrow \tan \tfrac{1}{2}v = \left(\frac{1+e}{1-e}\right)^{1/2} \tan \tfrac{1}{2}E.$$

Now reverse the roles of v and E in (6.28) – effectively the sign of e is changed – and so

$$\tan \tfrac{1}{2}E = \left(\frac{1-e}{1+e}\right)^{1/2} \tan \tfrac{1}{2}v \Rightarrow \frac{dE}{dv} = \frac{(1-e^2)^{1/2}}{(1+e\cos v)}.$$

Hence

$$\frac{dv}{dE}\cdot\frac{dE}{dv} = 1 = \frac{(1-e^2)}{(1-e\cos E)(1+e\cos v)}.$$

On using (6.24), this indicates that

$$r = \frac{a(1-e^2)}{1+e\cos v}. \tag{6.29}$$

This will be recognized as the standard equation for an ellipse in polar coordinates, verifying Kepler's first law. It is now possible to identify the constant e with the ellipse's eccentricity and a with its semimajor axis.

(ii) Refer back to Fig. 6.2, and consider the movement of the planet from P to P' during a short interval of time dt. The area described by the radius vector during this interval is approximately that of the triangle SPP'. The base of this triangle may be regarded as $SP = r$, and then the height will be $V_T\, dt$, where V_T is the transverse velocity. Consequently, the rate of description of area is $\tfrac{1}{2}rV_T$, which by (6.16) is $\tfrac{1}{2}h$. This justifies Kepler's second law.

(iii) The third law is contained in what was the definition of the mean motion n, equation (6.25). Since n is the constant rate of increase of the mean anomaly, it is also the mean rate of increase of the other two anomalies. As each of these increases by 2π in one orbital period, T say, it follows that

$$n = \frac{2\pi}{T}. \tag{6.30}$$

Consequently, using the definition of μ, (6.25) may be written as

$$\frac{a^3}{T^2} = \frac{G}{4\pi^2}(m_1 + m_2). \tag{6.31}$$

This general result relates the two masses to their period of revolution and the size of their relative orbit. It is the basis of mass determination in astronomy. Kepler's third law is derived by applying (6.31) to the orbits of the planets round the sun. In this case, m_2 will be the planet's mass and m_1 that of the sun. Writing (6.31) as

$$\frac{a^3}{T^2} = \frac{Gm_1}{4\pi^2}\left(1 + \frac{m_2}{m_1}\right),$$

it is seen that the right-hand side will be approximately the same for each of the planets, since $m_2 \ll m_1$. Kepler's third law – which states that a^3 is proportional to T^2 – is then deduced. Unlike the other two laws, however, it is only approximately correct, since it neglects the planet's mass in comparison with that of the sun. For the largest planet, Jupiter, this amounts to an error of one part in 10^3.

6.4   The solution of Kepler's equation

The orbit of a planet round the sun is usually specified by six orbital elements. These are constants of the motion, which will be formally introduced in the next chapter, although four of them (a, e, τ, ω) have already appeared in the analysis. The two remaining orbital elements are not needed for now, since they determine the plane of the orbit in space, which, for simplicity, has been made the x–y plane in the coordinate system. The six orbital elements completely determine the orbit. If they are known, the planet's position may be computed for any time, as we shall demonstrate in this section. Moreover, a knowledge of the orbital elements (and the two masses) provides values for the constants n and h through equations (6.23) and (6.25). Let us assume, therefore, that all the constants that have occurred in the analysis are known and that we wish to calculate the planet's position (r, v) at a particular time t,

First compute the mean anomaly from the equation

$$M = n(t - \tau).$$

The next step is to determine the eccentric anomaly E from Kepler's equation. This, however, does not give an explicit formula for E but rather the algebraic equation

$$f(E) = E - e \sin E - M = 0. \tag{6.32}$$

The form of the equation suggests that the solution for E might be obtained by a method of successive approximation $(E = E_0, E_1, \ldots)$, using the recurrence relation

$$E_{n+1} = M + e \sin E_n. \tag{6.33}$$

This is very easy to use, and an obvious initial approximation is $E_0 = M$.

The method is not recommended, however, as convergence is normally slow. Using (6.33) numerically is equivalent to the analytic procedure of expressing E as a power series in e. Certainly this may be expected to converge since $e < 1$, but, unless e is very small, a large number of terms must be included. A far better procedure is to solve (6.32) by the Newton–Raphson method, which places no practical restriction on the size of the eccentricity.

Let E_0 be a suitably chosen first approximation to the true value E. Suppose, in fact, that $E = E_0 + \Delta E$. Then $f(E_0 + \Delta E) = 0$. Expand this in a Taylor series, which truncated after two terms will give

$$f(E_0) + \Delta E f'(E_0) \simeq f(E_0 + \Delta E) = 0.$$

Consequently, we deduce an approximate formula for the correction to E_0 as

$$\Delta E \approx \Delta E_0 = -\frac{f(E_0)}{f'(E_0)}. \tag{6.34}$$

This is the Newton–Raphson formula. Normally $E_1 = E_0 + \Delta E_0$ will be an improved approximation to the eccentric anomaly. The process may be repeated with E_1 replacing E_0 and thereafter as often as is necessary. In most cases only a few iterations are required as convergence is rapid. It is easily seen that the error arising from the truncation of the Taylor series is $O(\Delta E^2)$. Consequently, once convergence has begun, each iteration roughly doubles the number of correct significant figures.

Applying the Newton–Raphson formula (6.34) to equation (6.32) leads to the iterative formula

$$E_{n+1} = E_n + \frac{(M - E_n + e \sin E_n)}{(1 - e \cos E_n)}, \tag{6.35}$$

more complicated certainly than (6.33), but usually well worth the extra trouble. Up to this point we have been working entirely in radians, and (6.35) is in these units. If the anomalies are expressed in degrees, however, Kepler's equation requires a slight modification which in turn modifies the iterative formula. This now becomes

$$E_{n+1} = E_n + \frac{[M - E_n + (180°e/\pi) \sin E_n]}{(1 - e \cos E_n)}. \tag{6.36}$$

The effective use of either (6.35) or (6.36) only requires an initial choice E_0 which is a reasonably close approximation. For small values of the eccentricity, the choice $E_0 = M$ is fine, the choice $E_0 = M + e \sin M$ even better. When the eccentricity is not small, however, such initial choices are not particularly close, and some improvement may well be required. A neat graphical method of selecting an accurate value for E_0 is presented in

Duffett-Smith (1981), pp. 122–4. Where the solution of Kepler's equation has been programmed, however, such an elaborate procedure is not essential, for the iterative process is very stable and will ultimately converge from either of the standard choices of E_0 mentioned above. Pathological cases can still occasionally arise (cf. Problem 6.6) and these must be resolved by *ad hoc* analysis.

Once the numerical procedure outlined above has been completed, the value of E is used to calculate (r, v) from equations (6.24) and (6.28), yielding

$$r = a(1 - e \cos E)$$

$$v = 2 \tan^{-1} \left[\left(\frac{1+e}{1-e} \right)^{1/2} \tan \tfrac{1}{2} E \right]. \tag{6.37}$$

The planet's position is, therefore, satisfactorily computed for time t. The section is concluded with an informative, if untypical, example.

Example. Halley's comet is in an orbit round the sun of period $T = 76.0081$ years and eccentricity $e = 0.9673$. Calculate its radius vector and true anomaly one year after perihelion.

First calculate the mean anomaly at the required time as

$$M = \frac{360°}{76.0081} = 4°.736\,337.$$

Let us try this as our first approximation for E and iterate using (6.36). The results are

$$E_0 = 4°.736\,337$$
$$E_1 = 131°.843\,196$$
$$E_2 = 79°.682\,604$$
$$E_3 = 54°.983\,412$$
$$E_4 = 44°.067\,435$$
$$E_5 = 41°.494\,423$$
$$E_6 = 41°.355\,533$$
$$E_7 = 41°.355\,514$$
$$E_8 = 41°.355\,514.$$

Clearly our initial choice for E_0 was inadequate, but, in the event, not hopelessly so. A better choice for E_0 could have saved the first four iterations.

The next step is to use equations (6.37). It is then found that

$$r = 0.273\,916\,9\,a,$$
$$v = 142°.280\,116.$$

Finally, we need the value of the semimajor axis, which may be obtained from Kepler's third law. Expressed in astronomical units and years, this is $a^3 = T^2$, leading to

$$a = 17.943\,476 \text{ AU},$$

and

$$r = 4.915\,022 \text{ AU}.$$

The accuracy of the data does not justify the number of decimal places which have been retained. Guard figures have been stated so that a reader wishing to perform the calculation may have a detailed check.

6.5 The equation of the centre

We now derive a series expansion for the true anomaly in terms of the orbital eccentricity and the mean anomaly. This formula, which is known as the equation of the centre, will be useful in some later investigations. It can give approximate results for small eccentricities, but it should be realized that the standard method of relating v to M is that just described in section 6.4.

The angular momentum equation (6.12) may be expressed in terms of the anomalies as

$$r^2 \frac{\mathrm{d}v}{\mathrm{d}M} = \frac{h}{n}.$$

Using (6.23), (6.25) and (6.29), this gives

$$\frac{(1-e^2)^{3/2}}{(1+e\cos v)^2} \frac{\mathrm{d}v}{\mathrm{d}M} = 1.$$

Consequently, since v and M are zero together, we can express the mean anomaly as the integral

$$M = \int_0^v \frac{(1-e^2)^{3/2}}{(1+e\cos x)^2} \,\mathrm{d}x. \tag{6.38}$$

For small values of e, the integrand may be expressed as a power series, giving, correct to order e^2, the formula

$$M = \int_0^v (1 - 2e\cos x + \tfrac{3}{2}e^2 \cos 2x \ldots) \,\mathrm{d}x.$$

This is integrated term by term to yield

$$M = v - 2e\sin v + \tfrac{3}{4}e^2 \sin 2v + O(e^3). \tag{6.39}$$

Unfortunately, this formula is the wrong way round for practical purposes, since M rather than v is the natural independent variable. The formula may, however, be inverted as follows. To zero order in e, clearly $v = M$. Substituting this in the second term on the right-hand side yields

$$v = M + 2e \sin M + O(e^2).$$

This in turn is used on the right-hand side of (6.39), giving

$$v = M + 2e \sin (M + 2e \sin M) - \tfrac{3}{4}e^2 \sin 2M + O(e^3).$$

With the usual small-angle approximations, it is found that

$$v - M = 2e \sin M + \tfrac{5}{4}e^2 \sin 2M + O(e^3). \tag{6.40}$$

The difference between the true and the mean anomaly is known as the *equation of the centre*. It is the angle between the planet's true heliocentric direction and the direction it would have if its angular velocity were uniform. The equation of the centre is derived to order e^2 in equation (6.40). The method described above may be extended to derive higher-order terms, but the algebra becomes increasingly laborious.

6.6   Components of a planet's velocity

The transverse component V_T of a planet's velocity was given in equation (6.16). This may, by (6.29), be rewritten as

$$V_T = V_0(1 + e \cos v), \tag{6.41}$$

where

$$V_0 = \frac{h}{a(1 - e^2)}. \tag{6.42}$$

The transverse velocity may, therefore, be split into two parts, a transverse velocity of constant magnitude V_0, and one of variable magnitude $eV_0 \cos v$. Now, as illustrated in Fig. 6.3, the true anomaly is the angle between the major axis CA and the radius vector. Draw a line PG through the planet's position parallel to the minor axis CB. The transverse direction PF makes

Figure 6.3

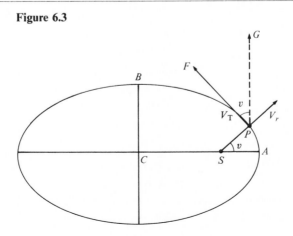

an angle v with this line. The variable part of the transverse velocity may, therefore, be regarded as the transverse component of a constant velocity eV_0 in the direction PG – that is, parallel to the minor axis of the ellipse.

The relevance of this decomposition of the transverse velocity will be seen when we consider the radial component of the planet's velocity, V_r, say. By equation (6.15), we have

$$V_r^2 + V_T^2 = \mu(2/r - 1/a).$$

On using (6.29), (6.23) and (6.42), this may be expressed as

$$V_r^2 + V_T^2 = V_0^2(1 + 2e \cos v + e^2).$$

Finally, the transverse component is removed by (6.41) to leave the result that

$$V_r = eV_0 \sin v. \tag{6.43}$$

Note that the positive square root is taken, since the radial velocity is positive in the outward half of the orbit, between perihelion and aphelion corresponding to the range $0 < v < \pi$.

Referring to Fig. 6.3, the interpretation of equation (6.43) is clear; the planet's radial velocity may be regarded as the radial component of a constant velocity eV_0 in the direction PG. Coupling this with the earlier resolution of the transverse component, we may state the following important theorem:

A planet's orbital velocity may be resolved into a transverse component of constant magnitude $V_0(\equiv h/(a(1 - e^2)))$ and a constant velocity of magnitude eV_0 parallel to the minor axis of the ellipse.

This theorem will be useful in later chapters as a means of detailing the effects of the earth's orbital motion on the apparent position and the apparent radial velocity of a source.

6.7   Elliptic and hyperbolic orbits

The principal formulae for elliptic motion have been summarized in Table 6.1. The argument that leads to these equations is contained mainly in sections 6.2 and 6.3. The starting point of the argument was the introduction of the parameter a in equation (6.14), viz.

$$C = -\mu/a, \tag{6.14}$$

where C was a constant of integration proportional to the total energy of the system. Only later was a identified with the orbit's semimajor axis. Since we were considering bound (elliptic) orbits, C was negative and a positive, and certainly the subsequent analysis bore this in mind. Nothing in the argument, however, actually requires a to be positive, and the formulae in

Table 6.1. *Formulae for an elliptic orbit*

Kepler's three laws

(i) $r = \dfrac{a(1-e^2)}{1+e\cos v}$

(ii) $r^2\dot{v} = h$

(iii) $n^2a^3 = \mu \equiv G(m_1 + m_2)$

where

$$h^2 = \mu a(1-e^2)$$

Position formulae

Kepler's equation:

$$n(t-\tau) = M = E - e\sin E$$
$$r = a(1-e\cos E)$$
$$\tan \tfrac{1}{2}v = \left(\frac{1+e}{1-e}\right)^{1/2}\tan \tfrac{1}{2}E$$

Velocity formula

$$V^2 = \mu\left(\frac{2}{r}-\frac{1}{a}\right)$$

Table 6.1 will describe unbound orbits if a is allowed to be negative. It is then seen that $e > 1$, since h^2 must be positive. The first equation then indicates that the orbit is a hyperbola with eccentricity e, but semimajor axis $-a$. Moreover, for consistency, it is necessary that the parameter n and the anomalies E and M should all be imaginary. This is clearly inconvenient.

A better procedure, for dealing with unbound orbits, is to redefine the recalcitrant parameters by writing $-a$ for a, in for n, iM for M, and $-iE$ for E. The resulting formulae then involve only real quantities and are given in Table 6.2. The reader may prefer to derive these formulae by making the appropriate adaptations to the argument expressed in full in sections 6.2 and 6.3. A complete rederivation is hardly required, however.

The position and velocity formulae may be used in the same way as their elliptic equivalents. Kepler's equation, in particular, is solved by the Newton–Raphson method, and the resulting value of E allows the body's position (r, v) to be determined. Kepler's three laws are modified somewhat. The first now states that the orbit is a hyperbola, the second is unchanged, while the third cannot be stated in the conventional manner, since the period of a hyperbolic orbit is not defined. Even so, the equation $n^2a^3 = \mu$ indicates that there is still a relationship, and a usable one, involving the size of the orbit, the timing – if not the period – and the masses of the two bodies involved.

Table 6.2. *Formulae for a hyperbolic orbit*

Kepler's three laws

$$(i) \qquad r = \frac{a(e^2 - 1)}{1 + e \cos v}$$

$$(ii) \quad r^2 \dot{v} = h$$

$$(iii) \quad n^2 a^3 = \mu \equiv G(m_1 + m_2)$$

where

$$h^2 = \mu a(e^2 - 1)$$

Position formulae

Kepler's equation:

$$n(t - \tau) = M = e \sinh E - E$$

$$r = a(e \cosh E - 1)$$

$$\tan \tfrac{1}{2} v = \left(\frac{e + 1}{e - 1} \right)^{1/2} \tanh \tfrac{1}{2} E$$

Velocity formula

$$V^2 = \mu \left(\frac{2}{r} + \frac{1}{a} \right)$$

6.8 Comparison of general relativity with Newtonian theory

While Newtonian gravitation gives an adequate description of planetary motion for most purposes, there are some effects that can only be accounted for by the General Theory of Relativity. The remainder of this chapter is devoted to a discussion of the two-body problem in general relativity and to establishing the differences from Newtonian theory. These differences are comparatively minor, in spite of the vast differences in the conceptual foundations of the two theories. As we shall see, near identical differential equations arise from quite distinct premises.

As explained in section 3.5, the sun's gravitational field is represented by Schwarzschild spacetime, with a Riemannian metric, expressible in the form

$$ds = g_{\mu\nu} \, dx^\mu \, dx^\nu. \tag{6.44}$$

A planet is treated as a free particle, and its world line is, therefore, an ordinary geodesic, whose differential equations will be

$$\frac{d}{ds} \left(g_{\lambda\nu} \frac{dx^\nu}{ds} \right) - \frac{1}{2} \frac{\partial g_{\mu\nu}}{\partial x^\lambda} \frac{dx^\mu}{ds} \frac{dx^\nu}{ds} = 0. \tag{6.45}$$

Here the interval s is being used as the curve parameter; this is the planet's proper time, that is, the time as measured by an observer on the planet. Notice, too, that, in this treatment, the mass of this particular planet, and, indeed, of all the planets, is being ignored when compared with the mass of the sun.

Let us start by examining the metric in its simplest form, namely (3.46), restated here as

$$ds^2 = \left(1 - \frac{2m}{r}\right)dt^2 - \frac{1}{c^2}\left[\frac{dr^2}{(1 - 2m/r)} + r^2\,d\theta^2 + r^2\sin^2\theta\,d\phi^2\right] \quad (6.46)$$

where

$$m = GM_\odot/c^2 \approx 1.5 \text{ km.} \quad (6.47)$$

It must be remembered that coordinates are only labels and so r should not be rashly identified with the planet's distance from the sun, nor should t be immediately identified with recorded time. This is more closely related to the interval s, but the coordinate time does provide an important standard. It is a label, but a special one, since t in (6.46) is that choice of time coordinate which expresses the metric in a completely static form, i.e. all $\partial g_{\mu\nu}/\partial t = 0$.

The individual components of the metrical tensor may be identified from equation (6.46). They are all zero, except the diagonal components, which are

$$g_{11} = \frac{-1}{c^2(1 - 2m/r)}, \quad g_{22} = -\frac{r^2}{c^2}, \quad g_{33} = -\frac{r^2\sin^2\theta}{c^2},$$

$$g_{44} = 1 - 2m/r. \quad (6.48)$$

These may now be inserted in the differential equations for the geodesics (6.45).

Consider first the θ equations, i.e. set $\lambda = 2$ in (6.45). Only two terms survive, giving

$$\frac{d}{ds}\left(g_{22}\frac{d\theta}{ds}\right) - \frac{1}{2}\frac{\partial g_{33}}{\partial\theta}\left(\frac{d\phi}{ds}\right)^2 = 0.$$

From (6.48), it follows that

$$\frac{d}{ds}\left(r^2\frac{d\theta}{ds}\right) - r^2\sin\theta\cos\theta\left(\frac{d\phi}{ds}\right)^2 = 0.$$

It is convenient to select a particular solution of this differential equation, namely

$$\theta = \tfrac{1}{2}\pi. \quad (6.49)$$

This means that the planet is moving in the equatorial plane of the spherical polar coordinate system. The selection of this particular solution amounts to no more than a deft choice of coordinate axes.

Setting $\lambda = 3$ and 4 in equations (6.45) also gives immediate results. Since all components of the metrical tensor are independent of both ϕ and t, these two equations clearly reduce to

$$\frac{d}{ds}\left(g_{33}\frac{d\phi}{ds}\right) = 0 = \frac{d}{ds}\left(g_{44}\frac{dt}{ds}\right).$$

These in turn will integrate to give the very important results that

$$r^2 \frac{d\phi}{ds} = h, \tag{6.50}$$

$$\left(1 - \frac{2m}{r}\right)\frac{dt}{ds} = k, \tag{6.51}$$

h and k being constants of integration. The first of these equations is reminiscent of equation (6.12), and indeed (6.50) and (6.51) are the general relativistic expressions for the conservation of angular momentum and energy respectively.

It is not necessary to make use of the remaining – and most complicated – geodesic differential equation. This would give the equivalent of the radial component of the equation of motion. Instead, we substitute the results of (6.49), (6.50) and (6.51) into the metric, which is itself an integral of the geodesic differential equations. After some simplification, this gives a first-order differential equation, namely

$$\left(\frac{dr}{ds}\right)^2 = c^2(k^2 - 1) + \frac{2mc^2}{r} - \frac{h^2}{r^2}\left(1 - \frac{2m}{r}\right). \tag{6.52}$$

Now guided by the notation used in the Newtonian case, let us introduce the parameters μ and a through the following two definitions. Set

$$\mu = GM_\odot = mc^2, \tag{6.53}$$

and define a length a by the equation

$$c^2(k^2 - 1) = -\mu/a. \tag{6.54}$$

Equation (6.52) can then be rewritten as

$$\left(\frac{dr}{ds}\right)^2 + \frac{h^2}{r^2} = \mu\left(\frac{2}{r} - \frac{1}{a}\right) + \frac{2mh^2}{r^3}. \tag{6.55}$$

This last equation should be compared with its Newtonian equivalent, namely (6.15). All the terms match up except the last, and, since $m \ll r$, this term is clearly much smaller than any of the others. Certainly some care is required in the interpretation of r, and s is the planet's proper time rather than the absolute Newtonian time. Even so, it is seen that differences between the two theories will be slight for $m \ll r$, and that Newtonian theory clearly gives a very close first approximation to the predictions of general relativity. The differences between the two theories are examined in detail in the next section.

6.9 The Kepler solution in general relativity

To simplify equation (6.55), introduce a parameter e with a definition that is suggested by Newtonian theory, viz.

$$h^2 = \mu a (1 - e^2).\tag{6.56}$$

The parameters a and e were the orbit's semimajor axis and eccentricity in the Newtonian case, but now such precise, as opposed to approximate, identifications cannot be made. The differential equation (6.55) can now be expressed in the form

$$\left(\frac{dr}{ds}\right)^2 = -\frac{\mu f(r)}{ar^3},\tag{6.57}$$

where $f(r)$ is the cubic equation, given by

$$f(r) = r^3 - 2ar^2 + a^2(1 - e^2)r - 2ma^2(1 - e^2),\tag{6.58}$$

For $a > 0$ and $0 < e < 1$, this cubic will have three positive roots, r_1, r_2 and r_3, say. Let us assume that $r_1 < r_2 < r_3$. Then it is easily seen from (6.58) that $r_1 \approx 2m$, $r_2 \approx a(1 - e)$, and $r_3 \approx a(1 + e)$. Equation (6.57) shows that $f(r)$ cannot be positive, so r is restricted to one of the ranges $r_2 \leqslant r \leqslant r_3$ or $0 \leqslant r \leqslant r_1$. The first of these concerns us, since it represents an approximately elliptic orbit – the second represents a body falling into a black hole!

It is only necessary to work correct to order m/a. For $a \sim 1$ AU, $m/a \sim 10^{-8}$, and so higher powers of this ratio may be safely neglected. Let us write

$$r_2 = \alpha(1 - \eta)$$
$$r_3 = \alpha(1 + \eta).\tag{6.59}$$

Then, correct to order m/a, it is found from (6.58) that

$$\alpha = a - m,$$

$$\eta = e\left[1 + \frac{m}{ae^2}(1 + e^2)\right],\tag{6.60}$$

$$r_1 = 2m.$$

In view of the limits that are imposed on r, it is convenient to introduce an 'eccentric anomaly' by setting

$$r = \alpha(1 - \eta \cos E).\tag{6.61}$$

It is then found that

$$f(r) = -\alpha^2 \eta^2 \sin^2 E \ (r - r_1) = (r - r_1)(dr/dE)^2.$$

When this is substituted into (6.57), that differential equation becomes

$$\left(\frac{\mu}{a}\right)^{1/2} ds = r\left(1 - \frac{r_1}{r}\right)^{-1/2} dE.\tag{6.62}$$

Define the mean motion n in the same way as before, writing

$$n^2 a^3 = \mu.\tag{6.63}$$

Then, using (6.60) and (6.61), the binomial expansion of (6.62) yields

$$n \, ds = [1 - (1 - m/a)\eta \cos E] \, dE. \tag{6.64}$$

This will integrate to give a Kepler equation. If the planet's proper time at perihelion is s_0, it is

$$n(s - s_0) = E - \varepsilon \sin E, \tag{6.65}$$

where ε is a modified 'eccentricity' defined by

$$\varepsilon = \eta(1 - m/\alpha). \tag{6.66}$$

Equation (6.65) can be used in the same way as the standard Kepler equation (6.26). For a given proper time s, a numerical solution is derived for E, and this is then used in (6.61) to calculate the radius vector. The only modification is that different 'eccentricities' are used in different equations. Actually it has not yet been demonstrated that the orbit is elliptical in any precise sense. In fact, this is a property of a different coordinate system which is considered in the next section. We shall defer until then any consideration of how the angular coordinate ϕ is related to the eccentric anomaly.

The Kepler equation that has been derived has a mean anomaly proportional to proper time. It is possible to convert this into an equation that involves the coordinate time. Using (6.51), equation (6.64) can be rewritten as

$$n \, dt/k = (1 - 2m/r)^{-1}(1 - \varepsilon \cos E) \, dE.$$

After applying the binomial theorem and using (6.61), this is expressed, with sufficient accuracy, as

$$n \, dt/k = (1 + 2m/\alpha - \varepsilon \cos E) \, dE.$$

This, too, will integrate to give an alternative Kepler equation. It will involve a modification to the mean motion and yet another modification to the eccentricity. They are now defined as

$$n' = \frac{n}{k}\left(1 + \frac{2m}{\alpha}\right)^{-1} = n\left(1 - \frac{3m}{2\alpha}\right), \tag{6.67}$$

and

$$\varepsilon' = \varepsilon\left(1 + \frac{2m}{\alpha}\right)^{-1} = \eta\left(1 - \frac{3m}{\alpha}\right). \tag{6.68}$$

Then, if t_0 is the coordinate time corresponding to perihelion, the new Kepler equation is simply

$$n'(t - t_0) = E - \varepsilon' \sin E. \tag{6.69}$$

Notice that, although the mean motion and eccentricity were modified, the eccentric anomaly is the same. So equation (6.61) allows the radius

vector to be determined for a particular instant of coordinate time, once
(6.69) has been solved in the usual way.

Equations (6.65) and (6.69), taken together, allow the proper and
coordinate times to be related. The relationship can be made slightly more
explicit by eliminating the term E between the two equations to yield

$$t - t_0 = (n/n')(s - s_0) + (\varepsilon - \varepsilon') \sin E/n'.$$

Define the orbital period P by the equation

$$P = 2\pi/n. \tag{6.70}$$

This corresponds to the Newtonian formalism and is, in fact, the proper
period – as measured by the planet. The coordinate period is slightly longer.
The result then follows that

$$t - t_0 = \left(1 + \frac{3m}{2\alpha}\right)(s - s_0) + \frac{m\eta \sin E}{\alpha\pi} P. \tag{6.71}$$

6.10   The relativistic advance of perihelion

The Schwarzschild form of the metric (6.46), although much the
simplest, is not the form that relates most naturally to the interpretation of
positional observations. As previously explained in section 3.6, the
isotropic form of the metric (3.52) is preferred. This involves only a change
to a new radial coordinate, $\bar{r}$, say, which is related to r by the equation

$$r = \bar{r}(1 + m/2\bar{r})^2. \tag{6.72}$$

The other coordinates are unaltered.

The two radial coordinates only differ to order m. In fact, to this order,
equation (6.72) reduces to

$$r = \bar{r} + m.$$

The new radial coordinate may, therefore, be expressed in terms of the
eccentric anomaly as

$$\bar{r} = \tilde{\alpha}(1 - \tilde{\eta} \cos E), \tag{6.73}$$

where, from (6.61) and (6.60),

$$\tilde{\alpha} = \alpha - m = a - 2m, \tag{6.74}$$

and

$$\tilde{\eta} = \eta(1 - m/\alpha)^{-1} = \eta(1 + m/\alpha). \tag{6.75}$$

Equation (6.73) may be used to determine $\bar{r}$ once E has been calculated from
either of the two Kepler equations established in the last section.

As might be expected, the semimajor axis and eccentricity of the planet's
orbit depend on the coordinate system that is being employed. These terms
are being used very loosely, however, since the shape of the orbit has yet to

be derived. This involves relating the angular coordinate ϕ to $\bar{r}$ (or r). This can be done through the eccentric anomaly.

Consider the differential equation for ϕ (6.50). After eliminating r and s through (6.61) and (6.65), it becomes

$$\frac{d\phi}{dE} = \frac{h}{n\alpha^2} \frac{(1 - \varepsilon \cos E)}{(1 - \eta \cos E)^2}.$$

The parameters h and n are removed by (6.56) and (6.63) to give

$$\frac{d\phi}{dE} = \frac{a^2}{\alpha^2} (1 - e^2)^{1/2} \frac{(1 - \varepsilon \cos E)}{(1 - \eta \cos E)^2}. \tag{6.76}$$

An exact integral is possible, but very complicated, and is not justified, since (6.76) is only correct to order m/a.

The complication in (6.76) arises because of the three distinct values of the eccentricity. In the Newtonian case they were all the same, and the corresponding equation reduced to the standard integral (6.27). The best procedure is, therefore, to reduce (6.76) to this standard form, while maintaining accuracy only to order m/a as follows. Introduce a new eccentricity η^*, which is yet to be determined, and which differs from the others only to order m/a. Then

$$\frac{d\phi}{dE} = (1 - e^2)^{1/2} \frac{a^2}{\alpha^2} \frac{[1 - \eta^* \cos E + (\eta^* - \varepsilon) \cos E]}{[1 - \eta^* \cos E + (\eta^* - \eta) \cos E]^2}.$$

Expanding by the binomial theorem, this gives, after a little reduction,

$$\frac{d\phi}{dE} = \frac{a^2(1 - e^2)^{1/2}}{\alpha^2(1 - \eta^* \cos E)} \left[1 + \frac{(2\eta - \varepsilon - \eta^*) \cos E}{(1 - \eta^* \cos E)} + O(m/a)^2 \right].$$

The parameter η^* is then chosen so that the second term in square brackets vanishes. It follows from (6.66) and (6.75) that

$$\eta^* = \bar{\eta}. \tag{6.77}$$

It is natural, therefore, to express (6.76) entirely in terms of parameters associated with the barred coordinate system. After some simplification, involving (6.60) and (6.75), equation (6.76) will reduce, correct to order $m/\bar{\alpha}$, to

$$\frac{d\phi}{dE} = \left[1 + \frac{3m}{\bar{\alpha}(1 - \bar{\eta}^2)} \right] \frac{(1 - \bar{\eta}^2)^{1/2}}{(1 - \bar{\eta} \cos E)}.$$

The standard integral (6.27) is now applied to yield

$$\phi - \phi_0 = 2 \left[1 + \frac{3m}{\bar{\alpha}(1 - \bar{\eta}^2)} \right] \tan^{-1} \left[\left(\frac{1 + \bar{\eta}}{1 - \bar{\eta}} \right)^{1/2} \tan \tfrac{1}{2} E \right], \tag{6.78}$$

where, of course, ϕ_0 is the value of ϕ at the perihelion $E = 0$.

This completes the determination of the planet's position. For a particular instant of time, either coordinate or proper, the appropriate Kepler equation – (6.65) or (6.69) – allows the computation of the eccentric anomaly. Substituting this value of E into (6.73) and (6.78) allows the planet's position $(\tilde{r}, \phi)$ to be determined in the isotropic coordinate system. If equation (6.61) is used instead, the planet's position is established in the standard Schwarzschild coordinate system.

The description of the planet's orbit is simpler in the isotropic coordinate system, since the same value of the eccentricity parameter occurs in the expressions for $\tilde{r}$ and ϕ – (6.73) and (6.78). Define a true anomaly v as

$$v = \left[1 + \frac{3m}{\tilde{\alpha}(1-\tilde{\eta}^2)} \right]^{-1} (\phi - \phi_0). \tag{6.79}$$

Then equation (6.78) gives

$$\tan \tfrac{1}{2}v = \left(\frac{1+\tilde{\eta}}{1-\tilde{\eta}} \right)^{1/2} \tan \tfrac{1}{2}E.$$

Comparison with the corresponding Newtonian equations then shows that the planet's orbit is represented by the curve

$$\tilde{r} = \frac{\tilde{\alpha}(1-\tilde{\eta}^2)}{1+\tilde{\eta}\cos v}. \tag{6.80}$$

This can be described as an ellipse of semimajor axis $\tilde{\alpha}$ and eccentricity $\tilde{\eta}$, but, as (6.79) shows, it is not a fixed ellipse, but one that is rotating with constant angular velocity.

Let ω be the argument of perihelion of the orbital ellipse, simply the value of ϕ at perihelion. Then, for the perihelion corresponding to $E=0$, equation (6.78) indicates that $\omega = \phi_0$. One period later, however, at the next perihelion, E will have increased by 2π, and the corresponding value of ϕ, given by (6.78), is

$$\phi = \phi_0 + 2\pi \left[1 + \frac{3m}{\tilde{\alpha}(1-\tilde{\eta}^2)} \right].$$

Consequently, in one orbital period – perihelion to perihelion – there is an increase $\Delta\omega$ in the argument of perihelion given by

$$\Delta\omega = \frac{6\pi m}{\tilde{\alpha}(1-\tilde{\eta}^2)}. \tag{6.81}$$

This prediction constitutes the first classical test of general relativity. It was noted by Einstein himself that this accurately accounted for a discrepancy between the computed and the observed motions of the planet Mercury, a discrepancy which had been recognized since the time of Le Verrier but which could not be explained in terms of perturbations by the major planets. The advance of perihelion for Mercury amounts to 43″

per century. It is much smaller for the other planets, since $\Delta\omega$ is inversely proportional to the planet's semimajor axis. Moreover, $\Delta\omega$ in (6.81) is the perihelion advance *per revolution*, and so the longer orbital periods of the other planets militate against the detection of this relativistic effect. It has not been possible, therefore, to isolate the effect so convincingly for any of the other planets in the solar system. Striking evidence of the effect has, however, been presented by the binary pulsar.

The orbit of a planet has been described in the isotropic coordinate system. The description in the standard Schwarzschild coordinate system is rather more complicated. The secular effect of the advance of perihelion is still given by (6.81), but the rotation of the orbital ellipse is not strictly uniform. For this reason, among others, the isotropic coordinate system has been preferred.

The general relativistic solution of the two-body problem makes modification to Kepler's three laws. The modification to the first law has been noted – the planet's orbit is a uniformly rotating ellipse in the isotropic coordinate system. The second law is also modified, for the coordinate angular momentum in the isotropic coordinate system is not conserved. The equivalent conservation law is equation (6.50). The non-conservation is completely allowed for by using a different eccentricity in the Kepler equation. For completeness we note the change in the third law. We still have $n^2a^3 = \mu$, but n is the mean motion in proper time, and a is not exactly the semimajor axis in any coordinate system. If isotropic coordinates and coordinate time are used, the result is $n'^2\tilde{a}^3 = \mu(1 - 9m/\tilde{a})$.

It should be remembered that the analysis of the last two sections is only correct to order $m/\tilde{a}$. This is quite sufficient for all applications to the solar system. It has been assumed throughout that the 'eccentricity' has a typical value in the range $0 < e < 1$. The analysis will break down at certain points if the eccentricity is within $0(m/\tilde{a})$ of the ends of this range. Since $m/\tilde{a} \sim 10^{-8}$, this is no serious disadvantage. In fact, exact solutions can be derived for circular orbits and for rectilinear radial motion (cf. Problems 6.8 and 6.9).

Problems

6.1 The ellipse $x^2/a^2 + y^2/b^2 = 1$ has the parametric solution

$x = a \cos E,$

$y = b \sin E,$

which provides a geometrical definition of the eccentric anomaly. Show that the application of Kepler's second law and performing the integration of the area in terms of E leads to Kepler's equation.

6.2 Prove that the orbital velocity of a planet may be expressed as

$$V^2 = \frac{\mu}{a}\left(\frac{1+e\cos E}{1-e\cos E}\right).$$

Show further that this velocity is inclined at an angle ψ to the radius vector, where

$$\tan\psi = \frac{(1-e^2)^{1/2}}{e\sin E}.$$

6.3 The planet Mars moves round the sun in an orbit of semimajor axis 1.5236 AU and eccentricity 0.0934. On 1984 January 1.0 its heliocentric distance was 1.6609 and decreasing. Calculate the date of the next perhelion passage.

6.4 A spacecraft is launched from a parking orbit of radius 6625 km with a velocity of 11.55 km s^{-1} in a horizontal direction. Calculate its geocentric distance 24 hours later.

(Geocentric gravitational constant $GM_{\oplus} = 3.986 \times 10^{14}$ m^3 s^{-2}.)

6.5 A minor planet is moving in the ecliptic in an orbit of semimajor axis 3.129 AU and eccentricity 0.1341. Calculate the increase in its heliocentric longitude in the year following perihelion.

6.6 Consider the numerical solution of the algebraic equation

$$f(E) = E - e\sin E - M = 0.$$

The Newton–Raphson method is used with a starting value $E_0 = M$. Show that this leads to a correction ΔE which is greatest for $M = \cos^{-1} e$. Show further that this maximum value is $e(1-e^2)^{-1/2}$ and deduce that this choice of E_0 will be unsatisfactory for large eccentricities at small values of M.

6.7 Show that in the circumstances envisaged in the last question, a more suitable first estimate for E is

$$E_0 = \sqrt[3]{6M}.$$

6.8 Prove that the coordinate and proper orbital periods of a body moving in a circular orbit of radius r in the standard Schwarzschild metric are given exactly by

$$T_c = 2\pi\left(\frac{r^3}{mc^2}\right)^{1/2}$$

and

$$T_p = 2\pi\left(\frac{r^3}{mc^2}\right)^{1/2}\left(1 - \frac{3m}{r}\right)^{1/2}.$$

6.9 A body is released from rest at radial coordinate r_0 in the standard Schwarzschild metric. Prove that it falls into the singularity in a proper time T_s given by

$$T_s = \frac{\pi}{c} \left(\frac{r_0^3}{8m} \right)^{1/2}.$$

6.10 Assuming that Mercury moves in an orbit of semimajor axis 0.3871 AU and eccentricity 0.2056, calculate the linear displacement of its perihelion position predicted by general relativity over one orbital period.

7

Planetary and satellite orbits

7.1 The orbit in space

The results that were established in chapter 6 will now be applied to the orbits of bodies within the solar system. These can be divided into two main classes, namely heliocentric orbits, relevant to planets, comets, asteroids and interplanetary space-probes, and satellite orbits – whether natural or artificial. The discussion is mainly in Newtonian terms, although some reference will be made to the relativistic modifications spelled out in sections 6.8 to 6.10.

A heliocentric orbit is uniquely specified by its *orbital elements* which are best defined with reference to the celestial sphere, shown in Fig. 7.1. This is a heliocentric celestial sphere – S is the sun – and the orbit, say, of a planet, is shown as the dotted curve within the celestial sphere. Let P' be the position of the planet at a general time t, and let A' be its position at the time of perihelion. Join SP' and SA' to cut the celestial sphere in the points P and A respectively. Then the planet's orbital plane intersects the celestial sphere in the great circle LAP. This great circle will cross the ecliptic ΥLH in two points which are referred to as *nodes*. Only one of them L is shown in the diagram. This is the *ascending node* corresponding to the point where the planet passes from south to north of the ecliptic. The *descending node* is, of course, the diametrically opposite point.

The ecliptic provides a suitable reference plane for a heliocentric orbit. The planet's orbital plane is determined with reference to the ecliptic by two orbital elements (Ω, i) defined by

$$\Omega = \Upsilon L \tag{7.1}$$
$$i = ALH.$$

The parameter Ω is the *longitude of the ascending node* and i is the *orbital inclination* (to the ecliptic). These two parameters define not only the orbital plane, but also the sense of the orbital motion. While Ω can take any value between 0 and 360°, i is confined to the first two quadrants. Consistent with the usage introduced in section 2.5, $i < 90°$ corresponds to direct orbits and $i > 90°$ to retrograde orbits.

The remaining orbital elements have all been introduced in chapter 6. The orientation of the orbit within its plane is specified by the *argument of perihelion ω*, where

$$\omega = LA. \tag{7.2}$$

The size and shape of the orbit are given by the semimajor axis a and eccentricity e respectively. Finally, the timing of the orbit is specified by the time of perihelion τ. The six parameters $(a, e, \tau, i, \Omega, \omega)$ are what are usually taken as the orbital elements for a heliocentric orbit, although certain variants are possible, and, sometimes, even essential.

Consider the earth's orbit for example. This is a special case, in that $i = 0$ by definition. The node is indeterminate and neither Ω nor ω can be defined. Consequently, the orientation of the earth's orbit in the ecliptic cannot be specified in terms of the orbital elements as defined above. Difficulties will arise, in practice, for heliocentric with small, albeit non-zero, inclinations to the ecliptic. This includes most of the major planets.

Figure 7.1. The orbital elements i, Ω and ω of a planet.

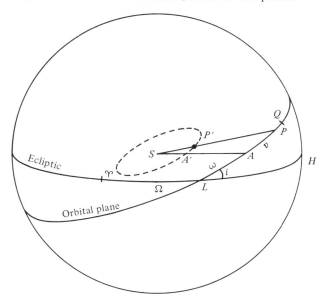

The difficulty is overcome by replacing ω with another parameter. One defines the *longitude of perihelion*, denoted by ϖ, as

$$\varpi = \Omega + \omega. \tag{7.3}$$

It is a slight misnomer in that ϖ is not exactly the ecliptic longitude of the planet at the time of perihelion, as the name might suggest. Rather, it is the broken arc from Υ to P measured in part along the ecliptic and in part along the orbital plane, i.e. $\varpi = \Upsilon L + LA$.

The arc AP, in Fig. 7.1, is the true anomaly v. Suppose we define the point Q on the orbital great circle to move with constant angular velocity n and to coincide with the planet's direction at perihelion. Then $AQ = n(t - \tau) = M$. The broken arc from Υ to Q ($\Upsilon L + LQ$) is termed the *mean longitude* of the planet at time t. Denoting it by L, we have

$$L = M + \varpi. \tag{7.4}$$

As an alternative to giving the time of perihelion, it is quite normal practice to state the mean longitude L_0 for some epoch, τ_0, say. Then the mean anomaly, at a general time t, is given by

$$M = n(t - \tau_0) + L_0 - \varpi. \tag{7.5}$$

Yet another possibility is to give the mean anomaly itself for the epoch.

Six orbital elements are needed to determine a two-body orbit provided the masses of the bodies are known. For heliocentric orbits, the sun's mass dominates, but the masses of the major planets, at least, cannot be ignored. The product $n^2 a^3$ is not exactly constant from one planet to another. To allow for this, n is usually treated as an additional orbital element. This is, in any case, a great convenience for computing the mean anomaly, so the practice is extended to minor bodies of the solar system as well. The following elements are given in the *Astronomical Almanac* for the major planets – a, n, e, i, Ω, ϖ, L_0. The tabulation is slightly different for the asteroids; the argument of perihelion and the mean anomaly at the epoch are given instead of the last two longitudes.

The orbital elements have been defined above for heliocentric orbits. Some modification is desirable for satellite orbits, as the ecliptic is not a suitable reference plane. The equatorial plane, defined by the planet's rotation, is the natural substitute. For the most part, the differences in the definitions and the terminology are fairly obvious. Consider an artificial satellite, for example, in geocentric orbit. The elements a, n and e are defined exactly as before, while τ is the *time of perigee*. Moreover, i is the *orbital inclination* to the *equator*, and the ascending node is the intersection of the orbital plane with the equator. The element ω is, then, the *argument of perigee* and Ω becomes the *right ascension of the ascending node*.

7.2   Calculation of an ephemeris

Suppose that the orbital elements of a planet, or some other body, are known, and that it is intended to predict its position in the sky for some particular date. The first step must be to compute the planet's heliocentric position. Consider the heliocentric celestial sphere of Fig. 7.2, in which K represents the pole of the ecliptic, ♈ the equinox, and U the point on the ecliptic with ecliptic longitude $90°$. The points L, A, P are defined as in Fig. 7.1. Take a set of rectangular axes (ξ, η, ζ) in the directions $S♈, SU, SK$. Then, referred to these axes, the planet has position vector $\mathbf{r}$ given by

$$\mathbf{r}=(\xi, \eta, \zeta)=r(\cos ♈P, \cos UP, \cos KP), \tag{7.6}$$

where r is the planet's radius vector at the time in question.

Now the arc $LP = v + \omega$, from definitions of the last section. Moreover, $♈L = \Omega$, $LU = 90° - \Omega$ and $LK = 90°$. In addition, note the three spherical angles: $♈LP = 180° - i$, $ULP = i$, and $KLP = 90° - i$. So the three direction cosines occurring in (7.6) may be evaluated by applying the cosine formula to spherical triangles $L♈P$, LUP and LKP. The results are

$$\xi = r[\cos (v + \omega) \cos \Omega - \sin (v + \omega) \sin \Omega \cos i]$$
$$\eta = r[\cos (v + \omega) \sin \Omega + \sin (v + \omega) \cos \Omega \cos i] \tag{7.7}$$
$$\zeta = r \sin (v + \omega) \sin i.$$

Figure 7.2

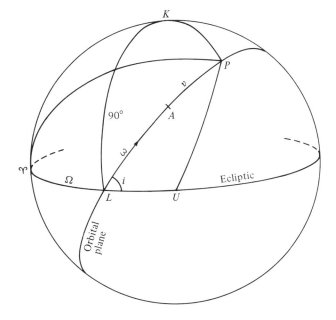

To make use of equations (7.7), it is necessary to know the relevant values of r and v, in addition to the orbital elements. The methods of section 6.4 make this possible. One successively derives M, say, from equation (7.5), E from the solution of Kepler's equation, and then r and v follow from equations (6.37).

The next step is to convert from ecliptic to equatorial coordinates. By (2.17), the position vector of P, referred to equatorial axes, is $\mathbf{r} = (x, y, z)$, where

$$x = \xi$$
$$y = \eta \cos \varepsilon - \zeta \sin \varepsilon \qquad (7.8)$$
$$z = \eta \sin \varepsilon + \zeta \cos \varepsilon,$$

and ε is the obliquity of the ecliptic.

These equations give the required heliocentric coordinates of the planet. The final step is to derive the geocentric position vector of the planet, $\mathbf{r}'$, say. Let $\mathbf{R} = (X, Y, Z)$ be the heliocentric position vector of the earth, expressed in rectangular equatorial coordinates. The components of this vector are available in section C of the *Astronomical Almanac*, where their negatives are tabulated as the geocentric rectangular coordinates of the sun. Using this information, we can derive $\mathbf{r}'$ as

$$\mathbf{r}' = \mathbf{r} - \mathbf{R}. \qquad (7.9)$$

Now let ρ be the geocentric distance of the planet, and let (α, δ) be its geocentric right ascension and declination. It then follows, from (7.9), that

$$x - X = \rho \cos \delta \cos \alpha$$
$$y - Y = \rho \cos \delta \sin \alpha \qquad (7.10)$$
$$z - Z = \rho \sin \delta.$$

Consequently the final results are

$$\alpha = \tan^{-1}\left(\frac{y - Y}{x - X}\right)$$
$$\delta = \sin^{-1}\left(\frac{z - Z}{\rho}\right), \qquad (7.11)$$

where

$$\rho^2 = (x - X)^2 + (y - Y)^2 + (z - Z)^2.$$

The coordinates (α, δ) derived in this way are *geometric* coordinates. They do not correspond exactly to the geocentric coordinates discussed in chapter 4, that is, the apparent coordinates of the planet. The latter correspond to what would be observed by a hypothetical observer at the earth's centre, moving with the earth's orbital velocity. Here, however, no aberrational effect has been included. Before any comparison with observation is possible, we must allow for annual aberration, caused by the earth's motion, and apply a light-time correction – to recognize the fact that, by the time the planet is observed, it will have moved from the location it occupied when the observed radiation was emitted. These two effects, annual aberration and the light-time correction, *together* are called *planetary aberration* which will be discussed in chapter 8. To sum up, the *geometric place* of equation (7.11) is the apparent place less planetary aberration.

7.3   Planetary masses

In the calculation just described, lengths will be expressed in astronomical units, since the planet's semimajor axis and the earth's coordinates are normally given in these units. To facilitate the calculation of the mean anomaly, the mean motion n is conventionally expressed in degrees per day. In these units, equation (6.25) becomes

$$n^2a^3 = \left(\frac{180}{\pi}\right)^2 G(M+m). \tag{7.12}$$

The units of mass, however, and, by implication, the gravitational constant, are still unspecified.

The IAU system of astronomical constants was revised in 1976 to produce a new system which has been used in preparing almanacs for the year 1984 onwards. In this revision, astronomical units of mass, length and time were defined as follows:

   (i) The astronomical unit of mass is the mass of the sun.
  (ii) The astronomical unit of time is the day, defined as 86 400 SI seconds.
 (iii) The astronomical unit of length is defined so that the constant of gravitation G is equal to k^2 in these units, where k is the Gaussian constant, cf. (4.57), given by

$$k = 0.017\,202\,098\,950. \tag{7.13}$$

Inserting the appropriate value of G in (7.12) will give a numerical result for the sum of the orbiting masses as

$$M+m = 1.029\,418\,n^2a^3. \tag{7.14}$$

Explicitly, the units in (7.14) are solar masses and AU, with n expressed in degrees per day.

Let us apply this equation to a planetary orbit, using data from the *Astronomical Almanac*. The following values are extracted for the planet Jupiter for the epoch 1984, January 21:

$$n = 0°.083\,089\,45$$
$$a = 5.204\,878.$$

Inserting these in (7.14) gives

$$M + m = 1.000\,955.$$

The effect of Jupiter's mass is evident.

Planetary masses are often expressed in reciprocal form as the ratio of the sun's mass to that of the planet. Since $M = 1$, this gives for Jupiter

$$m^{-1} = 1047.$$

The mass of any planet is detectable, in principle, by applying (7.14) to its heliocentric orbit. The results will lack precision, however, due to the dominance of the sun's mass in each case. For a small planet like Mars, for example, one can deduce little more than that its mass is virtually negligible, certainly less than one-millionth of a solar mass.

Equation (7.14) may be used for any two-body orbit, however. The classical method of determining a planet's mass is to apply this equation to the orbit of one of the planet's satellites. The equation then provides the sum of the masses of the planet and the satellite, with the latter usually negligible. The satellite's orbit will normally be less accurately determined than a heliocentric orbit, and it is this that now limits the precision of the method. The mean motion will be known to a very high precision, since satellite periods are usually only a matter of days, and observations of the satellite will have extended over a large number of orbital periods. By contrast, the semimajor axis is poorly determined, for this must be deduced from the small angular separation between the planet and its satellite. Study of natural satellite orbits usually allows a planet's mass to be determined to an accuracy of about one part in 10^4. The only exception is the planet Pluto, for which little more than an order of magnitude estimate is possible. In the following example the mass of the planet Mars is determined.

Example. The satellite Deimos moves round the planet Mars in an orbit of period $T = 1^{\text{d}}.262\,440\,7$ and semimajor axis $a = 2.346 \times 10^4$ km. Calculate the mass of Mars.

The mean motion n is computed, in the correct units, as $n = 360°/T$. The semimajor axis must be converted into AU, using

$$1\,\text{AU} = 1.496 \times 10^8\,\text{km}.$$

Equation (7.14) then yields

$$M + m = 3.228 \times 10^{-7}\,\text{solar masses.}$$

Here M is now the mass of Mars, m that of Deimos. Estimates of the mass of Deimos, derived from its size, indicate that $m/M < 10^{-8}$. It may, therefore, be safely neglected, giving the reciprocal of Mars' mass in solar units as

$$M^{-1} = 3\,098\,000.$$

Interplanetary probes have provided an important refinement of the classical method of determining planetary masses. Spacecraft have been placed in orbit about both Venus and Mars, thereby becoming artificial satellites of these planets. The tracking of these spacecraft provides much more accurate orbital data. This allows the mass of Mars to be determined to about three additional significant figures beyond the value given above. In the case of Venus, which has no natural satellite, the improvement is even more dramatic, as previous estimates of its mass had to rely primarily on the size of the perturbations that Venus induced in the orbits of the other planets.

All the major planets out to and including Saturn have been visited by at least one interplanetary probe. In the great majority of cases, the probe has been in a fly-past orbit and has only remained in the planet's vicinity for a few days. An artificial satellite of the planet was not established. Even so, while it was in the vicinity of the planet, the probe was moving principally under the planet's gravitational influence, and the tracking of the probe, therefore, provided detailed and accurate orbital data for a body in a *hyperbolic* orbit about the planet. Equation (7.14) may be used virtually unchanged for a hyperbolic orbit. The parameter a is the semimajor axis of the hyperbola, and n is the rate of increase of the mean anomaly which is related to the probe's position through Kepler's equation, cf. Table 6.2 in section 6.7. Thus planetary masses are determined from the tracking of space-probe orbits, even during comparatively transient encounters. Moreover, the precision of these mass determinations is considerably better than that obtained from the study of the natural satellites.

The masses of all the planets, Mercury out to Saturn, have now been established to about six significant figures, in terms of the mass of the sun, from the analysis of spacecraft orbits. Shortly, in January 1986, it is expected that a considerable improvement will be achieved in our knowledge of Uranus' mass, when Voyager 2 encounters that planet. Only in the cases of Neptune and Pluto, will it then be necessary to rely on the classical method for determining a planet's mass.

7.4 Planetary perturbations

The orbit of a planet round the sun is not an exact ellipse owing to the perturbing effects of the other planets. Orbital elements cannot, therefore, be regarded as exact constants of the orbital motion. An instantaneous elliptic orbit may, however, be defined as the orbit that the planet is describing at that instant of time. This is referred to as the *osculating ellipse*. It is determined entirely by the planet's mass and its position and velocity at that instant. A graphic way of describing the osculating ellipse is to say that it is the orbit that the planet would follow if the perturbing effects of the other planets could be instantaneously 'switched off'.

The orbital elements of the osculating ellipse are termed the *osculating elements*. They will, of course, vary with time, but, since the masses of all the perturbing planets are small compared with the mass of the sun, the variations in the osculating elements will normally be small, at least over limited periods of time. The osculating elements may, therefore, be used in order to calculate the position of a planet over a short interval of time. These elements are given, at 40-day intervals, for all the major planets, including the earth, in the *Astronomical Almanac*.

Complex methods of celestial mechanics, quite beyond the scope of this book, have been developed to determine the variation in the osculating elements of a planet. First- and higher-order perturbation theories have been constructed which express the elements as a series of terms of the form

$$\eta = \eta_0 + \lambda_1 t + \lambda_2 t^2 + \text{periodic terms} + (t \times \text{periodic terms}). \quad (7.15)$$

Here η represents one of the orbital elements, and η_0 is its initial value. The terms are of three kinds – secular terms, periodic terms and mixed terms. There are no secular terms for the semimajor axis to second order, although they are present in all the other elements, normally with $\lambda_2 \ll \lambda_1$. Secular terms are most evident in the angular elements, like the longitude of the ascending node or the argument of perihelion. It is the normal practice in celestial mechanics to treat the modifications due to general relativity as a perturbation of the Newtonian two-body problem. As shown in section 6.10, general relativity gives rise to a secular term in the argument of perihelion. Even for Mercury, the size of this relativistic term is, in fact, less than 10 per cent of the total secular advance of perihelion, most of which is due to perturbations by the other planets.

Mean elements of a planet's orbit are defined by omitting the periodic terms from the osculating elements. They provide a reference orbit that is a more stable basis for calculating the perturbations. Mean elements are not used in the *Astronomical Almanac*, but formulae for them can be found in the *Explanatory Supplement*.

Largely similar methods are applied to satellite orbits. Perturbations caused by other satellites of the system will be similar to the planetary perturbations discussed above. There will, however, be an additional, and very important, perturbation by the sun. The treatment of this must be slightly different. The small quantity involved in the perturbation theory is no longer the ratio of planetary to solar mass, but will be the ratio of the distances of the satellite from its primary and from the sun. Solar perturbations are very important for the moon's orbit, as values quoted in the next section will show.

One other source of perturbation of a satellite's orbit is the oblateness of its primary planet. For the point mass approximation of the two-body problem, considered in chapter 6, is only completely valid for spherical bodies. The small quantity to be used in the perturbation theory then involves the ratio of the planetary radius to the size of the orbit. The effect of oblateness is particularly important for artificial satellites in low orbits about the earth. Moreover, for such orbits, atmospheric drag cannot be neglected and produces a secular decay, involving a steady reduction in the semimajor axis.

7.5   The orbit of the moon

Owing to the importance of solar perturbations, the ecliptic, rather than the equator, is the natural reference plane for the moon's orbit about the earth. The orbital elements of the moon are, therefore, defined in the same way as those for a planetary orbit, which were shown in Fig. 7.1. The only differences are that the centre of the celestial sphere (Fig. 7.3) should now be the earth E rather than the sun, and the pericentre of the orbit, the point A, is *perigee* rather than perihelion.

Figure 7.3

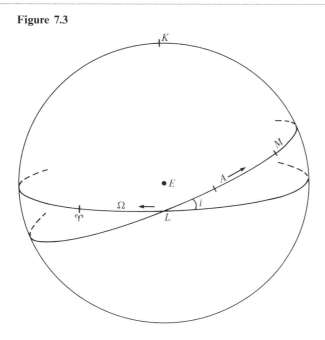

A mean lunar orbit is adopted, and the elements of this orbit – the mean elements – are defined with reference to the ecliptic and the mean equinox of date, and so they vary only secularly. In fact, three of the mean elements do not vary at all; they are

$$a = 384\,399.07 \text{ km},$$
$$e = 0.054\,900\,489, \tag{7.16}$$
$$i = 5°.145\,396\,4.$$

These values are derived from constants in the lunar theory. For example, a is consistent with the constant of lunar parallax P_0, quoted earlier in (4.53) (Murray, 1983).

The remaining mean elements, following the notation of the *Astronomical Almanac*, are given by

$$\Omega = 55°.204\,723 - 0°.052\,953\,78 \text{ d},$$
$$\Gamma' = 192°.917\,585 + 0°.111\,403\,61 \text{ d}, \tag{7.17}$$
$$L' = 18°.251\,908 + 13°.176\,396\,49 \text{ d}.$$

These are, referring to Fig. 7.3, respectively the longitude of the ascending node ΥL, the mean longitude of perigee ($\Upsilon L + LA$), and the mean longitude of the moon ($\Upsilon L + LM$). Notice, however, that M is not the true position of the moon, but its mean position, defined by the condition that AM is the mean anomaly. The expressions have been calculated for the year 1985, and d is the number of days that have elapsed since 1985 January 0, 0^{h}. One would expect the mean longitude to increase with the moon's mean motion of just over $13°$ per day, but (7.17) indicates that there are also significant secular changes in the longitudes of the node and perigee. In fact, the node regresses round the ecliptic, completing one circuit in 18.61 years. Perigee, on the other hand, moves directly round the mean orbit in a period of 8.85 years. The point A will not describe a closed curve on the celestial sphere in this interval, but it never departs from the ecliptic by more than the angle i ($\sim 5°$).

The phases of the moon, as seen from the earth, depend on the angular separation of the sun and the moon on the geocentric celestial sphere. It is convenient, therefore, to define the *mean elongation* of the moon from the sun as

$$D = L' - L, \tag{7.18}$$

that is, as the difference of the mean longitudes of the two bodies. When we speak of the mean longitude L of the sun, we are, of course, regarding the sun as being in orbit round the earth. This inverted perspective is justified – in this context at least. When the orbital theories of the moon and the sun are combined, it is found that, for 1985,

$$D = 98°.640\,537 + 12°.190\,749\,13\,d. \tag{7.19}$$

The lunar month can be defined in a number of ways. The most natural is the *synodic month*, which is the mean interval from one new moon to the next. This interval corresponds to an increase of $360°$ in the mean elongation D and may, therefore, be calculated from equation (7.19). An increase of $360°$ in the mean longitude L' defines the *tropical month*. This is not exactly the same as one revolution of the moon about the earth with respect to the fixed stars, which is called the *sidereal month*. The reason for the discrepancy is that the mean equinox itself has a small sidereal motion due to general precession – cf. equation (3.5).

There are two other fairly obvious definitions of the month suggested by (7.17). The mean interval from one perigee to the next is termed the *anomalistic month*, since it is the interval required for the moon's mean anomaly to increase by $360°$. Finally, the *Draconic month* is defined as the mean interval between successive passages through the node. This interval is important in predicting the occurrence of eclipses. The lengths of the anomalistic and Draconic months may be calculated by combining the rates of change given in equations (7.17). In fact, the lengths of the month corresponding to all five definitions, for 1985, are

$$\text{Synodic month} = 29^{\text{d}}.530\,589$$
$$\text{Tropical month} = 27^{\text{d}}.321\,582$$
$$\text{Sidereal month} = 27^{\text{d}}.321\,662 \tag{7.20}$$
$$\text{Anomalistic month} = 27^{\text{d}}.554\,550$$
$$\text{Draconic month} = 27^{\text{d}}.212\,221.$$

The discussion so far has concerned the mean orbit of the moon. The geocentric position calculated from the mean elements will differ from the apparent place of the moon due to periodic perturbations, most of which are due to the sun's influence. The problem of calculating the moon's apparent coordinates is extremely complex. The reader who wishes to understand the basic mechanics of the computation is referred to Sadler and Clemence (1954). The basic theory was developed by E. W. Brown at the beginning of the 20th century. It consists of forming expansions for the ecliptic longitude and latitude, and the horizontal parallax of the moon. Each of the series expansions consists of hundreds of terms; they are of the form

$$\lambda = L' + \sum a_{ijkl} \sin\,(iM' + jM + kF + lD)$$
$$\beta = \sum b_{ijkl} \sin\,(iM' + jM + kF + lD) \tag{7.21}$$
$$\sin P = \sin P_0 + \sum c_{ijkl} \cos\,(iM' + jM + kF + lD),$$

where i, j, k, l are integers. Further, M' and M are respectively the mean anomalies of the moon and the sun, F is the moon's mean argument from its node, i.e. $F = L' - \Omega$, while the mean elongation D has already been defined. The theory works out numerical values for the coefficients a_{ijkl}, etc. These depend on a number of small quantities, namely the eccentricities of the two mean orbits (the moon's and the sun's), the inclination i, and the ratio of the semimajor axes of the two mean orbits. The series solutions are, therefore, regarded as expansions in powers of four small parameters and may be expected to converge. These parameters are relevant to what is known as the main problem, the computation of the sun's direct influence on the orbit of the moon. Brown's theory goes on to consider other effects, principally the perturbing influence of the planets and of the figures of the earth and the moon. Each of these effects can be considered as a perturbation characterized by a small dimensionless parameter.

The first few terms in the series expansions for the moon's ecliptic longitude and latitude are given approximately by

$$\lambda = L' + 6°.29 \sin M' - 1°.27 \sin (M' - 2D) + 0°.66 \sin 2D$$
$$+ 0°.21 \sin 2M' - 0°.19 \sin M - 0°.11 \sin 2F, \tag{7.22}$$

$$\beta = 5°.13 \sin F + 0°.28 \sin (M' + F) - 0°.28 \sin (F - M')$$
$$- 0°.17 \sin (F - 2D). \tag{7.23}$$

The arguments in (7.22) and (7.23) may be derived, with the exception of M, from (7.17) and (7.19). They are, to sufficient accuracy,

$$\begin{aligned}
L' &= 18°.25 + 13°.176\,40\,\text{d} \\
M' &= 185°.33 + 13°.064\,99\,\text{d} \\
M &= 356°.93 + 0°.985\,60\,\text{d} \\
F &= 323°.05 + 13°.229\,35\,\text{d} \\
D &= 98°.64 + 12°.190\,75\,\text{d}.
\end{aligned} \tag{7.24}$$

As before, d is the number of days from 1985 January 0, 0^h.

Let us examine the terms in (7.22). Those with arguments M' and $2M'$ arise from the equation of the centre. That with argument $2F$ is due to the inclination of the moon's orbit to the ecliptic. None of these terms involves the solar perturbation, unlike the other three. The largest perturbation term, that with argument $(M' - 2D)$, is known as the *evection*. Its period is $31^d.8$. Its effect is to produce periodic variations in the osculating eccentricity and the longitude of perigee. The term with argument $2D$ is known as the *variation*. It is due to the fact that the sun's gravitational attraction on the moon varies during the synodic month, being strongest at new moon. Finally, the term with argument M is known as the *annual equation.* Its period is one year and it is due to the eccentricity of the earth's orbit round the sun.

Although all terms with amplitudes in excess of $0°.1$ have been included in (7.22) and (7.23), these equations can only give a rather crude ephemeris. When the largest terms that have been neglected are in phase, the total truncation error may be $10'$ in latitude, and twice that amount in longitude. Doubling the number terms in the series would produce a significant improvement, but, even then, errors of up to $3'$ would remain. A very precise ephemeris requires the inclusion of several hundred terms in each expansion.

The *Astronomical Almanac* gives the moon's apparent place in two forms. Ecliptic latitude and longitude are tabulated at daily intervals to an accuracy of $0°.01$. More accurate values are given of right ascension, declination and horizontal parallax – all expressed in degrees. These are again given at daily intervals, but, in addition, five polynomial coefficients are included for each entry which allow interpolated values to be rapidly calculated. The lunar theory, in its modern form, gives geometric coordinates referred to the mean equinox. The moon shares the earth's orbital motion round the sun, so annual aberration is not relevant, but the light-time correction and nutation are added to give the apparent place. The tabulated horizontal parallax, however, remains geometric.

Relativistic effects, at first sight, are not important for the moon's orbit. The Schwarzschild radius for the earth is given by

$$2m = 2GM_\oplus/c^2 \simeq 1 \text{ cm.} \tag{7.25}$$

Using (6.81), we may estimate the relativistic advance of perigee as $\sim 5 \times 10^{-5}$ arc seconds per revolution. This is, of course, totally swamped in the advance of perigee due to the solar perturbation. Now lunar theory was initially developed with coordinate determination as its prime objective. The determination of the moon's distance, the parallax expansion, was regarded as less fundamental, because positional coordinates could be measured more accurately and more directly than distances. Indeed, discrepancies in the moon's longitude provided the astronomical evidence of irregularities in the earth's rotation rate. Radar now provides a direct measurement of distance, however, and one that is considerably more accurate than that deduced from parallax observations. Moreover, laser ranging, using reflectors deposited on the moon's surface by the Apollo astronauts, has provided an exceptionally accurate means of determining the moon's distance – in fact, to an accuracy of about 1 cm. This is of the order of the earth's Schwarzschild radius, and, consequently, it is of the same order as the relativistic correction for distance, cf. (6.74). Lunar laser ranging has, therefore, provided new tests of general relativity and other post-Newtonian gravitational theories. The tests are broadly favourable to Einstein's theory.

7.6   Calculation of osculating elements

However perturbed an orbit may be, it can always be represented instantaneously by an osculating ellipse. This is uniquely determined by the relative position and velocity vectors at that instant, provided, of course, that the masses of the two bodies are known. For definiteness, consider a body in heliocentric motion. Let P be the point on the heliocentric celestial sphere defined by its position at a particular time t and let A be the body's direction at perihelion (Fig. 7.4). Further, let L be the ascending node, and N the position on the celestial sphere defined by the normal to the orbital plane. Finally, let the body's position and velocity vectors, at time t, be

$$\mathbf{r} = (\xi, \eta, \zeta)$$
$$\mathbf{V} = (V_\xi, V_\eta, V_\zeta). \tag{7.26}$$

We assume, for convenience, that these vectors are expressed in ecliptic rectangular coordinates, so that the coordinate axes are in the directions Υ, U and K. The notation is the same as in section 7.2, and so the three components of the position vector are given by equation (7.7). The problem

Figure 7.4

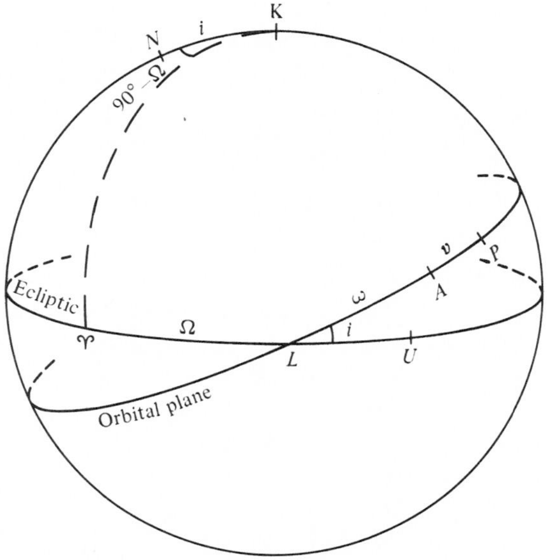

to be considered is how to convert the six items of data in (7.26) into six orbital elements.

The magnitudes of the two vectors in (7.26) may be computed and inserted in the velocity formula (6.15). This is inverted to yield the semimajor axis as

$$a = \frac{\mu r}{(2\mu - rV^2)}.$$
(7.27)

Remember that we are assuming that the masses are known, so μ is a known quantity.

Next compute the components and the magnitude of the vector $\mathbf{h}$, defined as

$$\mathbf{h} = \mathbf{r} \times \mathbf{V}.$$
(7.28)

This is the vector angular momentum per unit reduced mass. It is directed normal to the orbital plane, i.e. towards the point N on the celestial sphere. It is easily seen from Fig. 7.4 that this point N has ecliptic longitude $(\Omega - 90°)$ and ecliptic latitude $(90° - i)$. The direction cosines of $\mathbf{h}$ are, therefore, easily worked out, giving

$$h_\xi = h \sin \Omega \sin i$$
$$h_\eta = -h \cos \Omega \sin i$$
$$h_\zeta = h \cos i.$$
(7.29)

The orbital elements i and Ω are deduced as

$$\Omega = -\tan^{-1}\left(\frac{h_\xi}{h_\eta}\right)$$
$$i = \cos^{-1}\left(\frac{h_\zeta}{h}\right).$$
(7.30)

In applying the first of these equations, the principal value of the inverse tangent is taken if $h_\eta < 0$; otherwise add 180°. The magnitude of the vector $\mathbf{h}$ will give the eccentricity; for it follows, from (6.23), that

$$e = \left(1 - \frac{h^2}{\mu a}\right)^{1/2}.$$
(7.31)

Since both the semimajor axis and the eccentricity have now been calculated, the three anomalies may be determined. It is probably best to start with the eccentric anomaly, using (6.24) to deduce that

$$E = \pm \cos^{-1}\left(\frac{a - r}{ae}\right).$$
(7.32)

The plus sign corresponds to the body's outward journey, from perihelion to aphelion, when its radial velocity is positive. The sign of the radial velocity is that of the scalar product $\mathbf{r} \cdot \mathbf{V}$; so E is uniquely determined. The other two anomalies then follow from (6.26) and (6.28), that is

$$v = 2 \tan^{-1} \left[\left(\frac{1+e}{1-e} \right)^{1/2} \tan \tfrac{1}{2}E \right]. \tag{7.33}$$

$$M = E - e \sin E.$$

There is no ambiguity in either of these equations.

Let $\mathbf{n}$ denote the unit vector in the direction of the ascending node. It is seen from Fig. 7.4 that

$$\mathbf{n} = (\cos \Omega, \sin \Omega, 0), \tag{7.34}$$

and that $\mathbf{n} \cdot \mathbf{r} = r \cos LP = r \cos (v + \omega)$. We may, therefore, solve for $(v + \omega)$, as

$$v + \omega = \pm \cos^{-1} \left(\frac{\xi \cos \Omega + \eta \sin \Omega}{r} \right). \tag{7.35}$$

The ambiguity is easily resolved; the plus sign is to be taken when $\zeta > 0$. Since the true anomaly has already been calculated, equation (7.35) will allow the argument of perihelion to be determined. If the longitude of perihelion is preferred, simply use equation (7.3).

The timing of the orbit has already been established by computing the mean anomaly for time t. This can be converted to the mean longitude for this time by using (7.4). If, however, one really wants to obtain the time of perihelion τ explicitly, the definition of the mean anomaly is used to give

$$\tau = t - M/n. \tag{7.36}$$

The mean motion is, of course, prescribed by the condition that $n^2 a^3 = \mu$.

7.7   Preliminary orbit determination

The observational data on which orbit determination is based may take a number of different forms. Here we shall consider the classical problem of determining the orbit from three positional observations obtained within a short interval of time. This is the way that the preliminary orbit of a newly discovered asteroid or comet is obtained. The assumption is made that the mass of the body may be entirely neglected compared with the sun's mass. Each observation of the body's position gives two independent coordinates, usually a right ascension and declination. The intention is to derive, from the data of three such observations, the position and velocity vectors for one particular time. This is usually chosen as the time of the middle observation. Subsequently the methods of section 7.6 may be employed to derive the orbital elements.

Referring to Fig. 7.5, let S denote the sun, O the earth and P a newly discovered asteroid (or comet). Let $\mathbf{r}$ and $\mathbf{R}$ be the heliocentric position vectors of the asteroid and the observer respectively, and denote the vector $\overrightarrow{OP}$ by $\rho\mathbf{s}$, where $\mathbf{s}$ is a unit vector. A measurement of the asteroid's position gives this unit vector. The observation should be corrected for refraction, diurnal aberration and planetary aberration. The first two corrections present no problem. If the observation is made photographically, they are incorporated in the reduction of the photographic plate measurements, as explained in chapter 13. The correction for planetary aberration cannot be made *ab initio*, since it depends on the, as yet, unknown distance of the asteroid. The best procedure is to ignore this effect in the first instance in computing an orbit. The effect of planetary aberration may then be derived on the basis of this orbit, and the orbit itself is recomputed. One such iteration should suffice. A correction for geocentric parallax could be avoided if $\mathbf{R}$ is taken as strictly the position vector of the observer, rather than of the centre of the earth. The *Astronomical Almanac* gives the latter directly, but it may be adjusted by the methods of section 4.5. Alternatively, $\mathbf{R}$ can be taken as the tabulated position vector of the earth's centre, and geocentric parallax included in the same way as planetary aberration.

Suppose now that observations are made of the asteroid's position at three times $t_1 < t_2 < t_3$. Let $\mathbf{s}_1, \mathbf{s}_2, \mathbf{s}_3$ be the three resulting unit vectors. Denoting $(t_2 - t_1)$ by τ_1 and $(t_3 - t_2)$ by τ_3, we can expand $\mathbf{s}_1$ and $\mathbf{s}_3$ as Taylor series about time t_2, deriving

$$\begin{aligned} \mathbf{s}_1 &= \mathbf{s}_2 - \tau_1 \dot{\mathbf{s}}_2 + \tfrac{1}{2}\tau_1{}^2 \ddot{\mathbf{s}}_2 \\ \mathbf{s}_3 &= \mathbf{s}_2 + \tau_3 \dot{\mathbf{s}}_2 + \tfrac{1}{2}\tau_3{}^2 \ddot{\mathbf{s}}_2, \end{aligned} \tag{7.37}$$

when third-order terms are neglected. To justify this neglect, the time interval between the observations must be rather small, in fact, a few days. Equations (7.37) are now treated as simultaneous equations in $\dot{\mathbf{s}}_2$ and $\ddot{\mathbf{s}}_2$,

Figure 7.5

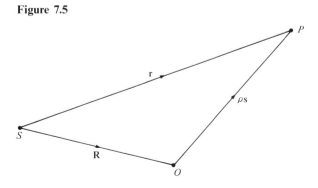

which may be solved to give

$$\dot{\mathbf{s}}_2 = \frac{\tau_3(\mathbf{s}_2 - \mathbf{s}_1)}{\tau_1(\tau_1 + \tau_3)} + \frac{\tau_1(\mathbf{s}_3 - \mathbf{s}_2)}{\tau_3(\tau_1 + \tau_3)}$$

$$\ddot{\mathbf{s}}_2 = \frac{2(\mathbf{s}_3 - \mathbf{s}_2)}{\tau_3(\tau_1 + \tau_3)} - \frac{2(\mathbf{s}_2 - \mathbf{s}_1)}{\tau_1(\tau_1 + \tau_3)}. \tag{7.38}$$

By using the three observations in this way, we have obtained the vector $\mathbf{s}$ and its first two derivatives all at time $t = t_2$. The suffix 2 will be dropped in what follows.

It is seen from the diagram that

$$\mathbf{r} = \rho\mathbf{s} + \mathbf{R}. \tag{7.39}$$

Notice that ρ is the only unknown on the right-hand side of this equation. Differentiate the equation twice with respect to time, deriving successively

$$\dot{\mathbf{r}} = \dot{\rho}\mathbf{s} + \rho\dot{\mathbf{s}} + \dot{\mathbf{R}}, \tag{7.40}$$

$$\ddot{\mathbf{r}} = \ddot{\rho}\mathbf{s} + 2\dot{\rho}\dot{\mathbf{s}} + \rho\ddot{\mathbf{s}} + \ddot{\mathbf{R}}. \tag{7.41}$$

Now the second equation gives the asteroid's acceleration, which is determined by the gravitational equation of motion. For definiteness, let us prescribe the units as astronomical units of length, mass and time, as defined in section 7.3. Then the equation of motion for the asteroid is

$$\ddot{\mathbf{r}} = -k^2\mathbf{r}/r^3.$$

Moreover, the equation of motion of the earth's orbit relative to the sun will, by (6.8), give

$$\ddot{\mathbf{R}} = -k^2(1+m)\mathbf{R}/R^3,$$

where m is the mass of the earth. Substitute these two results into equation (7.41) to derive, with the help of (7.39), the result that

$$(\ddot{\rho} + k^2\rho/r^3)\mathbf{s} + 2\dot{\rho}\dot{\mathbf{s}} + \rho\ddot{\mathbf{s}} = k^2\mathbf{R}[(1+m)/R^3 - 1/r^3]. \tag{7.42}$$

We want to derive an expression for ρ and eliminate its two derivatives from (7.42). So take the vector product of (7.42) with $\mathbf{s}$, followed by the scalar product with $\dot{\mathbf{s}}$, to give, after a little reduction,

$$\rho = k^2\left(\frac{(1+m)}{R^3} - \frac{1}{r^3}\right)\frac{[\mathbf{s}, \dot{\mathbf{s}}, \mathbf{R}]}{[\mathbf{s}, \dot{\mathbf{s}}, \ddot{\mathbf{s}}]}. \tag{7.43}$$

Everything on the right-hand side of this equation is known, with the exception of r. An estimate is, therefore, made for this quantity, so that a value of ρ may be calculated. Then r is obtained from the square of (7.39), namely

$$r^2 = \rho^2 + R^2 + 2\rho\mathbf{R}\cdot\mathbf{s}. \tag{7.44}$$

We iterate in this way between equation (7.43) and (7.44) until the process converges to a solution for ρ. Using this in equation (7.39) gives the planet's position vector at time t_2.

In order to derive the velocity vector, it is necessary to solve for $\dot{\rho}$ from (7.42). The procedure is similar. Take $((7.42) \times \mathbf{s}) \cdot \dot{\mathbf{s}}$ to obtain, after simplification,

$$\dot{\rho} = \frac{k^2}{2} \left(\frac{1}{r^3} - \frac{(1+m)}{R^3} \right) \frac{[\mathbf{s}, \ddot{\mathbf{s}}, \mathbf{R}]}{[\mathbf{s}, \dot{\mathbf{s}}, \ddot{\mathbf{s}}]}. \tag{7.45}$$

The value of $\dot{\rho}$ derived directly from this equation is then inserted in (7.40) to yield the velocity vector, again at time t_2. Since both $\mathbf{r}$ and $\dot{\mathbf{r}}$ are now known at the one time, the method of the previous section may be applied to obtain the orbital elements.

Three observations are the minimum necessary to derive a preliminary orbit. The elements obtained will be sensitive to observational errors. To minimize their effect, the observations should be as widely separated as possible. On the other hand, the separation is limited by the requirement that the truncation errors in the Taylor series (7.37) should not vitiate the result. Of course, an orbit may, in principle, be determined from three widely separated observations, but a different method must be developed.

The method of orbit determination will clearly be modified for different kinds of observational data. Positional observations of artificial satellites and space-probes, in particular, will often be supplemented by radar measurements of the range ρ, and the range-rate $\dot{\rho}$ is obtainable from the Doppler shift in the radar echo. One combined radar and positional observation is clearly sufficient to determine the body's position vector, and only one more is needed to give sufficient data for a complete orbit determination. The positional information, for satellites and space-probes, is frequently far less precise than the range and range-rate data. It is interesting to note that an orbit can, in fact, be derived solely from a knowledge of the range and range-rate at three separate instants of time. The method, however, is far more involved than that given above. Further details of methods of orbit determination will be found in Roy (1982), chapter 13.

Since dynamical principles are being applied to the observational data, it is necessary to ensure that the coordinates are referred to an inertial set of axes. Standard methods for the reduction of photographic plate measurements take care of this point by referring coordinates to a fixed equator and equinox, usually the mean equator and equinox of 1950.0 or 2000.0. If the positions were referred to the equator and equinox of date, this would mean that the coordinate system possessed a rotation due to precession, and the method is so sensitive to observational error that this would be sufficient to make nonsense of the derived orbit.

The analysis of this section is only intended to provide a preliminary orbit. The derived orbit makes an exact, as opposed to a statistical, fit with the observed coordinates. Once three observations of a newly discovered asteroid or comet are available, the preliminary orbit is obtained. Later, as additional observations become available, the provisional orbit is adapted to fit the whole corpus of observational data, but the fit will now be statistical. Once orbital elements are firmly established, they are used to calculate future ephemerides. Otherwise the asteroid would have to be rediscovered following each conjunction with the sun, when it is unobservable for a considerable period.

Problems

(Assume where needed that

1 AU $= 1.4960 \times 10^8$ km.

Obliquity of ecliptic $= 23° 26' 21''.4$.)

7.1 Neptune's satellite Triton has an orbital period of $5^d 21^h 02^m 39^s$ and a semimajor axis of 3.5525×10^5 km. Calculate the reciprocal mass of the planet, assuming that Triton's mass may be neglected.

7.2 A spacecraft is in a fly-past orbit round Venus with eccentricity 1.1898. At its closest approach its planetocentric distance is 8415 km and 49 minutes later this distance has increased to 19 998 km. Calculate the mass of Venus from this data.

7.3 The *Astronomical Almanac* provides the following orbital elements for the asteroid Ceres:

$i = 10°.606 \quad a = 2.7666$ AU

$\Omega = 80°.718 \quad n = 0°.214\,19\,\mathrm{d}^{-1}$

$\omega = 72°.890 \quad e = 0.0784$.

These are osculating elements for the date 1984 October 27.0 when the mean anomaly was $260°.117$. Calculate the heliocentric ecliptic longitude and latitude of Ceres for the date 1984 December 13.0.

7.4 The earth's heliocentric position vector is

$\mathbf{R} = (0.147\,528, 0.892\,916, 0.387\,161)$

on 1984 December 13.0. Calculate the geocentric right ascension and declination for Ceres at this date.

7.5 Show the heliocentric ecliptic latitude and longitude of a body in orbit about the sun are given by

$$\lambda = \Omega + \tan^{-1}[\cos i \tan (v + \omega)], \quad \beta = \sin^{-1}[\sin i \sin (v + \omega)],$$

where v is the true anomaly and the symbols for orbital elements have their usual meanings.

Show further that for small values of the inclination and eccentricity these equations become approximately

$$\lambda = M + \varpi + 2e \sin M - \tfrac{1}{4}i^2 \sin 2(M + \omega),$$

$$\beta \operatorname{cosec} i + e \sin \omega = \sin (M + \omega) + e \sin (2M + \omega).$$

7.6 Apply the equations established in the previous question to the moon's orbit. Identify the corresponding terms in equations (7.22) and (7.23), and verify that the coefficients are compatible with the elements quoted in equation (7.16).

7.7 Prove that the velocity vector **V** of a body in a heliocentric orbit may be expressed in rectangular ecliptic coordinates as

$$V_{\xi} = -\frac{na}{(1 - e^2)^{1/2}} \{[\sin (v + \omega) + e \sin \omega] \cos \Omega$$

$$+ [\cos (v + \omega) + e \cos \omega] \sin \Omega \cos i\}$$

$$V_{\eta} = -\frac{na}{(1 - e^2)^{1/2}} \{[\sin (v + \omega) + e \sin \omega] \sin \Omega$$

$$- [\cos (v + \omega) + e \cos \omega] \cos \Omega \cos i\}$$

$$V_{\zeta} = \frac{na}{(1 - e^2)^{1/2}} \{[\cos (v + \omega) + e \cos \omega] \sin i\}.$$

7.8 The minor planet Ceres has heliocentric position and velocity vectors

$$\mathbf{r} = (1.581\,990,\ 2.278\,502,\ -0.223\,537)\ \text{AU}$$

$$\mathbf{V} = (-8681.85,\ 5218.88,\ 1762.05)10^{-6}\ \text{AU d}^{-1}.$$

These are ecliptic coordinates for the date 1984 December 13.0. Calculate the orbital elements a, Ω, i, e, ω, τ.

(To recover the values quoted in question 7.3, it is necessary to use a slightly modified value of μ equal to $n^2 a^3$ based on the values given there. The values given in the Answers are based on $\mu = k^2$. Why are i and Ω unaffected?)

7.9 Prove that the method of section 7.7. breaks down for a body whose orbital inclination is zero.

8

Heliocentric and barycentric coordinates

8.1 Introduction

In chapter 4 the transformation from the observed position of a celestial object to its geocentric place was considered. The parallactic and aberrational corrections involved, although essential, are of comparatively little astronomical interest in themselves. The circumstances are markedly different with the transformation considered in this chapter, namely that from the geocentric origin to the heliocentric or barycentric one. The parallax correction, often referred to as stellar parallax, is crucial to the determination of stellar distance and to the whole of the astronomical distance scale beyond the solar system. Moreover, relativistic effects, neglected in chapter 4, should now be considered. They provide an important test of general relativity, and generally they should not be omitted from the reduction of high precision positional observations.

Most optical observations of stars and planets are made at elongations from the sun in excess of 30°. The general relativistic effects are then measured in milli-arc seconds. Many types of observation do not claim this order of precision, and, for these, the classical treatment is perfectly adequate and, of course, much simpler.

The classical treatment is developed in the earlier sections of this chapter. It treats as separate concepts, the effects of annual (or stellar) parallax, annual aberration and planetary aberration. We shall develop a fully relativistic treatment in the later sections of the chapter. In this, parallax and aberration – whether annual or planetary – cannot be so naturally distinguished. The treatment is, therefore, a synthesis along the lines discussed in section 3.6.

8.2 The motion of the earth

The parallax correction requires a knowledge of the heliocentric position vector of the earth; the aberrational correction depends on the earth's velocity vector. Each is computed from the earth's osculating elements by a process that is the inverse of section 7.6.

The position vector **r** is derived using the first part of the analysis of section 7.2. Setting $i = 0$ in equation (7.7) and using (7.3) yields, in ecliptic coordinates,

$$\mathbf{r} = [r \cos (v + \varpi), r \sin (v + \varpi), 0]. \tag{8.1}$$

The radius vector r and the true anomaly v will already have been calculated in terms of the eccentric anomaly, subsequent to the solution of Kepler's equation.

The velocity vector is most easily derived by applying the theorem of section 6.6. According to this the velocity vector can be written as

$$\mathbf{V} = \mathbf{V}_0 + \mathbf{V}_1, \tag{8.2}$$

where $\mathbf{V}_0$ is a transverse velocity of constant magnitude, and $\mathbf{V}_1$ is a constant velocity parallel to the minor axis.

The magnitudes of these two vectors are

$$V_0 = \frac{h}{a(1 - e^2)} = \left[\frac{\mu}{a(1 - e^2)} \right]^{1/2},$$
$$V_1 = e V_0. \tag{8.3}$$

Unit vectors parallel to $\mathbf{V}_0$ and $\mathbf{V}_1$ may be identified from equation (8.1). The transverse direction is obtained by replacing v with $v + 90°$, and the direction parallel to the minor axis is obtained by setting $v = 90°$. Equations (8.2) and (8.3), therefore, yield

$$\mathbf{V} = V_0[-\sin (v + \varpi) - e \sin \varpi, \cos (v + \varpi) + e \cos \varpi, 0]. \tag{8.4}$$

Osculating elements are tabulated in the *Astronomical Almanac* for all the major planets. The elements for the earth, however, strictly correspond to the centre of mass of the earth/moon system, for it is this point, rather than the earth's centre, that is describing the ellipse round the sun. In effect, the earth/moon system is regarded as a double planet.

The vectors given in (8.1) and (8.4) correspond to the earth/moon barycentre. In order to derive the position and velocity vectors of the true geocentric point, we must add that point's position and velocity vectors relative to the earth/moon barycentre. These may be derived from equations (6.3) and their derivatives. Consequently, if M and m are respectively the masses of the earth and the moon, and if $\mathbf{r}'$ and $\mathbf{V}'$ are the geocentric position and velocity vectors of the moon, then the required corrections are

$$\delta\mathbf{r} = -m\mathbf{r}'/(m + M)$$
$$\delta\mathbf{V} = -m\mathbf{V}'/(m + M). \tag{8.5}$$

This is a comparatively small correction. The earth/moon barycentre always lies within the earth's surface. The displacement δr produces effects, therefore, of slightly less importance than geocentric parallax. The velocity increment δV is about 12.5 m s^{-1}, which gives rise to an aberration effect of about 8.5 milli-arc seconds. Clearly, neither the positional nor the velocity effect can be totally ignored.

The distinction has already been made between the heliocentric and the barycentric origin. The relative orbit is normally considered when discussing planetary motions, and so the heliocentric origin is the natural one to use. It is not, however, unaccelerated and so it is not a suitable origin for an inertial frame of reference. The most practical origin to use with positional observations of stars and other bodies outside the solar system is the centre of mass of the whole system, which we refer to simply as the *barycentre*. A further correction is, therefore, required for the vectors $\mathbf{r}$ and $\mathbf{V}$ considered above.

Let M_i be the mass of the ith planet, together with its satellites, in order of increasing distance from the sun. So $M_3 = m + M$ in our earlier notation. Let $\mathbf{r}_i$ be the heliocentric position vector of the ith planet. Then the position vector of the barycentre with respect to the sun is $\mathbf{r}_0$, where

$$(1 + \sum M_i)\mathbf{r}_0 = \sum M_i\mathbf{r}_i, \tag{8.6}$$

the masses being expressed in solar units. The individual $\mathbf{r}_i$ must be derived from the heliocentric osculating elements of the planetary orbits.

The earth may now be referred to the barycentre instead of the sun. Combining equation (8.6), and its first derivative, with (8.2) and (8.5) yields the earth's barycentric position and velocity vectors as

$$\mathbf{R} = \mathbf{r} - \frac{m}{M_3}\mathbf{r}' - \sum\left[\frac{M_i\mathbf{r}_i}{(1 + \sum M_i)}\right]$$
$$\dot{\mathbf{R}} = \mathbf{V}_0 + \mathbf{V}_1 - \frac{m}{M_3}\dot{\mathbf{r}}' - \sum\left[\frac{M_i\dot{\mathbf{r}}_i}{(1 + \sum M_i)}\right]. \tag{8.7}$$

The components of $\mathbf{R} \equiv (X, Y, Z)$ and $\dot{\mathbf{R}} \equiv (\dot{X}, \dot{Y}, \dot{Z})$ are tabulated at daily intervals in the *Astronomical Almanac*. They are given as equatorial rectangular coordinates referred to the mean equator and equinox for the standard epoch J2000.0. The units are astronomical units for $\mathbf{R}$, and 10^{-9} AU per day for $\dot{\mathbf{R}}$.

The correction from heliocentric to barycentric coordinates is again a small one. Even if all the planets were lined up on one side of the sun at a particular instant, the sun's displacement from the barycentre would still be less than 0.01 AU. This would produce an effect near the limit of detectability in the parallax of the nearest stars. The sun's displacement is, of course, significant for a body within the solar system, but the heliocentric origin is acceptable, even preferable, in that context. The real importance in the change of origin resides in $\dot{\mathbf{R}}$. The major planets, principally Jupiter, produce a solar velocity relative to the barycentre which is usually in excess of 0.01 km s^{-1}. This, in turn, yields an aberrational effect of about 0″.01. Moreover, the barycentre is essential as the reference point for the precise timing of rapid periodic phenomena, like pulsar signals.

The barycentre will be used as the reference point in the ensuing discussion of parallax and aberration from the classical point of view. We shall revert to the heliocentric origin in the later relativistic discussion, so that a spherically symmetric spacetime may be considered. With the small change involved, the transfer from heliocentric to barycentric origin may validly be considered classically; the modifications that arise from the corrections considered in this section must be added to the relativistic results.

8.3   Annual parallax

Let X be the position of a celestial object, C the barycentre of the solar system, and E the earth's centre, all as shown in Fig. 8.1. Denote by $\mathbf{r}$ and $\mathbf{R}$ the barycentric position vectors of the object and the earth

Figure 8.1. Stellar parallax.

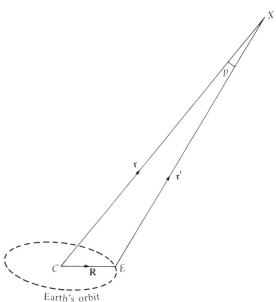

respectively, and let $\mathbf{r}'$ be the geocentric position vector of the object. Clearly

$$\mathbf{r} = \mathbf{r}' + \mathbf{R}. \tag{8.8}$$

If X is a body within the solar system, it is essential to use this equation in its exact form. For a star, however, some approximations may be made. The angle between the vectors $\mathbf{r}$ and $\mathbf{r}'$ is the annual parallax p. It will vary during the course of the year as the earth describes its orbit round the sun. Let E denote the elongation of the star from the sun, i.e. $E = C\hat{E}X$. Then, by the sine formula,

$$\sin p = (R/r) \sin E. \tag{8.9}$$

The *star's parallax* is formally defined as π, where

$$\sin \pi = 1/r, \tag{8.10}$$

r being expressed in astronomical units. This corresponds to the annual parallax for the standard conditions of $R = 1$ and $E = 90°$. It is shown later that, due to annual parallax, the star describes an approximate ellipse with semimajor axis π.

All stars have very small parallaxes; in no case does the stellar parallax exceed $1''$. Consequently, the distance of a star is given to high accuracy by

$$r = \pi^{-1}. \tag{8.11}$$

The practical unit of stellar distance is the *parsec*. It corresponds to a parallax of one arc second. If π is expressed in radians, then (8.11) gives the star's distance in AU, but, if π is expressed in arc seconds, the distance is given in parsecs. We have the following conversion table

$$
\begin{aligned}
1 \text{ parsec} &= 206\,265 \text{ AU} \\
&= 3.2616 \text{ light years} \\
&= 3.0857 \times 10^{13} \text{ km.}
\end{aligned}
\tag{8.12}
$$

Due to the smallness of a star's parallax, it is clearly permissible to work only to order R/r in equation (8.8). Let $\mathbf{s}$ and $\mathbf{s}'$ be unit vectors in the directions of $\mathbf{r}$ and $\mathbf{r}'$. Then the first-order parallax formula, equation (3.10), may be used to give

$$\mathbf{ds} = \mathbf{s}' - \mathbf{s} = r^{-1}\mathbf{s} \times (\mathbf{s} \times \mathbf{R}).$$

Using (8.11), this reduces to

$$\mathbf{ds} = \pi[(\mathbf{R} \cdot \mathbf{s})\mathbf{s} - \mathbf{R}]. \tag{8.13}$$

The components of $\mathbf{R}$ and $\mathbf{s}$, in rectangular equatorial coordinates, are, respectively

$$
\begin{aligned}
\mathbf{R} &= (X, Y, Z) \\
\mathbf{s} &= (\cos \alpha \cos \delta, \sin \alpha \cos \delta, \sin \delta).
\end{aligned}
\tag{8.14}
$$

Although we are working only to first order in the parallax, in other respects accuracy will be maintained. In particular, **R** is rigorously the barycentric position vector of the earth, as defined in the last section. The components of d**s** can be derived by differentiating the second of equations (8.14). Then the angular displacements (dα, dδ) may be obtained from (8.13), which has component equations

$$-\sin \alpha \cos \delta \, d\alpha - \cos \alpha \sin \delta \, d\delta = \pi[(\mathbf{R} \cdot \mathbf{s}) \cos \delta \cos \alpha - X],$$
$$\cos \alpha \cos \delta \, d\alpha - \sin \alpha \sin \delta \, d\delta = \pi[(\mathbf{R} \cdot \mathbf{s}) \cos \delta \sin \alpha - Y],$$
$$\sin \delta \, d\delta = \pi[(\mathbf{R} \cdot \mathbf{s}) \sin \delta - Z].$$

Combining the first two of these equations to eliminate dδ and using the third directly gives the final results:

$$d\alpha = (\pi/15) \sec \delta \, (X \sin \alpha - Y \cos \alpha)$$
$$d\delta = \pi(X \cos \alpha \sin \delta + Y \sin \alpha \sin \delta - Z \cos \delta).$$

(8.15)

The factor of $1/15$ has been inserted so that dα will be in time measure when π is expressed in arc seconds. It should be left out when π is expressed in radians.

8.4 Annual aberration

The treatment of annual aberration is very similar to that for parallax. Again, it is a classical first-order treatment. The earth's orbital velocity is about 30 km s^{-1}, which is a fraction 10^{-4} of the velocity of light. This indicates that the aberrational displacement will be of order 10^{-4} radians, or about 20″. Second-order effects ($\sim 10^{-8}$ radians) are not, therefore, entirely negligible in extremely high precision work. There is, however, no point in making the classical treatment more exact, since relativistic effects are also of this order.

The vector **r**′ in Fig. 8.1 corresponds to the direction in which the star would appear for an observer situated at E, but at rest with respect to the barycentre C. Let the direction of the star when aberration is allowed for be **s***. Then, using the first-order aberration formula (3.13), with the earth's velocity $\dot{\mathbf{R}}$ replacing $V\mathbf{n}$, yields the displacement

$$d\mathbf{s} = \mathbf{s}^* - \mathbf{s}' = c^{-1}\mathbf{s}' \times (\mathbf{s}' \times \dot{\mathbf{R}}).$$

It is sufficiently accurate to replace the unit vector **s**′ by **s**. The last equation may then be written as

$$d\mathbf{s} = c^{-1}(\dot{\mathbf{R}} - (\dot{\mathbf{R}} \cdot \mathbf{s})\mathbf{s}).$$

(8.16)

Now compare equation (8.16) with (8.13). The only differences are that **R** is replaced by its derivative and π by c^{-1}. Following the argument of the last section, we may, therefore, deduce the changes (dα, dδ) in a star's equatorial coordinates due to annual aberration, as

$$d\alpha = c^{-1} \sec \delta (\dot{Y} \cos \alpha - \dot{X} \sin \alpha)$$
$$d\delta = c^{-1} (\dot{Z} \cos \delta - \dot{X} \cos \alpha \sin \delta - \dot{Y} \sin \alpha \sin \delta). \tag{8.17}$$

The velocity of light c must be in the same units as the components of $\dot{\mathbf{R}}$. If the astronomical system of units are used, as in the *Astronomical Almanac*, velocities are expressed in AU per day. In these units

$$c = 173.14 \text{ AU d}^{-1}. \tag{8.18}$$

It has been tacitly assumed in the above treatment that the star itself has no velocity with respect to the solar system barycentre. This will not, of course, be true. Indeed, the star's velocity may exhibit itself through an observable proper motion. Correcting the observed position of a star for parallax and aberration, by the methods just considered, does not give the geometric barycentric position for the instant of observation, t_0 say. Rather it gives the position that the star occupied an interval τ years previously, where τ is the travel-time of the light between the star and the observer. Strictly speaking, to obtain the geometric position for time t_0, one should add on τ times the annual proper motion. This correction is known as *secular aberration*. It is, however, a correction that is never in practice attempted, due to the uncertainty in the star's distance and, therefore, in the value of τ. Instead, the contribution of secular aberration is absorbed into what are regarded as the star's barycentric coordinates.

The description of the light-time correction as an aberration will be justified in section 8.6. In fact, when both velocities are constant, there is no basic distinction between the effect of the observer's velocity (aberration) and the effect of the source's velocity (light-time correction) – only the relative velocity is significant. Although the light-time correction is ignored for a star, when the source is within the solar system, this correction can and must be made. The term *planetary aberration* is used to describe the total effect arising from both the earth's and the source's velocity. It is rigorously the correction from the apparent to the geometric position. By contrast, annual aberration, which has been the subject of this section, takes account only of the earth's motion with respect to the barycentre.

8.5   Low precision formulae for parallax and aberration

The displacement of a star's position on the celestial sphere, due to annual parallax, can never exceed an arc second. Consequently, in using equations (8.15), it is necessary to retain at most three significant figures in the components (X, Y, Z). In view of this, simpler expressions may be used for the earth's position vector $\mathbf{R}$. The distinction between the geocentric point and the earth/moon barycentre is never needed, and, for most stars, the distinction between the sun and the solar system barycentre can be dispensed with as well. The eccentricity of the earth's orbit is about 1/60, and so, provided the star's parallax is less than $0''.01$, it is sufficient to treat the earth as moving in a circular orbit round the sun.

The expressions are simpler if ecliptic, rather than equatorial coordinates, are used. Then, if $\lambda_\odot$ is the sun's true ecliptic longitude, $\lambda_\odot = -(v+\varpi)$, and equation (8.1) gives the earth's position vector approximately as

$$\mathbf{R}=(-\cos \lambda_\odot, -\sin \lambda_\odot, 0).\tag{8.19}$$

These components may be inserted in equation (8.15), but, since we are now working in ecliptic coordinates, (α, δ) must be replaced by (λ, β). The results are obtained as

$$d\lambda = -\pi \sec \beta \sin (\lambda-\lambda_\odot)$$
$$d\beta = -\pi \sin \beta \cos (\lambda-\lambda_\odot).\tag{8.20}$$

Now let x and y be components of the star's displacement in directions parallel and perpendicular to the ecliptic respectively, i.e. $x=d\lambda \cos \beta$, $y=d\beta$. In the course of one year the angle $(\lambda-\lambda_\odot)$ will take all values between 0 and $360°$. Eliminating this angle, we derive the apparent annual path traced out by the star on the celestial sphere as

$$\frac{x^2}{\pi^2}+\frac{y^2}{\pi^2 \sin^2 \beta}=1.\tag{8.21}$$

It is concluded that the annual path of a star due to parallax is an ellipse of semimajor axis π, semiminor axis $\pi \sin \beta$, and, therefore, of eccentricity $\cos \beta$. The major axis of this parallactic ellipse is parallel to the ecliptic.

Low precision aberrational formulae may be derived in a similar manner, that is, taking account only of the earth's heliocentric orbit and neglecting the fine distinction of barycentres. The ellipticity of the orbit may be included fairly easily by setting

$$\dot{\mathbf{R}}=\mathbf{V}_0+\mathbf{V}_1,\tag{8.22}$$

where $\mathbf{V}_0$ and $\mathbf{V}_1$ were defined in equation (8.3). Then, using equation (8.4), these velocities are given in ecliptic coordinates as

$$\mathbf{V}_0 = V_0(\sin \lambda_\odot, -\cos \lambda_\odot, 0)$$
$$\mathbf{V}_1 = eV_0(-\sin \varpi, \cos \varpi, 0).\tag{8.23}$$

The contributions of $\mathbf{V}_0$ and $\mathbf{V}_1$ may be considered separately. Inserting the components of $\mathbf{V}_0$ into (8.17) and converting to ecliptic coordinates yields

$$d\lambda = -\kappa \sec \beta \cos (\lambda-\lambda_\odot)$$
$$d\beta = \kappa \sin \beta \sin (\lambda-\lambda_\odot),\tag{8.24}$$

where κ is the dimensionless ratio V_0/c. This is known as the *constant of aberration* and, from (8.3), is given by

$$\kappa = \frac{k}{c}\left[\frac{1+m}{a(1-e^2)}\right]^{1/2}. \tag{8.25}$$

Here, k is the Gaussian constant, m the mass of the earth, and a and e are orbital elements of the earth's orbit. The constant of aberration is not an absolute constant, since the eccentricity does have a very small secular change. Using the IAU system of constants (1976), the value for J2000.0 is

$$\kappa = 20''.495\,52. \tag{8.26}$$

Examination of equation (8.24) indicates that annual aberration causes the star to describe an ellipse centred on its heliocentric position, with the major axis parallel to the ecliptic. The aberrational ellipse has semimajor axis κ and eccentricity $\cos\beta$.

The star's displacement due to the velocity $\mathbf{V}_1$ may be similarly derived from equation (8.17). The results are

$$d\lambda = \kappa e \sec\beta \cos(\varpi - \lambda)$$
$$d\beta = \kappa e \sin\beta \sin(\varpi - \lambda). \tag{8.27}$$

These are known as the *E*-terms. They are independent of the longitude of the sun, that is, they have no annual variation. Their magnitude, i.e. κe, is $0''.343$, and their effect is to displace the whole aberrational ellipse by the constant amount indicated by (8.27). They are not, however, strictly constant, but possess a slow secular variation as explained below.

The *E*-terms have an interesting, if slightly confusing, history. Prior to 1960 the aberrational displacement was computed along the lines just explained, no distinction, for example, being made between the heliocentric and barycentric origins. It was natural then not to make the corrections contained in (8.27) to the apparent places of the stars. Rather the *E*-terms were included in the so-called heliocentric positions. Both star catalogues and annual almanacs followed this practice. The mean places given in the star catalogues incorporated in the *E*-terms. The almanacs provided coefficients, for transforming from apparent to mean places, that ignored the *E*-terms and were based on (8.24) alone.

Increased precision required a modification of the almanac entries after 1960. The aberrational coefficients – they are discussed in chapter 12 – which are based on the earth's velocity needed to be computed using the full treatment outlined in section 8.2. The same star catalogues, however, were still in use, so the following compromise was devised. The coefficients were first computed rigorously from the earth's motion relative to the solar system barycentre. Then the *E*-terms were rather artificially extracted.

This procedure has not proved entirely satisfactory, however, due to the slow variations in the E-terms themselves. Firstly, both e and ϖ have secular variations due mainly to planetary perturbations. These affect both $d\lambda$ and $d\beta$ in (8.27). Further and more importantly, variations arise from precession. This produces a regular increase in the longitudes of all stars, latitudes are unaffected. This, it is true, leaves $d\lambda$ and $d\beta$ unchanged, since λ and ϖ increase at the same rate. The precessional displacements of the celestial pole and the equinox, however, do produce changes in the displacements in right ascension and declination ($d\alpha$, $d\delta$). It is these rather than their ecliptic equivalents that are relevant.

In 1976 the IAU decided to cut this Gordian knot and recommend that in future mean places should not include the E-terms. Further, it resolved that annual aberration should be computed rigorously from the earth's velocity relative to the solar system barycentre without any modification. The *Astronomical Almanac* from 1984 onwards gives aberrational coefficients computed in this way.

In chapter 12 we shall discuss in detail the reduction of the apparent place of a star to its mean place. The E-terms, following the new practice, are excluded from that discussion.

8.6   Planetary aberration

The term *planetary aberration* is conveniently used to describe the total aberrational effect on the position of a planet, or indeed any body in the solar system. It includes annual aberration – caused by the earth's velocity – together with the correction arising from the planet's movement during the interval, τ say, between the emission of the observed radiation and its reception. The effect of annual aberration is given by equation (8.16). The light-time correction leads to a similar expression, as we shall now demonstrate.

Suppose that an observation is made of a planet at the instant of time t. Let the points G, E and P – in Fig. 8.2 – represent the barycentre, the earth, and the planet respectively, all at time t. Let $\mathbf{r}$ and $\mathbf{R}$ be the barycentric position vectors of the planet and the earth, and denote the vector $\overrightarrow{EP}$ by $\rho\mathbf{s}$, where $\mathbf{s}$ is a unit vector. This vector gives the geometric direction of the planet at time t. Clearly

$$\mathbf{r} = \rho\mathbf{s} + \mathbf{R}. \tag{8.28}$$

The observed radiation does not emanate from P, but from the point P' where the planet was situated at time $t - \tau$. Denote the vector $\overrightarrow{EP'}$ by $\rho'\mathbf{s}'$. Then the unit vector $\mathbf{s}'$ gives the planet's direction corrected for annual aberration but not light-time. In this respect the notation is compatible with equation (8.16), provided s* is understood as the apparent direction of the planet. To sufficient accuracy we may write the annual aberrational displacement, from (8.16), as

$$\mathbf{s}^* - \mathbf{s}' = -c^{-1}\mathbf{s} \times (\mathbf{s} \times \dot{\mathbf{R}}).\tag{8.29}$$

Now the planet moves from the point P' to P in the time interval τ. So, if its acceleration within this interval is neglected, $\overrightarrow{P'P} = \tau\dot{\mathbf{r}}$. Consequently

$$\rho'\mathbf{s}' = \rho\mathbf{s} - \tau\dot{\mathbf{r}}.\tag{8.30}$$

This is the exact expression for the light-time correction. An explicit formula is derived by taking the vector product twice with $\mathbf{s}$; it is then found that

$$(\mathbf{s} \cdot \mathbf{s}')\mathbf{s} - \mathbf{s}' = -(\tau/\rho')\mathbf{s} \times (\mathbf{s} \times \dot{\mathbf{r}}).$$

Now $\rho' = c\tau$ by definition, and $(\mathbf{s} \cdot \mathbf{s}') \simeq 1$; so correct to first order

$$\mathbf{s}' - \mathbf{s} = c^{-1}\mathbf{s} \times (\mathbf{s} \times \dot{\mathbf{r}}).\tag{8.31}$$

This expression for the light-time correction closely parallels that for annual aberration (8.29). The sum of the two gives the correction for planetary aberration as

$$\mathbf{s}^* - \mathbf{s} = c^{-1}\mathbf{s} \times [\mathbf{s} \times (\dot{\mathbf{r}} - \dot{\mathbf{R}})].\tag{8.32}$$

Notice that the planetary aberrational displacement depends only on the relative velocity of the earth and the planet.

Figure 8.2

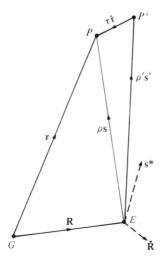

From equation (8.28) one obtains

$$\dot{\mathbf{r}} - \dot{\mathbf{R}} = \dot{\rho}\mathbf{s} + \rho\dot{\mathbf{s}},$$

which may be substituted into (8.32). Remembering that $\mathbf{s} \cdot \dot{\mathbf{s}} = 0$, since $\mathbf{s}$ is a unit vector, and putting $\rho \simeq \rho' = c\tau$, the very simple result emerges that

$$\mathbf{s}^* = \mathbf{s} - \tau\dot{\mathbf{s}}. \tag{8.33}$$

Planetary aberration is, therefore, a particularly easy correction to apply – simpler than either of its two constituent parts. For example, the apparent place of a planet is computed as follows. First the method described in section 7.2 is applied to yield the geocentric ephemeris (α, δ) and the geocentric distance ρ. These coordinates correspond to the vector $\mathbf{s}$. The value of τ is derived with sufficient accuracy as ρ/c. Then the apparent right ascension and declination (α^*, δ^*) are given by

$$\alpha^* = \alpha - \tau \, d\alpha/dt$$
$$\delta^* = \delta - \tau \, d\delta/dt. \tag{8.34}$$

The reverse problem of determining the planet's geometric barycentric place from its apparent place is more involved. It is bound up with the determination of the planet's orbit, as discussed in section 7.7. The main difficulty is that the light-time is not known *a priori*. An estimate must be made for it before one can make the planetary aberration correction. A slightly modified version of (8.33) will give $\mathbf{s}$ in terms of $\mathbf{s}^*$ as

$$\mathbf{s} = \mathbf{s}^* + \tau\dot{\mathbf{s}}^*, \tag{8.35}$$

but an even simpler procedure is to use the apparent coordinates unchanged, but to antedate the observation by an interval τ. The orbit is then computed from at least three observations, and this allows a new value of τ to be derived. The process is then repeated until it converges to a solution for τ.

The discussion of planetary aberration has, like the previous two sections, maintained only first-order accuracy in V/c. The planet's velocity will be of the same order as the earth's, and second-order terms in the planet's velocity were likewise neglected. However, one other approximation was made when it was assumed that any change in the planet's velocity during the interval τ could be neglected. Order of magnitude considerations indicate that the planet's acceleration (μ/r^2) can only produce second-order effects in a time interval τ. For, very roughly, $\tau \sim r/c$ and the planet's velocity $\sim (\mu/r)^{1/2}$. The treatment is, therefore, consistent. A valid second-order theory must be relativistic.

8.7   General relativistic light deflection

Basic formulae that define the apparent direction of a source in general relativity were established in section 3.6. The required procedure is described, in outline at least, on page 79. The first step is to determine the null geodesic that is the world line of the observed radiation. Consequently, we shall examine in this section the null geodesics of Schwarzschild spacetime, which describes the spherically symmetric gravitational field of the sun.

The analysis will, initially at least, closely parallel that given in section 6.8 where ordinary geodesics were considered. In particular, it is again simplest to study the spacetime in the standard form of the metric (6.46), and at a later stage to transform to the more physically significant isotropic form.

The differential equations for null geodesics, equations (3.36), are

$$\frac{d}{dp}\left(g_{\lambda\mu}\frac{dx^{\mu}}{dp}\right) - \frac{1}{2}\frac{\partial g_{\mu\nu}}{\partial x^{\lambda}}\frac{dx^{\mu}}{dp}\frac{dx^{\nu}}{dp} = 0. \tag{8.36}$$

This differs from the corresponding equation for ordinary geodesics (6.45) only in that the interval s is replaced by an affine parameter p. We may, therefore, repeat the analysis of section 6.8 and obtain integrals of the $\lambda = 2$, 3 and 4 equations in the form

$$\theta = \tfrac{1}{2}\pi,$$

$$r^2\frac{d\phi}{dp} = h, \tag{8.37}$$

$$\left(1 - \frac{2m}{r}\right)\frac{dt}{dp} = k.$$

The first of these equations simply means that the plane of the light ray is chosen as the equatorial plane of the coordinate system. The second and third involve the conservation of angular momentum and energy respectively. The affine parameter p, unlike the interval s, is not uniquely defined. There is no loss of generality, therefore, in taking

$$h = cq \tag{8.38}$$

$$k = 1.$$

Then, at large distances from the sun, p tends to the coordinate time. The significance of q is seen by taking the Newtonian approximation, i.e. set $m = 0$, so that $p = t$ and the second of equations (8.37) becomes $rV_T = cq$. The path of the photon is then a straight line that passes at a minimum distance q from the sun.

Instead of using the radial component of (8.36), substitute the results of

(8.37) into the metric to obtain

$$\left(\frac{ds}{dp}\right)^2 = 0 = k^2\left(1-\frac{2m}{r}\right)^{-1} - \frac{1}{c^2}\left[\frac{1}{(1-2m/r)}\left(\frac{dr}{dp}\right)^2 + \frac{h^2}{r^2}\right].$$

By (8.38), this reduces to

$$\frac{1}{c^2}\left(\frac{dr}{dp}\right)^2 = 1 - \frac{q^2}{r^2} + \frac{2mq^2}{r^3}. \tag{8.39}$$

Our main concern at present is to derive the path of the photons. It is convenient, therefore, to eliminate p and use ϕ as the independent variable. Set $u = r^{-1}$. Then by (8.37) and (8.38)

$$\frac{dr}{dp} = -qc\frac{du}{d\phi}, \tag{8.40}$$

and equation (8.39) will give

$$q^2\left(\frac{du}{d\phi}\right)^2 = 1 - q^2 u^2 + 2mq^2 u^3. \tag{8.41}$$

Now it will be sufficient to solve this equation correct to order m/q. The parameter m is about 1.5 km, while q will normally be of the order of an astronomical unit. Even for a grazing light ray or radio wave, q must exceed the solar radius ($\sim 7 \times 10^5$ km). So, in all circumstances, the ratio $m/q \leqslant 2 \times 10^{-6}$, and second-order effects are negligible.

The zero-order solution is easily derived by setting $m = 0$. Then (8.41) yields

$$\phi = \int \frac{q\,du}{(1-q^2 u^2)^{1/2}} = \sin^{-1} qu. \tag{8.42}$$

The constant of integration has been chosen equal to zero so that the ray is parallel to the x-axis (cf. Fig. 8.3), for (8.42) can be written as

$$y = r \sin \phi = \sin \phi / u = q. \tag{8.43}$$

The identification of Cartesian coordinates is a little arbitrary, but it is legitimate in the zero-order approximation.

To derive the first-order solution, let us write

$$qu = \sin \phi + (m/q)f. \tag{8.44}$$

Then substituting this into (8.41) and retaining only first-order terms yields, after a little reduction,

$$\cos \phi \, df/d\phi + f \sin \phi = \sin^3 \phi.$$

Multiplying by the integrating factor $\sec^2 \phi$, gives

$$\frac{d}{d\phi}(\sec \phi f) = \sin \phi \tan^2 \phi = \tan \phi \sec \phi - \sin \phi,$$

which will now integrate to give

$$f = 1 + \cos^2 \phi + A \cos \phi.$$

If the constant of integration A is chosen equal to zero, the ray is symmetrical about the y-axis. The first-order solution for u is now seen from (8.44) to be

$$u = q^{-1} \sin \phi + (m/q^2)(1 + \cos^2 \phi). \tag{8.45}$$

To interpret equation (8.45), notice that u, and therefore r, has a turning point on the y-axis when $\phi = \tfrac{1}{2}\pi$. This corresponds to the ray's closest approach to the sun, S, its perihelion, see Fig. 8.3. The light ray is curved, being concave to the sun. This is shown by the limiting values of ϕ as $r \to \infty$. Setting $u = 0$ in (8.45) gives, to sufficient accuracy,

$$\sin \phi = -2m/q.$$

So the directions of the asymptotes are not $\phi = 0$ and π as previously, but

$$\phi = -2m/q \quad \text{and} \quad \pi + 2m/q. \tag{8.46}$$

The total deflection of the ray $\Delta\phi$ in the sun's gravitational field is

$$\Delta\psi = 4m/q. \tag{8.47}$$

This deflection provides the second classical test of general relativity. Suppose that the source of the radiation is a star in the immediate vicinity of the sun. The parameter q is then a few solar radii, and in comparison both the star and the earth are considered as effectively at an infinite distance

---

Figure 8.3. Gravitational light deflection.

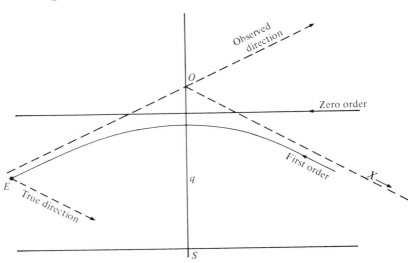

from the sun. They may be regarded as located on the asymptotes of Fig. 8.3. The asymptote EO gives the observed direction of the star, while the other asymptote OX is parallel to the direction in which the star would appear in the absence of any solar gravitational deflection. The angle EOX is equal to $180° - \Delta\psi$.

Stars can only be observed very close to the sun in the special circumstances of a total solar eclipse. The first observations were made of the effect of light deflection at the solar eclipse of 1919, and generally confirmatory results were obtained at that time and in subsequent eclipses. It is seen from equation (8.47) that the deflection is inversely proportional to q. Each star should be deflected from its 'normal' position by an amount proportional to its distance from the centre of the solar disc, the displacement being radially outwards. Star fields surrounding the eclipsed sun are, therefore, photographed, and the relative positions of the stars are measured from the photographic plate. These positions reveal the differential light deflection over the star field. For a ray at grazing incidence, the deflection is obtained by inserting the solar radius for q in (8.47) and is about $1''.75$.

While eclipse observations have established the existence of gravitational light deflection, they have never been able to provide convincing quantitative support for the General Theory of Relativity. Observations, essentially similar in principle, can, however, be made of the relative positions of radio sources near the sun, and an eclipse is no longer necessary. These observations have vastly superior accuracy and have established agreement with general relativity to better than one per cent, making the second test of general relativity as exact as the first. Such precise measurements of radio astrometry have an accuracy of the order of one milli-arc second. A more detailed formulation of the light deflection is required when working to such a high precision. In particular, it is insufficient to treat the earth as at an infinite distance from the sun. So we must develop expressions for the light ray's direction at a general point. The direction of the asymptote is insufficient.

Consider again the equation of the null geodesic (8.45). It is tempting to convert this to Cartesian coordinates by writing $x = r \cos \phi$, $y = r \sin \phi$, but this substitution is only meaningful in the isotropic coordinate system, since otherwise vectors in the x- and y-directions will not be orthogonal. The first step, therefore, is to replace the radial coordinate r with $\tilde{r}$ defined by equation (6.72). This is written with sufficient accuracy as

$$\tilde{r} = r - m. \tag{8.48}$$

Then acceptable quasi-Cartesian coordinates may be introduced as

$$x = \tilde{r} \sin \theta \cos \phi$$
$$y = \tilde{r} \sin \theta \sin \phi \qquad (8.49)$$
$$z = \tilde{r} \cos \theta.$$

The metric for this coordinate system is (3.53); so the non-zero components of the metrical tensor are given to order $m/\tilde{r}$, by

$$g_{11} = g_{22} = g_{33} = -\frac{1}{c^2}\left(1 + \frac{2m}{\tilde{r}}\right),$$
$$g_{44} = 1 - 2m/\tilde{r}. \qquad (8.50)$$

Now, by (8.48), $u = (\tilde{r} + m)^{-1}$. So the equation of the null geodesic (8.45) can now be expressed in the new coordinate system. After some reduction the result is most neatly expressed as

$$y = q - 2m\tilde{r}/q. \qquad (8.51)$$

8.8   The direction of the null geodesic

We require the tangent vector to the null geodesic. Since the ray is moving primarily in the negative x-direction, a suitable curve parameter is $-x/c$. The tangent vector is then given by

$$u^i = (-c, \ -c \ dy/dx, \ 0, \ -c \ dt/dx). \qquad (8.52)$$

It is found from (8.51) that

$$\frac{dy}{dx} = -\frac{2m}{q}\frac{d}{dx}(x^2 + y^2)^{1/2} \approx -\frac{2mx}{q\tilde{r}}.$$

The final component of the vector is obtained from the metric, remembering that the geodesic is null. Consequently (8.50) gives

$$\left(1 - \frac{2m}{\tilde{r}}\right)\left(\frac{dt}{dx}\right)^2 = \frac{1}{c^2}\left(1 + \frac{2m}{\tilde{r}}\right)\left(1 + O\left(\frac{m^2}{q^2}\right)\right).$$

Hence, to the accuracy that we are working, the required tangent vector is

$$u^i = \left(-c, \ \frac{2mcx}{q\tilde{r}}, \ 0, \ 1 + \frac{2m}{\tilde{r}}\right). \qquad (8.53)$$

This equation applies at any point of the ray path; it is of particular interest at the point of observation.

Equation (8.53) has been derived in a special coordinate system, with the coordinates axes chosen to minimize algebraic complexity. To be of general application, however, it should be expressed in conventional vector form. Now both the emission of the radiation and its observation are events in the spacetime. Observation hypothetically occurs at the earth's centre (X, Y, Z) at time T. So it is the event $(\mathbf{R}, T)$, where 3-vector notation is used as a shorthand. Similarly, the event of emission is represented by $(\mathbf{r}_e, T - \tau)$, where τ is the coordinate time of light-travel. It follows from (8.51) that in our special coordinate system

$$\mathbf{R} = (X, q - 2mR/q, 0)$$
$$\mathbf{r}_e = (x_e, q - 2mr_e/q, 0). \tag{8.54}$$

The method of analysis depends upon whether the source of the radiation is located within the solar system or not. The two cases are treated separately below.

(a)   *Body within the solar system*
        This case is the more complicated since the light deflection depends on the position of the source as well as that of the observer. We shall relate the tangent vector u^{λ} to the geocentric direction of the source, that is, the unit vector defined by

$$\rho \mathbf{s}_0 = \mathbf{r}_e - \mathbf{R}. \tag{8.55}$$

This is a formal definition of a 3-vector linking the source's position at emission to the observer's at the instant of observation. It is not a geometric direction, since no allowance for light-time is included. The quantity ρ is not strictly the physical distance between source and observer. It is seen from (8.54) that $\rho \approx x_e - X$, and so $\mathbf{s}_0$ has components given by

$$\mathbf{s}_0 = (1, \, 2m(R - r_e)/q\rho, \, 0). \tag{8.56}$$

The direction of the light-ray at the observer differs from this by a small amount. Let us, therefore, write

$$u^{\lambda} = [-c(\mathbf{s}_0 + \boldsymbol{\sigma}_0), \, 1 + 2m/R], \tag{8.57}$$

where, by equation (8.53),

$$\boldsymbol{\sigma}_0 = \left\{ 0, \, -\frac{2m}{q}\left(\frac{X}{R} + \frac{(R - r_e)}{\rho}\right), \, 0 \right\}. \tag{8.58}$$

It is necessary to express this vector $\boldsymbol{\sigma}_0$ in standard vector notation. Since it is in the y-direction, consider the essentially parallel vector $\mathbf{s}_0 \times (\mathbf{s}_0 \times \mathbf{R})$. To sufficient accuracy we find

$$\mathbf{s}_0 \times (\mathbf{s}_0 \times \mathbf{R}) = (0, \, -q, \, 0),$$
and so
$$q^2 = R^2 - (\mathbf{s}_0 \cdot \mathbf{R})^2. \tag{8.59}$$

Substituting these results into (8.58) allows $\boldsymbol{\sigma}_0$ to be expressed as

$$\boldsymbol{\sigma}_0 = \frac{2m}{(R^2 - (\mathbf{R} \cdot \mathbf{s}_0)^2)}\left(\frac{R - r_e}{\rho} + \frac{\mathbf{R} \cdot \mathbf{s}_0}{R}\right)\mathbf{s}_0 \times (\mathbf{s}_0 \times \mathbf{R}). \tag{8.60}$$

This rather complicated expression gives the light deflection quite generally in terms of the quasi-geometric direction of the source $\mathbf{s}_0$ and the position vectors of the source and the observer. Moreover, the tangent vector to the geodesic has, by (8.57), a similar general expression. Following the method of section 8.9, u^λ may be used to derive the apparent place of the source. The method of that section strictly applies to a star rather than to a body within the solar system, but the adaptations required are minimal.

(b)   *Source beyond the solar system*

This case is simpler, since the fact that $q < R \ll r_e$ allows considerable approximation. It is now convenient to relate the tangent vector to the heliocentric direction of the source. This is defined as **s**, where $\mathbf{r}_e = r_e \mathbf{s}$. It is also convenient to introduce the star's parallax π, which is equal to r_e^{-1}. Then, to first order in both m and π, equation (8.54) gives

$$\mathbf{s} = (1,\ q\pi - 2m/q,\ 0). \tag{8.61}$$

The null geodesic's tangent vector at the observer can now be written as

$$u^\lambda = [-c(\mathbf{s}+\boldsymbol{\sigma}),\ 1 + 2m/R]. \tag{8.62}$$

The vector $\boldsymbol{\sigma}$ is now the small correction due to light deflection and stellar parallax. From (8.53) and (8.61), it follows that

$$\boldsymbol{\sigma} = \left[0,\ \frac{2m}{q}\left(1 - \frac{X}{R}\right) - q\pi,\ 0\right].$$

Using the results already established in (8.59), it is found, after a little reduction that

$$\boldsymbol{\sigma} = \left\{\pi - \frac{2m}{R[R + (\mathbf{R}\cdot\mathbf{s})]}\right\}\mathbf{s} \times (\mathbf{s} \times \mathbf{R}). \tag{8.63}$$

This may be substituted into (8.62) to give a general expression for u^λ applicable in any isotropic coordinate system. This is an essential prerequisite for the rigorous determination of the star's apparent place according to general relativity.

Equation (8.63) indicates that the parallax displacement and the light deflection are in opposite directions; parallax displaces the star towards the sun, light deflection away from it. The parallax will dominate for nearby stars, but, for many stars, and particularly for radio sources, light deflection will be more important. It was shown in equation (8.9) that the parallax displacement is proportional to $\sin E$, E being the angular elongation from the sun. Reduction of (8.63) to scalar form shows that the light deflection is proportional to $\cot \frac{1}{2}E$. This factor may be large, as it is for eclipse observations and for the solar occultation of radio sources.

8.9   Relativistic derivation of the apparent place of a star

The analysis of sections 8.7 and 8.8 allows the completion of stage

(i) of the procedure laid down in section 3.6 (p. 79). The tangent vector u^i to the null geodesic at the observer is now known. Stage (ii) is to define unit vectors $I^\lambda, J^\lambda, K^\lambda, Q^\lambda$. From equation (3.58), correct to order m,

$$I^\lambda = [c(1 - m/R), 0, 0, 0], \tag{8.64}$$

with similar expressions for J^λ and K^λ. Moreover, to the same order,

$$Q^\lambda = (0, 0, 0, 1 + m/R). \tag{8.65}$$

This is the tangent vector to the world line of a stationary observer situated at the point (X, Y, Z), while $I^\lambda, J^\lambda, K^\lambda$ are the three coordinate directions for this observer.

Stage (iii) is to derive the equivalents for the geocentric observer. The tangent vector to this observer's world line may be written as

$$q^\lambda = \frac{dt}{ds} (\mathbf{V}, 1), \tag{8.66}$$

where $\mathbf{V}$ is the *coordinate velocity* of the observer, that is, $V_x = dX/dt$, etc. This is the most convenient and direct representation of q^λ. The vector $\mathbf{V}$ is not exactly the same as the 'physical velocity' which is normally defined as the coordinate velocity in the local Lorentz frame. They differ, however, only by a small amount of order Vm/R which is not significant in astrometry.

Since q^λ is a unit vector, dt/ds is determined from the metric to sufficient accuracy as

$$\frac{dt}{ds} = 1 + \frac{m}{R} + \frac{1}{2} \frac{V^2}{c^2}. \tag{8.67}$$

The unit tangent vector q^λ is now expressed in terms of the earth's position and velocity vectors. The last two terms on the right-hand side of (8.67) are of the same order of magnitude ($\sim 10^{-8}$). Higher-order terms are neglected, without acknowledgement, in the subsequent analysis.

The unit vector q^λ is now used to define unit vectors $i^\lambda, j^\lambda, k^\lambda$ which are in the directions of the observer's coordinate axes. The relevant equation to be applied is (3.65). After, admittedly, considerable reduction, it is found that

$$i^\lambda = \left[c\left(1 - \frac{m}{R}\right) + \frac{V_x^2}{2c}, \frac{V_x V_y}{2c}, \frac{V_x V_z}{2c}, \frac{V_x}{c} \right],$$

$$j^\lambda = \left[\frac{V_y V_x}{2c}, c\left(1 - \frac{m}{R}\right) + \frac{V_y^2}{2c}, \frac{V_y V_z}{2c}, \frac{V_y}{c} \right], \tag{8.68}$$

$$k^\lambda = \left[\frac{V_z V_x}{2c}, \frac{V_z V_y}{2c}, c\left(1 - \frac{m}{R}\right) + \frac{V_z^2}{2c}, \frac{V_z}{c} \right].$$

Stage (iv) of the procedure requires the implementation of equations (3.57) and (3.45). Then the apparent place of the source is given by the unit

3-vector $\mathbf{s}^*$ where

$$\mathbf{s}^* = (u^\lambda q_\lambda)^{-1}(u^\lambda i_\lambda, u^\lambda j_\lambda, u^\lambda k_\lambda). \tag{8.69}$$

The null geodesic's tangent vector is given by equation (8.62), in which $\boldsymbol{\sigma}$ is a small vector of order m. So, making use of (8.66) and (8.68), yields the result

$$\mathbf{s}^* = \frac{\left(1 + \dfrac{m}{R}\right)\mathbf{s} + \boldsymbol{\sigma} + \dfrac{\mathbf{V}}{c} + \dfrac{1}{2c^2}(\mathbf{V} \cdot \mathbf{s})\mathbf{V}}{1 + \dfrac{m}{R} + \dfrac{(\mathbf{V} \cdot \mathbf{s})}{c} + \dfrac{1}{2}\dfrac{V^2}{c^2}}. \tag{8.70}$$

Finally, expanding the denominator by the binomial theorem and substituting for $\boldsymbol{\sigma}$ from (8.63) gives the star's apparent place as

$$\mathbf{s}^* = \mathbf{s} + \left[\pi - \frac{2m}{R(R + \mathbf{R} \cdot \mathbf{s})}\right]\mathbf{s} \times (\mathbf{s} \times \mathbf{R})$$
$$- \frac{1}{c}\left(1 - \frac{\mathbf{V} \cdot \mathbf{s}}{c}\right)\mathbf{s} \times (\mathbf{s} \times \mathbf{V}) + \frac{1}{2c^2}\mathbf{V} \times (\mathbf{V} \times \mathbf{s}). \tag{8.71}$$

By studying the right-hand side of this equation we can recognize the significance of the different terms as follows: the vector $\mathbf{s}$ is the heliocentric place, the term in π is the annual parallax, the term in m is the light deflection, the first-order term in V/c is classical annual aberration, and the remaining terms are the second-order aberrational terms as given by special relativity (cf. equation (3.34)).

Inspection of equation (8.71) reveals that the effects of parallax, light deflection and aberration separate out from each other, although they are not separate concepts in a rigorous formulation within general relativity. The result is not surprising. In the analysis of this and the previous section the displacement from heliocentric to apparent place was expanded in a power series in the small quantities π, m/R and V/c. The largest possible cross-terms in this expansion would be of order $\pi V/c$ and mV/Rc, both of which were neglected in the treatment. It is concluded, therefore, that, to the accuracy we are working, a special relativistic treatment of aberration would be adequate.

8.10   Summary

This chapter has discussed the displacement of a celestial object from its heliocentric position to its apparent geocentric position. The formulation was entirely classical in the first six sections of the chapter, and it proved possible to derive explicit formulae for the changes in right ascension and declination $(d\alpha, d\delta)$. When relativistic effects are included, the formulae become much more complicated, and, while expressions for $d\alpha$ and $d\delta$ are not entirely out of the question, the vector formulation is much neater. The procedure for calculating the changes in right ascension and declination would be to convert the heliocentric place to a unit vector, apply the corrections of equations (8.71) to derive the unit vector corresponding to the apparent place, and finally convert back into right ascension and declination to obtain the new spherical coordinates.

Certain refinements were included in the classical treatment that have been omitted from the general relativistic analysis. These involved the distinction between the geocentric point and the earth/moon barycentre and also that between the heliocentric point and the barycentre of the solar system. They are small corrections for which second-order effects, either in parallax or aberration, are entirely negligible. They are, therefore, incorporated into the general relativistic analysis by modifying the position and velocity vectors of the earth, $\mathbf{R}$ and $\mathbf{V}$. Consequently, in equation (8.71), $\mathbf{R}$ and $\mathbf{V}$ become respectively the earth's isotropic coordinates and coordinate velocity components with the addition of barycentric corrections derived from classical considerations. All first- and second-order effects are then adequately covered.

The discussion of section 8.9 applied to a star, or certainly a body outside the solar system. For a body within the solar system, the procedure must be slightly different. Referring to case (a) in section 8.8, we replace $\mathbf{s}$ by $\mathbf{s}_0$ and $\boldsymbol{\sigma}$ by $\boldsymbol{\sigma}_0$. Hence, from equation (8.60), we derive the apparent place of a body within the solar system as

$$\mathbf{s}^* = \mathbf{s}_0 + \frac{2m}{[R^2 - (\mathbf{R} \cdot \mathbf{s}_0)^2]} \left[\frac{\mathbf{R} \cdot \mathbf{s}_0}{R} - \frac{(r_e - R)}{\rho} \right] \mathbf{s}_0 \times (\mathbf{s}_0 \times \mathbf{R})$$
$$- \frac{1}{c} \left(1 - \frac{\mathbf{V} \cdot \mathbf{s}_0}{c} \right) \mathbf{s}_0 \times (\mathbf{s}_0 \times \mathbf{V}) + \frac{1}{2c^2} \mathbf{V} \times (\mathbf{V} \times \mathbf{s}). \tag{8.72}$$

Note that in this formula $\mathbf{s}_0$ is the quasi-geometric geocentric direction and so no correction for parallax is included. The unit vector $\mathbf{s}_0$ must be related to the heliocentric direction through the vector equation (8.55). This displacement will not, in general, be small.

Both equations (8.71) and (8.72) express the apparent place in terms of a quasi-geometric place. These formulae can, however, be inverted comparatively easily. For example, the heliocentric place of a star is given, with sufficient accuracy, in terms of the apparent place by the formula

$$\mathbf{s} = \mathbf{s}^* - \left[\pi - \frac{2m}{R(R + \mathbf{R} \cdot \mathbf{s}^*)} \right] \mathbf{s}^* \times (\mathbf{s}^* \times \mathbf{R})$$
$$+ \frac{1}{c} \left(1 + \frac{\mathbf{V} \cdot \mathbf{s}^*}{c} \right) \mathbf{s}^* \times (\mathbf{s}^* \times \mathbf{V}) + \frac{1}{2c^2} \mathbf{V} \times (\mathbf{V} \times \mathbf{s}^*). \tag{8.73}$$

Equation (8.72) may be similarly inverted.

It should be borne in mind that the complicated relativistic formulae are only necessary in high-precision work. Frequently, the much simpler classical treatment is quite adequate.

Problems

8.1 A star has ecliptic longitude 270° and ecliptic latitude 45°. The total variation in the star's longitude due to annual parallax is 0″.8. What is the corresponding variation in the star's latitude? Determine the times of year at which maximum latitude and longitude occur and calculate the star's distance, on the assumption that the earth's orbit is a circle.

8.2 Prove that there are just two points on the celestial sphere for which the effect of annual aberration vanishes. Show that their equatorial coordinates are approximately given by

$$\alpha = -\tan^{-1} (\cos \varepsilon \cot \odot)$$

$$\delta = \pm \sin^{-1} (\sin \varepsilon \cos \odot)$$

where $\odot$ denotes the sun's true longitude.

8.3 Prove that the effect of annual parallax on a star's right ascension is greatest when the sun's longitude is approximately given by

$$\odot = 90° + \tan^{-1} (\tan \alpha \sec \varepsilon).$$

8.4. Prove that, in the isotropic form of the metric of Schwarzschild spacetime, the coordinate velocity of light is $c(1 - m^2/4r^2)$.

8.5 An observer's barycentric velocity V is directed towards the point O on the celestial sphere. Show that aberration will displace a star X along the great circle OX to a point X'. Prove that this effect is given exactly in special relativity by the formula

$$\tan \theta' = \frac{\tan \theta}{\gamma(1 + V \sec \theta/c)},$$

where $OX = \theta$, $OX' = \theta'$ and $\gamma = (1 - V^2/c^2)^{-1/2}$. Show that, for small velocities, this approximates to the classical formula

$$d\theta = -\frac{V}{c} \sin \theta.$$

8.6 Prove that a light ray in Schwarzschild spacetime characterized by the parameter q makes an angle ψ with the radial direction where

$$\sin \psi = \frac{q}{r}\left(1 - \frac{2m}{r}\right).$$

8.7 Prove that if R is the coordinate radius of the sun in the isotropic form of the metric, then the angular diameter D of the sun, observed from coordinate distance r, is given by

$$D = 2 \sin^{-1} \left[\frac{R}{r} \left(1 + \frac{2m}{R} - \frac{2m}{r} \right) \right].$$

8.8 Derive the expression in general relativity for the apparent direction of a body in the solar system in terms of its quasi-geometric direction $\mathbf{s}_0$, defined by

$$\rho \mathbf{s}_0 = \mathbf{r} - \mathbf{R}$$

where $\mathbf{r}$ and $\mathbf{R}$ are the position vectors of the body and observer respectively. Verify that your result is equivalent to equation (8.72).

8.9 Invert the result of the previous question to derive the quasi-geometric direction $\mathbf{s}_0$ of a body within the solar system in terms of its apparent direction $\mathbf{s}^*$.

9

Precession and nutation

9.1 Introduction

Detailed formulae are developed in this chapter for the changes in a star's coordinates that result from precession and nutation. These changes are not due to any movement on the celestial sphere of the star itself, but rather to the displacement of the celestial pole and the pole of the ecliptic which together define the coordinate system. A qualitative discussion of precession and nutation has already been given in section 3.2.

The discrimination between precession and nutation is arbitrary and is made purely for convenience in positional work. Long-term regular changes are called precession, while nutation is the short-term periodic displacement of the celestial pole about its mean position. The main part of the chapter discusses precessional effects, and consideration of nutation is deferred until section 9.8.

The annual rates of precession will first be established in the next three sections. They are sufficient to allow the determination of coordinate changes over short intervals of time, preferably less than one year. A more exact treatment is required when precessional changes over many years are being studied. So rigorous formulae are developed in sections 9.5 and 9.6. Thereafter approximate formulae are developed for application over long periods of time. These are sufficient for many purposes, being effectively second-order approximations.

It is assumed throughout this chapter that the stars themselves are fixed on the celestial sphere and that their coordinates only vary due to the precessional and nutational changes in the coordinate system.

9.2 Luni–solar precession

When treating luni–solar precession as a separate effect, only the mean motion of the celestial pole is included. The ecliptic is regarded as fixed. In Fig. 9.1, let P, K and Υ be respectively the mean celestial pole, the pole of the ecliptic and the mean equinox for some epoch t. Then $KP = \varepsilon$. A star at the point X will have equatorial coordinates (α, δ), where $PX = 90° - \delta$ and $KPX = 90° + \alpha$. Moreover, the star has ecliptic coordinates (λ, β), where $KX = 90° - \beta$ and $PKX = 90° - \lambda$.

The mean torque exerted on the earth by the moon and the sun is responsible for luni–solar precession. For reasons of symmetry the direction of this *mean* torque must lie both in the ecliptic and equatorial planes. It is, therefore, directed towards Υ. Now the instantaneous direction of the earth's angular momentum vector defines P. Consequently, the mean torque causes P to be displaced in a direction which is always perpendicular to the arc KP.

Since luni–solar precession is the dominant precessional effect, we can give an approximate description of precession by stating that the celestial pole describes a small circle $PP'R$ about the pole of the ecliptic. One circuit is completed in about 26 000 years. The annual rate of movement round this small circle ($\sim 50''$) is denoted by ψ. The above description is approximate because the point K is not itself fixed, due to the planetary precession considered in the next section. The approximation will hold good, however, over a short interval of time.

Figure 9.1. Luni–solar precession.

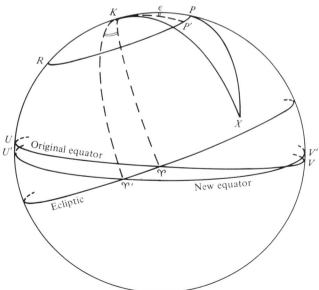

Suppose now that P' is the position of the celestial pole at time $t + \tau$. Then the angle $PKP' = \psi\tau$. The new equator is shown in Fig. 9.1 as $U'\Upsilon'V'$, the point Υ' where it intersects the ecliptic being the new equinox. Since $PK\Upsilon = P'K'\Upsilon' = 90°$, it follows that the arc $\Upsilon\Upsilon' = \psi\tau$. The equinox is, therefore, moving backwards round the ecliptic at a uniform rate ψ.

The coordinate changes for the star X are particularly simple in ecliptic coordinates. Since K is unmoved, there is no change in ecliptic latitude at all. The reference point for ecliptic longitude has been displaced along the ecliptic by $\psi\tau$, and so λ is increased by this amount. The changes in the ecliptic coordinates due to luni–solar precession are, therefore,

$$d\lambda = \psi\tau, \tag{9.1}$$

$$d\beta = 0.$$

The changes in the equatorial coordinates are more complicated. They are derived by considering the spherical triangle PKX, the parts of which have already been identified. We deduce from the cosine formula that

$$\sin\delta = \cos\varepsilon\sin\beta + \sin\varepsilon\cos\beta\sin\lambda,$$

in which only δ and λ vary as P is displaced to P'. Differentiation, therefore, yields

$$\cos\delta\,d\delta = \sin\varepsilon\cos\beta\cos\lambda\,d\lambda.$$

This may be simplified by applying the sine formula, which, curiously enough, gives

$$\cos\beta\cos\lambda = \cos\delta\cos\alpha. \tag{9.2}$$

The change in declination is then derived from (9.1) as

$$d\delta = \psi\tau\sin\varepsilon\cos\alpha. \tag{9.3}$$

The change in right ascension is obtained by differentiating equation (9.2). Using (9.1) and (9.3), this gives

$$\cos\delta\sin\alpha\,d\alpha = \psi\tau(\cos\beta\sin\lambda - \sin\varepsilon\sin\delta\cos^2\alpha).$$

The ecliptic coordinates on the right-hand side are eliminated by applying the analogue formula to the first term. The equation then reduces to

$$d\alpha = \psi\tau(\cos\varepsilon + \sin\varepsilon\sin\alpha\tan\delta). \tag{9.4}$$

Equations (9.3) and (9.4) are usually only applied over short intervals of time of the order of a year or less. They are first-order formulae derived from the geometrical representation of Fig. 9.1 which is itself approximate.

The annual rate of luni–solar precession is understood principally in terms of Newtonian dynamics with a small relativistic correction ($\sim 0''.02$) which is known as geodesic precession. It is found in general relativity that the inertial frame of reference in the neighbourhood of the orbiting earth has a slight rotation with respect to the heliocentric inertial frame. This is incorporated in the calculation of ψ. The value of ψ depends on a number of parameters, the dynamical figure of the earth, the obliquity of the ecliptic, the masses of the moon and the sun and their orbital elements. It is, in fact, directly proportional to cos ε, and, since the obliquity of the ecliptic has a slow secular change due to planetary precession, ψ is not strictly constant. Its value is given by

$$\psi = 50''.3878 + 0''.0049\, T, \tag{9.5}$$

where T is the time in centuries from the fundamental epoch J2000.0, i.e. $T = (t - 2000)/100$.

9.3   Planetary precession

The planets have a negligible influence on the direction of the earth's rotational axis. Planetary perturbations do, however, affect the earth's orbit round the sun. As explained in chapter 7, this means that the earth's orbital elements are subject to gradual change and, in particular, that the plane of the orbit, the ecliptic, is not fixed. The ecliptic is defined as the *mean* orbital plane of the earth/moon barycentre. As such it is not subject to any short-term periodic variations, but only to secular changes. The consequent variation in the coordinate system may, therefore, be described as a precession – *planetary precession*. There are, by definition, no nutational terms.

In this section, the celestial pole and the equator will be regarded as fixed. Let K and Υ (Fig. 9.2) be respectively the pole of the ecliptic and the equinox at some initial epoch, and let K' and Υ' be their positions a short time τ later. The original ecliptic is the great circle $U \Upsilon V$, and the new ecliptic is $U' \Upsilon' V'$. These two ecliptics will intersect in two diametrically opposite points, N and N', one of which is shown in Fig. 9.2.

The motion of the ecliptic may be described as a slow rotation of the reference plane about the axis NN'. The rate of this rotation ($\sim 0''.5$ per year) will be denoted by π. It then follows that the angle $\Upsilon N \Upsilon' = \pi\tau$. The position of the rotation axis is specified by its longitude (on the original ecliptic) and is usually denoted by Π. Then $\Upsilon N = \Pi$. It is fairly easy to see that N and N' are the poles of the arc KK' that the pole of the ecliptic describes on the celestial sphere.

The effects of planetary precession on the equatorial coordinates (α, δ) of a star X are particularly simple. Since the celestial pole is unaffected, there can be no change in the star's declination. Further, the equinox is displaced along the equator by the arc $\Upsilon \Upsilon'$. This will be denoted by $\lambda'\tau$, where λ' is known as the *annual rate of planetary precession*. Since the equinox is the reference for right ascension, it is clear that the right ascension of any star will be decreased by $\lambda'\tau$. Formally, therefore, the changes in a star's equatorial coordinates, due to planetary precession, are given by

$$d\alpha = -\lambda'\tau,$$
$$d\delta = 0. \tag{9.6}$$

The parameter λ' may be determined from the spherical triangle $\Upsilon\,\Upsilon'N$. Parts may be identified in this triangle as follows: As already noted, $\Upsilon N = \Pi$, $\Upsilon\,\Upsilon' = \lambda'\tau$, and $\Upsilon N\,\Upsilon' = \pi\tau$. Moreover, since the obliquity of the ecliptic is the angle between the equator and the ecliptic, it follows that $N\,\Upsilon\,\Upsilon' = \varepsilon$, while $\Upsilon\,\Upsilon'N = 180° - (\varepsilon + d\varepsilon)$. Notice that planetary precession causes a change in the obliquity of the ecliptic.

Applying the sine formula to triangle $\Upsilon\,\Upsilon'N$ yields

$$\sin\Pi\,\sin\pi\tau = \sin\lambda'\tau\,\sin(\varepsilon + d\varepsilon).$$

When τ is small, this gives, to sufficient accuracy,

$$\lambda' = \pi\,\sin\Pi\,\text{cosec}\,\varepsilon. \tag{9.7}$$

The change in the obliquity of the ecliptic is likewise derived from this triangle. From the four-parts formula

$$\cos\varepsilon\,\cos\lambda'\tau = \sin\lambda'\tau\,\cot\Pi + \sin\varepsilon\,\cot(\varepsilon + d\varepsilon),$$

Figure 9.2. Planetary precession.

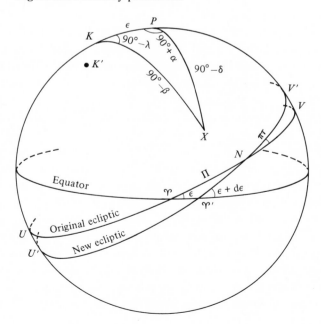

or

$$\sin (\varepsilon + d\varepsilon) \cos \varepsilon \cos \lambda'\tau - \cos (\varepsilon + d\varepsilon) \sin \varepsilon$$
$$= \sin \lambda'\tau \cot \Pi \sin (\varepsilon + d\varepsilon).$$

Making the usual small-angle approximations, this reduces, by (9.7), to

$$d\varepsilon = \pi\tau \cos \Pi. \tag{9.8}$$

The changes in a star's ecliptic coordinates are more complicated than those in their equatorial counterparts. The most direct way of obtaining them is to consider the spherical triangle KPX. The same triangle was encountered in section 9.2 and its parts were identified; they are labelled accordingly in Fig. 9.2. The changes $d\alpha$ and $d\varepsilon$ have already been established. By the cosine formula

$$\sin \beta = \cos \varepsilon \sin \delta - \sin \varepsilon \cos \delta \sin \alpha.$$

Taking differentials of this yields

$$\cos \beta \, d\beta = - (\sin \delta \sin \varepsilon + \cos \delta \cos \varepsilon \sin \alpha) \, d\varepsilon$$
$$- \sin \varepsilon \cos \delta \cos \alpha \, d\alpha.$$

The equatorial coordinates must be removed from the right-hand side by applying an analogue formula to the first term and the sine formula (9.2) to the second. This leads to the result

$$d\beta = \pi\tau \sin (\Pi - \lambda). \tag{9.9}$$

The effect on longitude may be discerned by differentiating equation (9.2) to obtain

$$\cos \beta \sin \lambda \, d\lambda = \cos \delta \sin \alpha \, d\alpha - \sin \beta \cos \lambda \, d\beta.$$

The factor $\cos \delta \sin \alpha$ is eliminated by another application of the analogue formula. Then, after a little simplification, it is found that

$$d\lambda = \pi\tau[\tan \beta \cos (\Pi - \lambda) - \sin \Pi \cot \varepsilon].$$

The effect of planetary precession on a star's ecliptic coordinates is, therefore, established as

$$d\lambda = - \lambda'\tau \cos \varepsilon + \pi\tau \tan \beta \cos (\Pi - \lambda),$$
$$d\beta = \pi\tau \sin (\Pi - \lambda). \tag{9.10}$$

Figure 9.2 is slightly misleading in one respect. The longitude of the rotation axis has, for purposes of illustration, been drawn as an acute angle. In fact, $\Pi \simeq 175°$. The point N is in reality close to the autumnal equinox on the reverse side of the diagram, and the arc KK' is close to the meridian of ecliptic longitude KP. These differences, however, in no way affect the formulae that have been deduced. Note, however, that λ' is positive as drawn, but the obliquity of the ecliptic is, in fact, decreasing with time.

The annual rate λ' of planetary precession has been expressed in terms of the two parameters π and Π. The latter are derived from the planetary perturbations that are studied by the methods of celestial mechanics. Neither of the parameters is strictly constant. The rotation rate π has a very small secular variation. The longitude Π, in addition to an intrinsic secular displacement, also has a precessional variation, due to the movement of the equinox. This parameter requires a rather more precise definition than that given earlier. This may be taken as the longitude (referred to the mean equinox of date) of the ascending node of the instantaneous ecliptic on its immediately preceding position.

The numerical values of the two parameters under discussion are

$$\Pi = 174°.8764 + 0°.9137\ T$$
$$\pi = 0''.4700 - 0''.0007\ T, \tag{9.11}$$

where T is the time in centuries from J2000.0. The annual rate of planetary precession and the rate of change of the obliquity of the ecliptic may be calculated from these values by using equations (9.7) and (9.8). It is found that

$$\lambda' = 0''.1055 - 0''.0189\ T$$
$$\varepsilon = 23°\ 26'\ 21''.45 - 46''.81\ T. \tag{9.12}$$

The data considered so far and the methods that have been employed are not sufficiently accurate to give any higher-order terms for ε. These must be derived by the rigorous methods of section 9.5.

9.4   General precession

The treatment of planetary precession given in the last section will appear somewhat artificial. The equator was treated as fixed, ignoring the fact that its displacement due to luni–solar precession is considerably greater than the displacement of the ecliptic that was being investigated. The treatment does, however, give useful results. *General precession*, which is the total effect arising from the combined movement of the equator and the ecliptic, may be considered as the superposition of luni–solar precession and planetary precession. Provided that the interval τ over which the displacement is being considered is sufficiently small, the principle of superposition will be valid. The effects of general precession are then derived simply by adding together the results derived separately in sections 9.2 and 9.3.

Let us consider first the general precessional changes in a star's equatorial coordinates (α, δ). Add the displacements of equations (9.3) and (9.4) to those given in (9.6) to derive

$$d\alpha = m\tau + n\tau \sin \alpha \tan \delta,$$
$$d\delta = n\tau \cos \alpha, \tag{9.13}$$

where we have introduced two new precessional constants, namely

$$m = \psi \cos \varepsilon - \lambda',$$
$$n = \psi \sin \varepsilon. \tag{9.14}$$

The constants m and n are referred to as the annual precessions in right ascension and declination respectively.

The ecliptic coordinates (λ, β) are treated similarly. Combining the results of equations (9.1) and (9.10) gives

$$d\lambda = p\tau + \pi\tau \tan \beta \cos (\Pi - \lambda),$$
$$d\beta = \pi\tau \sin (\Pi - \lambda), \tag{9.15}$$

where p is known as the annual general precession (in longitude). It is defined by the equation

$$p = \psi - \lambda' \cos \varepsilon. \tag{9.16}$$

Once again it is found that the annual rates m, n, and p are not entirely constant, but have slow secular variations. From their definitions and from the numerical values already quoted in equations (9.5), (9.11) and (9.12), it may be established that

$$p = 50''.2910 + 0''.0222\ T, \tag{9.17}$$

while

$$m = 3^s.074\,96 + 0^s.001\,86\ T$$
$$n = 1^s.336\,21 - 0^s.000\,57\ T \tag{9.18}$$
$$= 20''.0431 - 0''.0085\ T.$$

Frequently, the parameter n is quoted both in time and arc seconds, as above, since it is required for both right ascension and declination computations. On the other hand, m is required only for right ascension and p only for longitude; so the appropriate units are used.

In addition to affecting the coordinates of a celestial object, general precession produces changes in the orbital elements of a heliocentric orbit. The elements involved are the orbital inclination i, the longitude of the ascending node Ω, and the argument of perihelion ω. They are shown in Fig. 9.3 respectively as the angle ALN and the arcs ΥL and LA. In this diagram, ΥLN represents the ecliptic at some initial epoch, while $\Upsilon'LN$ is its position a short time τ (years) later. The great circle $L'LA$ is, of course, the orbital plane, and the point A represents the planet's position at perihelion.

It will be immediately recognized that $LL' = d\omega$, the increment in the argument of perihelion. Moreover, the spherical angle $LL'N = i + di$, the new orbital inclination. Now consider the node N which is common to both ecliptics. Referred to the original equinox Υ, its longitude is Π. Hence $LN = \Pi - \Omega$. Its longitude referred to the new equinox Υ' follows from (9.15). Since $\beta = 0$, $d\lambda = p\tau$. We, therefore, conclude that

$$L'N = \Pi - \Omega + p\tau - d\Omega.$$

The remaining two parts of spherical triangle $LL'N$ are readily seen to be $LNL' = \pi\tau$, and $L'LN = 180° - i$. For clarity, an expanded view of this triangle is shown in Fig. 9.4.

Now let us perform the following construction. Draw a small circle arc with pole N through the point L to intersect NL' in a point Q. Then $NL = NQ = \Pi - \Omega$, and so $L'Q = p\tau - d\Omega$. Moreover, the small circle arc LQ is of

---

Figure 9.3. Precessional changes in the orbital elements.

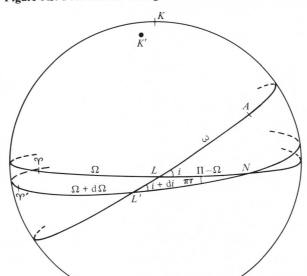

Figure 9.4. Expanded view of Fig. 9.3.

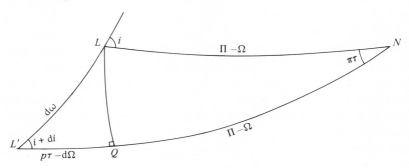

length $\pi\tau \sin (\Pi - \Omega)$. Now the small figure LQL' may be treated as a plane triangle right-angled at Q. It is then seen that

$$d\omega = LQ \operatorname{cosec} (i + di),$$

$$p\tau - d\Omega = LQ \cot (i + di).$$

Substituting for the small circle arc, we obtain, with sufficient accuracy, the two results

$$d\omega = \pi\tau \operatorname{cosec} i \sin (\Pi - \Omega),$$

$$d\Omega = p\tau - \pi\tau \cot i \sin (\Pi - \Omega). \tag{9.19}$$

The increment in the orbital inclination is obtained by applying the four-parts formula to spherical triangle $L'LN$. Then

$$-\cos d\omega \cos i = \sin d\omega \cot (\Pi - \Omega) - \sin i \cot (i + di),$$

i.e.

$$\sin i \cos (i + di) - \cos d\omega \cos i \sin (i + di)$$

$$= \sin d\omega \cot (\Pi - \Omega) \sin (i + di).$$

Since di and $d\omega$ are small quantities, this gives

$$\sin di + O(d\omega^2) = -d\omega \sin i \cot (\Pi - \Omega) + O(d\omega\, di).$$

Using the first result of (9.19) will yield the first-order result

$$di = -\pi\tau \cos (\Pi - \Omega). \tag{9.20}$$

It is worth reiterating that the formulae for general precession derived in this section are primarily intended for use over short intervals of time, normally less than one year. If used over longer intervals, precision must be sacrificed. The rigorous formulae developed in the next section will avoid this. In some ways they are not so flexible, but they are very powerful when used for the purpose they are intended, namely for the conversion of equatorial coordinates between epochs separated by many years.

9.5   Rigorous formulae

When it is necessary to compare observations that have been taken many years apart, it is normal practice to refer each set to a standard epoch. Until recently this has been the epoch 1950.0, although in future the epoch J2000.0 will be generally used (cf. section 10.5). Let us consider, therefore, the transformation of the equatorial coordinates of a star between a general epoch at time t and a standard epoch at time t_0.

Let P_0 and Υ_0 be the celestial pole and equinox at time t_0. The equator for this epoch is shown in Fig. 9.5 as $U_0\Upsilon_0V_0$. Also shown is the position of a star X with coordinates (α_0, δ_0) at this epoch. Now suppose that P is the new position of the celestial pole at time t. Denote the arc P_0P by θ_A. This is to be understood as the great circle arc joining these two points. So P_0P as shown in Fig. 9.5 does not represent the actual path of the pole from one point to the other, although it will be close to it. Now the first crude approximation to precession was to describe the pole as moving in a small circle about the pole of the ecliptic. This motion initially at least would be along the great circle $P_0 \Upsilon_0$. Consequently, the angle $PP_0 \Upsilon_0$ will be a small angle, which must tend to zero with the time interval $(t - t_0)$. Denote this angle $PP_0 \Upsilon_0$ by ζ_A. Now the right ascension $\alpha_0 = \Upsilon_0 P_0 X$. So, in the spherical triangle $PP_0 X$ we recognize that $PP_0 X = \alpha_0 + \zeta_A$, while $P_0 X = 90° - \delta_0$, as usual.

Now let $U \Upsilon V$ be the equator at time t, where Υ is the equinox at this new epoch. For the same reasons that it was anticipated that the angle $PP_0 \Upsilon_0$ would be small, it is expected that ΥPP_0 will be close to $180°$. So let us set $\Upsilon PP_0 = 180° + z_A$. As shown in Fig. 9.5, both ζ_A and z_A will be small positive angles (when $t > t_0$). They are in fact equal to first order in the time interval $(t - t_0)$. Now denote the star's coordinates by (α, δ) when referred to the new equator and equinox. Since $\alpha = \Upsilon PX$, it follows that $P_0 PX = 180° - (\alpha - z_A)$. Moreover, $PX = 90° - \delta$.

Now, in all, five parts have been identified in spherical triangle $P_0 PX$, namely $P_0 P = \theta_A$, $P_0 X = 90° - \delta_0$, $PX = 90° - \delta$, $PP_0 X = \alpha_0 + \zeta_A$, $P_0 PX =$

Figure 9.5. The precessional angles ζ_A, z_A, θ_A.

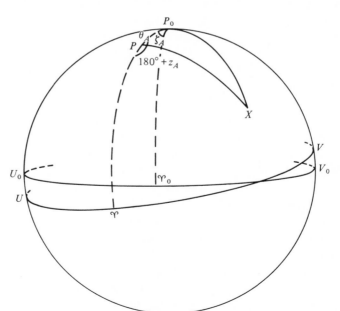

$180° - (\alpha - z_A)$. The spherical coordinates of the star at the two epochs may now be related by applying the standard trigonometric formulae to spherical triangle P_0PX. For example, (α, δ) are expressed in terms of (α_0, δ_0) and the three precessional angles by invoking in turn the analogue, sine and cosine formulae to give

$$\cos \delta \cos (\alpha - z_A) = \cos \theta_A \cos \delta_0 \cos (\alpha_0 + \zeta_A) - \sin \theta_A \sin \delta_0,$$
$$\cos \delta \sin (\alpha - z_A) = \cos \delta_0 \sin (\alpha_0 + \zeta_A), \qquad (9.21)$$
$$\sin \delta = \sin \theta_A \cos \delta_0 \cos (\alpha_0 + \zeta_A) + \cos \theta_A \sin \delta_0.$$

On the other hand, if the transformation is to be applied in the other direction, slightly different applications of the same three basic formulae will yield

$$\cos \delta_0 \cos (\alpha_0 + \zeta_A) = \cos \theta_A \cos \delta \cos (\alpha - z_A) + \sin \theta_A \sin \delta,$$
$$\cos \delta_0 \sin (\alpha_0 + \zeta_A) = \cos \delta \sin (\alpha - z_A), \qquad (9.22)$$
$$\sin \delta_0 = - \sin \theta_A \cos \delta \cos (\alpha - z_A) + \cos \theta_A \sin \delta.$$

The formulae in (9.21) and (9.22) are exact. No approximation of any kind has been made. In order to apply them, however, it is necessary to have the values of the three precessional angles ζ_A, z_A, and θ_A, which must be derived from precessional theory. They are computed in practice as power series in the time interval $(t - t_0)$ retaining terms up to and including the third order. The coefficients in the series do depend slightly on the initial epoch that has been chosen. If this is taken as J2000.0, the following apply

$$\zeta_A = 0°.640\,616\,1\ T + 0°.000\,083\,9\ T^2 + 0°.000\,005\,0\ T^3,$$
$$z_A = 0°.640\,616\,1\ T + 0°.000\,304\,1\ T^2 + 0°.000\,005\,1\ T^3, \qquad (9.23)$$
$$\theta_A = 0°.556\,753\,0\ T - 0°.000\,118\,5\ T^2 - 0°.000\,011\,6\ T^3.$$

In these formulae T is the interval $(t - t_0)$ expressed in Julian centuries of 36 525 days.

Using the standard epoch of J2000.0 will mean that $(t - t_0)$, and, therefore, T, will normally be negative, at least for the time being. All three parameters given in equations (9.23) will then also be negative. This contrasts with what was tacitly assumed in deriving equations (9.21) and (9.22) and with what is illustrated in Fig. 9.5. Although the diagram should be drawn differently, the resulting formulae are unchanged.

The previous standard epoch, now written as B1950.0, may also be used. The definitions of these two standard epochs and the different conventions associated with them are discussed fully in section 10.5. Using B1950.0

requires different formulae for ζ_A, z_A, and θ_A. They are

$$\zeta_A = 0°.640\,263\,3\ T + 0°.000\,083\,9\ T^2 + 0°.000\,005\,0\ T^3,$$

$$z_A = \zeta_A + 0°.000\,219\,7\ T^2, \tag{9.24}$$

$$\theta_A = 0°.556\,737\,6\ T - 0°.000\,118\,3\ T^2 - 0°.000\,011\,7\ T^3.$$

Now, however, T is the time interval subsequent to B1950.0 expressed in tropical centuries of 36 524.22 days. It should be noted that the old system used slightly different values of the precessional constants, and this is one reason for the discrepancy between equations (9.23) and (9.24).

The three precessional angles ζ_A, z_A, and θ_A define the positions of the pole P and the equinox γ (Fig. 9.5) with respect to their initial positions. They, therefore, determine the changes in a star's equatorial coordinates. They are not sufficient, however, to determine changes in ecliptic coordinates, since they do not locate the pole of the ecliptic. Referring to Fig. 9.5, it may be stated that the initial ecliptic pole K_0 lies on the great circle $P_0 U_0$ drawn at right angles to $\gamma_0 P_0$, and a similar statement may be made about its position at time t, but no more. To locate the ecliptic pole, and so define the ecliptic plane, it is necessary to know the obliquity of the ecliptic. This is derived from precessional theory as

$$\varepsilon = 23° 26' 21''.448 - 46''.815\ T - 0''.001\ T^2 + 0''.002\ T^3 \tag{9.25}$$

where T is the time in Julian centuries after the fundamental epoch J2000.0.

Ecliptic coordinates of a celestial object may be derived at any epoch by using equation (9.25). As shown in section 2.5, the transformation from equatorial to ecliptic coordinates is a rotation of axes through an angle ε. Equation (9.25) provides the value of this angle. The actual transformation from one epoch to another is best performed in equatorial coordinates.

9.6   Rotation matrices

The formulae derived in the last section are rigorous, but they are not in the most convenient form for computation. They can be improved by employing vector notation. Then transforming from the equator and equinox of one epoch to those of another becomes a matrix operation on a star's position vector.

Let $\mathbf{s}_0$ be a unit vector which specifies the star's position referred to the normal set of rectangular equatorial axes determined by the equator and equinox of the standard epoch t_0. We shall write $\mathbf{s}_0$ as a column vector which, in the notation of the previous section, is

$$\mathbf{s}_0 = \begin{pmatrix} x_0 \\ y_0 \\ z_0 \end{pmatrix} = \begin{pmatrix} \cos \delta_0 \cos \alpha_0 \\ \cos \delta_0 \sin \alpha_0 \\ \sin \delta_0 \end{pmatrix}. \tag{9.26}$$

Similarly let **s** be the star's direction referred to axes defined by the equator and equinox at time t. Then

$$\mathbf{s} = \begin{pmatrix} x \\ y \\ z \end{pmatrix} = \begin{pmatrix} \cos \delta \cos \alpha \\ \cos \delta \sin \alpha \\ \sin \delta \end{pmatrix}. \tag{9.27}$$

Now the transformation equations (9.21) may be expressed in terms of the components of these two vectors. The third equation of (9.21) is the simplest and, on expanding $\cos(\alpha_0 + \zeta_A)$ on the right-hand side, gives directly

$$z = \cos \zeta_A \sin \theta_A \, x_0 - \sin \zeta_A \sin \theta_A \, y_0 + \cos \theta_A \, z_0. \tag{9.28}$$

To derive expressions for x and y, it is necessary to combine the first two equations of (9.21) and to note that

$$x = \cos \delta \cos(\alpha - z_A) \cos z_A - \cos \delta \sin(\alpha - z_A) \sin z_A,$$

and

$$y = \cos \delta \cos(\alpha - z_A) \sin z_A + \cos \delta \sin(\alpha - z_A) \cos z_A.$$

Then substituting from (9.21), and expanding as necessary, will yield the two results

$$\begin{aligned}
x &= (\cos \zeta_A \cos z_A \cos \theta_A - \sin \zeta_A \sin z_A)x_0 \\
&\quad - (\cos \zeta_A \sin z_A + \sin \zeta_A \cos z_A \cos \theta_A)y_0 - \cos z_A \sin \theta_A z_0. \\
y &= (\sin \zeta_A \cos z_A + \cos \zeta_A \sin z_A \cos \theta_A)x_0 \\
&\quad + (\cos \zeta_A \cos z_A - \sin \zeta_A \sin z_A \cos \theta_A)y_0 - \sin z_A \sin \theta_A z_0.
\end{aligned} \tag{9.29}$$

Some of the coefficients in (9.29) have the appearance of cosine formulae, as indeed they are. For the coefficients of x_0, y_0, z_0 in equations (9.28) and (9.29) are the direction cosines of the new axes with respect to the old. So far the results that have been established will hardly appear to be an improvement on (9.21). They can be elegantly expressed in matrix notation, however, and, what is more important, this is particularly well suited to routine computation.

Writing (x_1, x_2, x_3) for the coordinates instead of (x, y, z), the results contained in (9.28) and (9.29) may be written as

$$x_i = P_{ij}x_{0j}, \tag{9.30}$$

using the dummy suffix notation. Here P_{ij} are the elements of the matrix of the coefficients that appear in (9.29) and (9.28). To be specific,

$$\begin{aligned}
P_{11} &= -\sin \zeta_A \sin z_A + \cos \zeta_A \cos z_A \cos \theta_A \\
P_{12} &= -\cos \zeta_A \sin z_A - \sin \zeta_A \cos z_A \cos \theta_A \\
P_{13} &= -\cos z_A \sin \theta_A
\end{aligned}$$

$$P_{21} = \sin \zeta_A \cos z_A + \cos \zeta_A \sin z_A \cos \theta_A$$
$$P_{22} = \cos \zeta_A \cos z_A - \sin \zeta_A \sin z_A \cos \theta_A \qquad (9.31)$$
$$P_{23} = -\sin z_A \sin \theta_A$$
$$P_{31} = \cos \zeta_A \sin \theta_A$$
$$P_{32} = -\sin \zeta_A \sin \theta_A$$
$$P_{33} = \cos \theta_A.$$

We shall write the transformation equations (9.30) in formal matrix notation as

$$\mathbf{s} = \mathbf{P} \, \mathbf{s}_0. \qquad (9.32)$$

The matrix $\mathbf{P}$ is called the *rotation matrix*. Examining the diagram Fig. 9.5, it can be seen that the coordinate transformation may be resolved into the product of three rotations, (i) a rotation through the angle $-\zeta_A$ about the initial z-axis (P_0), (ii) a rotation through an angle θ_A about the resulting y-axis, and (iii) a rotation through the angle $-z_A$ about the resulting and final z-axis (P).

Since the three precessional parameters ζ_A, z_A, and θ_A are all small angles, most of the elements of the rotation matrix are small quantities. The exceptions are the three diagonal elements which are close to unity. As one would expect, if the time interval $(t - t_0)$ tends to zero, the rotation matrix tends to the unit matrix; the transformation becomes an identity.

The reverse transformation, from epoch t to epoch t_0, follows at once from equation (9.32) as

$$\mathbf{s}_0 = \mathbf{P}^{-1} \, \mathbf{s}. \qquad (9.33)$$

It is not necessary to invert the matrix, however. The inverse transformation is obtained by replacing ζ_A, z_A, and θ_A by $-z_A$, $-\zeta_A$, and $-\theta_A$ respectively. When these changes are made in (9.31), it is found that the inverse of $\mathbf{P}$ is its transpose. Hence

$$\mathbf{s}_0 = \mathbf{P}^T \, \mathbf{s}. \qquad (9.34)$$

The rotational matrix is very straightforward to use. The elements do not normally need to be computed as they are given at daily intervals in the *Astronomical Almanac*. The tabulated elements differ slightly from the P_{ij} defined above as nutational effects are also included. The necessary modification is discussed in section 9.9.

9.7   Approximate formulae

The rigorous formulae that have been developed in the last two sections are clearly much more elaborate than the first-order formulae established in section 9.4. Although the matrix formulation is elegant, its application still involves considerable computation, and the first-order formulae give coordinate changes much more directly. The latter become imprecise, however, if applied over time intervals much in excess of one year. Nevertheless, it is possible to improve the precision of the first-order formulae while retaining their basic form. By taking into account the variations in the annual precessional rates over the time interval involved, they may be converted into similar formulae which are essentially correct to second order. By contrast, the so-called rigorous formulae can be made as precise as we wish by improving the precision of the expressions for the three angles ζ_A, z_A, and θ_A. The power series are truncated in practice after third-order terms. In their application, therefore, the rigorous formulae, whether used in the matrix formulation or not, are only correct to third order in the time interval.

The formulae that we wish to modify are equations (9.13), (9.15), (9.19) and (9.20). The modified forms are to be applied over a time interval τ of many years, where $\tau = t - t_0$ – as before, t is a general epoch, while t_0 is the standard epoch. The modification consists of substituting values for the precessional parameters which correspond to the midpoint of the interval τ. Similarly, where appropriate, the star's coordinates are estimated for this same time. Let us use the subscript m to denote a value corresponding to time $\frac{1}{2}(t_0 + t)$. Then new parameters are defined as follows:

$$
\begin{aligned}
M &= m_m(t - t_0) \\
N &= n_m(t - t_0) \\
a &= p_m(t - t_0) \\
b &= \pi_m(t - t_0) \\
c &= 180° - \Pi_m + \tfrac{1}{2}a \\
c' &= 180° - \Pi_m - \tfrac{1}{2}a.
\end{aligned}
\tag{9.35}
$$

The formulae just cited then give the following results for a star's coordinates, or a planet's orbital elements, at time t in terms of the values at the fundamental epoch:

$$
\begin{aligned}
\alpha &= \alpha_0 + M + N \sin \alpha_m \tan \delta_m \\
\delta &= \delta_0 + N \cos \alpha_m \\
\lambda &= \lambda_0 + a - b \tan \beta_0 \cos (\lambda_0 + c) \\
\beta &= \beta_0 + b \sin (\lambda_0 + c) \\
i &= i_0 + b \cos (\Omega_0 + c) \\
\Omega &= \Omega_0 + a - b \cot i_0 \sin (\Omega_0 + c) \\
\omega &= \omega_0 + b \operatorname{cosec} i_0 \sin (\Omega_0 + c).
\end{aligned}
\tag{9.36}
$$

We shall not derive each of these formulae separately but just consider the ones for the ecliptic coordinates (λ, β) as an example. It follows from (9.15) that, to the accuracy that we are working,

$$\lambda - \lambda_0 = p_m \tau + \pi_m \tau \tan \beta_m \cos (\Pi_m - \lambda_m),$$
$$\beta - \beta_0 = \pi_m \tau \sin (\Pi_m - \lambda_m).$$

Since the rotation rate π of the ecliptic is only about one-hundredth of the general precession rate p, or the rate of change of Π, its contribution to λ_m may be ignored within the trigonometric functions on the right-hand sides. It is sufficient, therefore, to replace $\tan \beta_m$ by $\tan \beta_0$, and λ_m by $\lambda_0 + \frac{1}{2} p_m \tau$ $(= \lambda_0 + \frac{1}{2} a)$. Using the definitions (9.35), the two relevant results in (9.36) are then obtained. The formulae given in the *Astronomical Almanac* differ slightly from those given here, but they maintain the same order of precision.

In order to derive the equatorial coordinates (α, δ), it is necessary to supply values for (α_m, δ_m) in addition to the initial values of right ascension and declination. With sufficient accuracy, one may write

$$\alpha_m = \alpha_0 + \frac{1}{2}(M + N \sin \alpha_0 \tan \delta_0) \tag{9.37}$$
$$\delta_m = \delta_0 + \frac{1}{2} N \cos \alpha_m.$$

One could iterate to improve on these, but it is not usually worthwhile.

The values of the parameters defined in equations (9.35) may be computed from the annual precessional rates already given in sections 9.2–9.4. They may be expressed as power series in the time interval of which only the first two terms are significant. If the fundamental epoch is taken as J2000.0, the following numerical results are obtained

$$M = 1°.281\,232\ T + 0°.000\,388\ T^2$$
$$N = 0°.556\,753\ T - 0°.000\,119\ T^2$$
$$a = 1°.396\,971\ T + 0°.000\,309\ T^2 \tag{9.38}$$
$$b = 0°.013\,056\ T - 0°.000\,010\ T^2$$
$$c = 5°.1236 + 0°.2416\ T$$
$$c' = 5°.1236 - 1°.1553\ T.$$

In these equations T is the time interval from J2000.0 expressed in Julian centuries.

Using these numerical expressions, equations (9.36) give the transformation equations for equatorial or ecliptic coordinates and orbital elements from the standard equinox of J2000.0 to the equinox at any time t. Often this will be the equinox of date. The reverse transformation is also often required, as a means of standardizing observations. The formulae (9.36) must then be inverted.

The smallness of b allows β and β_0, and i and i_0, to .be used interchangeably as arguments of the trigonometric functions on the right-hand sides of (9.36). Moreover, it is easily seen that, since $\lambda \simeq \lambda_0 + a$,

$$\lambda_0 + c \simeq \lambda + c'$$

and

$$\Omega_0 + c \simeq \Omega + c',$$

for the same reason. This is justified if second-order terms in b are to be neglected. Equations (9.36) may, therefore, be rewritten in a form suitable for reduction to the standard equinox as

$$\alpha_0 = \alpha - M - N \sin \alpha_m \tan \delta_m$$
$$\delta_0 = \delta - N \cos \alpha_m$$
$$\lambda_0 = \lambda - a + b \tan \beta \cos (\lambda + c')$$
$$\beta_0 = \beta - b \sin (\lambda + c') \tag{9.39}$$
$$i_0 = i - b \cos (\Omega + c')$$
$$\Omega_0 = \Omega - a + b \cot i \sin (\Omega + c')$$
$$\omega_0 = \omega - b \operatorname{cosec} i \sin (\Omega + c').$$

To sufficient accuracy, the mid-interval values (α_m, δ_m) are given by

$$\alpha_m = \alpha - \tfrac{1}{2}(M + N \sin \alpha \tan \delta)$$
$$\delta_m = \delta - \tfrac{1}{2}N \cos \alpha_m. \tag{9.40}$$

The approximate formulae that have been considered in this section may be expected to give a precision of better than an arc second over intervals of up to 50 years.

9.8 Nutation

The actual movement of the pole on the celestial sphere is very complicated, and it is for this reason that it is arbitrarily split into luni–solar precession and nutation. The latter is defined as the periodic variations in the position of the true pole about its mean position.

The dynamical theory of precession and nutation is extremely complex and far beyond the scope of this book. The following comments, however, will hopefully illustrate some of the basic principles involved. Let us consider first the torque exerted on the earth by the sun at a time when its right ascension and declination are $(\alpha_\odot, \delta_\odot)$. The magnitude of this torque is found for a simple rigid earth model to be proportional to $\sin 2\delta_\odot$. The torque vanishes, therefore, at the equinoxes. It may be represented by a vector $\mathbf{K}$ which will, from symmetry considerations, be perpendicular both to the earth–sun line and to the earth's axis of symmetry, the polar axis. This vector, therefore, lies in the equator in the direction of right ascension

$\alpha_\odot - 90°$. In fact, from the above considerations, the torque may be written in equatorial rectangular coordinates as

$$\mathbf{K} = K_0 \sin 2\delta_\odot [\cos(\alpha_\odot - 90°), \sin(\alpha_\odot - 90°), 0],$$

where K_0 is a constant, that is,

$$\mathbf{K} = 2K_0 \sin \delta_\odot (\sin \alpha_\odot \cos \delta_\odot, -\cos \alpha_\odot \cos \delta_\odot, 0). \qquad (9.41)$$

The effect of the various terms in (9.41) will become apparent if $(\alpha_\odot, \delta_\odot)$ are replaced by the sun's longitude. In Fig. 9.6, the equator, the ecliptic, and the plane of the moon's orbit are all shown; N is the ascending node of the moon's orbit on the ecliptic and M the ascending node on the equator. In the notation of section 7.5, $\Upsilon NM = i$, $\Upsilon N = \Omega$. Now let S be the position of the sun. If we neglect the eccentricity of the earth's orbit, $\Upsilon S = L$, the sun's mean longitude. Further, $N \Upsilon M = \varepsilon$. Since the sun's ecliptic latitude is zero, the standard transformation equations (2.19) will yield

$$\cos \alpha_\odot \cos \delta_\odot = \cos L$$
$$\sin \delta_\odot = \sin L \sin \varepsilon \qquad (9.42)$$
$$\sin \alpha_\odot \cos \delta_\odot = \sin L \cos \varepsilon.$$

Substituting this into (9.41) yields

$$\mathbf{K} = K_0 \sin \varepsilon [\cos \varepsilon (1 - \cos 2L), -\sin 2L, 0]. \qquad (9.43)$$

Figure 9.6

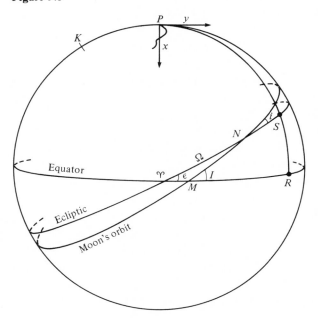

The earth's angular momentum vector is directed towards P. The torque is perpendicular to this direction, and so it cannot affect the magnitude of the earth's angular momentum, but only its direction, which means the location of the point P itself.

In discussing the movement of P, it is necessary to fix the coordinate system to a particular pole. So let $s = (x, y, z)$ be the position vector of the celestial pole referred to equatorial axes based on the mean pole and equinox of an initial epoch when $L = 0$. Then the components of K are proportional to $(dx/dt, dy/dt, 0)$. Integrating (9.43), the subsequent position of the pole may be expressed approximately as

$$x \operatorname{cosec} \varepsilon = \psi_1 \left[t - \frac{1}{2} \left(\frac{dL}{dt} \right)^{-1} \sin 2L \right]$$

$$y = \frac{1}{2} \psi_1 \left(\frac{dL}{dt} \right)^{-1} \tan \varepsilon \cos 2L, \tag{9.44}$$

$$z = 1,$$

where ψ_1 is a constant depending on the magnitudes of the angular momentum, the mean solar torque, and the obliquity of the ecliptic.

Now $x \operatorname{cosec} \varepsilon$ is the advance in the celestial pole's longitude, while y is an increment in the obliquity of the ecliptic. We recognize the term ψ_1 as the solar contribution (about one-third) to luni–solar precession. Furthermore, the existence of two nutational terms, one in longitude and one in obliquity, has been established. Each of the nutational terms has a period of six months. If t is expressed in years, then $dL/dt = 2\pi$, and equation (9.44) relates the amplitudes of the nutational terms to the solar precessional rate.

Further terms of purely solar origin arise when the annual variation of the earth's distance from the sun is taken into account. Treated to first order in the earth's orbital eccentricity, this couples with the precessional term to produce nutational terms of a period of one year. Furthermore, it couples with the nutational terms already established to produce two further nutational terms of periods one year and four months. A few higher-order solar terms also need to be included.

While the direct solar effect of precession and nutation may be dealt with fairly quickly, as outlined above, the lunar contribution is far more complicated. The discussion of section 7.5 described some of the intricacies of the lunar orbit, and these have correspondingly complex nutational effects. We shall only attempt below a rather heuristic derivation of the principal nutational effects.

Following the same argument that was applied to the sun, a study may be made in terms of the mean lunar orbit, neglecting its eccentricity. The lunar torque on the earth is then of the form

$$\mathbf{K}' = K_0' \sin I[\cos I(1 - \cos 2L), -\sin 2L, 0], \tag{9.45}$$

where $I \ (= S\hat{M}R)$ is indicated in Fig. 9.6 as the inclination of the orbit to the equator, and L is the moon's angular distance from the point M. Notice that the coordinate system (x', y', z'), say, that is being used in (9.45) is not the standard equatorial coordinate system. It is an equatorial system in that the z'-axis points to P, but the x'-axis will be directed towards the point M, not Υ. Inspection of (9.45) reveals a quasi-precessional term and nutational terms which will have a period of about 14 days, half the lunar month. These terms are not the largest nutational terms, however, as might at first sight be expected. They are, in fact, smaller than the principal solar terms. For, although the lunar torque is larger $(K_0' \simeq 2K_0)$, the factor $(dL/dt)^{-1}$ that arises upon integration is about 12 times smaller than its solar equivalent – cf. equation (9.44). The principal nutational terms are actually contained in what has been described as the quasi-precessional term. This must be examined in more detail.

The node N of the mean lunar orbit on the ecliptic regresses, completing one circuit in 18.6 years. The direction of the x'-axis, the point M, therefore oscillates with this period about the equinox, and the inclination I oscillates about a mean value ε. Since the inclination i of the moon's mean orbit to the ecliptic is small, the extent of these oscillations will also be small. The quasi-precessional term in the torque, therefore, contains small variations in both its magnitude and its direction. When referred to the standard rectangular equatorial axes, it is

$$\mathbf{K}_P = \tfrac{1}{2}K_0' \sin 2I(\cos \Upsilon M, \sin \Upsilon M, 0). \tag{9.46}$$

Let us evaluate this correct to first order in i. Applying the sine formula to spherical triangle ΥMN yields

$$\sin \Upsilon M \sin I = \sin \Omega \sin i.$$

Hence, to sufficient accuracy,

$$\sin \Upsilon M = i \sin \Omega \csc \varepsilon, \tag{9.47}$$

and, of course, $\cos \Upsilon M \approx 1$. Further, the four-parts formula will give

$$\cos \Upsilon M \cos \varepsilon = \sin \Upsilon M \cot \Omega + \sin \varepsilon \cot I.$$

Making use of the results already established, this can be rewritten as

$$\sin (I - \varepsilon) \approx \sin i \cos \Omega,$$

which, to first order, is

$$I = \varepsilon + i \cos \Omega. \tag{9.48}$$

These results for ΥM and I may now be substituted into (9.46). After a little manipulation, it is established that

$$\mathbf{K}_P = K_0'(\sin \varepsilon \cos \varepsilon + i \cos \Omega \cos 2\varepsilon, i \cos \varepsilon \sin \Omega, 0). \tag{9.49}$$

We now proceed to integrate this and derive this displacement of the pole relative to its initial position, when $\Omega = 0$, as

$$x \csc \varepsilon = \psi_1'[t + 2i \cot 2\varepsilon (d\Omega/dt)^{-1} \sin \Omega]$$
$$y = -\psi_1' i (d\Omega/dt)^{-1} \cos \Omega. \tag{9.50}$$

Here ψ_1' is a constant chosen so that it represents the precessional term in longitude. In (9.50) we can identify the lunar precession in longitude and nutational terms both in longitude and the obliquity of period 18.6 years. These are, in fact, the largest nutational terms, being over ten times larger than the six-month solar terms which are the next most significant.

The argument given above indicates how luni–solar precession and the most important nutational terms arise. Until fairly recently the theory of nutation in common use was based on a rigid model of the earth. The ratios between the amplitudes of the precessional and nutational terms in (9.44) and (9.50) are in agreement with that theory. A new theory of nutation was put forward, however, in 1980 and has been accepted by the IAU (1982). This theory is based on a more realistic model of the earth which is non-rigid and not even axially symmetric. In all, the 1980 theory of nutation contains 106 terms both in longitude and the obliquity, although some of the latter are negligible.

The displacement of the true celestial pole from the mean pole in longitude is denoted by $\Delta\psi$, its orthogonal displacement by $\Delta\varepsilon$. These are referred to respectively as the *nutation in longitude* and the *nutation in the obliquity*; they correspond to $x \csc \varepsilon$ and y in our earlier notation. The theory of nutation provides series expansions for $\Delta\psi$ and $\Delta\varepsilon$ of the form

$$\Delta\psi = \sum a_{ijklm} \sin (iM' + jM + kF + lD + m\Omega)$$
$$\Delta\varepsilon = \sum b_{ijklm} \cos (iM' + jM + kF + lD + m\Omega), \tag{9.51}$$

where $i, j, k, l,$ and m are integers. The quantities $M', M, F,$ and D were introduced in section 7.5 in connection with the moon's orbit. They are, respectively, the moon's mean anomaly, the sun's mean anomaly, the moon's mean argument from the node, and the moon's mean elongation from the sun. All change uniformly with time, as does the mean longitude of the node Ω. In the 1980 theory of nutation, the principal terms that have been discussed in this section are as follows:

$$\Delta\psi = -17''.1996 \sin \Omega - 1''.3187 \sin (2F - 2D + 2\Omega)$$
$$- 0''.2274 \sin (2F - 2\Omega)$$
$$= 9''.2025 \cos \Omega + 0''.5736 \cos (2F - 2D + 2\Omega)$$
$$+ 0''.0927 \cos (2F - 2\Omega). \tag{9.52}$$

The coefficient of the leading term in the nutation in the obliquity is often referred to as the *constant of nutation*. The coefficients in the two expansions have small secular variations, but these are only significant for about 15 of the 106 terms. The coefficients quoted in (9.52) correspond to the epoch J2000.0.

It is sometimes convenient to separate the 'long period' from the 'short period' nutational terms. By the long-period terms is meant those which do not depend upon the moon's mean longitude. There is, in fact, a clear division in the periods; the long-period terms all have periods in excess of 90 days, while no short-period term has a period greater than 35 days. The short-period terms summed separately are denoted by $d\psi$ and $d\varepsilon$.

In *The Apparent Places of the Fundamental Stars*, for example, right ascensions and declinations are given at 10-day intervals. This is too long to allow meaningful interpolation of short-period nutational effects, which are, therefore, excluded, that is the nutational corrections applied are $\Delta\psi - d\psi$ and $\Delta\varepsilon - d\varepsilon$. The full nutations $(\Delta\psi, \Delta\varepsilon)$ are tabulated at daily intervals in the *Astronomical Almanac*.

9.9   The effect of nutation on a star's coordinates

The nutation in longitude simply provides a periodic addition to luni–solar precession. As a result, the true ecliptic longitude of a star exceeds its mean longitude by $\Delta\psi$. The ecliptic latitude is unaffected. The effect of $\Delta\psi$ on a star's equatorial coordinates may be deduced immediately from equations (9.3) and (9.4) which give the corresponding changes for luni–solar precession. Consequently

$$d\alpha = \Delta\psi(\cos \varepsilon + \sin \varepsilon \sin \alpha \tan \delta)$$
$$d\delta = \Delta\psi \sin \varepsilon \cos \alpha. \tag{9.53}$$

The coordinate changes due to nutation in the obliquity will next be considered. In Fig. 9.7, P represents the mean pole and P' its displaced position allowing only for $\Delta\varepsilon$. Let us consider the standard spherical triangle KPX formed by the pole of the ecliptic, the celestial pole and the star. The parts of this triangle are indicated in the diagram in terms of the star's ecliptic coordinates (λ, β) and its equatorial coordinates (α, δ). The displacement of the pole from P to P' affects neither the side KX nor the angle PKX, so the ecliptic coordinates are completely unaltered. The change in ε will, however, affect both α and δ.

By the cosine formula in triangle PKX,

$$\sin \delta = \sin \beta \cos \varepsilon + \cos \beta \sin \varepsilon \sin \lambda. \tag{9.54}$$

Taking differentials in this equation will yield

$$\cos \delta \, d\delta = (-\sin \beta \sin \varepsilon + \cos \beta \cos \varepsilon \sin \lambda) \Delta\varepsilon,$$

since λ and β are unchanged. The bracket on the right-hand side may be simplified by the analogue formula, which gives

$$-\cos \delta \sin \alpha = \sin \beta \sin \varepsilon - \cos \beta \cos \varepsilon \sin \lambda, \tag{9.55}$$

hence

$$d\delta = \Delta \varepsilon \sin \alpha.$$

Now the sine formula gives the familiar result

$$\cos \alpha \cos \delta = \cos \lambda \cos \beta. \tag{9.56}$$

Taking differentials of this will now yield

$$\sin \alpha \cos \delta \, d\alpha + \cos \alpha \sin \delta \, d\delta = 0.$$

Substituting for $d\delta$ will then yield the increment in right ascension. So the final results are

$$d\alpha = -\Delta \varepsilon \cos \alpha \tan \delta$$
$$d\delta = \Delta \varepsilon \sin \alpha. \tag{9.57}$$

The first-order formulae of equations (9.53) and (9.57) are sufficiently accurate in almost every circumstance, since the two nutations are small angles. The total nutational effect is simply the superposition of the two equations. It is often more convenient, however, to express the results in rectangular coordinates.

Figure 9.7

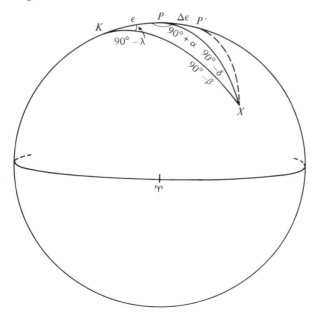

Let $\mathbf{s} = (x, y, z)$ be the unit vector defining the position of a star referred to the set of Cartesian axes defined by the mean pole and equinox. The components of this vector are given by (9.27). Furthermore, let $\mathbf{s}'$ be the same unit vector referred to the true equator and equinox. Now

$$\mathbf{s}' - \mathbf{s} = \mathbf{ds} = (dx, dy, dz). \tag{9.58}$$

Hence, from the differential form of (9.27)

$$\mathbf{s}' - \mathbf{s} = d\alpha \begin{pmatrix} -\sin \alpha \cos \delta \\ \cos \alpha \cos \delta \\ 0 \end{pmatrix} + d\delta \begin{pmatrix} -\cos \alpha \sin \delta \\ -\sin \alpha \sin \delta \\ \cos \delta \end{pmatrix}. \tag{9.59}$$

The increments $(d\alpha, d\delta)$ are obtained by adding together those given in (9.53) and (9.57). After a little reduction, the results are obtained in scalar form as

$$dx = -(y \cos \varepsilon + z \sin \varepsilon) \, \Delta\psi$$
$$dy = x \cos \varepsilon \, \Delta\psi - z \, \Delta\varepsilon \tag{9.60}$$
$$dz = x \sin \varepsilon \, \Delta\psi + y \, \Delta\varepsilon.$$

This last result may be expressed in matrix form as

$$\mathbf{s}' = \mathbf{N} \, \mathbf{s}, \tag{9.61}$$

where $\mathbf{N}$ is the *rotation matrix* for nutation, given by

$$\mathbf{N} = \begin{pmatrix} 1 & -\Delta\psi \cos \varepsilon & -\Delta\psi \sin \varepsilon \\ \Delta\psi \cos \varepsilon & 1 & -\Delta\varepsilon \\ \Delta\psi \sin \varepsilon & \Delta\varepsilon & 1 \end{pmatrix}. \tag{9.62}$$

This formulation allows nutation to be combined with precession in a simple way. Let $\mathbf{s}_0$ be the star's position vector on the celestial sphere referred to the equator and equinox of the standard epoch. Then, as shown in section 9.6, $\mathbf{s} = \mathbf{P} \, \mathbf{s}_0$, where $\mathbf{P}$ is the rotation matrix for precession. Hence the true rectangular coordinates of the star may be deduced from the matrix equation

$$\mathbf{s}' = \mathbf{N} \, \mathbf{s} = (\mathbf{N} \, \mathbf{P}) \, \mathbf{s}_0 = \mathbf{R} \, \mathbf{s}_0. \tag{9.63}$$

The rotation matrix $\mathbf{R}$ $(\equiv \mathbf{N} \, \mathbf{P})$ thus allows for both precession and nutation, and normally the two corrections are required simultaneously. The elements of the rotation matrix $\mathbf{R}$ are given at daily intervals in the *Astronomical Almanac*, so that the matrix elements for the precise date in question may be obtained by simple interpolation.

An alternative method for correcting for precession and nutation is considered in chapter 12. This uses Besselian Day Numbers and enables corrections for parallax and aberration to be applied at the same time. This method is a first-order method, and so corrections for precession are only applied over a short interval of time. Essentially the midpoint of the year is used as the reference epoch instead of the standard epoch.

The use of rotation matrices, is by contrast, rigorous, at least as far as precession is concerned. The nutation matrix, however, as given in (9.62) is only correct to first order in $\Delta\psi$ and $\Delta\varepsilon$. In corroboration of this, it may be noted that the rows and columns of (9.62) are not exactly unit vectors. The errors involved in using this form of the rotation matrix are likely to be about $0''.001$, which for most purposes is acceptable. If desired, however, an exact form of the nutational rotation matrix may be derived by considering the problem in ecliptic coordinates.

The transformation from the mean to the true pole involves (i) the increase of a star's longitude by $\Delta\psi$, and (ii) an increase in the obliquity of the ecliptic by $\Delta\varepsilon$. Now equations (9.54), (9.55) and (9.56) give expressions in ecliptic coordinates for the equatorial coordinates z, $-y$, and x respectively. Consequently, the coordinates of $\mathbf{s}'$ are given rigorously as

$$x' = \cos\beta \cos(\lambda + \Delta\psi)$$
$$y' = -\sin\beta \sin(\varepsilon + \Delta\varepsilon) + \cos\beta \cos(\varepsilon + \Delta\varepsilon) \sin(\lambda + \Delta\psi) \qquad (9.64)$$
$$z' = \sin\beta \cos(\varepsilon + \Delta\varepsilon) + \cos\beta \sin(\varepsilon + \Delta\varepsilon) \sin(\lambda + \Delta\psi).$$

Now expand $\cos(\lambda + \Delta\psi)$ and $\sin(\lambda + \Delta\psi)$ on the right-hand sides and identify the rectangular coordinates (x, y, z) as they occur. The results are

$$x' = \cos\Delta\psi\, x - \cos\varepsilon \sin\Delta\psi\, y - \sin\varepsilon \sin\Delta\psi\, z$$
$$y' = \cos(\varepsilon + \Delta\varepsilon) \sin\Delta\psi\, x$$
$$\qquad + [\cos\varepsilon \cos(\varepsilon + \Delta\varepsilon) \cos\Delta\psi + \sin\varepsilon \sin(\varepsilon + \Delta\varepsilon)]y$$
$$\qquad + [\sin\varepsilon \cos(\varepsilon + \Delta\varepsilon) \cos\Delta\psi - \cos\varepsilon \sin(\varepsilon + \Delta\varepsilon)]z \qquad (9.65)$$
$$z' = \sin(\varepsilon + \Delta\varepsilon) \sin\Delta\psi\, x$$
$$\qquad + [\cos\varepsilon \sin(\varepsilon + \Delta\varepsilon) \cos\Delta\psi - \sin\varepsilon \cos(\varepsilon + \Delta\varepsilon)]y$$
$$\qquad + [\sin\varepsilon \sin(\varepsilon + \Delta\varepsilon) \cos\Delta\psi + \cos\varepsilon \cos(\varepsilon + \Delta\varepsilon)]z.$$

The elements of the rotation matrix N may then be identified.

Problems

9.1 A star (α, δ) lies on the ecliptic and has longitude λ. Prove the changes in right ascension, declination and longitude over a short interval of time due to luni–solar precession are related by

$\cos^2\delta \cot\alpha\, \Delta\alpha = \cot\delta\, \Delta\delta = \cot\lambda\, \Delta\lambda.$

9.2 The Pole Star has right ascension $2^{\mathrm{h}} 15^{\mathrm{m}} 54^{\mathrm{s}}.6$ and declination $89° 11' 39''$ for the epoch J1984.5. Use the first-order formulae (9.13) to calculate its coordinates for J1985.0.

9.3 Calculate the three precessional angles ζ_A, z_A and θ_A for the transformation from the standard epoch J2000.0 to the epoch J1985.0.

9.4 Prove that over a short interval of time τ

$$m\tau = \zeta_A + z_A$$
$$n\tau = \theta_A.$$

9.5 Prove that the stars whose declinations are unchanged by precession over a certain period T from the standard epoch lie on a great circle whose poles are the points $\alpha = \zeta_A$, $\delta = -\frac{1}{2}\theta_A$ and $\alpha = 12^h + \zeta_A$, $\delta = \frac{1}{2}\theta_A$.

9.6 The Pole Star has equatorial coordinates $(2^h 31^m 46^s.3, 89° 15' 50''.6)$ at the standard epoch J2000.0. Use the approximate formulae of section 9.7 to compute its coordinates at the epoch J1985.0. Compare your result with that obtained in question 9.2. Which set of values is more accurate?

9.7 Check your answer to the previous question by using the rigorous formulae to derive the Pole Star's right ascension and declination for J1985.0.

9.8 A reference star has coordinates (α_0, δ_0) at the standard epoch. A nearby object is measured t years later at displacements $\Delta\alpha$, $\Delta\delta$ from the reference star. Show that its coordinates (α, δ) at the standard and epoch are

$$\alpha = \alpha_0 + (1 - e \tan \delta_0) \Delta\alpha - f \sec^2 \delta_0 \Delta\delta$$
$$\delta = \delta_0 + \Delta\delta + f \Delta\alpha,$$

where

$$e = -(nt + \Delta\psi \sin \varepsilon) \cos \alpha - \Delta\varepsilon \sin \alpha,$$
$$f = (nt + \Delta\psi \sin \varepsilon) \sin \alpha - \Delta\varepsilon \cos \alpha.$$

9.9 Show that the constant of nutation is approximately related to the annual rate of luni–solar precession by the equation

$$N = \frac{18.61mi}{2\pi(r^3 + m)}$$

where m is the moon's mass in solar units and r the radius of its orbit in AU.

9.10 Using the coordinates quoted in question 9.6, determine the epoch when the Pole Star is closest to the celestial pole. Calculate its declination at that time.

9.11 Show that all the non-diagonal elements of the rotation matrix for precession are small and that the diagonal elements are close to unity. In particular prove that

$$P_{12} + P_{21} = 2 \sin^2 \tfrac{1}{2}\theta_A \sin (\zeta_A - z_A)$$

and that, therefore, the magnitudes of P_{12} and P_{21} differ only in the fourth order.

10

Time

10.1 Astronomical and physical concepts of time

The ultimate nature of time is a deep and uncertain philosophical question, but fortunately not one that it is necessary to grapple with in an introductory text on positional astronomy. It is sufficient to regard time as being measured out and defined by the repetition of events. These may be simply the ticking of a clock, or the passage of the sun across the observer's meridian, or something more subtle. Although some time-bases must be considered as more fundamental than others in the overall context of physical science, their independence must be appreciated before any attempt is made to reconcile them and establish relationships between them. Six different time-bases are used in astronomy that are qualitatively distinct, leading to the following time-scales:

(i) Sidereal time
(ii) Solar time
(iii) Dynamical time
(iv) Atomic time
(v) Proper time
(vi) Coordinate time

Formal definitions will be given in this chapter of these time-scales, their variants, and the conventions associated with each of them. Before embarking on this, however, it is important that the underlying principles should be understood. This section is, therefore, devoted to making some general introductory comments about each of the time-scales listed above.

(i) *Sidereal time*

Sidereal time is the hour angle of the equinox. Apart from the small influence of precession, it depends entirely on the rotation of the earth. It is now recognized that the earth's rotation rate is not completely uniform when compared with other more fundamental time-scales. The rate possesses certain quasi-periodic irregularities in addition to a slow secular decrease. These irregularities produce a corresponding unevenness in the passage of sidereal time.

(*ii*)   *Solar time*
        Solar time is defined by the hour angle of the sun. This is subject to comparatively gross irregularities, which were discussed in section 2.6, and consequently a mean solar time must be introduced. The sun's hour angle depends on two things – on the rotation of the earth and on the sun's right ascension, which changes with the earth's orbital motion. Consequently, solar time, quite apart from its gross, but removable irregularities, is a hybrid, being dependent on two distinct and unconnected periodic phenomena, namely the earth's diurnal rotation and its annual revolution round the sun. It has been astronomical practice to define mean solar time in two distinct ways in order to remove this source of confusion. Universal time (UT) is made totally dependent on the rotation of the earth, while ephemeris time, considered shortly, is a dynamical time-scale that depends only on the earth's orbital motion.

UT may be defined in terms of the Greenwich hour angle (GHA) of a fictitious body which we shall call the *Universal Mean Sun* (UMS). In order that the day should begin at midnight the definition is

$$UT = 12^h + GHA \ UMS. \tag{10.1}$$

The Universal Mean Sun is prescribed to move round the equator at a uniform rate *in the time-scale under discussion*. Both sidereal and UT are based on the rotation of the earth. Although the units of time (the second, minute, hour, day) in the two systems differ, they are in a constant ratio. We may state, therefore, that the right ascension of the Universal Mean Sun increases uniformly with sidereal time. Put rather dramatically, but, I hope, vividly – if the earth's rotation were to stop, both sidereal time and universal time would stand still, and the progress of the Universal Mean Sun round the equator would also halt, even though the right ascension of the true sun would continue to increase as the earth maintained its orbital motion. In fact, a tiny discrepancy has built up of some $2''$ between the right ascension of the Universal Mean Sun and the value that would be deduced from the mean motion of the true sun.

(*iii*)   *Dynamical time*
        Dynamical time means the independent variable that appears in the gravitational equations of motion. The mean anomalies of unperturbed

planetary orbits, therefore, increase uniformly in dynamical time. In Newtonian physics this time-scale is regarded as absolute. This time-scale has in the past been defined in Newtonian terms with *ad hoc* relativistic corrections having the same status as purely Newtonian perturbations.

Ephemeris time (ET) is a dynamical time-scale expressed in solar form. It may be defined in terms of the hour angle of another fictitious body which we shall call the Ephemeris Mean Sun (EMS). This is prescribed to move round the mean equator at a uniform rate that corresponds to the mean motion of the true sun. One cannot, however, use the Greenwich hour angle of this body to define a dynamical time-scale since the sidereal direction of this meridian depends on the rotation of earth. Another standard meridian is, therefore, defined, the *ephemeris meridian*, which is essentially the direction that the Greenwich meridian would have if the earth's rotation were strictly uniform. The ephemeris meridian is not, therefore, fixed to the earth's surface but has a slow eastward drift. The hour angle with reference to this meridian is termed the *ephemeris hour angle* (EHA), and ET may be defined by an equation similar to (10.1) as

$$ET = 12^h + EHA \, EMS. \tag{10.2}$$

ET was introduced in 1960 as the argument used in the ephemerides published in the annual almanacs. The difference between UT and ET is denoted by ΔT, in the sense:

$$\Delta T = ET - UT. \tag{10.3}$$

This quantity cannot be given accurately in advance since it depends on irregularities in the earth's rotation which are not predictable in detail. Dynamical theory, however, allows heliocentric positions of the planets to be determined in advance for any ET. Moreover, the coordinates of the geocentric point may be similarly so determined. Hence geocentric ephemerides may be computed with ET as the independent argument.

(iv)   *Atomic time*

Atomic clocks provide the most accurate and consistent means of measuring time. International Atomic Time (TAI) was introduced in 1972, although atomic time-scales were available for many years before that. Consequently it has been possible to extrapolate TAI back as far as 1956. The fundamental unit of TAI is the SI second, which is defined in terms of the frequency of radiation resulting from a particular hyperfine transition in the caesium atom.

There is a conceptual difference, therefore, between atomic and dynamical time. However, provided the constants of nature are absolute, the two time-scales should be rigidly related. Moreover, the definitions of the second in TAI and ET, although formally independent, are in fact numerically matched, so that there is no more than an offset between the two time-scales. The distinction between the two time-scales would be significant if it transpired, for example, that the gravitational constant G had a variation over a cosmological time-scale. This is suggested in some unorthodox cosmologies; dynamical time would then be slowing down with respect to atomic time.

In classical physics, time is an absolute quantity, and the distinction between atomic and dynamical time is only a matter of their measurement and definition. In general relativity, however, the difference becomes more important as we naturally distinguish between proper and coordinate time.

(v)   *Proper time*

The world line of the earth in general relativity is an ordinary geodesic. The proper time s along this geodesic was used in section 6.9 as the curve parameter. The physical significance of this parameter is that it is the time that would be measured by an observer on the world line. It is time for an earth-bound observer. Consequently, time measured on any clock on the earth will be a proper time. In particular, atomic time will be a form of proper time.

(vi)   *Coordinate time*

It was also shown in section 6.9 that the proper time had periodic variations in relation to the coordinate time t. It is true that the proper time is algebraically the most convenient curve parameter to use in solving the geodesic differential equations, in this sense it is used as an independent dynamical variable. It is unsuitable as the basis for an overall dynamical time-scale, however, as a different proper time-scale is defined for each heavenly body. It is only the earth's proper time that is measurable. The coordinate time is the obvious alternative, since it may be applied globally. While the time coordinate in general relativity is not uniquely defined, a particular choice may be made. In static spherically symmetric Schwarzschild spacetime, the choice is obvious. It is the time coordinate that has been used in all forms of the metric quoted in chapters 3, 6 and 8.

In 1976 the IAU recommended the introduction of two new time-scales to replace ET and to recognize the importance of relativistic effects. The IAU did not, however, wish to endorse general relativity unequivocally as the definitive relativistic theory, since other post-Newtonian theories of gravity were, and are, in the process of being studied and tested. Consequently only one of these time-scales is defined in an uncomprising manner.

This is *Terrestrial Dynamical Time* (TDT) which was introduced from 1977 January 1. It is based on the SI second, but its zero point is chosen so that it is effectively a continuation of ET. It is intended for use as the time-scale for apparent geocentric ephemerides. Although no assumption is made about a theory of gravity, within the context of general relativity this time-scale is the proper time on the earth and is measured by atomic clocks.

The second time-scale is Barycentric Dynamical Time (TDB). This is intended for use in the equations of motion of planetary bodies referred to the barycentre of the solar system. This time-scale is not uniquely defined but will depend on the particular post-Newtonian gravitational theory that is adopted. Even so, a restriction is imposed on TDB in that it is required that it differ from TDT by only periodic discrepancies. This restriction can be met in any gravitational theory. In general relativity, TDB will be the coordinate time *t* mentioned above with an appropriately chosen zero point and scaling factor.

10.2   Sidereal and solar time

The definition of sidereal time has already been given. It is very simple. For example, Greenwich sidereal time is given by

$$GST = GHA \, \Upsilon. \tag{10.4}$$

In equation (10.4), Υ always refers to an equinox of date, but nutation may or may not be included. If, in this equation, the true equinox is intended, the sidereal time is termed *apparent sidereal time*. If, however, Υ is the mean equinox, (10.4) gives *mean sidereal time*.

The difference between the two sidereal times is referred to as the *equation of the equinoxes*. It is easily seen from equation (9.53) and (9.57) that

$$\text{Equation of the equinoxes} = \Delta\psi \cos \varepsilon. \tag{10.5}$$

Apparent sidereal time is not used as a time-scale, but is required for meridian observations. In what follows we shall be concerned only with mean sidereal time which is a uniform time-scale, apart from irregularities in the earth's rotation.

The mean sidereal day differs slightly from the earth's rotation period owing to precession. The right ascension of a star situated at the equinox would have an annual increase (cf. (9.4)) of $\psi \cos \varepsilon$. This corresponds to a daily rate of $0^s.0084$, and so the earth's rotation period exceeds the mean sidereal day by this amount.

Precession also affects the length of the year, or, to be precise, the length of the tropical year. For, like the month, the year may be defined in a number of ways. The tropical year is the most important, being the mean interval required for the sun's longitude to increase by 360°. It is the interval from one vernal equinox to the next, and is, therefore, the interval between identical seasonal dates. The civil calendar is matched to this length of the year; the occurrence of leap years, at a rate of 97 every 400 years in the Gregorian calendar, gives an average length of the calendar year of $365^d.2425$ which is very close to the tropical year. By contrast, the older Julian calendar had a leap year every four years, and so the interval of $365^d.25$ is termed the *Julian year*. The Julian century is exactly 36 525 days.

The period of the earth's revolution round the sun with respect to the fixed stars is termed the *sidereal year*. Due to planetary perturbations, this differs slightly from the interval from one perihelion to the next, which is called the *anomalistic year*. Finally, one may define the *eclipse year* as the interval between successive passages of the mean sun through the ascending node of the moon's mean orbit. This differs appreciably from the other lengths of the year and determines the average frequency of solar and lunar eclipses.

The several definitions of the year gives the following lengths

$$\text{Julian year} = 365^d.25$$
$$\text{Tropical year} = 365^d.2422$$
$$\text{Sidereal year} = 365^d.2564 \tag{10.6}$$
$$\text{Anomalistic year} = 365^d.2596$$
$$\text{Eclipse year} = 346^d.6201.$$

These values are subject to minute secular changes, which for most purposes are too small to be significant.

The tropical year is important in defining the relationship between solar and sidereal time. For any object X, one may write

$$\text{GMST} = \text{GHA } X + \text{RA } X, \tag{10.7}$$

provided the right ascension is measured from the mean equinox. The point X may now be identified with the universal mean sun, so that using (10.1) we have

$$\text{GMST} = \text{UT} - 12^h + \text{RA UMS}.$$

Now the right ascension of the universal mean sun increases by 24^h in one tropical year. Hence, if β is the reciprocal of one tropical year in days, it follows that

$$\text{GMST} = T_0 + (1 + \beta)\,\text{UT} \tag{10.8}$$

where T_0 is a constant.

This, of course, assumes that the universal mean sun's right ascension increases uniformly with UT. This is part of its definition. The concept of the mean sun was first introduced by Newcomb (1895) at a time when the variability of the earth's rotation rate was not measurable. Newcomb had no need, therefore, to distinguish between the universal and ephemeris mean suns. The fictitious body which he considered moved uniformly (apart from slight secular effects) in both sidereal and dynamical time. He gave an expression for the right ascension of his mean sun which effectively defined that body. This expression was later retained to define the universal mean sun with the coefficients unchanged but with the argument now explicitly defined as UT, as in equation (10.8). Following the 1976 revision of astronomical constants, slight modifications were introduced into this expression; the resulting relationship (IAU, 1983) between UT and sidereal time is as follows:

$$\begin{aligned} \text{GMST at } 0^h \text{ UT} \\ = 6^h\,41^m\,50^s.54841 \\ + 8\,640\,184^s.812\,866\ T_u + 0^s.093\,104\ T_u^2 \\ - 6^s.2 \times 10^{-6}\ T_u^3. \end{aligned} \tag{10.9}$$

Here UT should more properly be written as UT1 which is defined in the next section, and T_u is the interval in Julian centuries from 2000 January 1, 12^h UT1.

The final terms in (10.9) are due to secular change in the tropical year which is mainly caused by a variation in the rate of precession. It has nothing to do with variations in the earth's rate of axial rotation. Neglecting this secular change and dividing the coefficient of T_u by 36 525, we derive the excess of the mean solar day over the mean sidereal day in sidereal time. In fact

$$1 \text{ mean solar day} = 24^h\,03^m\,56^s.55537 \text{ of mean sidereal time}$$

while $\tag{10.10}$

$$1 \text{ mean sidereal day} = 23^h\,56^m\,04^s.09053 \text{ of mean solar time.}$$

10.3   Ephemeris and universal time

The fictitious mean sun introduced by Newcomb (1895) is conceptually closer to the ephemeris mean sun. The right ascension of the latter is simply the sun's mean longitude L. A dynamical time-scale can be defined directly in terms of this quantity without recourse to the ephemeris meridian used in equation (10.2). In fact the formal definition of ET followed this procedure. Newcomb's expression for the geometric mean longitude L of the sun, namely

$$L = 279°\,41'\,48''.04 + 129\,602\,768''.13\ T + 1''.089\ T^2, \tag{10.11}$$

was taken as the definition of ET T, expressed in Julian centuries. The epoch from which T is measured is, by definition, 1900 January 0, 12^h ET.

The above expression for the sun's mean longitude defines the units of ephemeris time through the coefficient of T. Neglecting the small secular variations, this coefficient gives the increase in the sun's mean longitude in a period of 36 525 ephemeris days. Since L increases by $360°$ (or $1\,296\,000''$) in one tropical year, one may derive the tropical year in ephemeris days, or indeed in ephemeris seconds. An ephemeris day is defined as 86 400 ephemeris seconds.

The primary unit of ET is the topical year for 1900.0, which may be subdivided into N ephemeris seconds, where

$$N = \frac{1\,296\,000 \times 36\,525 \times 86\,400}{129\,602\,768.13}$$

i.e.

$$N = 31\,556\,925.9747. \tag{10.12}$$

The ephemeris second is formally defined as the fraction $1/N$ of the tropical year of 1900.0.

In what follows we shall continue to neglect the term in T^2 in (10.11). Expressed in time measure this equation is of the form

$$\text{RA EMS} = T_0' + \beta\,\text{ET}. \tag{10.13}$$

where β is again the reciprocal of the tropical year in days. For both the universal and the ephemeris mean sun, it is convenient to use apparent rather than geometric right ascension. The necessary aberrational correction is a constant which may be absorbed into T_0'.

The ephemeris meridian is introduced as a means of comparing ET with UT. Since ET itself is formally defined by (10.11), the ephemeris hour angle and consequently the ephemeris meridian are determined by equation (10.2). In Fig. 10.1, the ephemeris meridian and the Greenwich meridian are shown respectively as the great circles PHQ and PGQ. The points E and U represent the ephemeris and universal mean suns, while S is the true sun. The arc HG will be denoted by λ_E; it is the longitude of the ephemeris meridian east of Greenwich.

Now by (10.1) and (10.2),

$$\text{HE} = \text{ET} - 12^h$$
$$\text{GU} = \text{UT} - 12^h. \tag{10.14}$$

The arc ΥG is the Greenwich sidereal time, similarly ΥH is termed the *ephemeris sidereal time* (EST), being the sidereal time on the ephemeris meridian. Clearly

$$\text{EST} = \text{GST} + \lambda_E. \tag{10.15}$$

Now, since EST = EHA EMS + RA EMS, equations (10.2) and (10.13) will yield

$$\text{EST} = T_0' - 12^{\text{h}} + (1 + \beta)\, \text{ET}. \tag{10.16}$$

This has a very similar form to equation (10.8).

Now, at some epoch in the past, thought to have been in 1902, the universal and ephemeris mean suns coincided in position. At this instant the ephemeris meridian coincided with the Greenwich meridian. If therefore both ephemeris and Greenwich sidereal times are measured from this instant, it follows from (10.8) and (10.16) that $T_0 = T_0' - 12^{\text{h}}$. Subtracting the two equations then yields with the aid of (10.15) and (10.3)

$$\lambda_E = (1 + \beta)\, \Delta T \tag{10.17}$$
$$= 1.002\,738\, \Delta T.$$

The difference in the right ascensions of the two mean suns is the arc UE. Using (10.14), this is seen to be $\lambda_E - \Delta T$. Hence

$$\text{RA EMS} - \text{RA UMS} = \beta\, \Delta T. \tag{10.18}$$

The measurement of ET requires observations of the sun's longitude at least in principle. Since it is a dynamical time-scale, however, the longitudes of other bodies in the solar system may also be used. In practice, ET was

---

Figure 10.1 The ephemeris and Greenwich meridians.

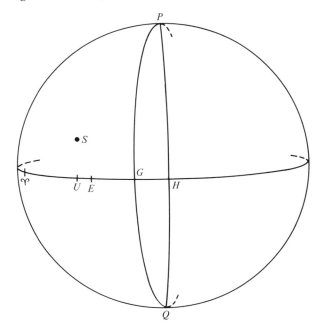

derived principally from observations of the moon, since its longitude varies most rapidly. A considerable body of observations must be reduced before an effective knowledge of ET is obtained. This means that the ET of a particular observation will only be known some time after the event.

By contrast, the determination of UT is almost immediate. Meridian observations of the stars provide a direct measurement of sidereal time which may be converted to UT through equation (10.9). The time difference ΔT is only known later, however, when the ET has become available.

It is necessary to distinguish between several different forms of UT. The conversion of the observed sidereal time to universal time through equation (10.9) involves the station's longitude. This longitude will be accurately known with respect to the pole of figure. Let us represent it by λ_0. On the other hand, let λ be the longitude with respect to the instantaneous pole of rotation, following the conventions of section 5.5. The relationship between the two longitudes is given in equation (5.29).

Since λ_0 is a station constant, the reduction from sidereal to UT may be carried out immediately using this value of the longitude. The resulting UT is denoted by UT0. Owing to polar motion, the value of UT0 is site dependent and does not provide a global standard. For this reason the reduction should be carried out using the station's instantaneous value of λ. This version of UT is called UT1. It is the form that is used when relating UT to other time-scales. Whenever the symbol UT is used without qualification it is UT1 that is implied. Using (5.29) gives

$$UT1 = UT0 - (u_x \sin \lambda_0 + u_y \cos \lambda_0) \tan \phi_0, \qquad (10.19)$$

where ϕ_0 is the observing station's geocentric latitude and (u_x, u_y) are the instantaneous pole's coordinates expressed in time measure, i.e. $(\frac{1}{15}x, \frac{1}{15}y)$ in the notation of section 5.5. The coefficients (u_x, u_y) are not available at the time of observation, they are published subsequently by the *International Polar Motion Service* and by the *Bureau International de l'Heure*.

A further refinement of UT is possible. UT1 exhibits irregularities due to variations in the earth's rotation rate. These variations are not predictable in detail, but contain recognizable seasonal periodicities. When these are removed a new time-scale called UT2 is derived. Both UT1 and UT2 are independent of the observer's position on the surface of the earth.

10.4   Modern dynamical time-scales

It has long been felt that there are unsatisfactory features in the time-scale of ET. It has been replaced as the basic astronomical time-scale system in the annual almanacs for the year 1984 onwards. The new time-scale, terrestrial dynamical time (TDT), has, however, been designed to maintain effective continuity with ephemeris time. So our earlier discussion of ET is still relevant and explains the origins of the new time-scale.

The principal objections to ET are that it is artificial and that it is not immediately available. It is a dynamical time-scale, of the form needed for ephemerides, but it is based on pre-relativistic dynamics. Moreover, from a Newtonian point of view, it is essentially no different from atomic time, which can be obtained more readily and with considerably greater precision.

The ephemeris second was defined, c.f. equation (10.12), as the fraction N^{-1} of the tropical year for 1900.0. The fundamental unit in International Atomic Time is the SI second, which is defined as 9 192 631 770 periods of the radiation corresponding to the transition between the two hyperfine levels in the ground state of the atom caesium 133. No systematic difference has been established between these two time units. The relationship between ET and atomic time is taken as

$$ET = TAI + 32^s.184. \tag{10.20}$$

The astronomical time-scale for ephemerides has been brought formally closer to TAI with the introduction of TDT in 1977. This new time-scale was not in general use prior to 1984. In TDT the fundamental unit is the SI second, exactly as in atomic time. All that is needed to relate the two time-scales is an exact value for the offset. This is stated in the form

$$1977 \text{ January } 1 \ 0^h \text{ TAI} = 1977 \text{ January } 1.000\,372\,5 \text{ TDT}. \tag{10.21}$$

This is the same value of the offset as in (10.20), in order to maintain continuity with ET.

The definition of UT was not changed in the revision of the time-scales, at least in principle. Slight changes were, however, introduced in the numerical values of coefficients due to a redefinition of the equinox. In the new system, UT1 is effectively defined by its relation to GST given in equation (10.9). The strict correspondence between UT and sidereal time is certainly useful, but there is a price to be paid for it. For, as (10.18) indicates, the right ascension of the universal mean sun differs from the mean longitude of the true sun. It is true that the discrepancy is very small, being only about $0^s.15$. But it is increasing with time, and eventually a step modification will have to be made, or the basis of UT will have to be redefined.

Broadcast time signals are based on Coordinated Universal Time (UTC). This is not strictly a solar time-scale at all, but a modified form of atomic time. UTC differs from TAI by an integral number of seconds, and it is always maintained with $0^s.9$ of UT1. This is achieved by the occasional introduction of additional seconds, called leap seconds. In principle, seconds could also be omitted from UTC if this was necessary, but, in fact, TAI is gaining about one second a year on UT1 so the adjustment has always been the introduction of a leap second. It is arranged that this takes place, when required, at the end of either June or December, so the last day of the month is extended by one second. At the beginning of 1983, for example, the difference between TAI and UTC was 21^s; the last leap second had been introduced at 1982, June 30.

TDT is intended as a time-scale for geocentric ephemerides. Many observations should, however, be referred to the solar system barycentre rather than the geocentric point. Just as a transformation for the positional coordinates was considered in chapter 8, so the time corresponding to some observed event should be transformed to the time at which the event would be observed by a hypothetical barycentric observer. This is particularly important for rapid periodic phenomena such as pulsar signals, and short-period variables and binary systems. It is essential that the reception of the light or radio signals from such objects should be referred to a common unaccelerated origin.

The geometry, which is very simple, is illustrated in Fig. 10.2, in which G is the barycentric and E the geocentric point. The vector $\overrightarrow{GE} = \mathbf{R}$, and the components (X, Y, Z) of this vector were discussed in section 8.2. Suppose a star is in the direction given by the unit vector $\mathbf{s}$. The distance of the star is being treated as infinite, as parallax is unlikely to be significant in this problem. Let t be the time of observation at the earth and $t + \tau$ the time of

Figure 10.2

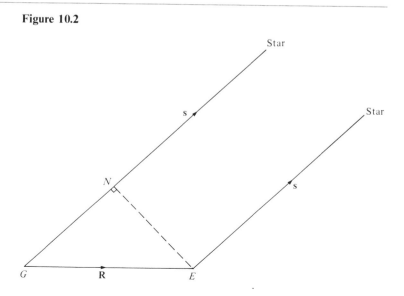

arrival of the light signal at G. Then clearly $c\tau = GN = \mathbf{R} \cdot \mathbf{s}$. Expressing $\mathbf{s}$ in terms of the stars equatorial coordinates (α, δ) yields

$$c\tau = X \cos \alpha \cos \delta + Y \sin \alpha \cos \delta + Z \sin \delta. \qquad (10.22)$$

This gives the difference between terrestrial and barycentric time at least for observations of bodies outside the solar system. The components (X, Y, Z) will be expressed in AU and so using (8.18) we find

$$\tau = 0^{\mathrm{d}}.005\ 775\ 6(X \cos \alpha \cos \delta + Y \sin \alpha \cos \delta + Z \sin \delta).$$
$$(10.23)$$

For a body within the solar system one is unlikely to require this form of barycentric correction. The dynamical theory of planetary motion must, however, be formulated in what is known as Barycentric Dynamical Time (TDB). When relativistic effects are included there are subtle differences between TDB and TDT which will be considered in section 10.7. Moreover, the calculation of the total light-time is important.

Suppose, for example, a planet is observed at time t and that the radiation was emitted at time $t - \tau$. In Fig. 10.3, G again represents the barycentre, E is the earth's position at time t and P that of the planet at time $t - \tau$. Since PE is simply $c\tau$, it follows that

$$c^2\tau^2 = r^2 + R^2 - 2\mathbf{r} \cdot \mathbf{R}. \qquad (10.24)$$

The relativistic correction to this equation is derived in section 10.8.

Figure 10.3

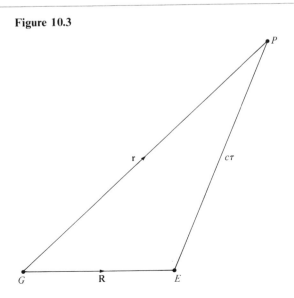

10.5   The Julian and the Besselian year

This section is concerned with defining and describing a number of astronomical conventions associated with time. First of all, one minor point: when specifying the time of an event, the astronomical practice is to state the time intervals (year, month, day, etc.) in order of decreasing size. For example, whereas one might casually speak of something happening at noon on 23rd September 1985, the correct astronomical designation is 1985 September 23 12^h. Alternatively, this can be written as 1985 Sept 23.5. Moreover, the astronomer does not feel constrained to adhere the conventional number of days in a month and could, in principle, refer to New Year's Day as December 32, and Hogmanay as January 0. For example, the instant 1985 December 31 18^h might be written as 1986 January $0^d.75$. These conventions can be applied in any of the time-scales discussed earlier.

There is an alternative method of specifying a particular instant of astronomical time that is used in a dynamical time-scale. This dispenses altogether with the calendar date and uses only the year, but in fractional form, e.g. 1985.1672. Two systems are in use, the older one is based on the Besselian year, while the new system uses the Julian year. It is important to recognize in each system how the instant at which the year begins is defined. In neither case is it the instant that heralds in 'Auld Lang Syne' – not even at Greenwich.

The length of the Besselian year is the interval required for the ephemeris mean sun's right ascension to increase by 24^h. This may be identified with the tropical year, although there is a subtle distinction between the two, but it amounts to only $0^s.148\,T$ where T is the time in centuries from 1900. So, if secular effects are ignored, the Besselian year is identified with a tropical year of length $365^d.2422$. The instant at which a particular Besselian year begins is defined as the instant when the sun's mean longitude is exactly 280°, or

$$\text{RA EMS} = 18^h\,40^m. \tag{10.25}$$

This is always close to the beginning of the calendar year. The fundamental epoch, which is written as B1900.0, is 1900 January $0^d.813$ ET, while the standard epoch B1950.0 is exactly 50 tropical years or 18 262.110 days later. This interval exceeds 50 ordinary years of 365 days by $12^d.110$. So, remembering that 1900 was *not* a leap year, it is seen that

$$\text{B1950.0} \equiv 1950\ \text{January}\ 0^d.923\ \text{ET}$$
$$\equiv 1949\ \text{December}\ 31\ 22^h\,09^m\ \text{ET}. \tag{10.26}$$

The Besselian epoch for any subsequent instant may be computed by dividing the interval that has elapsed in days by 365.2422.

Calculations are made rather simpler if the Julian system is used. The date is now expressed as a fraction of a Julian year of $365\frac{1}{4}$ days. The fundamental epoch, written as J2000.0, is

$$J2000.0 \equiv 2000 \text{ January } 1^d.5 \text{ TDB.} \tag{10.27}$$

With this definition the Julian epoch of any instant may be calculated. For comparison let us compute J1950.0. It will be exactly $18\,262^d.5$ prior to the fundamental epoch, which, allowing for 12 leap years, gives

$$J1950.0 \equiv 1950 \text{ January } 1^d.0.$$

The new system of Julian epoch was introduced with the 1976 IAU revision of astronomical constants (IAU, 1977). At the same time the old Besselian epoch was redefined in terms of the new time-scale. The definition of the Besselian year was simplified by making it equal in length to the tropical year of B1900.0. Precise formulae for the Julian epoch and the redefined Besselian epoch are given below.

For some observations it is convenient to express the time entirely in days. This can be done by using the *Julian date* (JD) which is formally defined as the time in days, including the fraction, that has elapsed from the epoch 4713 BC January $1^d.5$. The reasons for this apparently extraordinary choice of epoch are rather obscure and involve the confluence of several chronological cycles. The explanation may be found in the *Explanatory Supplement*, p. 431. Thus the JD of the standard epochs discussed earlier are found to be

$$\begin{aligned}
B1900.0 &= JD\ 241\ 5020.313 \\
B1950.0 &= JD\ 243\ 3282.423 \\
J2000.0 &= JD\ 245\ 1545.0.
\end{aligned} \tag{10.28}$$

The JD was originally defined using UT, and the epoch was taken as Greenwich mean noon on 1st January 4713 BC. A similar definition made in ET was called the Julian ephemeris date. We have purposely not been specific in defining JD so that any time-scale may be used; the term JD is no longer reserved for UT. If there is any possibility of misunderstanding, the time-scale in use must be stated explicitly. Usually this is not necessary, and it will probably be clear to the reader that the first two time-scales in (10.28) are ET, while the third is TDB. The starting epoch for JD has the same formal definition in every time-scale, but will not correspond to the same instant of time. Indeed, little is known about the relationship between the different time-scales in the remote past.

The relationships between the Julian epoch or the Besselian epoch and

JD may be stated as follows

$$\text{Julian epoch} = \text{J2000.0} + (\text{JD} - 2\,451\,545)/365.25$$
$$\text{Besselian epoch} = \text{B1900.0} + (\text{JD} - 2\,415\,020.313\,52)/365.242\,198\,781$$

$$(10.29)$$

The simplicity and exactness of the new Julian system will be evident.

The integral part of the JD is called the *Julian Day Number*. It is seldom necessary to calculate this, since the JD is tabulated for each day of the year in the *Astronomical Almanac*. It must be stressed that the Julian day begins at noon and not at midnight. This is a point that is easily overlooked, since it is contrary to all other astronomical conventions.

Occasionally a modified Julian date (MJD) is used, which is defined as

$$\text{MJD} = \text{JD} - 240\,000.5. \qquad (10.30)$$

The MJD reverts to the normal practice of starting the day at midnight and is the time in days from 1858 Nobember 17.0.

10.6   Ephemeris transit

The distinction between apparent and mean solar time has already been discussed in section 2.6. The difference between the two is called the *equation of time E*, defined in the sense,

$$E = \text{Apparent solar time} - \text{mean solar time.} \qquad (10.31)$$

This may be expressed in terms of right ascensions as

$$E = \text{RA UMS} - \text{RA} \odot. \qquad (10.32)$$

Low-precision formulae are given in the *Astronomical Almanac* for calculating this quantity, but no attempt is made to tabulate E to high precision for the following reason.

The right-hand side of (10.32) contained two quantities that are defined in different time-scales. The quantity RA UMS is known for any UT, while the right ascension of the true sun may be calculated only for an ET. Since the difference of the two time-scales ΔT is not known in advance, the equation of time cannot be computed with precision.

The difficulty is removed if the *equation of ET, E**, is used instead. This is defined as

$$E^* = \text{RA EMS} - \text{RA} \odot. \qquad (10.33)$$

It follows from equation (10.18) that E^* exceeds the equation of time by $\beta\,\Delta T$.

The quantity that is tabulated in the *Astronomical Almanac* is the *ephemeris transit* of the sun. This is simply the ET of the sun's transit across the ephemeris meridian. Write (10.33) as

$$E^* = \text{EHA} \odot - \text{EHA EMS}$$
$$= \text{EHA} \odot - \text{ET} + 12^h.$$

The first term on the right-hand side vanishes at ephemeris transit, and so

$$\text{Ephemeris transit} = 12^h - E^*. \tag{10.34}$$

The *Astronomical Almanac* also gives the ephemeris transits for the moon – at both upper and lower transit – the major planets and some minor planets.

Let us now consider the calculation of E^*. In Fig. 10.4, $F \Upsilon G$ represents the mean equator and $A \Upsilon S$ the ecliptic. Suppose S is the position of the sun at time t and A its position at perigee corresponding to time τ. The point E represents the ephemeris mean sun at time t. By definition, $\Upsilon E = L$ the sun's mean longitude. The true sun's right ascension is $\Upsilon PS = \alpha$, say. So by (10.33)

$$E^* = L - \alpha. \tag{10.35}$$

It is necessary to derive separate expressions for L and α.

Let ϖ_0 be the sun's longitude at perigee. This is in fact about 282° and so $A \Upsilon = 360° - \varpi_0$. Suppose now that M is the mean anomaly of the sun (or earth) at time t. This is calculated as $n(t - \tau)$ where n is the mean motion. The

Figure 10.4

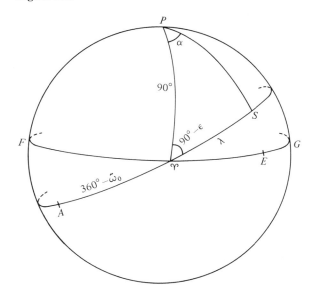

mean longitude is $M - A\Upsilon$, consequently

$$L = M + \varpi_0. \tag{10.36}$$

Now the arc AS is the true anomaly v. This is obtained from the mean anomaly in the usual way, i.e. solve Kepler's equation to derive the eccentric anomaly and then apply equation (6.28). It is then found that the longitude of the true sun is

$$\lambda = \Upsilon S = v + \varpi_0 \tag{10.37}$$

which may now be calculated.

The final step is to derive α from the spherical triangle $P\Upsilon S$ in which we may identify the four consecutive parts as shown in the diagram. An application of the four-parts formula then gives

$$\tan \alpha = \cos \varepsilon \tan \lambda. \tag{10.38}$$

The equation of ET is now given by (10.35) to (10.38) as

$$E^* = M + \varpi_0 - \tan^{-1} \left[\cos \varepsilon \tan (v + \varpi_0) \right]. \tag{10.39}$$

10.7 Proper and coordinate time

It is necessary to make the distinction in general relativity between proper time and coordinate time. The time measured by an earth-bound observer will be the proper time s, the interval along the observer's world line. This was identified as an ordinary geodesic in Schwarzschild spacetime and was investigated in section 6.9.

The coordinate time, although not directly measurable, is the more satisfactory dynamical time-scale. It may be used as the independent variable, or curve parameter, for the world line of any body in a heliocentric orbit. It must be remarked that by the coordinate time we mean that particular choice of time coordinate which is made in the two forms of the metric that have been considered, namely (3.46) and (3.52). This choice of time coordinate t allows the metric to be written in a time-independent form, and this property is sufficient to define it uniquely apart from a zero point and a constant scaling factor.

The relationship between proper and coordinate time was established in equation (6.71). Using (6.70), we find to sufficient accuracy that

$$t - t_0 = \left(1 + \frac{3}{2}\frac{m}{a}\right)(s - s_0) + \frac{2m}{a}\frac{e \sin E}{n}. \tag{10.40}$$

Here, a, e, n are respectively the earth's semimajor axis, eccentricity and mean motion and t_0 and s_0 are the coordinate and proper times of perihelion. Working to order m, the distinction between relativistic and classical values of the orbital elements is not important.

Now let us identify the proper time with the terrestrial time-scale (TDT) and attempt to construct from the coordinate time a global dynamical time-scale, the TDB mentioned earlier. Only periodic variations are to be allowed between the two time-scales. These are contained in the second term of (10.40). We therefore adopt as TDB a coordinate time-scale T such that

$$T - s_0 = \left(1 + \frac{3}{2}\frac{m}{a}\right)^{-1}(t - t_0).$$ (10.41)

Working still to order m, this gives

$$\text{TDB} = \text{TDT} + \frac{2m}{a}\frac{e \sin E}{n}.$$ (10.42)

It only remains to substitute the values for the orbital elements and for the Schwarzschild radius. Inserting the following values

$$2m = 2.956 \text{ km}$$
$$a = 1.496 \times 10^8 \text{ km}$$
$$e = 0.016\,71$$
$$n = 1.991 \times 10^{-7} \text{ radian s}^{-1}$$

gives the result as

$$\text{TDB} = \text{TDT} + 0^s.001\,658 \sin E.$$ (10.43)

If desired, the eccentric anomaly may be eliminated from (10.43) by expressing $\sin E$ as a power series in M and e. Kepler's equation may be solved by the recurrence relation given in equation (6.33). Then neglecting terms in e^3, we find

$$e \sin E = e \sin (M + e \sin M)$$
$$\simeq e \sin M + \tfrac{1}{2}e^2 \sin 2M.$$

Substituting this into (10.42) leads to the final numerical result

$$\text{TDB} = \text{TDT} + 0^s.001\,658 \sin M + 0^s.000\,014 \sin 2M.$$ (10.44)

10.8   The computation of light-time – radar delays

The derivation of the light travel time from the source to the observer is relevant in any observation of a body within the solar system. Usually the positional accuracy is not such as to warrant more than a classical treatment. The high precision of radar and laser ranging, however, which is independent of any angular measurement, make relativistic effects not only detectable but significant. The measured radar delay between transmission of the pulse and reception of the echo is, of course, a proper time. It can only be determined, however, once the coordinate time interval required by the pulse to reach the target and return has been established.

This will involve the study of the null geodesics that are the world lines of the outward signal and the returning echo.

Null geodesics in Schwarzschild spacetime were derived in section 8.7 and their directional properties were studied in section 8.8. We need to extend the analysis of case (a) from the latter section, to derive the time properties of a null geodesic linking a target at position vector $\mathbf{r}_e$ with an earth-bound observer with position vector $\mathbf{R}$. We shall concentrate on the return leg of the pulse's path, but the analysis of the outward leg is essentially similar.

Figure 10.5 shows the ray-path from the target S to the observer at O. The coordinate system is a system of isotropic coordinates that has been selected so that the ray-path lies in the x–y 'plane' (i.e. $z=0$) symmetrically about the y-axis. The diagram is very similar to Fig. 8.3. Then by equation (8.54) the two position vectors may be written as

$$\mathbf{R}=\left(X, q-\frac{2m}{q}R, 0\right)$$

$$\mathbf{r}_e=\left(x_e, q-\frac{2m}{q}r_e, 0\right). \tag{10.45}$$

As in (8.55), introduce a geocentric direction $\mathbf{s}_0$ and a geocentric distance ρ by the formal definition

$$\rho\mathbf{s}_0=\mathbf{r}_e-\mathbf{R}. \tag{10.46}$$

Figure 10.5. The null geodesic linking the radar target S to the observer O.

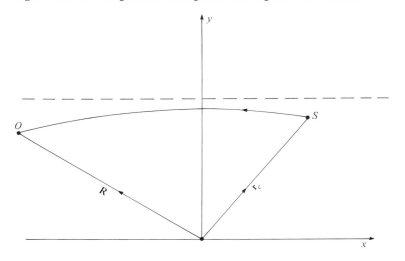

Then

$$\rho^2 = (x_e - X)^2 + \frac{4m^2}{q^2} (r - R)^2$$

and so, correct to order m, we have

$$\rho = (x_e - X). \tag{10.47}$$

Then the light-time t_0 determined classically is simply

$$t_0 = \rho/c = (x_e - X)/c. \tag{10.48}$$

The relativistic coordinate time, on the other hand, may be obtained by equating the fourth components of the tangent vector to the null geodesic as they are expressed in equations (8.52) and (8.53).

This gives the differential equation

$$\frac{dt}{dx} = -\frac{1}{c}\left(1 + \frac{2m}{r}\right). \tag{10.49}$$

It will be sufficient to substitute r correct only to zero order (in m), i.e. $r = (x^2 + q^2)^{1/2}$, since the form of the position vectors given in (10.45) applies also at any intermediate point. Integrating equation (10.49) then gives the coordinate travel time of the light as

$$t = t_0 + \frac{2m}{c} \int_X^{x_e} \frac{dx}{(x^2 + q^2)^{1/2}}$$

This yields the final result

$$t = t_0 + \frac{2m}{c}\left(\sinh^{-1}\frac{x_e}{q} - \sinh^{-1}\frac{X}{q}\right). \tag{10.50}$$

The factor $2m/c$ is about 10 μs, but under certain circumstances the time delay $(t - t_0)$ may be much greater than this. Suppose the target is a planet at superior conjunction, then the parameter q, expressed in AU, will be small. For a sun-grazing ray, q is one solar radius which is 0.0046 AU. Then $\sinh^{-1}(X/q)$ is about -10, and $\sinh^{-1}(x_e/q)$ will be of the same order of magnitude but opposite sign. When both the outward and the return path of the pulse are allowed for the total radar delay may exceed 400 μs. Measurements of such radar delays constitute an important test of general relativity – sometimes referred to as the fourth classical test. The precision attainable is similar to the second test to which is is related. For the second test measures the bending of the ray-path, while the fourth test measures the concomitant delay.

It only remains to express the result (10.50) in a general vector form that is independent of a special choice of coordinate system. To sufficient accuracy, we have

$$\begin{aligned} x_e &= \mathbf{r}_e \cdot \mathbf{s}_0 \\ X &= \mathbf{R} \cdot \mathbf{s}_0 \end{aligned} \tag{10.51}$$

while by equation (8.59)

$$q^2 = [R^2 - (\mathbf{s}_0 \cdot \mathbf{R})^2].\tag{10.52}$$

Moreover, it is easy to see that $\mathbf{R}$ may be replaced in this equation by $\mathbf{r}_e$, giving

$$q^2 = [r_e^2 - (\mathbf{s}_0 \cdot \mathbf{r}_e)^2].\tag{10.53}$$

It is usual to write the inverse hyperbolic sines in logarithmic form. Since X is negative, we have

$$-\sinh^{-1}\frac{X}{q} = \ln\left[-X + (X^2 + q^2)^{1/2}\right] - \ln q.$$

Using (10.51) and (10.52), this becomes

$$-\sinh^{-1}\frac{X}{q} = \ln(R - \mathbf{R} \cdot \mathbf{s}_0) - \ln q.$$

Proceeding similarly we find, using (10.53) that, since x_e is positive,

$$\sinh^{-1}\frac{x_e}{q} = \ln(r_e + \mathbf{r}_e \cdot \mathbf{s}_0) - \ln q.$$

Using these two expressions, the formula for the radar delay (10.50) gives

$$t - t_0 = \frac{2m}{c}\ln\left[\frac{(r_e + \mathbf{r}_e \cdot \mathbf{s}_0)(R - \mathbf{R} \cdot \mathbf{s}_0)}{q^2}\right],$$

which, by (10.52), reduces to

$$t - t_0 = \frac{2m}{c}\ln\left[\frac{(r_e + \mathbf{r}_e \cdot \mathbf{s}_0)}{(R + \mathbf{R} \cdot \mathbf{s}_0)}\right].\tag{10.54}$$

The factor $2m/c$ is 9.86 µs.

Equation (10.54) only gives the radar delay over the return journey of the pulse. The delay in the outward journey may be obtained similarly, but a little care is necessary in adapting the formulae. Neglecting the change in the observer's position during the travel time of the pulse, the total delay is simply twice that given in (10.54). It is left as an exercise to the reader to prove that this total delay may be written in the form

$$t - t_0 = \frac{4m}{c}\ln\left[\frac{(r_e + R + \rho)}{(r_e + R - \rho)}\right].\tag{10.55}$$

Problems

10.1 Prove that the Greenwich mean sidereal time is given at *any* epoch T as

GMST $= 18^h.697\,374\,6 + 879\,000^h.051\,336\,9\ T$

where T is the interval in Julian centuries from the standard epoch. Why is formula (10.9) to be preferred?

10.2 Given that:

the tropical year $= 365^{\mathrm{d}}.2422$

the sidereal year $= 365^{\mathrm{d}}.2564$

the anomalistic year $= 365^{\mathrm{d}}.2596$

determine the approximate amount and sign of annual precession and the motion of perihelion.

10.3 Find the dates that correspond to the two epochs B1985.1672 and J1985.1672.

10.4 Prove that the equation of ephemeris time may be approximately written as

$$E^* = y \sin 2L - 2e \sin M + 4ey \sin M \cos 2L$$
$$- \tfrac{1}{2}y^2 \sin 4L - \tfrac{5}{4}e^2 \sin 2M,$$

where M is the sun's mean anomaly, L is its mean longitude, e is the earth's orbital eccentricity and $y = \tan^2 \tfrac{1}{2}\varepsilon$.

10.5 Prove that the rate of change of the equation of ET is negative at the two solstices, assuming $e = 0.016\,75$. Show that this implies that the earliest sunrise occurs before midsummer's day and the latest sunset after it.

10.6 Prove that the UT of apparent transit of the sun exceeds the tabulated ephemeris transit by $\beta\,\Delta T$.

10.7 A radar pulse is transmitted from the earth at a time when its heliocentric position vector is $\mathbf{R}_1$. The radar echo is received from a target at position vector $\mathbf{r}$ and when the echo is received the earth's position vector is $\mathbf{R}_2$. Show the coordinate time interval between transmission and reception is

$$\tau = \frac{\rho_1 + \rho_2}{c} + \frac{2m}{c} \ln \left| \frac{(r + R_1 + \rho_1)(r + R_2 + \rho_2)}{(r + R_1 - \rho_1)(r + R_2 - \rho_2)} \right|$$

where

$$\rho_1 = (r^2 + R_1{}^2 - 2\mathbf{r} \cdot \mathbf{R}_1)^{1/2},$$

and

$$\rho_2 = (r^2 + R_2{}^2 - 2\mathbf{r} \cdot \mathbf{R}_2)^{1/2}.$$

11

Proper motion and radial velocity

11.1   Components of stellar motion

This chapter will discuss the motion of a star with respect to the sun and its observational consequences. The discussion does not include the determination of the reference frame, which must depend on the statistical interpretation of observations of many stars. For the present it is assumed that a reference frame, centred on the solar system barycentre, can be defined which is approximately inertial. The motions of individual, mainly nearby, stars within such a reference frame will be dealt with here.

Suppose that a star has a velocity $\mathbf{V}$ with respect to the sun, strictly speaking the barycentre, and that it is observed in the direction $\mathbf{s}$. This velocity may be resolved into a radial component V_r and a transverse velocity $\mathbf{V}_T$, where

$$\mathbf{V} = V_r\mathbf{s} + \mathbf{V}_T. \tag{11.1}$$

It immediately follows that

$$V_r = \mathbf{V} \cdot \mathbf{s}$$

and

$$\mathbf{V}_T = \mathbf{s} \times (\mathbf{V} \times \mathbf{s}). \tag{11.2}$$

In principle, the radial component can always be measured from the Doppler shift in the star's spectrum. The transverse velocity cannot be directly measured. Its direction is known from the positional displacement $\dot{\mathbf{s}}$ of the star, its proper motion, but its magnitude can only be determined if the star's distance from the sun is already known. Parallax measurements provide this information for a minority of stars, but in most cases the parallactic displacement is too small to be measured with any certainty. While parallax is an annual periodic phenomenon, proper motion is a secular effect that may be allowed to build up over long intervals to a measurable displacement.

In Fig. 11.1, let X be the position of the star and S that of the sun, which will be treated as fixed. Now the velocity of the star with respect to the sun will change with time, but only very slowly. The time-scale of variation will be of the order of a period of galactic rotation $(\sim 2 \times 10^8$ years). Consequently, it is a realistic, and very useful, approximation to treat the star's velocity with respect to the sun as a constant. The path of the star is then represented by the straight line AX being described with a constant velocity V. The point A is the foot of the perpendicular from the sun to the star's path. Denote the star's distance from the sun XS by r and the angle XSA by θ; these are simply polar coordinates of X referred to the sun as pole and SA as initial line.

For definiteness, let us state that r is measured in kilometres and θ in radians, while the time t is measured in years. It is convenient, however, to express velocities in km s^{-1}. In these units the radial and transverse velocities are given by

$$V_r = V \sin \theta = \frac{1}{n}\frac{\mathrm{d}r}{\mathrm{d}t}, \tag{11.3}$$

$$V_T = V \cos \theta = \frac{1}{n}r\frac{\mathrm{d}\theta}{\mathrm{d}t}, \tag{11.4}$$

where n is the number of seconds in one year, namely 3.156×10^7.

A star's *annual proper motion* is defined as its total angular displacement in one year on the celestial sphere referred to a fixed equator and equinox. It

---

Figure 11.1. A star's velocity relative to the sun.

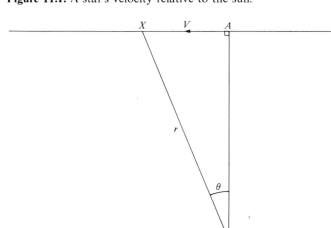

is normally measured in arc seconds yr^{-1}. In these units

$$\mu = \frac{d\theta}{dt} \operatorname{cosec} 1''. \tag{11.5}$$

The star's distance may be expressed in terms of its parallax π. Again, this is normally measured in arc seconds. So, using (8.11), the distance r in km is given by

$$r = a\pi^{-1} \operatorname{cosec} 1'', \tag{11.6}$$

where a is the astronomical unit expressed in km, i.e. $a = 1.496 \times 10^8$. Then, (11.4) may be written as

$$V_T = \frac{a}{n}\frac{\mu}{\pi}, \tag{11.7}$$

which gives the numerical result that

$$V_T = 4.74 \frac{\mu}{\pi} \text{ km s}^{-1}. \tag{11.8}$$

The star's annual proper motion μ may be resolved into components in right ascension and declination, respectively written as μ_α and μ_δ. These components give the annual rates of change of the star's right ascension and declination. In Fig. 11.2, X represents the position of the star on the celestial sphere at one instant and X' its position dt years later, so that $XX' = \mu\, dt$. If P is the north celestial pole, the angle PXX', denoted by ϕ, is called the

---

Figure 11.2. The position angle ϕ of a star's proper motion.

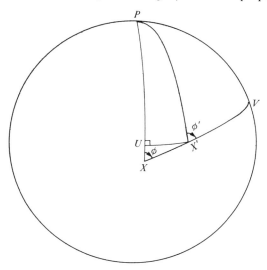

position angle of the proper motion. It is measured in the direction shown in the diagram and can take any value between 0 and 360°.

Construct a small circle arc, with pole P, through the point X' to cut PX in U. Then if (α, δ) are the equatorial coordinates of X and $(\alpha + d\alpha, \delta + d\delta)$ those of the point X', it is easily seen that

$$UX' = d\alpha \cos (\delta + d\delta)$$

$$UX = d\delta.$$

The small figure UXX' may be approximated to a plane triangle right-angled at U. Consequently we find to first order that

$$d\alpha \cos \delta = \mu \, dt \sin \phi$$

$$d\delta = \mu \, dt \cos \phi.$$

The components of proper motion are simply the derivations $d\alpha/dt$ and $d\delta/dt$. In practical units, however, μ_α will be expressed in time seconds yr^{-1} and μ_δ in arc seconds yr^{-1}. Hence

$$\mu_\alpha = \tfrac{1}{15}\mu \sin \phi \sec \delta$$
$$\mu_\delta = \mu \cos \phi. \tag{11.9}$$

The above analysis has not yet made use of the assumption that the star's velocity is a constant. In particular, equations (11.8) and (11.9) would apply in any event. This assumption is, however, required to deduce the rates of change of the components μ_α and μ_δ in the next section. One consequence of the assumption that the star is moving in a straight line is that its path on the celestial sphere will be a great circle. (The star's path and the sun together define a plane which intersects the celestial sphere in a great circle.)

Let V be the point on the celestial sphere which is determined by the direction of the star's velocity. Then V is clearly on the great circle arc XX' produced. Denote the angle $PX'V$ by ϕ' – this is the position angle of the star's proper motion at time $t + dt$ and may legitimately be written as

$$\phi' = PX'V = \phi + d\phi. \tag{11.10}$$

11.2   Intrinsic changes in proper motion

The components of proper motion are first derivatives of the star's coordinates. Computing the star's motion over an interval t as $(\mu_\alpha t, \mu_\delta t)$ is equivalent to truncating a Taylor series after the first-order terms. For nearby stars with large proper motions this is insufficient, and at least the second-order terms must be included. These are expressed as the derivatives of μ_α and μ_δ, which are derived below. In this derivation we maintain a fixed equator and equinox so that the changes in μ_α and μ_δ are due entirely to the motion of the *star* on the celestial sphere. We refer to these changes as the intrinsic changes in the proper motion components. The components also vary due to precession, but such changes are excluded from consideration here.

Taking the derivatives of μ_z and μ_δ from equation (11.9) yields the following two formulae

$$\frac{\mathrm{d}\mu_z}{\mathrm{d}t} = \frac{1}{15}\frac{\mathrm{d}\mu}{\mathrm{d}t}\sin\phi\sec\delta + \frac{1}{15}\mu\cos\phi\sec\delta\frac{\mathrm{d}\phi}{\mathrm{d}t}$$

$$+ \frac{1}{15}\mu\sin\phi\sec\delta\tan\delta\frac{\mathrm{d}\delta}{\mathrm{d}t} \tag{11.11}$$

$$\frac{\mathrm{d}\mu_\delta}{\mathrm{d}t} = \frac{\mathrm{d}\mu}{\mathrm{d}t}\cos\phi - \mu\sin\phi\frac{\mathrm{d}\phi}{\mathrm{d}t}.$$

The time derivatives of ϕ and δ in these equations are in circular measure, but otherwise practical units are being used. Moreover,

$$\frac{\mathrm{d}\delta}{\mathrm{d}t} = \mu_\delta\sin 1''. \tag{11.12}$$

So, using the results of (11.9), a little simplification may be effected to give

$$\frac{\mathrm{d}\mu_z}{\mathrm{d}t} = \frac{1}{15}\frac{\mathrm{d}\mu}{\mathrm{d}t}\sin\phi\sec\delta + \frac{1}{15}\mu_\delta\sec\delta\frac{\mathrm{d}\phi}{\mathrm{d}t}$$

$$+ \mu_z\mu_\delta\tan\delta\sin 1'' \tag{11.13}$$

$$\frac{\mathrm{d}\mu_\delta}{\mathrm{d}t} = \frac{\mathrm{d}\mu}{\mathrm{d}t}\cos\phi - 15\mu_z\cos\delta\frac{\mathrm{d}\phi}{\mathrm{d}t}.$$

In order to use these formulae, it will be necessary to obtain rates of change of both the total proper motion and its position angle.

Refer back to Fig. 11.2. Denoting the coordinates of X' by (α', δ'), the following parts may be identified in spherical triangle PXX': $PX = 90° - \delta$, $PX' = 90° - \delta'$, $PXX' = \phi$, $PX'X = 180° - \phi'$. The sine formula applied to this triangle yields

$$\cos\delta\sin\phi = \cos\delta'\sin\phi'.$$

This implies that, as the stars move along the great circle $XX'V$, the quantity $\cos\delta\sin\phi$ is conserved, i.e.

$$\frac{\mathrm{d}}{\mathrm{d}t}(\cos\delta\sin\phi) = 0,$$

yielding

$$\frac{\mathrm{d}\phi}{\mathrm{d}t} = \tan\phi\tan\delta\frac{\mathrm{d}\delta}{\mathrm{d}t}.$$

From equations (11.9) and (11.12), this can be written as

$$\frac{\mathrm{d}\phi}{\mathrm{d}t} = 15\mu_z\sin\delta\sin 1''. \tag{11.14}$$

The rate of change of μ is known as the *perspective acceleration*. It is derived by differentiating equation (11.4) which then gives

$$-V \sin \theta \frac{d\theta}{dt} = \frac{1}{n} \frac{dr}{dt} \frac{d\theta}{dt} + \frac{r}{n} \frac{d^2\theta}{dt^2}.$$

The derivatives of θ are eliminated by using (11.5) and its first derivative. Further simplification is provided by (11.3) to give the result that

$$\frac{d\mu}{dt} = -\frac{2n\mu V_r}{r}.$$

Substituting for r from (11.6) then gives the perspective acceleration as

$$\frac{d\mu}{dt} = -\frac{2n}{a} V_r \mu \pi \sin 1''. \tag{11.15}$$

In numerical form, this is

$$\frac{d\mu}{dt} = -0.422 V_r \mu \pi \sin 1''. \tag{11.16}$$

The expressions that have been derived for $d\mu/dt$ and $d\phi/dt$ may now be inserted into (11.13). After a little simplification the final results are obtained as

$$\frac{d\mu_\alpha}{dt} = -0.422 V_r \pi \mu_\alpha \sin 1'' + 2\mu_\alpha \mu_\delta \tan \delta \sin 1''$$

$$\frac{d\mu_\delta}{dt} = -0.422 V_r \pi \mu_\delta \sin 1'' - 225\mu_\alpha^2 \sin \delta \cos \delta \sin 1''. \tag{11.17}$$

Notice that, in these expressions, μ_α and μ_δ are measured respectively in time and arc seconds per year, π is in arc seconds, and V_r is in km s^{-1}.

The derivatives given in (11.17) will only be required if the proper motion is comparatively large, a circumstance that will arise for a nearby high-velocity star. They clearly allow the star's displacement to be computed with greater precision. Suppose, for example, that the star has equatorial coordinates (α, δ) and proper motion components (μ_α, μ_δ) at some initial epoch. Then the star's coordinates t years later will be (α', δ') where

$$\alpha' = \alpha + \left(\mu_\alpha + \tfrac{1}{2} t \frac{d\mu_\alpha}{dt} \right) t$$

$$\delta' = \delta + \left(\mu_\delta + \tfrac{1}{2} t \frac{d\mu_\delta}{dt} \right) t. \tag{11.18}$$

This is sufficiently accurate for almost all stars over intervals of the order of 100 years or less. It is only in quite pathological cases that the second derivatives of the proper motion need to be included.

The analysis of this section is based on the assumption that the star's velocity relative to the sun is constant. In fact, this assumption permits an exact solution for the star's displacement as the following considerations show (cf. also Problem 11.2).

Suppose $s = (x, y, z)$ is the unit vector in the direction of the star. Then the proper motion may be considered as the vector μ where

$$\mu = \dot{s} = \frac{d}{dt}(\cos \alpha \cos \delta, \sin \alpha \cos \delta, \sin \delta). \tag{11.19}$$

The three components of μ may be readily expressed in terms of μ_z and μ_δ. Maintaining the system of 'practical units' that has been used so far, the components of μ in arc seconds are found to be

$$\mu_x = -15 \sin \alpha \cos \delta \, \mu_z - \cos \alpha \sin \delta \, \mu_\delta$$
$$\mu_y = 15 \cos \alpha \cos \delta \, \mu_z - \sin \alpha \sin \delta \, \mu_\delta \tag{11.20}$$
$$\mu_z = \cos \delta \, \mu_\delta.$$

The transverse velocity vector is related to μ by the vector form of equation (11.7), namely

$$V_T = \frac{a\mu}{n\pi}. \tag{11.21}$$

Consequently the total space velocity of the star is given, from (11.1), by

$$V = V_r s + \frac{a}{n\pi}\mu. \tag{11.22}$$

Let r be the initial position vector of the star, i.e. $r = rs$, and let r' be the position vector t years later. Since V is a constant vector, it follows that

$$r' = rs + Vnt.$$

By (11.6), this can be written in the form

$$r' = ks*$$

where k is a constant (equal to $a\pi^{-1} \operatorname{cosec} 1''$) and the $s*$ is a near unit vector given by

$$s* = s\left(1 + V_r \frac{\pi t}{4.74} \sin 1''\right) + \mu t \sin 1''. \tag{11.23}$$

Completing the normalization of $s*$ will give the unit vector in the direction of the star at the later epoch. The validity of this analysis is, of course, limited to the accuracy of the underlying assumption that the star's velocity V with respect to the sun is constant.

11.3   Precessional changes in proper motion

In the previous section the intrinsic changes in a star's proper motion were investigated. These arose from two related causes, (i) the change in μ itself (the perspective acceleration), and (ii) changes in how μ is decomposed into its two components as the star becomes displaced over the surface of the celestial sphere. It must be clearly understood that while the displacement was taking place the coordinate system was regarded as fixed. The star's right ascension and declination and its proper motion components were all referred to a fixed equator and equinox. No precessional effects were considered. They are considered now.

Whatever the date of an observation or an ephemeris, the coordinates of a star may be referred to the equator and equinox of any epoch, not necessarily the epoch of date. The same is true of the proper motion components. The problem considered in this section is, therefore, how to transform these components from one precessional epoch to another.

The most precise way of doing this, and analytically the most direct, is to apply the rotation matrix discussed in section 9.6. Suppose μ_0 is the proper motion vector referred to the equator and equinox of the epoch t_0, and that μ is this same vector but referred to equator and equinox of time t. Then, by analogy with equation (9.32)

$$\mu = \mathbf{P}\,\mu_0, \tag{11.24}$$

where $\mathbf{P}$ is the rotation matrix relating the two epochs. The vectors μ_0 and μ are related to the corresponding components in right ascension and declination by equation (11.20) and its inverse, which is

$$\mu_\alpha = \tfrac{1}{15}(1-z^2)^{-1}(x\mu_y - y\mu_x)$$
$$\mu_\delta = (1-z^2)^{-1/2}\mu_z. \tag{11.25}$$

Although this method is analytically attractive, even elegant, it may involve unnecessarily lengthy computation, particularly if the elements of the rotation matrix need to be calculated *ab initio*. The proper motion components are themselves small quantities and their precessional rates of change are, in reality, second-order quantities. It is not necessary, therefore, to maintain as full a precision in calculating their precessional rates of change as one would in calculating the precessional changes in a star's coordinates. The approximate formulae derived in section 9.4 will normally suffice.

Consider again the two equations (11.11) which give the rates of change of the proper motion components. Since they are merely the derivatives of the definitions of (μ_α, μ_δ), they are valid whatever the cause of the change. To obtain precessional rates of change, it is only necessary to substitute appropriate expressions for the three derivatives of the right-hand sides. Clearly, in this case,

$$\frac{d\mu}{dt} = 0, \tag{11.26}$$

since precession cannot affect the total proper motion, only its direction. Moreover, by (9.13),

$$\frac{d\delta}{dt} = n \cos \alpha \sin 1'',$$

(11.27)

where n is expressed in arc seconds.

The precessional rate of change of the position angle ϕ is established as follows. Let P and P' be the two positions of the pole (Fig. 11.3) at epochs separated by a short interval dt. The position of the star is X and its proper motion is along the great circle XV. Clearly the precessional increment in ϕ over the interval dt is $d\phi = P'XP$. Now let (α, δ) and (α', δ') be, respectively, the star's coordinates referred to the poles P and P' and the corresponding equinoxes. Then $P'X = 90° - \delta'$. Due to luni–solar precession the celestial pole is describing a small circle about the pole of the ecliptic K. So the length of the arc PP' is given by

$$PP' = \psi \sin \varepsilon \, dt = n \, dt.$$

Strictly this is the length of small circle arc, but over the short interval involved we may treat it as the great circle arc PP'. Moreover, $KPP' = 90°$ and so $P'PX = \alpha$. Now apply the sine formula to spherical triangle $PP'X$ to obtain

$$\sin d\phi \cos \delta' = \sin (n \, dt) \sin \alpha.$$

Figure 11.3. Precessional change in position angle.

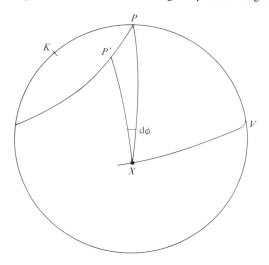

Proceeding to the limit as dt tends to zero, this gives

$$\frac{\mathrm{d}\phi}{\mathrm{d}t} = n \sin \alpha \sec \delta \sin 1'', \tag{11.28}$$

since n is expressed in arc seconds.

The results of equations (11.26), (11.27) and (11.28) may now be substituted into the right-hand sides of equation (11.11). The precessional rates of change of (μ_z, μ_δ) are then given as

$$\frac{\mathrm{d}\mu_z}{\mathrm{d}t} = n\left(\mu_z \cos \alpha \tan \delta + \frac{\mu_\delta}{15} \sin \alpha \sec^2 \delta\right) \sin 1''$$

$$\frac{\mathrm{d}\mu_\delta}{\mathrm{d}t} = -15n\mu_z \sin \alpha \sin 1''. \tag{11.29}$$

These equations use practical units of time seconds for μ_z and arc seconds for μ_δ and n.

Only luni–solar precession was included in the derivation. This is justified, however, since planetary precession affects only K leaving the celestial pole unaffected. It can clearly have no effect either on the star's declination or position angle.

If the proper motion components are required at several times, referred on each occasion to the equator and equinox of date, it will be necessary to apply both the intrinsic and the precessional rates of change. Under these circumstances the total rates of change are obtained by adding together the right-hand sides of equations (11.17) and (11.29). This is discussed further in the next chapter.

The precessional constant n that appears in (11.29) is about $20''$ which is larger than the proper motion of any star. It is clear, therefore, that whenever both rates of change are to be included, it is the precessional rate that will be the more significant. When deriving precessional changes over a long time interval it may be necessary to improve the accuracy of (11.29). This may be achieved by using the value of n corresponding to the midpoint of the time interval. The greatest accuracy, however, is achieved by dispensing with (11.29) altogether and using the rotation matrix, as explained earlier.

11.4   Barycentric radial velocity

Radial velocity is not strictly a positional measurement, but it is clearly related to proper motion. Moreover, it is affected by viewpoint, so that it is necessary to distinguish between the topocentric, geocentric and barycentric radial velocities. Since radial velocity is independent of distance, the distinction is always significant. In what follows we shall denote the topocentric, geocentric and barycentric radial velocities by V_r'', V_r' and V_r respectively.

The radial velocity is deduced from a measurement of the Doppler displacement or red-shift, z, say. Suppose that v_0 is the observed frequency of a spectral line, and v is the unshifted frequency of this line as provided by a laboratory source. Then the red-shift z is given by the equation

$$1+z=\frac{v}{v_0}. \tag{11.30}$$

The observed red-shift z is related to the topocentric radial velocity V_r'' by the classical Doppler formula, which is

$$z=\frac{V_r''}{c}. \tag{11.31}$$

This formula is only valid for small red-shifts corresponding to radial velocities much less than the speed of light. Large red-shifts are encountered in extragalactic sources and a relativistic interpretation is essential; this will be discussed later. Equation (11.31) is, however, usually adequate for the reduction of spectroscopic measurements of ordinary stars. Notice the convention for radial velocity is that velocities of recession are treated as positive.

A Doppler measurement, therefore, yields the topocentric radial velocity of a star. This must be reduced to the barycentre of the solar system before any dynamical interpretation of the phenomena is possible. Suppose that the star is in the direction of the unit vector $\mathbf{s}$ and that $\mathbf{v}'$ is the observer's velocity with respect to the earth's centre, while $\mathbf{v}$ is the velocity of this geocentric point with respect to the barycentre of the solar system. Then the geocentric and barycentric radial velocities are given by

$$\begin{aligned} V_r' &= V_r'' + \mathbf{v}' \cdot \mathbf{s} \\ V_r &= V_r' + \mathbf{v} \cdot \mathbf{s}. \end{aligned} \tag{11.32}$$

The velocity $\mathbf{v}'$ was considered in section 4.8 in connection with diurnal aberration. It is directed towards the point E in Fig. 11.4, the east point of the horizon. If X is the star's position (H, δ) on the celestial sphere and P is the north celestial pole, then $PE = 90°$, $PX = 90° - \delta$, and spherical angle $XPE = 90° + H$. The magnitude of $\mathbf{v}'$ is given in equation (4.67). To sufficient accuracy, this gives

$$v' = 0.465 \cos \phi' \text{ km s}^{-1}, \tag{11.33}$$

where ϕ' is the observer's geocentric latitude.

Now applying the cosine formula to spherical triangle PXE will yield

$$\mathbf{v}' \cdot \mathbf{s} = v' \cos XE = -v' \cos \delta \sin H.$$

So by (11.32) and (11.33) we have the geocentric radial velocity expressed in km s^{-1} as

$$V_r' = V_r'' - 0.465 \cos \phi' \cos \delta \sin H. \tag{11.34}$$

The equatorial components of the velocity **v** of the earth with respect to the solar system barycentre are tabulated in the *Astronomical Almanac* as $\mathbf{v} = (\dot{X}, \dot{Y}, \dot{Z})$. The units in which these components are given are AU day^{-1}. Applying the necessary conversion factor, therefore, the barycentric radial velocity is given directly by (11.32), in km s^{-1}, as

$$V_r = V_r' + 1731.5(\dot{X} \cos \alpha \cos \delta + \dot{Y} \sin \alpha \cos \delta + \dot{Z} \sin \delta). \tag{11.35}$$

The earth's velocity may be resolved into two velocities, one transverse and one parallel to the minor axis of its orbit. Writing

$$\mathbf{v} = \mathbf{V}_0 + \mathbf{V}_1, \tag{11.36}$$

these two velocities are given by equation (8.23) in ecliptic coordinates. Moreover, the magnitudes of these two velocities are expressible in terms of the constant of aberration. In fact,

$$\begin{aligned} V_0 &= \kappa c \\ V_1 &= e\kappa c, \end{aligned} \tag{11.37}$$

where e is the eccentricity of the earth's orbit.

Figure 11.4

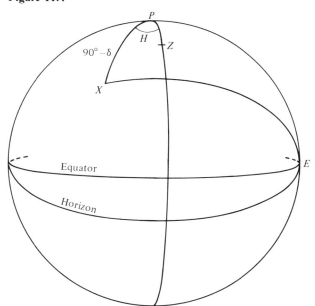

Let (λ, β) be the star's ecliptic coordinates. Then

$$\mathbf{s} = (\cos \lambda \cos \beta, \sin \lambda \cos \beta, \sin \beta).$$

The barycentric radial velocity follows from (11.32), (8.23) and (11.37). The result is

$$V_r = V_r' + \kappa c \cos \beta [\sin (\lambda_\odot - \lambda) + e \sin (\lambda - \varpi)]. \qquad (11.38)$$

In this equation κ is, of course, expressed in circular measure.

When numerical values are substituted for the constant of aberration and the orbital elements the formula in km s^{-1} becomes

$$V_r = V_r' + 29.79 \cos \beta \sin (\lambda_\odot - \lambda) + 0.50 \cos \beta \sin (\lambda - \varpi). \qquad (11.39)$$

The precision of this formula is inferior to that of (11.35), since no distinction has been drawn between the heliocentric point and the barycentre. Nevertheless it is sufficiently accurate for many purposes.

The small final term in (11.39) is analogous to the E-term of annual aberration. It is virtually constant and exhibits no annual variation. In some circumstances it may be omitted. For example, when deriving the orbit, a spectroscopic binary (a problem to be considered in chapter 19), one is primarily interested in the variation in the star's radial velocity. Clearly, the annual variation must be extracted to reveal the true orbital variation, but the constant term will be of no consequence. It will affect only the derived value of the radial velocity of the entire binary system. On the other hand, if stars' radial velocities are being studied as a means of investigating stellar motions or galactic dynamics, then the omission of the final term will introduce systematic errors of the order of $\frac{1}{2}$ km s^{-1} which depend on latitude and longitude. In these circumstances it should be included.

11.5 Pulsar timing

The radial velocity of every star will exhibit the annual variation contained in the second term of (11.39). The amplitude and phase of this variation depend on the star's position. Since the angle $(\lambda_\odot - \lambda)$ spans all four quadrants in the course of one year, equation (11.39) indicates that the star's observed radial velocity, after correction for the small diurnal variation, will oscillate about its mean value with an amplitude of $29.79 \cos \beta$ km s^{-1}. Moreover, maximum radial velocity is attained when the longitude of the sun is $\lambda + 90°$.

This annual variation in radial velocity has been used as a means of determining the positional coordinates of radio pulsars. The spectra of these radio sources do not exhibit spectral lines as such, but the pulse itself is a periodic phenomenon, and its repetition frequency is subject to the Doppler shift. The intrinsic pulse repetition frequency (PRF) cannot be determined without knowing the pulsar's velocity with respect to the sun. The mean value, $\tilde{v}$, say, only gives the frequency that would be measured from the heliocentric point. The variations about this mean, however, are due to the earth's orbital velocity, and so, in the notation of the previous section, we may write

$$\frac{\tilde{v}}{v_0} - 1 = z = -\frac{\mathbf{V}_0 \cdot \mathbf{s}}{c}. \tag{11.40}$$

Here, v_0 is the observed PRF and it is assumed that correction for the diurnal effect has already been made. The effect of the constant component V_1 of the earth's orbital motion is, of course, absorbed into the mean PRF $\tilde{v}$.

It follows from (11.40) and (11.38) that

$$\frac{\tilde{v}}{v_0} = 1 - \kappa \cos \beta \sin (\lambda_\odot - \lambda). \tag{11.41}$$

The amplitude of the variation then allows $\cos \beta$ and hence β to be determined, while the time of year corresponding to minimum v_0 has $\lambda_\odot = \lambda + 90°$. The longitude is, therefore, derived as well. The determination of β becomes imprecise near the ecliptic. The position of a pulsar may by this means be determined from observations with a simple antenna system without recourse to high-resolution radio techniques. Pulse timings can be made so precisely that a positional accuracy in excess of one arc second is possible. The method has been outlined in terms of simple elliptic motion of the earth about the sun. The precision of the measurements, however, necessitates a reduction of observations that is based on (11.35) rather than (11.38). The distinction between the barycentre and the centre of the sun will be significant.

The positions of pulsars may be determined either from the analysis of the pulse repetition rate as described above or from a study of the pulse arrival times at the earth. The two methods are in fact closely related. The difference in the arrival time of the pulse at the earth and at the barycentre may be determined from equation (10.22). Comparing this equation with (11.35), a strong similarity will be seen. The radial velocity formula is essentially the derivative of the time difference formula. The relationships of the Doppler shift to time will be clearer after it has been considered from a relativistic standpoint in the next section.

11.6   The red-shift in general relativity

In general relativity it is not entirely natural to distinguish between a Doppler shift, due to the relative motion of the source and the observer, and a gravitational red-shift, due loosely to a gravitational potential difference between the sites of emission and reception. Certainly the distinction is frequently made, but it is to some extent arbitrary. For the relative velocity of a source widely separated from the observer is not uniquely defined, but depends on the coordinate system. What is invariantly defined, however, is the red-shift.

To fix ideas, let us consider the emission of a sequence of pulses with repetition frequency v, but the same analysis may be applied to a spectrum line. Let v_0 be the observed pulse repetition frequency. Suppose that the event of emission of a particular pulse has coordinates (x_e, y_e, z_e, t_e), while the event of observation of that same pulse is (x_0, y_0, z_0, t_0). Although the particular choice of coordinate system need not affect the argument, it will be convenient to use isotropic quasi-Cartesian coordinates in the Schwarzschild metric, and so the argument is couched in the notation of particular coordinates. The two events of emission and observation are connected by a null geodesic of the spacetime. Consequently, their coordinates cannot all be arbitrarily assigned. For example, if the four coordinates of emission are specified, the three space coordinates of the observer can be arbitrarily selected, but not the time of observation. This is determined by the null geodesic. In more general terms we may say that the null geodesic is a functional relationship between the two sets of coordinates which may be written as

$$f(x_e, y_e, z_e, t_e, x_0, y_0, z_0, t_0) = 0 \tag{11.42}$$

Suppose next that $(x_e + dx_e, y_e + dy_e, z_e + dz_e, t_e + dt_e)$ are the coordinates of the event of emission of the immediately following pulse of the sequence, and that this pulse is observed by the same observer at the event $(x_0 + dx_0, y_0 + dy_0, z_0 + dz_0, t_0 + dt_0)$. The time between the emission of the two pulses as determined at the source is simply the interval, ds_e, say, between the two events of emission. Similarly, the observed time lapse between the reception of the two pulses is ds_0, the interval between the two events of observation. It, therefore, follows that the emitted and observed frequencies are given by

$$v = \frac{1}{ds_e}$$
$$v_0 = \frac{1}{ds_0}. \tag{11.43}$$

Now let $\mathbf{V}_e$ and $\mathbf{V}_0$ be respectively the 3-vectors that give the coordinate velocities of the source and the observer. Then

$$\mathbf{V}_e = \left(\frac{dx_e}{dt_e}, \frac{dy_e}{dt_e}, \frac{dz_e}{dt_e}\right)$$
$$\mathbf{V}_0 = \left(\frac{dx_0}{dt_0}, \frac{dy_0}{dt_0}, \frac{dz_0}{dt_0}\right). \tag{11.44}$$

Moreover, if we differentiate equation (11.42) totally with respect to t_e, the following result is obtained, in the fairly obvious notation,

$$\frac{\partial f}{\partial t_e} + \mathbf{V}_e \cdot \text{grad}_e \, f + \left(\frac{\partial f}{\partial t_0} + \mathbf{V}_0 \cdot \text{grad}_0 \, f \right) \frac{dt_0}{dt_e} = 0. \tag{11.45}$$

This last equation allows the ratio $dt_0 : dt_e$ to be established, that is, the ratio of coordinate times between pulses at emission and reception. What is required, however, is a similar relationship between the proper time intervals, for it follows, from (11.30) and (11.43), that

$$1 + z = \frac{ds_0}{ds_e}. \tag{11.46}$$

The final stage of the argument is to relate the proper time intervals to their coordinate time counterparts. This is done through the metric. For example, if the metric is the Schwarzschild metric in its isotropic quasi-Cartesian form, equation (3.53), one derives

$$\left(\frac{ds_e}{dt_e} \right)^2 = \frac{\left(1 - \dfrac{m}{2r_e} \right)^2}{(1 + m/2r_e)^2} - \frac{V_e^2}{c^2} \, (1 + m/2r_e)^4, \tag{11.47}$$

with a similar expression for ds_0/dt_0. Use of the metric in this way will, from (11.46), establish a relationship of the form

$$1 + z = h \frac{dt_0}{dt_e}, \tag{11.48}$$

where h is a function of the coordinates and the coordinate velocities of the source and the observer. The red-shift may then be derived by using equation (11.45). The form of (11.46) shows that it is an invariant quantity independent of the coordinate system. It contains both Doppler and gravitational effects.

Let us now return to the particular case of Schwarzschild spacetime. To sufficient accuracy, equation (11.47) yields

$$\frac{ds_e}{dt_e} = 1 - \frac{m}{r_e} - \frac{1}{2} \frac{V_e^2}{c^2}.$$

Consequently (11.48) will take the form

$$1 + z = \left[1 + \frac{m}{r_e} - \frac{m}{r_0} + \frac{1}{2} \frac{(V_e^2 - V_0^2)}{c^2} \right] \frac{dt_0}{dt_e}. \tag{11.49}$$

Notice that the bracket on the right-hand side of (11.49) differs from unity only to second order, so the first-order Doppler effect is contained in the derivative dt_0/dt_e.

The radial velocity of the source cannot be convincingly defined in an invariant manner. It is possible, however, to define a coordinate radial velocity, V_r, say. Writing $\mathbf{r}_e = (x_e, y_e, z_e)$ and $\mathbf{r}_0 = (x_0, y_0, z_0)$, a coordinate distance ρ is defined, as in (10.46), by

$$\rho \mathbf{s}_0 = \mathbf{r}_e - \mathbf{r}_0, \tag{11.50}$$

$\mathbf{s}_0$ of course being a unit vector. Then the coordinate radial velocity may be taken as

$$V_r = \frac{\mathrm{d}\rho}{\mathrm{d}t_e}. \tag{11.51}$$

The relationship (11.42) between the coordinates of emission and observation has already been established in the analysis of section 10.8. With some slight modifications, mainly in notation, equation (10.55) will give

$$t_0 = t_e + \frac{\rho}{c} + \frac{2m}{c} \ln\left(\frac{r_e + r_0 + \rho}{r_e + r_0 - \rho}\right). \tag{11.52}$$

The final term in this equation is already of second order, so its derivative is third order and may, therefore, be neglected. Differentiating

$$\frac{\mathrm{d}t_0}{\mathrm{d}t_e} = 1 + \frac{V_r}{c}, \tag{11.53}$$

to sufficient accuracy. This is exactly what one would expect classically. Finally combining this result with (11.49) gives the general relativistic red-shift in Schwarzschild spacetime as

$$z = \frac{V_r}{c} + m\left(\frac{1}{r_e} - \frac{1}{r_0}\right) + \frac{1}{2}\frac{(V_e^2 - V_0^2)}{c^2}. \tag{11.54}$$

This is, of course, only correct to second order in the velocity, and equivalently first order in the Schwarzschild radius. The first term on the right-hand side of (11.54) is the classical Doppler effect, the second is the gravitational red-shift and the third is the second-order Doppler effect.

Equation (11.54) was developed by considering a series of pulses and gives the Doppler displacement in the pulse repetition frequency. It may also be applied to a natural spectral line or to the radio frequency of a space-borne radio beacon. For the latter, all the terms may in some circumstances be significant.

The formula has been based on the Schwarzschild spacetime and so is immediately applicable to interplanetary probes. With reference to pulsar timings, it should be noted that the gravitational red-shift in the pulsar's own gravitational field is not included. It must be assumed that this is constant and is absorbed into the frequency of emission v_e. For this reason, equation (11.54) is not adequate for studying a binary pulsar. Further modelling is then required.

As a final example, let us apply (11.54) to solar spectral lines. Then r_e is one solar radius and $r_e \ll r_0$. To one per cent accuracy, at least, we may omit the term in m/r_0 and the second-order Doppler term to derive

$$z = \frac{V_r}{c} + \frac{m}{r_e}. \tag{11.55}$$

This formula then predicts that solar spectrum lines should exhibit a variable Doppler shift due to the radial component of the orbital motion and a systematic gravitational red-shift. The Doppler shift may have either sign and has an amplitude of about 0.5 km s^{-1}. The gravitational red-shift is equivalent to a velocity 0.64 km s^{-1} which is well within the compass of traditional measurements in solar spectroscopy. This is the third classical test of general relativity. Solar measurements, however, have been less convincing than might have been expected. The anticipated gravitational red-shift is largely obscured by systematic effects arising from convection in the solar atmosphere. The gravitational red-shift has, however, been unambiguously established by laboratory measurements (Pound and Snider, 1965). It is less important to general relativity's credibility *per se* than the other tests, since almost every post-Newtonian theory predicts an identical gravitational red-shift.

11.7   Systematic effects in proper motion and radial velocity

Measurement of a star's annual parallax is the only direct and unambiguous way of determining a star's distance. It is the foundation of the astronomical distance scale. This scale must, however, extend to distances which are far outwith the range of parallax determination. Beyond the sun's immediate vicinity, the distance of stars are deduced by statistical inference rather than direct measurement. This is a much subtler and more uncertain process. The distance scale is established through a long and elaborate chain of argument, in which the validity of each link must depend on all that has preceded it. The argument must make use of whatever observational evidence is available and proper motion and radial velocity play an important part. The former is an important distance indicator for stars in the sun's neighbourhood, but beyond the range of parallax measurement. Radial velocity provides a useful supplement, even in this context. For the most distant extragalactic sources, it is the only quantitative means that remains to give an indication of distance, however uncertain.

Routine proper motion measurements of a whole star field may be effected from a small number of astrographic plates taken over a considerable interval of time. Parallax measurements require a more concentrated effort and will often yield a null measurement. Radial velocity measurements involve studying stars singly so that the stellar spectrum may be compared with a laboratory standard. The proper motion measurement is, therefore, normally the starting point. It will provide the first indication of stars that are suitable candidates for parallax measurement.

All the stars in the solar neighbourhood, including the sun, are in orbital motion about the centre of the galaxy. The proper motion of a particular star arises from its *relative* transverse velocity with respect to the sun. This is usually much less than the velocity of either the star or the sun with respect to the galactic centre. What is being observed is essentially a differential effect. The local standard of rest (LSR) was defined in section 3.2 as the frame of reference moving about the galactic centre with the mean velocity of stars in the solar neighbourhood. Suppose that the sun has a velocity U with respect to the LSR. In the course of one year this amounts to a displacement of nU km, in the notation of section 11.1, and this displacement is reflected in the proper motion of a star. Now a displacement of a kilometres (1 AU) produces a star's annual parallax π. By analogy with (11.6), a displacement nU km is regarded as defining a *secular parallax h* given, in arc seconds, by

$$r = nU\, h^{-1} \operatorname{cosec} 1''. \tag{11.56}$$

The secular parallax gives the annual displacement due to solar motion, of a star whose direction is orthogonal to the direction of the sun's motion. From (11.6) and (11.56) one deduces that

$$h = \frac{nU}{a}\,\pi = \frac{U}{4.74}\,\pi. \tag{11.57}$$

That part of a star's proper motion that arises from the sun's, rather than its own, velocity with respect to the LSR is known as its (secular) *parallactic motion*. Let us denote the components of the parallactic motion by (P_χ, P_δ) and the total parallactic motion by μ_p. The point A on the celestial sphere (Fig. 11.5) to which the solar motion is directed is called the *solar apex*. Let (A, D) be its equatorial coordinates, and let (α, δ) be those of a star X. Now the parallactic motion of the star will be along the great circle AX. The angle χ, shown in Fig. 11.5, is its position angle. Hence from (11.9)

$$P_\chi = \tfrac{1}{15}\mu_p \sin\chi \sec\delta$$
$$P_\delta = \mu_p \cos\chi. \tag{11.58}$$

Neglecting the star's own motion with respect to the LSR, it is clear that its transverse velocity relative to the sun is $U \sin AX$ or $U \sin \lambda$, say, so by (11.8)

$$\mu_p = \frac{U\pi \sin\lambda}{4.74}. \tag{11.59}$$

Using this and (11.57), equation (11.58) may be expressed as

$$P_\chi = \tfrac{1}{15}h \sin \lambda \sin \chi \sec \delta$$
$$P_\delta = h \sin \lambda \cos \chi. \qquad (11.60)$$

Alternatively, equation (11.60) may be expressed in terms of the right ascension and declination of the solar apex. In spherical triangle PAX, we identify the following parts $PA = 90° - D, PX = 90° - \delta, AX = \lambda, A\hat{P}X = \alpha - A, P\hat{X}A = 180° - \chi$. The angles λ and χ are then eliminated from (11.60) by applications of the sine and analogue formulae, to give the results

$$P_\chi = \tfrac{1}{15}h \sin (\alpha - A) \cos D \sec \delta$$
$$P_\delta = h[\cos D \sin \delta \cos (\alpha - A) - \sin D \cos \delta]. \qquad (11.61)$$

The components P_χ and P_δ are systematic effects that will be present in every star's proper motion. A similar systematic effect is to be expected in the radial velocity. Let us denote that part of the radial velocity of a star that arises from the solar motion by V_p. Clearly $V_p = -U \cos \lambda$. Applying the cosine formula to spherical triangle PAX then yields

$$V_p = -U[\sin D \sin \delta + \cos D \cos \delta \cos (\alpha - A)]. \qquad (11.62)$$

Methods for the determination of the solar velocity U and its direction (A, D) are considered in chapter 14. Once they are known, the components of parallactic motion are given by (11.61) in terms of the star's secular

Figure 11.5

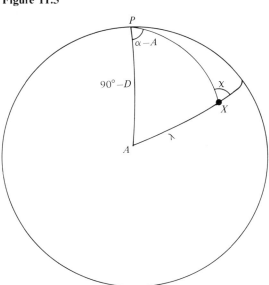

parallax h. The parallactic motion of a particular star may well be obscured by the star's peculiar motion, but statistical analysis of a homogeneous group can reveal its presence in a meaningful way. If the secular parallax can be determined, the trigonometric parallax, and hence the star's distance, may be derived from (11.57). Radial velocity measurements provide independent information through (11.62), which constitutes, if not a rigorous check, at least confirmatory evidence to reduce the persistence of reasonable doubt in the statistical process.

11.8   The cosmological red-shift

Radial velocity measurements become particularly important in observational cosmology. For very distant extragalactic sources, whether they are normal galaxies, peculiar galaxies, or quasars, observational data is of necessity rather sparse. Three important observable parameters that can be associated with a source are red-shift, apparent magnitude and angular diameter. Each can be used as a distance indicator. The apparent brightness and angular diameter of source should decrease in a known way with increasing distance, assuming that the luminosity and physical diameter of source are known. Even here, however, there are important relativistic modifications to the interpretation of the observations. The red-shift is interpreted as due to the overall expansion of the universe, and it, too, must be treated relativistically. Cosmological models of the universe can make predictions of how the three parameters are related for identical sources. It is our intention to relate the red-shift to the essential features of the simplest, but the most widely adopted, cosmological models.

Homogeneous isotropic models in general relativity possess a metric which may be expressed as

$$ds^2 = dt^2 - \frac{[R(t)]^2}{c^2} \left(\frac{dr^2}{1 - kr^2} + r^2\, d\theta^2 + r^2 \sin^2 \theta\, d\phi^2 \right). \qquad (11.63)$$

This is known as the Robertson–Walker form of the metric. The constant k assigns the curvature of the space, and there is no loss of generality in restricting it to values $k = \pm 1$ or 0. The coordinate system is regarded as being centred on the observer. It is a 'co-moving' coordinate system so that the spatial coordinates (r, θ, ϕ) are constant for any material particle and, therefore, for any source of radiation. The expansion of the universe is contained in the variation of the *scale factor* R which is a function of the time t only. The era in the past at which R was zero corresponds to the big bang. The function $R(t)$ is established from Einstein's field equations after the model parameters such as the average density of the universe and its rate of expansion at the present time have been inserted.

The red-shift z is derived by the method that was developed in section 11.6. Let us consider radiation emitted from the point (r_e, θ_e, ϕ_e) at time t_e and received at the origin at present time t_0. The world line of this radiation is a radial null geodesic, which from (11.63) will have differential equation

$$\frac{dr}{dt} = -\frac{c(1-kr^2)^{1/2}}{R(t)}.$$

Consequently, the time of emission and of observation are related by the integral equation

$$\int_{t_e}^{t_0} \frac{dt}{R(t)} = \frac{1}{c} \int_0^{r_e} \frac{dr}{(1-kr^2)^{1/2}}. \tag{11.64}$$

Now the red-shift is related by (11.46) to proper time intervals. But, since both the source and the observer are at rest in the coordinate system, (11.63) indicates that

$$1+z = \frac{ds_0}{ds_e} = \frac{dt_0}{dt_e}. \tag{11.65}$$

Now the right-hand side of (11.64) is independent of time and so differentiating with respect to t_e will yield

$$\frac{1}{R(t_0)} \frac{dt_0}{dt_e} - \frac{1}{R(t_e)} = 0.$$

Finally substituting into (11.65) will give the red-shift as

$$1+z = \frac{R(t_0)}{R(t_e)}. \tag{11.66}$$

Thus the red-shift in distant sources arises from the change in the scale factor of the universe between the time of emission of the radiation and the time of its reception. This is rather different from a straightforward Doppler shift. It is probably best to regard the red-shift simply as a property of the spacetime, rather than attempt to attribute it to kinematical, or even gravitational, causes. The more distant sources will have larger red-shifts since the travel time of the radiation is longer and so the universal scale factor had a smaller value when the radiation we observe now was emitted.

Proper motion is not measurable in extragalactic sources. A more specific comment to this effect has already been made at the very end of section 3.2. There are, however, observed structural changes within some very distant radio sources that are analogous to the proper motion, although a proper motion of the source as an entity may not be detectable. The only way that the distance of a quasar may be estimated involves regarding the observed red-shift as of wholly cosmological origin and interpreting it in terms of the expansion of the universe. When the distance is known, the structural changes may be translated into transverse velocities using the standard proper motion equation (11.8). The velocities deduced in this way are sometimes very large, so that doubt has been cast on the distance estimate and even on the principle of a cosmological interpretation of the red-shift. In several cases the velocities deduced exceed the speed of light. While these superluminal velocities may be difficult to understand, they are not in contradiction of any relativistic principle as the following argument will show.

Consider a source S at a distance r from the observer O, as shown in Fig. 11.6. Let us suppose that the source is moving with velocity V in a direction that makes an angle ψ with the inward radial direction SO. Then the radial velocity is $-V \cos \psi$. Let S be the position of the source at time t and S' its position a short interval dt later. Then $SS' = V \, dt$. Finally, let $d\theta$ be the angle subtended by SS' at the observer. Then

$$r \, d\theta = V \sin \psi \, dt. \tag{11.67}$$

Now let t_0 be the time at which the radiation emitted from S is observed at O. The true transverse velocity is $r \, d\theta / dt$ or $V \sin \psi$ from (11.67). What is immediately deduced as the transverse velocity from proper motion, or

Figure 11.6

structural changes, is not $r \, \mathrm{d}\theta/\mathrm{d}t$, but $r \, \mathrm{d}\theta/\mathrm{d}t_0$. Since $r = c(t_0 - t)$, it follows from differentiation that

$$\frac{\mathrm{d}r}{\mathrm{d}t} = -V \cos \psi = c\left(\frac{\mathrm{d}t_0}{\mathrm{d}t} - 1\right).$$

Consequently, the immediately deduced transverse velocity is

$$V_T = r \frac{\mathrm{d}\theta}{\mathrm{d}t_0} = \frac{V \sin \psi}{1 - V \cos \psi/c}. \tag{11.68}$$

This can be superluminal even if the true velocity is less than the velocity of light. For a fixed value of V the maximum value of V_T is found from (11.68) to correspond, not to $\psi = 90°$, but to $\psi = \cos^{-1} V/c$. This maximum value of V_T is then found to be γV, which is superluminal if V exceeds $c/\sqrt{2}$.

Problems

11.1 Show that the assumption that a star's velocity with respect to the sun is constant implies that its radial velocity must always increase in an algebraic sense. Show further that, in conventional units, the rate of change is

$$\frac{\mathrm{d}\rho}{\mathrm{d}t} = 4.74 \frac{\mu^2}{\pi} \sin 1''.$$

11.2 The following data apply to Barnard's star for the epoch 1950.0:

$\alpha = 17^\mathrm{h} 55^\mathrm{m} 40^\mathrm{s}$ $\delta = 4° 33'$

$\mu = 10''.25$ $\phi = 356°$

$\rho = -108 \ \mathrm{km \ s^{-1}}$ $\pi = 0''.546$.

Determine the epoch when the star's radial velocity will be zero. Further, calculate the star's position (referred to the 1950 equator and equinox), its proper motion and parallax at this epoch.

11.3 A star has position vector (x_0, y_0, z_0) at the standard epoch. Show that, if the perspective acceleration is neglected, its position vector t years later has components given by

$$x = x_0 + \mu_x t - \tfrac{1}{2}(\mu t)^2 x_0, \text{ etc.}$$

where the components of the vector $\boldsymbol{\mu}$ and its magnitude are expressed in circular measure. Under what circumstances would this be a good approximation?

11.4 A star has equatorial coordinates (α, δ) and its proper motion has position angle ϕ, all referred to the mean equator and equinox of date. Show that the position angle referred to the standard equator and equinox is ϕ_0, which is given exactly by the equation

$$\tan(\phi - \phi_0) = \frac{\sin(\alpha - z_A)\sin\theta_A}{\cos\delta\cos\theta_A + \sin\delta\sin\theta_A\cos(\alpha - z_A)}$$

where z_A and θ_A are the precessional angles defined in section 9.5.

11.5 Calculate the gravitational red-shift for the sodium line at 5896 Å in the sun's spectrum, assuming that the sun's radius is 6.96×10^5 km.

11.6 A pulsar's period is found to vary during the course of the year between the limits $0^s.7144600$ and $0^s.7145774$. The maximum period occurs when the longitude of the sun is $335°$. Calculate the pulsar's right ascension and declination. (Take $\varepsilon = 23°\,26'.5$, $\kappa = 20''.496$.)

11.7 Prove that it is always possible to adjust the radial coordinate and the scale factor R in the Robertson–Walker metric so that it takes the form (11.63) with $k = \pm 1$ or 0.

11.8 In a homogeneous isotropic model of the universe, the 'horizon' is determined by the condition that the red-shift tends to infinity. This normally occurs at a finite value of the radial coordinate r. Show that this implies that the angular diameter of sources, of the same physical extent, will have a minimum value at a finite red-shift.

11.9 Justify the statements that have been made in the final paragraph of section 11.8.

12

Mean and apparent coordinates

12.1 Terminology

The coordinates of a star are subject to many causes of variation. The most important are proper motion, precession and nutation, aberration and parallax. These effects have been separately discussed in chapters 11, 9 and 8. It is the intention of this chapter to derive formulae that combine in a meaningful and practical way the positional effects arising from these disparate causes.

Suppose that the coordinates of a star are given as a right ascension and a declination. If this information is to be unambiguous, it is necessary to have stated, or at least understood, three additional pieces of information. We must ask the following questions:

(i) To what *date* do these coordinates refer?

(ii) What is the centre of the celestial sphere on which these coordinates are defined?

(iii) How have the *equator* and *equinox* been chosen on the celestial sphere?

The date referred to in the first question is the date of the observation or of the ephemeris. Until this is known, no allowance can be made for the proper motion of the star.

The centre of the celestial sphere defines the origin of the coordinate system that is being used. There are several possibilities, of which the most important are the barycentre of the solar system and the geocentric point. The transference from one to the other involves correction for annual aberration and annual parallax. The much smaller relativistic light deflection and second-order aberrational terms are omitted from consideration in most of this chapter.

In practice, the equator and equinox will be chosen either as a *mean* equator and equinox or as a *true* equator and equinox for some epoch. In the former case only precessional effects are allowed for, while in the latter nutation is included as well. In either event the epoch may or may not be the same as the date of observation referred to in question (i).

Clearly the possibilities for defining coordinate systems are legion. Standardization is desirable, and it is convenient to introduce the definitions and notations listed below.

(*a*) *Mean place*
The mean coordinates (α_1, δ_1) of a star are its coordinates on the barycentric celestial sphere referred to the mean equator and equinox of date. (This means that the epoch of the equator and equinox is the same as the date of observation.)

The mean coordinates of a star only vary due to precession and proper motion, all other effects being excluded by the definition.

(*b*) *True place*
The true coordinates (α_2, δ_2) of a star are its coordinates on the barycentric celestial sphere referred to the true equator and equinox of date.

True coordinates introduce nutation but are in fact rarely used. They generally represent no more than a link in the transformation from the mean place to the apparent place of a star defined below.

(*c*) *Apparent place*
The apparent coordinates of a star (α, δ) are its coordinates on the geocentric celestial sphere referred to the true equator and equinox of date.

The apparent place involves further corrections for annual aberration and parallax. It differs from the observed position only in that purely local effects of refraction and diurnal aberration have not been included.

The definitions of the mean, true and apparent place of a star do not, of course, exhaust all the possibilities for a star's coordinates, but they are sufficiently comprehensive for most purposes. We will, however, formally add one further definition for completeness.

(*d*) *Standard mean place*
The standard mean coordinates (α_0, δ_0) of a star are its mean coordinates at the date of a standard epoch.

The standard epoch will usually be either B1950.0 or J2000.0. The catalogued position of a star is a standard mean place. We shall take this as our starting point and examine in the ensuing sections how the coordinates (α_0, δ_0) may be converted to apparent coordinates (α, δ). It will be recalled that chapter 5 contained a discussion of how meridian observations may be reduced to yield apparent coordinates. The present discussion will, therefore, complete the chain of argument that allows catalogued positions to be compared with current observations. The reverse argument is used in the computation of star catalogues.

The *Astronomical Almanac* recommends and provides data for two different methods of reduction. The first, developed in the next three sections, works directly in terms of right ascension and declination. It combines the corrections for the different effects in a set of Almanac entries, known as *Besselian Day Numbers*. The second method uses a vectorial approach and works in terms of the rectangular coordinates. It will be described in section 12.5.

12.2   The annual and secular variations

Let us suppose that the apparent coordinates of a star are required for a date $(t + \tau)$ years after the standard epoch, where $(t - \frac{1}{2})$ is an integer chosen so that the remaining fraction τ is in the range $-\frac{1}{2} < \tau \leqslant \frac{1}{2}$. Then, as a first step, the mean coordinates are calculated for a time t years after the standard epoch. These coordinates (α_1, δ_1) may be derived by a Taylor expansion from the standard epoch. Retaining only the first three terms, this gives

$$\alpha_1 = \alpha_0 + \left(\frac{d\alpha}{dt}\right)_0 t + \frac{1}{2}\left(\frac{d^2\alpha}{dt^2}\right)_0 t^2$$

$$\delta_1 = \delta_0 + \left(\frac{d\delta}{dt}\right)_0 t + \frac{1}{2}\left(\frac{d^2\delta}{dt^2}\right)_0 t^2.$$

(12.1)

The subscript zero, of course, indicates that the coordinate or the derivative is evaluated at the standard epoch.

A comprehensive star catalogue will give not just the standard mean coordinates (α_0, δ_0) but their first two derivatives as well. The first derivatives in (12.1) are referred to as the *annual variations in right ascension and declination*. Since only precession and proper motion are involved, the annual variations may, by (9.13), be expressed as

$$\left(\frac{d\alpha}{dt}\right)_0 = m + \tfrac{1}{15}n \sin \alpha_0 \tan \delta_0 + \mu_\alpha$$

$$\left(\frac{d\delta}{dt}\right)_0 = n \cos \alpha_0 + \mu_\delta.$$

(12.2)

It is, of course, to be understood that proper motion components and the precessional constants are evaluated for the standard epoch. The catalogue entries for the annual variations in right ascension and declination will be in time seconds yr^{-1} and arc seconds yr^{-1} respectively. Again it is assumed that these units are used on the right-hand sides of (12.2); n is expressed in arc seconds.

The second derivatives in (12.1) are very small quantities. The catalogue entries are, therefore, slightly modified, and the second derivatives are given in the form of secular variations (s_α, s_δ) in right ascension and declination. These are defined as the rates of change per century of the corresponding annual variation, i.e.

$$s_\alpha = 100 \left(\frac{d^2\alpha}{dt^2}\right)_0$$

$$s_\delta = 100 \left(\frac{d^2\delta}{dt^2}\right)_0.$$
(12.3)

The computation of the secular variations is rather involved, but it is something that is only undertaken in the compilation of a star catalogue. Differentiating equations (12.2) yields, in the appropriate units,

$$s_\alpha = 100 \left[\frac{dm}{dt} + \frac{1}{15}\frac{dn}{dt} \sin \alpha_0 \tan \delta_0 + \frac{d\mu_\alpha}{dt} \right.$$

$$+ n \cos \alpha_0 \tan \delta_0 \left(\frac{d\alpha}{dt}\right)_0 \sin 1''$$

$$\left. + \tfrac{1}{15}n \sin \alpha_0 \sec^2 \delta_0 \left(\frac{d\delta}{dt}\right)_0 \sin 1'' \right]$$

$$s_\delta = 100 \left[\frac{dn}{dt} \cos \alpha_0 + \frac{d\mu_\delta}{dt} - 15n \sin \alpha_0 \left(\frac{d\alpha}{dt}\right)_0 \sin 1'' \right].$$
(12.4)

Even these equations disguise the extent of the computation, since the rates of change of the proper motion components involve many separate terms. As explained in chapter 11, the proper motion components have both intrinsic and precessional rates of change. Both are needed in equations (12.4), so the derivatives of μ_α and μ_δ in these equations should be replaced by the sum of the right-hand sides of (11.17) and (11.29). Clearly the computation of the secular variations is a daunting task.

By contrast, the routine use of star catalogue entries is straightforward. Equation (12.1) may be rewritten in terms of the secular variations as

$$\alpha_1 = \alpha_0 + t \left[\left(\frac{d\alpha}{dt}\right)_0 + \frac{s_\alpha t}{200} \right]$$

$$\delta_1 = \delta_0 + t \left[\left(\frac{d\delta}{dt}\right)_0 + \frac{s_\delta t}{200} \right].$$
(12.5)

These equations can be improved by the inclusion of a third-order term and some star catalogues include such an entry. Once again, the formal expressions of the third derivatives are very complicated, but, once evaluated, they are easy to incorporate in the Taylor expressions.

12.3   A star's true place

The coordinates (α_1, δ_1) derived in the last section give the mean place of the star for the midpoint of the year that is closest to the date actually required. The next step is to calculate the star's true place (α_2, δ_2) for the date in question. This will require further correction for both precession and proper motion, but only over the short interval τ, and the inclusion of nutation. Suppose we write

$$\alpha_2 = \alpha_1 + \Delta\alpha_1$$
$$\delta_2 = \delta_1 + \Delta\delta_1. \tag{12.6}$$

Then, on using equations (9.53), (9.57) and (12.2), it is seen that

$$\Delta\alpha_1 = \tfrac{1}{15}m\tau + \tfrac{1}{15}n\tau \sin \alpha_1 \tan \delta_1 + \mu_x\tau$$
$$+ \frac{\Delta\psi}{15} (\cos \varepsilon + \sin \varepsilon \sin \alpha_1 \tan \delta_1) - \frac{\Delta\varepsilon}{15} \cos \alpha_1 \tan \delta_1,$$

$$\Delta\delta_1 = n\tau \cos \alpha_1 + \mu_\delta\tau + \Delta\psi \sin \varepsilon \cos \alpha_1 + \Delta\varepsilon \sin \alpha_1. \tag{12.7}$$

Here $\Delta\psi$ and $\Delta\varepsilon$ are the nutations in longitude and the obliquity at the date in question. The factor $1/15$ occurs in the expression for $\Delta\alpha_1$, since the two nutations are normally given in arc seconds, and to simplify the discussion both m and n are now expressed in arc seconds.

It is convenient to rearrange the terms in (12.7) to derive a form of these equations that is suitable for computation. In particular, it is desirable to separate those factors that depend upon the date – namely τ, $\Delta\psi$ and $\Delta\varepsilon$ – from those that depend upon the star's coordinates. To this end, we first eliminate the obliquity of the ecliptic. From the definitions of m and n (9.14), we may write

$$\cos \varepsilon = \frac{m + \lambda'}{\psi}$$
$$\sin \varepsilon = \frac{n}{\psi}. \tag{12.8}$$

Substituting these expressions into equations (12.7) yields, after a little reduction, the results

$$\Delta\alpha_1 = n\left(\tau + \frac{\Delta\psi}{\psi}\right) \frac{1}{15} \left(\frac{m}{n} + \sin \alpha_1 \tan \delta_1\right)$$
$$- \frac{\Delta\varepsilon}{15} \cos \alpha_1 \tan \delta_1 + \frac{\lambda' \Delta\psi}{15\psi} + \tau\mu_x \tag{12.9}$$

$$\Delta\delta_1 = n\left(\tau + \frac{\Delta\psi}{\psi}\right) \cos \alpha_1 + \Delta\varepsilon \sin \alpha_1 + \tau\mu_\delta.$$

Now make the following definitions, which amount really to a short-hand notation:

$$A = n\left(\tau + \frac{\Delta\psi}{\psi}\right)$$

$$B = -\Delta\varepsilon \qquad (12.10)$$

$$E = \frac{\lambda' \Delta\psi}{\psi}$$

$$a = \frac{1}{15}\left(\frac{m}{n} + \sin \alpha_1 \tan \delta_1\right)$$

$$b = \tfrac{1}{15} \cos \alpha_1 \tan \delta_1$$

$$a' = \cos \alpha_1 \qquad (12.11)$$

$$b' = -\sin \alpha_1.$$

Then the increments in right ascension and declination can be written in a much neater form. By (12.6), the true coordinates of the star are given as

$$\alpha_2 = \alpha_1 + \tau\mu_\alpha + Aa + Bb + E$$

$$\delta_2 = \delta_1 + \tau\mu_\delta + Aa' + Bb'. \qquad (12.12)$$

The quantities A, B and E are known as *Besselian Day Numbers*. They are independent of the stars' coordinates but vary rapidly with date. They are tabulated at daily intervals in the *Astronomical Almanac* together with the value of τ. It will be seen from their definitions that A and B are expressed in arc seconds, while E is given in time measure. The values of the Besselian Day Numbers and the fraction of a year τ are based on the midpoint of the year in question.

The quantities a, b, a', b' are called *Besselian star constants*. They are not strictly constant, since the star's coordinates and m and n change slowly with time. If these variations were neglected, they could be calculated once and for all using the standard mean place of the star. They could then be incorporated within the star catalogue. In practice, this is found to be insufficiently accurate, and they must be computed afresh for the year in question. They are only constant, therefore, in the sense that they are independent of the fraction τ.

Equations (12.12) may be written in an alternative form that does not make use of star constants. Instead the star's right ascension and declination are used directly. Using (12.10) to simplify equations (12.9) leads to the result

$$\Delta\alpha_1 = \frac{1}{15}\frac{mA}{n} + E + \frac{1}{15}(A \sin \alpha_1 + B \cos \alpha_1) \tan \delta_1 + \tau\mu_\alpha$$

$$\Delta\delta_1 = A \cos \alpha_1 - B \sin \alpha_1 + \tau\mu_\delta. \qquad (12.13)$$

Now introduce the quantities f, g and G defined as follows:

$$f = \frac{1}{15}\frac{mA}{n} + E$$

$$g \sin G = B \qquad (12.14)$$

$$g \cos G = A.$$

The true right ascension and declination of the star will then be given by

$$\alpha_2 = \alpha_1 + \tau\mu_x + f + \tfrac{1}{15}g \sin (G + \alpha_1) \tan \delta_1$$
$$\delta_2 = \delta_1 + \tau\mu_\delta + g \cos (G + \alpha_1). \qquad (12.15)$$

The quantities f, g and G are known as *Independent Day Numbers*. As may be seen from their definitions, f will be expressed in time, g in arc seconds, while G is an angle which is most conveniently expressed in time measure. The Independent Day Numbers are at first sight simpler to use than the Besselian Day Numbers, since there is no need to compute the stars' constants. If the latter are available, however, Besselian Day Numbers are preferred as the calculation is then simpler. The publication of Independent Day Numbers was discontinued in the *Astronomical Almanac* in 1981, although they are still given in some other national ephemerides.

12.4   A star's apparent place

The apparent coordinates (α, δ) of a star will differ from the true coordinates, just derived, due to the presence of annual aberration and parallax. Denote the coordinate increments by $(\Delta\alpha, \Delta\delta)$, i.e. set

$$\Delta\alpha = \alpha - \alpha_2$$
$$\Delta\delta = \delta - \delta_2. \qquad (12.16)$$

These increments are related to the components of the earth's position and velocity vectors by equations (8.15) and (8.17). In fact

$$\Delta\alpha = \frac{1}{15}\left(\frac{\dot{Y}}{c} - \pi Y\right)\cos \alpha \sec \delta - \frac{1}{15}\left(\frac{\dot{X}}{c} - \pi X\right)\sin \alpha \sec \delta$$

$$\Delta\delta = -\left(\frac{\dot{Y}}{c} - \pi Y\right)\sin \alpha \sin \delta - \left(\frac{\dot{X}}{c} - \pi X\right)\cos \alpha \sin \delta \qquad (12.17)$$

$$+ \left(\frac{\dot{Z}}{c} - \pi Z\right)\cos \delta.$$

In these equations, π is the star's parallax and the component velocities and the velocity of light must be expressed in the same units; the ratios are then converted from radians into arc seconds. If the astronomical system of units is to be used, consistent with the tabulation of $(\dot{X}, \dot{Y}, \dot{Z})$ in the *Astronomical Almanac*, then the value of c is given by equation (8.18).

Equations (12.17) will be recast in terms of the Day Numbers and stars' constants defined below, since the equations clearly invite a treatment similar to that of the last section. Before making the necessary definitions, however, there are several points that should be noted. First of all, the coordinates of the star on the right-hand side cannot be inserted as the apparent coordinates (α, δ) as indicated, since these have not yet been determined. It would be most natural to insert the true coordinates (α_2, δ_2). This is impracticable, however, as these coordinates are to be absorbed into star constants which should be independent of date. Actually it will be sufficiently accurate to use the mean coordinates (α_1, δ_1) corresponding to the midpoint of the year.

Similar considerations apply to the components of the earth's position and velocity vectors. Ideally, they should be referred to the true equator and equinox of date, but once again the mean equator and equinox for the midpoint of the year is adequate. These components are absorbed into new Day Numbers which are tabulated in the *Astronomical Almanac* at daily intervals. There is no reason in principle why the true equator and equinox should not be used. In practice, however, the correction for parallax and aberration is made simultaneously with the corrections considered in section 12.3, and, it is more orderly to use the mean equator and equinox throughout. Moreover, the difference will be a small second-order effect which is covered by the second-order Day Numbers mentioned later in this section.

The parallax correction in (12.17) is peculiar to each star and so it cannot be included in the Day Numbers. The components (X, Y, Z) of the earth's position vector are tabulated in the *Astronomical Almanac* referred to the standard mean equator and equinox. Since the parallax correction is much smaller than that due to aberration, it is sufficiently accurate to use this tabulated information.

Equations (12.17) may be simplified by eliminating the z-components of the earth's position and velocity vectors. If we assume that the earth lies exactly in the plane of the ecliptic at longitude λ, then its position vector can be written

$$\mathbf{R} = R(\cos \lambda, \sin \lambda \cos \varepsilon, \sin \lambda \sin \varepsilon),$$

consequently

$$Z = Y \tan \varepsilon$$
$$\dot{Z} = \dot{Y} \tan \varepsilon. \tag{12.18}$$

These equations are not exactly satisfied, since the coordinates are referred to the barycentre, not the sun, and mean rather than true coordinates are being used. The error involved in (12.18) is not, however, significant, and this equation allows the number of Day Numbers required to be reduced from three to two. Making this substitution in equation (12.17), and using (α_1, δ_1) on the right-hand sides, then gives the following results

$$\Delta\alpha = \tfrac{1}{15} \cos \alpha_1 \sec \delta_1 \left(\frac{\dot{Y}}{c} - \pi Y\right) - \tfrac{1}{15} \sin \alpha_1 \sec \delta_1 \left(\frac{\dot{X}}{c} - \pi X\right)$$

$$\Delta\delta = (\tan \varepsilon \cos \delta_1 - \sin \alpha_1 \sin \delta_1)\left(\frac{\dot{Y}}{c} - \pi Y\right) \qquad (12.19)$$

$$- \cos \alpha_1 \sin \delta_1 \left(\frac{\dot{X}}{c} - \pi X\right).$$

Now we make the following definitions

$$C = \frac{\dot{Y}}{c}$$

$$\qquad (12.20)$$

$$D = -\frac{\dot{X}}{c}$$

$$c = \tfrac{1}{15} \cos \alpha_1 \sec \delta_1$$
$$d = \tfrac{1}{15} \sin \alpha_1 \sec \delta_1$$
$$\qquad (12.21)$$
$$c' = \tan \varepsilon \cos \delta_1 - \sin \alpha_1 \sin \delta_1$$
$$d' = \cos \alpha_1 \sin \delta_1.$$

The quantities C and D are new Besselian Day Numbers, which are tabulated in arc seconds at daily intervals in the *Astronomical Almanac*. The quantities c, d, c', d' are further Besselian star constants. Equations (12.19) can now be written in the much neater and more convenient form

$$\Delta\alpha = (C - \pi Y)c + (D + \pi X)d$$
$$\Delta\delta = (C - \pi Y)c' + (D + \pi X)d'. \qquad (12.22)$$

An alternative formulation is again possible in terms of Independent Day Numbers. Suppose that we write

$$h \sin H = C$$
$$h \cos H = D \qquad (12.23)$$
$$i = C \tan \varepsilon.$$

Then, by (12.21),

$$Cc + Dd = \tfrac{1}{15}h \sin (H + \alpha_1) \sec \delta_1$$
$$Cc' + Dd' = h \cos (H + \alpha_1) \sin \delta_1 + i \cos \delta_1. \qquad (12.24)$$

Although these equations appear simpler at first sight, they do not include the parallax correction which necessitates the use of star constants.

As already mentioned, current practice is to use the Besselian rather than the Independent Day Numbers.

When the star's parallax is sufficiently small, it is possible to absorb it into the star constants by making the approximation that the earth's orbit is circular. The earth's position and velocity vectors may then be written as

$$\mathbf{R} = (-\cos \lambda_\odot, \ -\sin \lambda_\odot, \ 0),$$

$$\dot{\mathbf{R}} = \kappa c (\sin \lambda_\odot, \ -\cos \lambda_\odot, \ 0)$$

where κ is the constant of aberration.

The vectors are here expressed in ecliptic, not equatorial, coordinates. Rotating the axes through an angle ε about the x-axis will express the two vectors in the required equatorial system as

$$\mathbf{R} = (-\cos \lambda_\odot, \ -\sin \lambda_\odot \cos \varepsilon, \ -\sin \lambda_\odot \sin \varepsilon)$$

$$\dot{\mathbf{R}} = \kappa c (\sin \lambda_\odot, \ -\cos \lambda_\odot \cos \varepsilon, \ -\cos \lambda_\odot \sin \varepsilon). \tag{12.25}$$

It is important to notice that, although an approximation is being made to the earth's velocity, this approximation is only to be used to simplify the parallax correction. The accurate treatment of aberration is maintained.

It may be deduced by inspection of the components in (12.25) that

$$X = \frac{\dot{Y} \sec \varepsilon}{\kappa c} = \frac{C \sec \varepsilon}{\kappa}$$

$$Y = \frac{-\dot{X} \cos \varepsilon}{\kappa c} = \frac{D \cos \varepsilon}{\kappa} \tag{12.26}$$

when the definitions of the Besselian Day Numbers (12.20) are used. Now substituting for X and Y in (12.22) allows these equations to be written as

$$\Delta \alpha = C \left(c + \frac{\pi d \sec \varepsilon}{\kappa} \right) + D \left(d - \frac{\pi c \cos \varepsilon}{\kappa} \right)$$

$$\Delta \delta = C \left(c' + \frac{\pi d' \sec \varepsilon}{\kappa} \right) + D \left(d' - \frac{\pi c' \cos \varepsilon}{\kappa} \right). \tag{12.27}$$

Clearly equations (12.27) are of the form

$$\Delta \alpha = C c_1 + D d_1$$

$$\Delta \delta = C c_1' + D d_1', \tag{12.28}$$

where c_1, d_1, c_1', d_1' are the star constants redefined to include the contribution of stellar parallax. Inserting values for the obliquity of the ecliptic and the constant of aberration, they are given by

$$c_1 = c + 0.0532 \, d\pi$$

$$d_1 = d - 0.0448 \, c\pi$$

$$c_1' = c' + 0.0532 \, d'\pi \tag{12.29}$$

$$d_1' = d' - 0.0448 \, c'\pi.$$

These modified star constants are calculated once and for all for use in (12.28). The operation of these equations is obviously speedier than the more exact (12.22). The error in the approximation involved may be considered negligible if $\pi < 0''.2$.

In deriving the apparent place of a star there is no need to obtain the true place explicitly. The corrections that have been considered in this and the previous section may be applied in one fell swoop, thereby relating the apparent place directly to the mean place (α_1, δ_1) for the midpoint of the year. Thus combining the results of equations (12.22) and (12.12) gives the apparent place of the star as

$$\alpha = \alpha_1 + \tau\mu_\alpha + Aa + Bb + (C - \pi Y)c + (D + \pi X)d + E$$
$$\delta = \delta_1 + \tau\mu_\delta + Aa' + Bb' + (C - \pi Y)c' + (D + \pi X)d'. \tag{12.30}$$

These are the formulae that are used in practice, the true place is never derived.

Equations (12.30) are first-order formulae which are accurate enough for most purposes. Even so, the Day Number may be extended to include some second-order effects, which are most likely to be significant at high declinations. It is not intended to elaborate on the second-order effects here; they are fully discussed in Woolard and Clemence (1966), p. 320, and, in less detail, in the *Explanatory Supplement*. The practice recommended in the *Astronomical Almanac* is to add a term $J \tan^2 \delta_1$ to right ascension and $J' \tan \delta_1$ to declination in equations (12.30). The quantities J and J' are called second-order Day Numbers. Unlike the first-order Day Numbers, they are not independent of the star's coordinates, but they are tabulated in the *Astronomical Almanac* as functions of date and of right ascension. They do not give all the second-order terms, but only an approximation to those that are likely to be most significant. The nature of the approximation is such that separate tables for J and J' are required for stars at northern and southern declinations.

The use of second-order Day Numbers is really an attempt to reduce systematic errors which become significant at high declinations. This is due to the singularities in the coordinate system of right ascension and declination at the celestial poles. The difficulty can be avoided by a vector treatment of the whole problem, like that developed in the next section. In short, this treatment is recommended whenever the first-order formulae of (12.30) are inadequate.

The E-terms in annual aberration were discussed in section 8.5. Since it had been the practice to incorporate these in the catalogued positions of the star, the constant velocity responsible for them was, until 1984, extracted from $\dot{X}$ and $\dot{Y}$ prior to the calculation of the aberrational Day Numbers.

This procedure has now been discontinued and C and D are now defined strictly in terms of the earth's barycentric velocity. The definitions in (12.20) correspond to the new practice.

12.5   Vector derivation of a star's apparent place

In converting the standard mean place to the apparent place, the last three sections have examined what may be called the Day Number technique. It uses the mean place for the nearest midpoint of a year as a go-between. Day Numbers provide what are essentially first-order corrections from that point. We shall now consider the alternative technique which makes the necessary corrections in a different and perhaps more logical order. The method is exact, and generally to be preferred, although it does involve more elaborate computation; a programmed calculation is essential. The *Astronomical Almanac* describes this method as the rotation matrix technique and illustrates it with a worked example. In deriving the formulae on which this technique is based, it will clearly be convenient to use compatible notation and similar units to those employed in the *Astronomical Almanac*.

The starting point is the same as before – the star's standard mean place (α_0, δ_0), and the apparent place (α, δ) for a date t years later is being sought. All the intervening work, however, is performed in rectangular coordinates using a system based on the mean equator and equinox of the standard epoch. Allowance is made for proper motion, parallax and aberration so that geocentric rectangular coordinates are derived for the required date, but still referred to the original coordinate axes. The final step is to transform to the coordinates based on the true equator and equinox of date, thereby allowing for precession and nutation.

Let $\mathbf{s}_0$ be the unit vector determined by the standard mean place, i.e.

$$\mathbf{s}_0 = (\cos \alpha_0 \cos \delta_0, \sin \alpha_0 \cos \delta_0, \sin \delta_0). \qquad (12.31)$$

This vector will first be modified to allow for the purely geometric effects of proper motion and parallax. In Fig. 12.1, G represents the solar system barycentre and E the position of the earth for the date of observation. The points X_0 and X_1 are the quasi-geometric positions of the star (secular aberration is not corrected for) for the standard epoch and the date of observation respectively. The four points will not in general be coplanar, and it is not intended that the diagram should suggest this. Let $\mathbf{r}_0$, $\mathbf{r}_1 + \mathbf{R}$ and $\mathbf{R}$ be the barycentric position vectors of the points X_0, X_1 and E. Then we may write

$$\mathbf{r}_0 = r_0 \mathbf{s}_0$$
$$\mathbf{r}_1 = r_1 \mathbf{s}_1 \qquad (12.32)$$
$$\mathbf{R} = (X, Y, Z).$$

The components of **R** may be extracted from the *Astronomical Almanac*.

Now suppose that the star has a velocity **V** with respect to the barycentre. It follows from Fig. (12.1) that

$$\mathbf{r}_1 = \mathbf{r}_0 + \mathbf{V}t - \mathbf{R}. \tag{12.33}$$

Since, as explained in section 11.1, it may be assumed that the velocity **V** is constant.

The catalogue entries will provide the following information: the standard mean place (α_0, δ_0), the proper motion components (μ_α, μ_δ), the parallax π and the radial velocity V_r (in km s^{-1}) all at the standard epoch. It is convenient to express distances in astronomical units, since the components of **R** are tabulated in these units. Then, if π is expressed in radians, $r_0 = \pi^{-1}$. Equation (12.33) may, therefore, be rewritten as

$$r_1 \pi \mathbf{s}_1 = \mathbf{s}_0 + \mathbf{m}t - \pi \mathbf{R}, \tag{12.34}$$

where **m** is the star's space motion vector given by

$$\mathbf{m} = \pi \mathbf{V}.$$

Figure 12.1

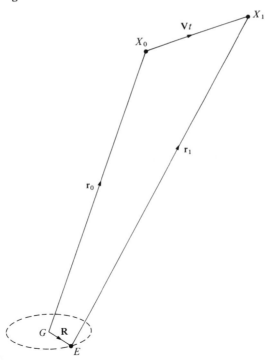

This vector must be expressed in radians per year. To achieve this, the velocity $\mathbf{V}$ which was given in km s^{-1} by equation (11.22), should be in AU yr^{-1}. Making the necessary conversion, one derives

$$\mathbf{m} = \boldsymbol{\mu} + \frac{\pi}{4.74} V_r \mathbf{s}, \tag{12.35}$$

provided $\boldsymbol{\mu}$ is expressed in radians per year.

On the other hand, if the appropriate practical units are used for μ_z, μ_δ and π, the required components of $\mathbf{m}$ are given by

$$m_x = \left(-15\mu_z \sin \alpha_0 \cos \delta_0 - \mu_\delta \cos \alpha_0 \sin \delta_0 + \frac{\pi V_r}{4.74} \cos \alpha_0 \cos \delta_0 \right) \sin 1''$$

$$m_y = \left(15\mu_z \cos \alpha_0 \cos \delta_0 - \mu_\delta \sin \alpha_0 \sin \delta_0 + \frac{\pi V_r}{4.74} \sin \alpha_0 \cos \delta_0 \right) \sin 1''$$

$$m_z = \left(\mu_\delta \cos \delta_0 + \frac{\pi V_r}{4.74} \sin \delta_0 \right) \sin 1''. \tag{12.36}$$

Once these components have been worked out, the entire right-hand side of (12.34) may be evaluated. Normalizing, the resulting vector gives $\mathbf{s}_1$ – the quasi-geometric direction of the star for the date of observation. The correction for proper motion and parallax is complete.

The next step is to correct for aberration. How this is done depends on whether a classical treatment is regarded as adequate. If it is, the argument is very simple. The observed radiation has a velocity vector $\mathbf{V}_1$ with respect to the barycentre, given by

$$\mathbf{V}_1 = -c\mathbf{s}_1.$$

With respect to the earth the radiation's velocity is $\mathbf{V}_2$, where

$$\mathbf{V}_2 = -c\mathbf{s}_1 - \dot{\mathbf{R}}.$$

The components of $\dot{\mathbf{R}}$ may be extracted from the *Astronomical Almanac* where they are tabulated in AU day^{-1}. Expressing the velocity of light in these units – cf. (8.18) – the direction of the star, corrected for aberration, is given by the unit vector $\mathbf{s}_2$, where

$$\frac{V_2}{c} \mathbf{s}_2 = \mathbf{s}_1 + 0.005\,775\,6\,\dot{\mathbf{R}}. \tag{12.37}$$

The right-hand side of (12.37) is not a unit vector, but it may be normalized to yield $\mathbf{s}_2$.

A slight modification is required if the catalogued position has already been corrected for the E-terms of aberration. The velocity, $\mathbf{V}_E$, say, responsible for these terms must be subtracted from $\dot{\mathbf{R}}$ before (12.37) is applied. Referred to the mean equator and equinox for 1950.0, this velocity has components given by

$$V_E = (-0.000\,281, -0.000\,055, -0.000\,024). \tag{12.38}$$

This modification should not be required for more recent star catalogues based on the epoch J2000.0.

It is now desired to refer the vector s_2 to axes corresponding to the true equator and equinox of date. The rotation matrix R_M, say, incorporating both precession and nutation, was defined in section 9.9. Its elements are tabulated at daily intervals in the *Astronomical Almanac*. The unit vector s that corresponds to the star's apparent place is given simply by

$$s = R_M s_2. \tag{12.39}$$

Finally the apparent place (α, δ) is readily derived from this vector, since

$$s = (\cos \alpha \cos \delta, \sin \alpha \cos \delta, \sin \delta). \tag{12.40}$$

The method just described is exact within the terms of the underlying classical assumptions. The classical treatment of aberration is not valid beyond the first order in $|\dot{R}|/c$. Still, the precision of the method may be improved by using the relativistic formulae that were developed in section 8.9. The effect of stellar parallax is already adequately allowed for in the present treatment, so this should be excluded from (8.71). The remaining terms will then give the effects of relativistic aberration and light deflection. Equation (8.71) is used to replace (12.37). With the appropriate changes in notation, it will become

$$s_2 = s_1 - \frac{2m}{R(R + R \cdot s_1)} s_1 \times (s_1 \times R)$$

$$- \frac{1}{c}\left(1 - \frac{\dot{R} \cdot s_1}{c}\right) s_1 \times (s_1 \times \dot{R}) + \frac{1}{2c^2} \dot{R} \times (\dot{R} \times s_1). \tag{12.41}$$

This is the only change required in the above vector treatment to include relativistic effects. The unit vector s_1 is derived as before, and the rotation matrix is applied to s_2 to obtain the apparent place.

Equation (12.41), as it stands, presents an extremely complicated formula. Fortunately, considerable simplification is possible. First of all express m and c in the same units as the tabulated components of R and $\dot{R}$, namely AU and AU per day respectively. Then

$$m = 9.87 \times 10^{-9}$$

$$c^{-1} = 0.005\,775\,6. \tag{12.42}$$

Due to the extreme smallness of m it is sufficiently accurate to set $R = 1$ in the denominator of the second term on the right-hand side of (12.41). Expanding all the triple vector products and rearranging terms, (12.41) may be rewritten as

$$s_2 = s_1 \left[1 - \frac{2m(s_1 \cdot R)}{1 + s_1 \cdot R} - \frac{s_1 \cdot \dot{R}}{c} + \frac{(s_1 \cdot \dot{R})^2}{c^2} - \frac{1}{2} \frac{|\dot{R}|^2}{c^2} \right]$$
$$+ \frac{\dot{R}}{c} \left(1 - \frac{1}{2} \frac{\dot{R} \cdot s_1}{c} \right) + \frac{2m}{1 + s_1 \cdot R} R. \tag{12.43}$$

This equation in itself is little simplification of (12.41). However, suppose that we divide (12.43) through by the coefficients of s_1 in that equation, which we shall call γ. Then, if only second-order terms are retained, the right-hand side simplifies significantly and equation (12.43) reduces to

$$\gamma^{-1} s_2 = s_1 + \frac{\dot{R}}{c} \left(1 + \frac{1}{2} \frac{\dot{R} \cdot s_1}{c} \right) + \frac{2m}{1 + R \cdot s_1} R$$

Using (12.42), this may be expressed in numerical form as

$$\gamma^{-1} s_2 = s_1 + 5.7756 \times 10^{-3} \dot{R} (1 + 2.9 \times 10^{-3} \dot{R} \cdot s_1)$$
$$+ \frac{1.974 \times 10^{-8}}{(1 + R \cdot s_1)} R. \tag{12.44}$$

The right-hand side of (12.44) may be computed and normalized to derive the vector s_2. Correction for precession and nutation will then give the apparent place.

The technique described in this section is preferred to the Day Number technique whenever high precision is required. The inclusion of relativistic effects necessitates the use of the vector method, since adaptation of the Day Number technique would involve the definition of new, and untabulated, Besselian Day Numbers.

12.6 The apparent place of a planet

The derivation of the apparent place of an object that is inside the solar system differs from the method just considered. Proper motion is replaced by orbital motion and the source's velocity can no longer be treated as constant.

The problem that will be considered is the following: Given a barycentric ephemeris of a planet for a particular time t, calculate the apparent place of the planet at this time.

Suppose that the barycentric ephemeris is given in the form of rectangular coordinates (x_0, y_0, z_0) referred to the standard mean equator and equinox, so that the barycentric position vector of the planet at time t is

$$r_0 = r_0 s_0 = (x_0, y_0, z_0). \tag{12.45}$$

This vector is shown as $\overrightarrow{GP}$ in Fig. 12.2. Let E be the earth's position at this time and let R be its barycentric position vector. Finally let P' be the position of the planet at the time the radiation was emitted, and denote the vector $\overrightarrow{EP'}$ by $\rho s'$.

The classical treatment is very simple, amounting to a transformation to geocentric coordinates followed by the correction for planetary aberration prior to the reduction to the true equator and equinox of date.

The geocentric position vector of P is first derived as

$$\mathbf{r}_1 = \mathbf{r}_0 - \mathbf{R}. \tag{12.46}$$

This is, of course, exact. To correct for planetary aberration, it is necessary to know the light-time τ. Strictly, this should be calculated as ρ/c, but it is sufficiently accurate in the classical treatment to use the approximation

$$\tau = \frac{|\mathbf{r}_1|}{c}. \tag{12.47}$$

The effects of planetary aberration were considered in section 8.6, where it was shown that the apparent coordinates at time t are equivalent to the geometric coordinates at time $t - \tau$. This correction gives the complete first-order effect arising from both the planet's and the earth's velocity. There is no need to reduce $\mathbf{r}_1$ to a unit vector at this stage, but an apparent geocentric position vector of the planet may be defined as

$$\mathbf{r}_2 = \mathbf{r}_1 - \tau \dot{\mathbf{r}}_1. \tag{12.48}$$

The components of $\dot{\mathbf{r}}_1$ are obtained from the derivative of equation (12.46).

Figure 12.2

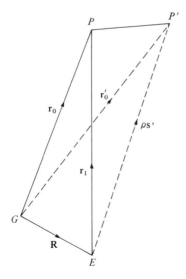

After all, we may presume that the gravitational dynamical analysis that furnished the original barycentric ephemeris $\mathbf{r}_0$ (and $\mathbf{R}$) will also provide the derivatives. If the astronomical system of units is being used, so that velocities are expressed in AU day^{-1}, equation (12.48) may by (12.42) and (12.47) be expressed as

$$\mathbf{r}_2 = \mathbf{r}_1 - 0.005\,775\,6\,r_1(\dot{\mathbf{r}}_0 - \dot{\mathbf{R}}). \tag{12.49}$$

The apparent place will be obtained when this vector $\mathbf{r}_2$ is referred to the true equator and equinox of date. Applying the rotation matrix, R_M say, the apparent place corresponds to the unit vector $\mathbf{s}$ where

$$r\mathbf{s} = \mathbf{r} = \mathbf{R}_M\mathbf{r}_2. \tag{12.50}$$

The components of $\mathbf{r}$ are, therefore, computed from (12.50) and normalizing this vector gives the required apparent place of the planet at time t. The transformation to angular coordinates is performed in the usual way.

Some considerable modification to this method is required when relativistic effects are to be included. Planetary aberration is no longer a neat correction and is not, in fact, a useful concept. We shall consider its two parts separately, namely those due to the planet's and to the earth's velocity. This means that the vector $\overrightarrow{EP'}$ will be used in place of the vector $\overrightarrow{EP}$ and then annual aberration is applied to the former. First of all we must derive the position of P'.

The analysis is valid as far as equation (12.46); equation (12.47) gives a first, but inadequate, approximation to τ. Calling this approximation τ_0, the point P' has barycentric position vector approximately given by

$$\mathbf{r}_0' = \mathbf{r}_0 - \tau_0\dot{\mathbf{r}}_0.$$

We may, therefore, compute an improved value of τ as

$$\tau_1 = \frac{|\mathbf{r}_0 - \tau_0\dot{\mathbf{r}}_0 - \mathbf{R}|}{c}. \tag{12.51}$$

This is a sufficiently good approximation to the true light travel time. It has only been defined classically and it is perhaps surprising that the relativistic correction to the light-time is not needed. As shown in section 10.8, however, the relativistic correction is of order m/c – cf. equation (10.55) – and this can give rise only to what is effectively a third-order aberrational effect.

This value of the light-time may now be used to recompute the position vector of P', but the next term in the Taylor expansion must be included to achieve second-order precision. Hence

$$\mathbf{r}_0' = \mathbf{r}_0 - \tau_1\dot{\mathbf{r}}_0 + \tfrac{1}{2}\tau_1^2\ddot{\mathbf{r}}_0.$$

The second derivative can be regarded as the planet's acceleration due to the sun's gravitational attraction, that is, to sufficient accuracy

$$\ddot{\mathbf{r}}_0 = -\frac{GM_\odot}{r_0{}^3}\,\mathbf{r}_0.$$

So the position vector of the point P' is given by

$$\mathbf{r}_0' = \mathbf{r}_0 - \tau_1\dot{\mathbf{r}}_0 + \frac{1}{2}\frac{mc^2\tau_1{}^2}{r_0{}^3}\,\mathbf{r}_0. \tag{12.52}$$

It is easily seen that the magnitude of the final term of (12.52) will, in general, be of order m. It is, therefore, a second-order term which should be included in the relativistic treatment. The geocentric position vector of P' may clearly be written as

$$\rho\mathbf{s}' = \mathbf{r}_0 - \mathbf{R} - \tau_1\dot{\mathbf{r}}_0 + \frac{1}{2}\frac{mc^2\tau_1{}^2}{r_0{}^3}\,\mathbf{r}_0. \tag{12.53}$$

Using (12.51) the components of this vector may be calculated and the vector itself normalized to give $\mathbf{s}'$. The magnitude ρ may also be determined.

The treatment so far has not been specifically relativistic. Sufficient accuracy has, however, been maintained to enable us to identify $\mathbf{s}'$ with the quasi-geometric direction $(\mathbf{s}_0)$ used in section 8.8, case (a). That vector was first defined in equation (8.55).

Allowance must now be made for relativistic annual aberration and the light deflection, using (8.72). With the appropriate change of notation, this will give the following result

$$\mathbf{s}_2 = \mathbf{s}' + \frac{2m}{[R^2 - (\mathbf{R}\cdot\mathbf{s}')^2]}\left(\frac{\mathbf{R}\cdot\mathbf{s}'}{R} - \frac{(r_0'-R)}{\rho}\right)\mathbf{s}'\times(\mathbf{s}'\times\mathbf{R})$$
$$-\frac{1}{c}\left(1 - \frac{\dot{\mathbf{R}}\cdot\mathbf{s}'}{c}\right)[\mathbf{s}'\times(\mathbf{s}'\times\dot{\mathbf{R}})] + \frac{1}{2c^2}\dot{\mathbf{R}}\times(\dot{\mathbf{R}}\times\mathbf{s}'). \tag{12.54}$$

This equation may be simplified by the method of reduction that was applied to equation (12.41). Multiplying out the three triple vector products and setting $R = 1$, we derive a vector $\gamma^{-1}\mathbf{s}_2$ given by

$$\gamma^{-1}\mathbf{s}_2 = \mathbf{s}' + 5.7756\times10^{-3}\,\dot{\mathbf{R}}(1 + 2.9\times10^{-3}\,\dot{\mathbf{R}}\cdot\mathbf{s}')$$
$$+ \frac{1.974\times10^{-8}}{1-(\mathbf{R}\cdot\mathbf{s}')^2}\left[\frac{(r_0'-R)}{\rho} - \mathbf{R}\cdot\mathbf{s}'\right]\mathbf{R}. \tag{12.55}$$

This equation enables us to calculate the components of $\gamma^{-1}\mathbf{s}_2$, and hence, by normalization, those of $\mathbf{s}_2$ itself. This gives the apparent direction of the planet, but still referred to the original axes corresponding to the standard mean equator and equinox. Multiplying by the rotation matrix will refer this vector to the true equator and equinox of date. The apparent place of the planet is then derived as a right ascension and declination by converting to angular coordinates.

12.7   Star catalogues

It was demonstrated in chapter 5 that the meridian circle has the capability of measuring the positional coordinates of a star quite independently of any assumed or previously measured stellar positions. The term 'fundamental' is applied to such measurements, in contrast to less-exacting relative measurements. Catalogues recording the results of such measurements are described as fundamental catalogues. We must distinguish, however, between two kinds of fundamental catalogue. Observational catalogues give the positions of stars determined by one observatory over a comparatively short period of time. They essentially give the positions corresponding to one epoch. A fundamental general catalogue, on the other hand, is constructed from the combination of many observational catalogues prepared at many observatories over an extended period.

The fundamental (general) catalogue will give the mean places of selected stars together with the rates of change of the stars' coordinates due to precession and proper motion – the annual and secular variations. The catalogue defines a reference system. Although we frequently speak of referring a star's coordinates to a particular equator and equinox as if this great circle and ♈ were recognizable in some way in the sky, the stars themselves are the only observable realities. It is the catalogued star positions that define the equator and equinox. The mean equator and equinox at the standard epoch are defined implicitly by the right ascensions and declinations of the stars in the fundamental catalogue, and the catalogued annual and secular variations allow the definition to be extended to other epochs.

The reference system established in this way is, of course, a rotating reference system – rotating due to precession. The moving equinox of this system is referred to as the *catalogue equinox*. It is implicitly defined as the equatorial zero point for right ascension. This point could, in principle, be chosen arbitrarily, but it has been traditionally chosen to coincide as closely as possible with the intersection of the earth's instantaneous orbit with celestial equator. The latter is referred to as the *dynamical equinox*. A distinction, albeit a fine one, must be made between the idealization of the dynamical equinox and its practical realization as the zero point in the fundamental reference system. The difference in right ascension between the two is called the *equinox correction*. A constant equinox correction does not invalidate the catalogue reference frame in any way, but a varying correction implies that precession or some other dynamical effect is being imprecisely allowed for.

A non-rotating reference system must be referred to the fixed equator and equinox of the standard epoch. This is defined for the date of the standard epoch itself by the catalogue positions, but it is only defined for other dates if the proper motions of the stars are given. Disentangling the contributions of proper motion and precession within the annual and secular variations involves assuming a rate of precession. This is normally included in the fundamental catalogue so that the fundamental reference system is defined both as a moving reference system and as a fixed reference system for all dates.

The fixed reference frame defined by the fundamental catalogue will be called the *stellar reference frame*. Ignoring the residual inaccuracies in the catalogue, this is defined by the fundamental catalogue in an entirely self-consistent manner. Ideally, the stellar reference frame should be inertial, but the extent to which this ideal is achieved must be measured by comparison with reference frames established in other ways, using either the dynamics of the solar system, or of the Galaxy, or alternatively extragalactic objects. Such alternative reference frames will be introduced in later chapters.

By international agreement under the aegis of the IAU, the fundamental reference system that has been in use since 1963 is that defined by the Fourth Fundamental Catalogue (FK4 – Fricke and Kopff, 1963). This was prepared by the Astronomisches Rechen-Institut at Heidelberg and contains 1535 stars. It is the basis of star positions in the *Astronomical Almanac*. The same institute also publishes an annual almanac entitled *The Apparent Places of the Fundamental Stars* (APFS), which gives apparent coordinates of all 1535 stars at 10-day intervals.

The observations on which the FK4 catalogue is based were completed in the 1950s. The standard epoch of the catalogue is B1950.0. Subsequent observations provide the basis of a new fundamental catalogue (Fricke, 1980), the FK5 catalogue which is also being prepared by the Heidelberg Institute. This is based on the standard epoch J2000.0. Revision of the positions of the FK4 stars was completed in 1983 and it is intended to include in many additional fainter stars. It was found that no significant revision of the FK4 equator was necessary, but the equinoxes in the two fundamental catalogues do not precisely match. Consequently, right ascensions derived from the two systems differ systematically as follows

$$\alpha_{FK5} = \alpha_{FK4} + 0^{s}.0775 + 0^{s}.085 \, T, \tag{12.56}$$

where T is the interval in Julian centuries from J2000.0. This means that proper motions of stars in the two systems also differ. In fact

$$(\mu_{\alpha})_{FK5} = (\mu_{\alpha})_{FK4} + 0^{s}.000\,85. \tag{12.57}$$

This difference amounts to only $1''.275$ per century. It is intended that the FK5 system should be available for use in annual almanacs from 1984 onwards.

The stars in the fundamental catalogues can be used as the basis for relative astrometry by classical methods. It is a much simpler matter to measure the right ascension and declination of a star relative to the fundamental system with, for example, a meridian circle. General catalogues, of a non-fundamental nature, are published which give the positions and proper motions of large numbers (10^4–10^5) of stars, going down to a much fainter magnitude. These stars provide a secondary reference system. They are spread over the celestial sphere at a density such that several such stars may be expected to fall on a photographic plate taken in any region of the sky. The known coordinates of these stars may then be used to derive the positions of other stars on the plate from purely relative measurements. The new spherical principles that are involved in the reduction of photographic plate measurements are the subject matter of the next chapter.

Problems

12.1 Show that the points on the celestial sphere for which the annual variations in right ascension and declination reduce to the components of proper motion are displaced from the poles of the ecliptic by an angle $(\lambda' \sin \varepsilon)/\psi$.

12.2 Show that the annual variations may be written in rectangular coordinates as

$$\frac{dx}{dt} = -my - nz + \mu_x$$

$$\frac{dy}{dt} = mx + \mu_y$$

$$\frac{dz}{dt} = nx + \mu_z.$$

12.3 A (fictitious) star has right ascension 6^h and declination zero at the standard epoch. It has no detectable radial velocity and its annual proper motion at this epoch is $6''$ with zero position angle. Show that in these circumstances the secular variations reduce to

$$s_x = 100 \left(\frac{dm}{dt} + 0.4n \sin 1'' \right),$$

$$s_\delta = -1500mn \sin 1'',$$

when m and n are respectively expressed in time and arc seconds per year.

12.4 Calculate the equatorial coordinates of the star of the previous question 25 years before the standard epoch. (Use the values of m and n given in equation (9.18).)

12.5 The star γ Ursae Minoris has mean equatorial coordinates $(15^h 20^m 44^s.7,\ 71°53'21'')$ for the epoch 1984.5. Calculate its Besselian star constants for use in the year 1984. cf. (9.12) and (9.18) for data.)

12.6 Prove that, if the distinction between the solar system barycentre and the sun is ignored and the E-terms are omitted, then the Besselian star constants for use in the year 1984 (cf. (9.12) and (9.18)

$$C^2 \sec^2 \varepsilon + D^2 = \kappa^2,$$

where κ is the aberrational constant.

12.7 The mean coordinates (α_1, δ_1) of a reference star are known for the epoch corresponding to the midpoint of the year. At a certain date measurements indicate that the apparent place of another star is removed from that of the reference star by amounts $(\Delta\alpha, \Delta\delta)$. Show that the mean place of this star, may, provided proper motion and parallax can be ignored, be written in terms of Independent Day Numbers as

$$\alpha_1{}^* = \alpha_1 + \Delta\alpha \sin 1''[1 - g \cos (G + \alpha_1) \tan \delta_1 - h \cos (H + \alpha_1) \sec \delta_1]$$
$$- \frac{\Delta\delta}{15} \sin 1''[g \sin (G + \alpha_1) \sec^2 \delta_1 + h \sin (H + \alpha_1) \sec \delta_1 \tan \delta_1],$$
$$\delta_1{}^* = \delta_1 + 15 \Delta\alpha \sin 1''[g \sin (G + \alpha_1) + h \sin (H + \alpha_1) \sin \delta_1]$$
$$+ \Delta\delta \sin 1''[1 - h \cos (H + \alpha_1) \cos \delta_1 + i \sin \delta_1].$$

12.8 Verify that equation (12.41) may be reduced to the more practical form of (12.44).

12.9 Show that, if the gravitational light deflection is isolated from the other relativistic effects, it may be approximately written as

$$\Delta\alpha = 0^s.136 \times 10^{-3} \cos \delta_\odot \sec \delta \sin (\alpha - \alpha_\odot) \operatorname{cosec}^2 \tfrac{1}{2}E,$$
$$\Delta\delta = 2''.04 \times 10^{-3}[\sin \delta \cos \delta_\odot \cos (\alpha - \alpha_\odot) - \cos \delta \sin \delta_\odot] \operatorname{cosec}^2 \tfrac{1}{2}E$$

where E is the object's elongation from the sun. Show further that

$$\cos E = \sin \delta \sin \delta_\odot + \cos \delta \cos \delta_\odot \cos (\alpha - \alpha_\odot).$$

12.10 Deduce equation (12.55) from equation (8.72) for the apparent direction of a body within the solar system.

13

Astrographic plate measurements

13.1 Central projection

Although detailed discussions of observational techniques are certainly not appropriate to this book, some contact with the basic principles underlying the more important observational methods is essential. Thus chapter 5 was concerned with the direct and fundamental measurement of a star's coordinates. This chapter considers the photographic method of determining a star's position. This can, of course, only be a relative position, for it is necessary to assume that the celestial coordinates of some stars on the photographic plate are known *a priori*. The measurements of photographic astrometry must be underpinned by a satisfactory framework of reference stars. The assumed positions of these depend, ultimately at least, on measurements of the type considered in chapter 5.

As in that earlier chapter, the discussion will concentrate on the geometrical aspects of the observational process, purely technical problems receiving only a cursory acknowledgement. The central geometrical problem to be considered in astronomical photography is the mapping of the celestial sphere on to the plane surface of a photographic plate. To a high degree of approximation, this may be regarded as the simple central projection, shown in Fig. 13.1. The diagram illustrates the essential features of an astronomical refractor when used for photography. Similar considerations to those developed below will apply with a reflecting telescope. The details of the optical system will be different, but the celestial sphere is mapped on to the photographic plate in mathematically an identical manner provided the focal surface of the telescope is plane. More significant differences arise in Schmidt reflectors where the photographic plate or film must be bent to fit the curved focal surface. Even in this case, however, much of the subsequent analysis may be used after the appropriate corrections have been applied. The simplified geometry of the idealized refractor in Fig. 13.1 should be considered as representative of many more complex systems.

In Fig. 13.1 the line CO is the optical axis of the telescope; C is the centre of the objective, and O is the point where the axis intersects the photographic plate. This plate is positioned in the focal plane of the telescope, and consequently CO is normal to the surface of the plate.

It is convenient to treat C formally as the centre of the celestial sphere. Produce the optical axis OC to intersect the celestial sphere in the point A. Clearly a star in the direction of A would be imaged on the photographic plate at O. In reality, this point will not be marked in any way, but the coordinates of the point A are required for the reduction of photographic plate measurements. For reasons that will be apparent from Fig. 13.1, the point A is called the *tangential point* on the celestial sphere.

Now let X be the position of a star on the celestial sphere. Join the line CX and produce it in both directions. Let CX produced intersect the photographic plate at S; in the other direction it intersects the tangential plane, which is drawn to touch the celestial sphere at A, at the point T. Since

Figure 13.1 Central projection.

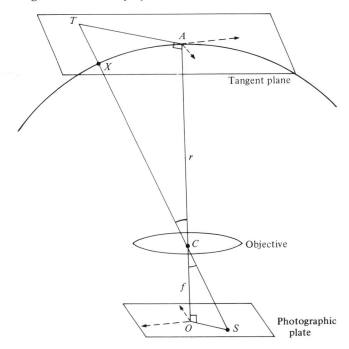

C is the centre of the objective, any light ray through this point, and, in particular, CX, is undeviated. The image of the star X is, therefore, formed at S.

Consider next the two triangles OCS and ACT. Since the line AT lies in the tangent plane, $T\hat{A}C = 90°$. Moreover, since SO lies in the focal plane, $S\hat{O}C$ is also a right angle. Furthermore, $T\hat{C}A = S\hat{C}O$, and so the triangles OCS and ACT are similar. It is then seen that the process of imaging stars on to the photographic plate is similar to a central projection of the stars on to the tangent plane to the celestial sphere at A. This geometrical similarity may be expressed in more concrete form as follows.

Let r be the radius of the celestial sphere and f the focal length of the telescope, i.e. $CA = r$, $OC = f$. Since the triangles ACT and OCS are similar, it follows that

$$AT = \frac{r}{f} OS. \tag{13.1}$$

Moreover, since AT is parallel to SO, this result may be extended to give the vector equivalent

$$\overrightarrow{AT} = -\frac{r}{f}\overrightarrow{OS}. \tag{13.2}$$

In spherical astronomy, the radius of the celestial sphere is usually taken as unity. Following this convention, we set $r = 1$, or, equivalently, measure lengths in the tangent plane in units of the radius of the celestial sphere. Let (ξ, η) be the coordinates of the point T referred to suitably chosen orthogonal axes in the tangent plane with the point A as origin. Clearly these axes will image into antiparallel axes through O on the photographic plate. As the diagram illustrates, the handedness of the axes as viewed from the emulsion side of the plate, the side nearest the objective, will be the same as the handedness of the equivalent axes as viewed from the exterior of the celestial sphere. Viewed in this way, the normal practice in plate measurement, the negative will give a reversed view of the sky, but will correspond to normal diagrams of the celestial sphere.

In principle, at least, the axes on the plate may be realized and the point O may be identified. The coordinates of the star's image may be measured with respect to these axes. Let (X, Y) be the coordinates of the point S. Initially these will be expressed in some arbitrary unit of measurement, but suppose that they are converted so that both X and Y are expressed in terms of the focal length of telescope. Then, with the idealizations implicit in this paragraph, equation (13.2) will take the simple form

$$\xi = X$$
$$\eta = Y. \tag{13.3}$$

The coordinates (ξ, η) are often referred to as the *tangential coordinates* of the star. They depend on the choice of the tangential point A and its spherical coordinates. If the latter are assumed, the definition of the tangential coordinates is purely formal and unambiguous. In practice the assumed position of A will be in error, and this will produce systematic differences between the computed tangential coordinates and what is measured on the photographic plate. The process of plate measurement will also be subject to systematic errors, quite apart from the observational errors that occur in any measurements. At a later stage it will be necessary to discuss systematic causes of discrepancy between the tangential and the measured coordinates, particularly those of a geometrical nature. Before doing so, however, we shall refine the definition of the tangential coordinates and derive relationships between them and the equatorial coordinates of a star.

13.2   Standard coordinates

The *standard coordinates* of a star are defined as the tangential coordinates (ξ, η) referred to axes in the directions of increasing right ascension and declination respectively. These two directions are chosen correctly orientated with respect to the mean equator of a standard epoch. By referring all stars' coordinates to the standard equator and equinox, any involvement with precession and nutation is avoided.

The ξ- and η-axes are shown in Fig. 13.2 for the tangential point A. Suppose that this point is in the direction of the unit vector s_A and has right ascension and declination (A, D) referred to the standard equator and equinox. Then

$$s_A = (\cos D \cos A, \cos D \sin A, \sin D), \tag{13.4}$$

referred to the usual rectangular equatorial axes through the centre C of the celestial sphere.

Denote the unit vector in the direction of the celestial pole P by $\mathbf{k}$, i.e.

$$\mathbf{k} = (0, 0, 1). \tag{13.5}$$

Let $\mathbf{I}$ and $\mathbf{J}$ be unit vectors respectively parallel to the ξ and η axes. Then the three unit vectors $\mathbf{I}$, $\mathbf{J}$ and s_A define a right-hand set of axes. Converting from equatorial to standard coordinates, therefore, involves a rotation of the coordinate axes which may be effected by applying a rotation matrix (cf. Problem 13.4). It is simpler, however, to work in terms of angular coordinates for the present.

The vector $\mathbf{I}$ is perpendicular to both $\mathbf{k}$ and s_A; it is, therefore, in the direction of $\mathbf{k} \times \mathbf{s}_A$. Since that vector has a magnitude cos D, we conclude that

$$\mathbf{I} = \sec D \mathbf{k} \times \mathbf{s}_A$$

and

$$\mathbf{J} = \mathbf{s}_A \times \mathbf{I} = \sec D[\mathbf{k} - (\mathbf{k} \cdot \mathbf{s}_A)\mathbf{s}_A].$$

(13.6)

These two vectors may readily be expressed in component form by using (13.4) and (13.5) to give

$$\mathbf{I} = (-\sin A, \cos A, 0)$$

$$\mathbf{J} = (-\sin D \cos A, -\sin D \sin A, \cos D).$$

(13.7)

Now the standard coordinates of a star X refer not to its position on the celestial sphere, but to the point T in which the line CX intersects the tangent plane. Suppose that the star has right ascension and declination (α, δ) and lies in the direction of unit vector $\mathbf{s}$, so that

$$\mathbf{s} = (\cos \delta \cos \alpha, \cos \delta \sin \alpha, \sin \delta).$$

(13.8)

Then the position vector of T is of the form $\lambda\mathbf{s}$, where λ is a scalar. The value of λ is easily derived from the requirement that T should lie in the tangent plane. This means that TA is perpendicular to CA or, in vector form,

$$(\lambda\mathbf{s} - \mathbf{s}_A) \cdot \mathbf{s}_A = 0.$$

Hence

$$\lambda^{-1} = \mathbf{s} \cdot \mathbf{s}_A.$$

(13.9)

By (13.4) and (13.8)

$$\lambda^{-1} = \sin D \sin \delta + \cos D \cos \delta \cos (\alpha - A).$$

(13.10)

Figure 13.2. The standard coordinates (ξ, η).

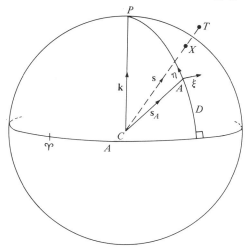

The standard coordinates (ξ, η) of the star are derived by taking scalar products of T's position vector with $\mathbf{I}$ and $\mathbf{J}$, that is,

$$\xi = \lambda \mathbf{I} \cdot \mathbf{s}$$
$$\eta = \lambda \mathbf{J} \cdot \mathbf{s}. \tag{13.11}$$

Making substitution from (13.7), (13.8) and (13.10), we derive the final results as

$$\xi = \frac{\cos \delta \sin (\alpha - A)}{\sin D \sin \delta + \cos D \cos \delta \cos (\alpha - A)}$$
$$\eta = \frac{\cos D \sin \delta - \sin D \cos \delta \cos (\alpha - A)}{\sin D \sin \delta + \cos D \cos \delta \cos (\alpha - A)}. \tag{13.12}$$

Equations (13.12) will allow the standard coordinates of a star of known position (α, δ) to be calculated for an assumed position of the tangential point A. The photographic plate will contain a number of reference stars whose positions are known *a priori*. The standard coordinates of each of these may be derived from the above equations. A relationship is then set up between standard coordinates and the coordinates measured on the plate – under ideal circumstances, they would be identical. Measured coordinates of objects on the plate of unknown position may then be transformed into standard coordinates. The final step is to transform these standard coordinates into a right ascension and declination for each object of unknown position. This requires the inverse of (13.12), which will now be derived.

After some rearrangement, the second of equations (13.12) may be written as

$$\cot \delta \cos (\alpha - A) = \frac{\cos D - \eta \sin D}{\eta \cos D + \sin D}.$$

When this is substituted into the first equation, a little simplification will yield

$$\cot \delta \sin (\alpha - A) = \frac{\xi}{\eta \cos D + \sin D}.$$

These two results will allow $(\alpha - A)$ and δ to be derived separately. Explicitly, it is found that

$$\tan (\alpha - A) = \frac{\xi}{\cos D - \eta \sin D}$$
$$\tan \delta = \frac{\sin D + \eta \cos D}{\cos D - \eta \sin D} \cos (\alpha - A). \tag{13.13}$$

A number of points should be emphasized in connection with standard coordinates.

(i) Standard coordinates are defined formally in terms of the celestial sphere. They depend only on the star's position and that of the adopted tangential point.

(ii) The star's coordinates (α, δ) are referred to a standard mean epoch. They will differ from the standard mean place only through the inclusion of annual parallax and proper motion.

(iii) The origin's right ascension and declination (A, D) are referred to the standard equator and equinox, and the ξ and η axes are correctly aligned with respect to the standard equator.

(iv) Ideally, the standard coordinates are realizable as measured coordinates on the plate. In practice, there will be systematic differences. Since the plate will cover at most a few degrees of the sky, we expect both ξ and η to be small.

13.3   Centring error

The position of the tangential point on the celestial sphere is determined by the direction of the telescope's optical axis. In order to calculate the standard coordinates, it is necessary to estimate this position either from the telescope setting, or, more likely, from an inspection or preliminary measurement of the photographic plate. The error in the assumed position of the tangential point will be called the *centring error*.

Suppose that (A, D) are the right ascension and declination of the true tangential point, but that errors (dA, dD) have been made in estimating this point's coordinates. It is intended to derive the corresponding changes $(d\xi, d\eta)$ in a star's coordinates. These will depend on the star's position; that is, its standard coordinates. Although it is proposed to neglect second-order terms in (dA, dD), the treatment is not restricted to only first-order terms in (ξ, η).

Using the notation of the last section, the tangential point is defined by the vector $\mathbf{s}_A$ and the ξ and η directions by the vectors $\mathbf{I}$ and $\mathbf{J}$. All these are unit vectors, and so each is orthogonal to its derivative. From (13.4) the change in $\mathbf{s}_A$ is found to be

$$\mathrm{d}\mathbf{s}_A = \mathrm{d}A(-\cos D \sin A, \cos D \cos A, 0)$$
$$+ \mathrm{d}D(-\sin D \cos A, -\sin D \sin A, \cos D).$$

On using (13.7), this simplifies to

$$\mathrm{d}\mathbf{s}_A = \cos D \, \mathrm{d}A \, \mathbf{I} + \mathrm{d}D \, \mathbf{J}. \tag{13.14}$$

The same process applied to the two equations in (13.7) leads to the results

$$\mathrm{d}\mathbf{I} = \sin D \, \mathrm{d}A \, \mathbf{J} - \cos D \, \mathrm{d}A \, \mathbf{s}_A$$
$$\mathrm{d}\mathbf{J} = -\sin D \, \mathrm{d}A \, \mathbf{I} - \mathrm{d}D \, \mathbf{s}_A. \tag{13.15}$$

Consider next the parameter λ given by equation (13.9). Taking differentials, one derives

$$d\lambda = -\lambda^2 \mathbf{s} \cdot d\mathbf{s}_A.$$

Substituting from (13.14) and (13.11) will then give

$$d\lambda = -\lambda(\cos D \; dA \; \xi + dD \; \eta). \tag{13.16}$$

With these preliminary results we may now determine the required changes in ξ and η. Differentiating equations (13.11) will yield

$$d\xi = d\lambda \; \mathbf{I} \cdot \mathbf{s} + \lambda \; d\mathbf{I} \cdot \mathbf{s}$$
$$d\eta = d\lambda \; \mathbf{J} \cdot \mathbf{s} + \lambda \; d\mathbf{J} \cdot \mathbf{s}.$$

Substituting for the differentials on the right-hand side from (13.15) and (13.16), and making use of (13.11) and (13.9), leads to the result

$$d\xi = -\cos D \; dA + \sin D \; dA \; \eta - \cos D \; dA \; \xi^2 - dD \; \xi\eta$$
$$d\eta = -dD - \sin D \; dA \; \xi - dD\eta^2 - \cos D \; dA \; \xi\eta. \tag{13.17}$$

The errors (dA, dD) in the assumed tangential point are, of course, unknown and must ultimately be determined from comparison of the computed standard coordinates with measurements of the reference stars. This comparison will be complicated by the presence of other effects still to be considered. For now, let us note the form of equations (13.17); it is

$$d\xi = c_1 + b_1\eta + \xi(c_1\xi + f_1\eta)$$
$$d\eta = f_1 - b_1\xi + \eta(c_1\xi + f_1\eta), \tag{13.18}$$

where b_1, c_1 and f_1 are all constants.

The leading terms (c_1, f_1) which are independent of the star represent the change in origin of the coordinate system. The first-order terms in (ξ, η), which have coefficient b_1, represent a small rotation of the coordinate axes. The reduction of photographic plate measurements is considerably simplified if the second-order terms, sometimes known as the *tilt terms*, in the standard coordinates can be neglected. This is likely to be in order near the centre of the plate. An error of one arc minute in the tangential point will give rise to second-order errors in the standard coordinates equivalent to about 0″.02 at a distance of 1° from the optical axis. In many circumstances such an error may be acceptable and the second-order terms are then omitted. They should, however, be included for high-precision astrometry particularly with wide fields. It will be clear from (13.17) that the maximum values of the second-order terms are directly proportional to the displacement of the true tangential point from its adopted position and to the square of the angular radius of the field of view.

13.4   Refraction and annual aberration

The influence of precession and nutation is wholly removed by defining standard coordinates with reference to the standard equator and equinox. Other positional effects, however, are not so simply dealt with. Proper motion and parallax cannot be removed. The standard mean places of the reference stars, must be adjusted to include these two effects. The safest way of proceeding would seem to involve making similar adjustments for refraction and aberration as well. The process is rather laborious, however, and would lead to the determination of coordinates of 'target' stars which would contain these effects. It is better, therefore, to try to absorb these corrections into the process of reduction of the plate measurements.

Neither refraction nor aberration is a property of the stars themselves, rather of the observer and his environment. All stars in the same area of the sky are affected in nearly the same way. It might be expected, therefore, that refraction and aberrational effects could be removed in the same way as precession and nutation. This is only partly true. The precessional and nutational coordinate changes are equivalent to pure rotations of the celestial sphere. In so far as refraction, or aberration, can be approximated by a small rotation of the celestial sphere, they can be removed and will not affect the relative positions of the stars on the photographic plate. In reality, however, both refraction and aberration are displacements along a pencil of great circles, and, while a small rotation may account for the main part of their effect, small residual displacements of a differential nature will remain. Before considering the particular effects of atmospheric refraction and annual aberration, we shall derive the changes in a star's standard coordinates that arise from a general displacement.

As before, let $\mathbf{s}$ be the unit vector specifying the star's direction. As shown in section 1.7, a general displacement on the celestial sphere can be represented as

$$\mathrm{d}\mathbf{s} = k\mathbf{s} \times (\mathbf{s} \times \mathbf{s}_0), \tag{13.19}$$

where $\mathbf{s}_0$ is the vertex of the displacement, i.e. zenith for refraction and the apex of the earth's motion for aberration. The parameter k may be regarded as a constant in our treatment of aberration, providing a very useful simplification. This is not possible for refraction, since the displacement is proportional to the tangent rather than to the sine of the star's distance from the vertex. It will be more convenient to use (13.19) in its expanded form, giving

$$\mathrm{d}\mathbf{s} = k[(\mathbf{s} \cdot \mathbf{s}_0)\mathbf{s} - \mathbf{s}_0]. \tag{13.20}$$

Differentiating equations (13.11) will yield the changes in the standard coordinates of the star as

$$d\xi = d\lambda \ \mathbf{I} \cdot \mathbf{s} + \lambda \mathbf{I} \cdot d\mathbf{s}$$
$$d\eta = d\lambda \ \mathbf{J} \cdot \mathbf{s} + \lambda \mathbf{J} \cdot d\mathbf{s}.$$

Substitute for d$\mathbf{s}$ from (13.20), noting from (13.9) that

$$d\lambda = -\lambda^2 \mathbf{s}_A \cdot d\mathbf{s}.$$

This yields

$$d\xi = -k\lambda^2(\mathbf{I} \cdot \mathbf{s})[(\mathbf{s} \cdot \mathbf{s}_0)(\mathbf{s} \cdot \mathbf{s}_A) - (\mathbf{s}_A \cdot \mathbf{s}_0)]$$
$$+ k\lambda[(\mathbf{s} \cdot \mathbf{s}_0)(\mathbf{I} \cdot \mathbf{s}) - \mathbf{I} \cdot \mathbf{s}_0],$$

with a similar expression for dη. Now using equations (13.11) and (13.9), these two expressions reduce to

$$d\xi = k\lambda[-(\mathbf{I} \cdot \mathbf{s}_0) + \xi(\mathbf{s}_A \cdot \mathbf{s}_0)]$$
$$d\eta = k\lambda[-(\mathbf{J} \cdot \mathbf{s}_0) + \eta(\mathbf{s}_A \cdot \mathbf{s}_0)].$$

(13.21)

The scalar products occurring in (13.21) are constants, independent of the particular star. The parameter λ, however, was originally defined as the magnitude of the position vector of the point T in Figs. 13.1 and 13.2. Referred to the set of axes defined by the unit vectors $\mathbf{I}, \mathbf{J}, \mathbf{s}_A$, the point T has coordinates $(\xi, \eta, 1)$. Hence

$$\lambda = (1 + \xi^2 + \eta^2)^{1/2}.$$

(13.22)

Substituting this into (13.21), while maintaining only second-order accuracy in the standard coordinates, will give the following results

$$d\xi = -k(\mathbf{I} \cdot \mathbf{s}_0) + k(\mathbf{s}_A \cdot \mathbf{s}_0)\xi - \frac{k}{2}(\mathbf{I} \cdot \mathbf{s}_0)(\xi^2 + \eta^2)$$

$$d\eta = -k(\mathbf{J} \cdot \mathbf{s}_0) + k(\mathbf{s}_A \cdot \mathbf{s}_0)\eta - \frac{k}{2}(\mathbf{J} \cdot \mathbf{s}_0)(\xi^2 + \eta^2).$$

(13.23)

We now proceed to make application of these formulae to annual aberration and refraction in turn.

(i)   *Annual aberration*

The adopted form of d$\mathbf{s}$ in (13.19) implies that only first-order (classical) aberration is being considered. If it is necessary to include second-order aberration, the attempt would not be made to incorporate the aberrational correction in the plate reduction, which is the goal of the present analysis. Instead the equatorial coordinates of each reference star would be converted for aberration – and for light deflection – by applying the method of section 12.5, leaving out only the final application of the rotation matrix. In the classical treatment, the parameter k is a constant determined by the earth's velocity.

Equations (13.23), therefore, have the form

$$d\xi = c_2 + a_2\xi + \tfrac{1}{2}c_2(\xi^2 + \eta^2)$$
$$d\eta = f_2 + a_2\eta + \tfrac{1}{2}f_2(\xi^2 + \eta^2), \tag{13.24}$$

where a_2, c_2 and f_2 are constants. There are obvious intentional similarities in notation between this equation and (13.18), but, unlike the constants in that equation, each of the coefficients in (13.24) can in principle be calculated. This is not necessary, however. The zero and first-order coefficients become absorbed with similar terms arising from other positional and instrumental effects. The total coefficients for terms of these orders are determined empirically by the best fit between the measured and standard coordinates of the reference stars. Individual contributions are not required.

The situation is slightly different for the second-order terms. Their contribution, which is very small, is easily estimated by noting that k is essentially the aberrational constant ($\sim 20''$). It is found that, one degree from the optical axis, the maximum value of the second-order aberrational terms is only about $0''.006$. Consequently, these terms are often ignored, particularly if the plate measurements cover only a restricted field. If they have to be included, this is best done as a correction to the standard coordinates prior to comparison with the measured position. The coefficients c_2 and f_2 must then be worked out, but this need not be done to any great precision owing to the extreme smallness of the second-order correction.

From (13.24) it is seen that (c_2, f_2) are the zero-order changes in the standard coordinates, i.e. the displacements $(d\alpha \cos \delta, d\delta)$ due to annual aberration at the plate centre. It is, therefore, seen from (8.17) that

$$c_2 = \frac{1}{c}(\dot{Y}\cos A - \dot{X}\sin A)$$

$$f_2 = \frac{1}{c}(\dot{Z}\cos D - \dot{X}\sin D\cos A - \dot{Y}\sin D\sin A). \tag{13.25}$$

Diurnal aberration is treated similarly. It will lead to a correction of the same form as (13.24). Since the diurnal effect is much smaller, however, second-order terms are usually ignored and the whole effect is absorbed into the reduction of the plate measurements.

(ii)   *Atmospheric refraction*
    The unit vector $\mathbf{s}_0$ now points to zenith and its components are given, in equatorial form, by

$$\mathbf{s}_0 = (\cos \phi \cos T, \cos \phi \sin T, \sin \phi). \tag{13.26}$$

Here ϕ is the observer's astronomical latitude and T is the local sidereal time of the observation. It will be convenient to introduce the following notation:

$$\xi_0 = \frac{\mathbf{I} \cdot \mathbf{s}_0}{\mathbf{s}_A \cdot \mathbf{s}_0}$$

$$\eta_0 = \frac{\mathbf{J} \cdot \mathbf{s}_0}{\mathbf{s}_A \cdot \mathbf{s}_0}. \tag{13.27}$$

Then (ξ_0, η_0) can be calculated. Formally they are the standard coordinates of the zenith point. As such they are of order unity – they are not small quantities. Moreover, introducing $\lambda_0 = (\mathbf{s}_0 \cdot \mathbf{s}_A)^{-1}$, equations (13.21) may be expressed in the form

$$d\xi = \frac{k\lambda}{\lambda_0}(-\xi_0 + \xi)$$

$$d\eta = \frac{k\lambda}{\lambda_0}(-\eta_0 + \eta). \tag{13.28}$$

The simplest representation of refraction, that considered in section 4.1, treats the effect as a reduction of zenith distance proportional to tan z. Had it been proportional to sin z, the parameter k could have been treated as a constant. Now, however, it must be written as

$$k = -k_0 \sec z = -k_0(\mathbf{s} \cdot \mathbf{s}_0)^{-1}, \tag{13.29}$$

where k_0 is constant which will depend slightly on atmospheric conditions. This simple treatment is useful at small zenith distances. Near the horizon a more elaborate formulation of refraction is necessary, leading to further algebraic complication. Here we shall make the assumption that k_0 may be treated as constant over the extent of the plate.

The star's direction $\mathbf{s}$ may be expressed as

$$\mathbf{s} = \lambda^{-1}(\mathbf{s}_A + \xi\mathbf{I} + \eta\mathbf{J}),$$

so, by (13.27), the parameter k is now given as

$$k = -k_0\lambda\lambda_0(1 + \xi\xi_0 + \eta\eta_0)^{-1}. \tag{13.30}$$

Then, by (13.22), equation (13.28) reduces to

$$d\xi = k_0 \frac{(1 + \xi^2 + \eta^2)}{(1 + \xi\xi_0 + \eta\eta_0)}(\xi_0 - \xi)$$

$$d\eta = k_0 \frac{(1 + \xi^2 + \eta^2)}{(1 + \xi\xi_0 + \eta\eta_0)}(\eta_0 - \eta). \tag{13.31}$$

To make practical use of these formulae, it is necessary to expand them in power series in the standard coordinates. Applying the binomial theorem yields the following results correct to second order

$$d\xi = k_0[\xi_0 - (1 + \xi_0^2)\xi - \xi_0\eta_0\eta + \xi_0(2 + \xi_0^2)\xi^2$$
$$+ \eta_0(1 + 2\xi_0^2)\xi\eta + \xi_0(1 + \eta_0^2)\eta^2]$$
$$d\eta = k_0[\eta_0 - \xi_0\eta_0\xi - (1 + \eta_0^2)\eta + \eta_0(1 + \xi_0^2)\xi^2$$
$$+ \xi_0(1 + 2\eta_0^2)\xi\eta + \eta_0(2 + \eta_0^2)\eta^2].$$

$$(13.32)$$

Zero- and first-order terms are incorporated into the reduction of the plate measurements and need not be explicitly derived. The second-order corrections, however, must be made to the reference stars individually. The coordinates (ξ_0, η_0) are obtained from (13.12) as

$$\xi_0 = \frac{\cos \phi \sin H}{\sin D \sin \phi + \cos D \cos \phi \cos H}$$
$$\eta_0 = \frac{\cos D \sin \phi - \sin D \cos \phi \cos H}{\sin D \sin \phi + \cos D \cos \phi \cos H},$$

$$(13.33)$$

where H is the hour angle of the tangential point. The magnitude of these second-order terms is likely to be only slightly larger than the equivalent terms for aberration, since $k_0 \sim 60''$. They will, however, increase rapidly at large zenith distances since (ξ_0, η_0) tend formally to infinity on the horizon. Third-order terms can be included in such circumstances, but a more realistic refraction law must also be used. This involves allowing the parameter k_0 to depend on (ξ, η). The algebra then becomes exceedingly complicated, and we shall investigate the matter no further. The second-order treatment (13.32) is certainly adequate at moderate zenith distances.

13.5 The plate constants

The determination of the positions of objects from astrographic plate measurements becomes possible once a relationship has been set up between standard and measured coordinates. This relationship is based on the measured positions of reference stars of known position. It is assumed that the measured positions are expressed as rectangular coordinates (x, y) in units of the focal length of the telescope. To be more precise, however, the comparison is made between these measured coordinates and the computed, as opposed to the true, standard coordinates.

The differences between computed and true standard coordinates were studied in the previous two sections. They arise from the centring error, annual and diurnal aberration, and atmospheric refraction. Combining the expression for the differences arising from each cause, the total difference between true and computed standard coordinates will take the form

$$d\xi = a_1\xi + b_1\eta + c_1$$
$$d\eta = d_1\xi + e_1\eta + f_1, \qquad (13.34)$$

where each of the coefficients $a_1, \ldots, f_1$ is constant but unknown. Equations (13.34) ignore the second-order terms. These are certainly small and will be omitted from the argument of this section for simplicity and clarity. If this neglect is not justified, the second-order terms for aberration and refraction may be incorporated in the computed standard coordinates; only the tilt terms are unknown. Their addition to the right-hand side of (13.34) requires two further parameters, as the form of (13.18) indicates. They complicate the problem, but do not make it insolvable. More reference stars are required.

Since all the coefficients in (13.34) are small (of the order of $1'$, or less), it does not matter whether the true or computed coordinates are used on the right-hand side. For definiteness, let us denote the computed standard coordinates by (ξ, η), and their true values by (X, Y).

Then we may also write equations (13.34) as

$$\xi - X = a_1 X + b_1 Y + c_1$$
$$\eta - Y = d_1 X + e_1 Y + f_1. \qquad (13.35)$$

The measured coordinates (x, y) will also differ from (X, Y) due to a number of instrumental effects. We shall list and comment on the geometrical ones, but assume that otherwise the imaging process is ideal. In Fig. 13.3, the point O is the intersection of the optical axis with

Figure 13.3

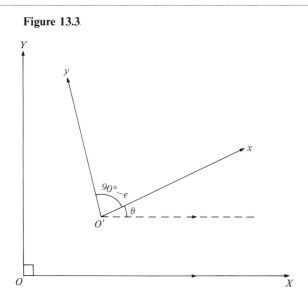

photographic plate. It is the origin of the (X, Y) coordinate system that identifies with the true standard coordinates. The point O', on the other hand, is the origin of the measured coordinate system. The principal instrumental errors are as follows:

(a) *Displacement of origin.* This will clearly lead to constant differences (x_0, y_0) between the measured and the true coordinates.

(b) *Error of orientation.* The true axes OX, OY are correctly aligned for the standard epoch, by definition. The x- and y-axes of the measured coordinate system will, however, be inclined at a small angle θ to the desired directions.

(c) *Non-perpendicularity of axes.* The true axes are strictly orthogonal, but the axes of the measured coordinate system will be oblique, through a small misalignment ε of one axis with respect to the other.

(d) *Scale errors.* The measured coordinates are to be expressed in units of the telescope's focal length. This indicates a calibration with an inevitable scale error. There will probably be separate scale errors in the x- and y-coordinates.

When each of the above effects is examined separately it is found to produce a discrepancy between true and measured coordinates that is of linear form. Consequently their aggregate effect may be expressed as

$$x - X = a_2 X + b_2 Y + c_2$$
$$y - y = d_2 X + e_2 Y + f_2, \tag{13.36}$$

where the coefficients $a_2, \ldots, f_2$ are constants. There is, however, one other instrumental effect that should be mentioned, namely:

(e) *Error of tilt.* The plate surface should ideally be perpendicular to the optical axis. There will in practice be a small angle of inclination between the normal to the plate and the optical axis. This error does not produce a discrepancy of the form (13.36) but a purely quadratic discrepancy which has exactly the same form as the tilt terms – cf. (13.18). The two sets of tilt terms may, therefore, be combined, if their inclusion is essential. For the present we assume this is not the case and regard equation (13.36) as adequate.

We can now see how the computed standard coordinates (ξ, η) relate to the measured coordinates (x, y). Combining equations (13.35) and (13.36) will give

$$\xi - x = aX + bY + c$$
$$\eta - y = dX + eY + f, \tag{13.37}$$

where $a = a_1 - a_2$, etc. The quantities a, b, c, d, e, f are known as *plate constants.* They arise from a combination of causes that have been considered in this and the previous two sections. The particular causes are not a matter of operational concern. A relationship between standard and measured coordinates of the form (13.37) is postulated, and the plate constants are then determined empirically by comparing standard and measured coordinates for the reference stars. Individual contributions to the plate constants are never investigated.

The plate constants will all be small quantities – typically of the order of one arc minute. The treatment has been first order in these quantities. While the neglect of terms which are $O(aX^2)$ may be questionable, terms $O(a^2 X)$ will be in the submilliarcsecond range and may certainly be neglected. Replacing (X, Y) on the right-hand side either with the computed standard coordinates or the measured coordinates will introduce differences of the latter order. We are free, therefore, to use whichever is more convenient, normally the measured coordinates, since they are available both for reference and target stars. We shall, therefore, rewrite (13.37) using measured coordinates on the right-hand side, giving

$$\xi - x = ax + by + c$$
$$\eta - y = dx + ey + f. \tag{13.38}$$

13.6   Principles of plate reduction

In this section the procedure for determining the right ascension and declination of a single target object of unknown position is considered. It is intended to describe the principles, rather than practice. Replacements, improvements and short-cuts may be devised in particular problems. The method is based on the simple linear relationship between measured and standard coordinates stated in equation (13.38).

Suppose that there are N reference stars on the photographic plate whose mean places can be obtained for a star catalogue. Corrections for proper motion and annual parallax, appropriate to the date of the observation, are applied to these mean places. This will provide right ascensions and declinations for each of the reference stars, which we shall denote by (α_i, δ_i), $i = 1, \ldots, N$.

It is assumed that previous knowledge of the equipment parameters allows the position of the optical axis on the photographic plate to be located at least approximately. This point is adopted as the origin of a coordinate system of measurement. We shall further assume that the desired orientation of the coordinate axes is known *a priori* with sufficient accuracy. If not, the reference stars' positions may be used to realign the coordinate axes. Once the measured coordinate system has been settled, measurements are made of all the relevant objects on the photographic plate. Suppose (x_i, y_i), $i = 1, \ldots, N$, are the measured coordinates of the reference stars and (x, y) those of the target object.

An estimate must now be made of the equatorial coordinates (A, D) of the tangential point. Using the measured coordinates of some of the reference stars, one may, for example, perform an inverse linear interpolation to derive the equatorial coordinates of the origin. If this is not sufficiently accurate, the resulting centring error may produce significant tilt terms. The estimate can, if necessary, be improved at a later stage. Let us assume that a sufficiently close estimate is, in fact, successfully made. Using these coordinates (A, D) and the equatorial coordinates of the reference stars, the standard coordinates (ξ_i, η_i), $i = 1, \ldots, N$, are now computed for each reference star from equations (13.12).

Both standard and measured coordinates have now been obtained for the reference stars. Equations (13.38), therefore, provide $2N$ equations of condition on the plate constants, namely

$$
\begin{aligned}
ax_i + by_i + c &= \xi_i - x_i \\
dx_i + ey_i + f &= \eta_i - y_i.
\end{aligned}
\qquad i = 1, \ldots, N
\qquad (13.39)
$$

Clearly, the minimum number of reference stars required to solve for the plate constants is three. In practice, more will be available, and, in order to make use of all the data, the problem is solved by the method of least squares.

It follows from (13.39) that the solution for a, b and c will decouple from that for d, e, f. The solutions for a, b, c are obtained from the normal equations, provided by the method of least squares, which are

$$
\begin{aligned}
a \sum x_i^2 + b \sum x_i y_i + c \sum x_i &= \sum x_i(\xi_i - x_i) \\
a \sum x_i y_i + b \sum y_i^2 + c \sum y_i &= \sum y_i(\xi_i - x_i) \\
a \sum x_i + b \sum y_i + cN &= \sum (\xi_i - x_i).
\end{aligned}
\qquad (13.40)
$$

The summations extend over the N reference stars. This form of the normal equations attributes equal weight to each reference star. A similar set of normal equations gives the solutions for d, e and f; in fact, the coefficients on the left-hand sides are identical.

If the solutions for (c, f) are unacceptably large, this is probably an indication of a large centring error. The coordinates of the tangential point may be adjusted, so that it corresponds more closely with the origin of the measured coordinate system. The standard coordinates of the reference stars may then be recomputed and the whole process of determining the plate constants repeated. Such an iterative process is only advantageous, if sufficient confidence can be placed in the assumed location of the optical axis on the photographic plate.

If tilt terms are to be included in the analysis, two further parameters are required in addition to the plate constants. The minimum number of reference stars needed for a definitive solution is four rather than three. Again, in practice, the method of least squares is used. The problem is complicated not just by the presence of additional parameters, requiring further normal equations, but also by the fact that the two sets of normal equations no longer decouple.

Leaving aside these complications, let us assume that the linear treatment is adequate and that the plate constants have been satisfactorily determined. The standard coordinates of the target object are now derived from its measured coordinates by equation (13.38). Its equatorial coordinates (α, δ) then follow from the application of equations (13.13). The significance of these coordinates is slightly different depending on whether the target object is within the solar system or not. Let us consider separately the representative cases of a star and a planet. In each case (α, δ) means the coordinates derived from plate measurements by the method just considered.

(i) **Star.** Geocentric parallax is entirely negligible for all stars. Aberration and refraction have been automatically corrected for in the reduction of the plate measurements. Their first-order contributions are included in the plate constants. It is true that, if second-order effects cannot be ignored, then an adjustment must be made for these in the derived position of the target star. If, however, the plate has been approximately centred on this object, second-order effects can be safely ignored. Furthermore, the use of standard coordinates insures that the derived coordinates will be referred to the standard mean equator and equinox. Consequently (α, δ) will differ from the standard mean place (α_0, δ_0) only through the presence of annual parallax and proper motion.

If the observation is made t years after the standard epoch, then it follows, from (8.15), that the derived place and the standard mean place are related by

$$\alpha = \alpha_0 + \mu_\alpha t + \frac{\pi}{15} \sec \delta (X \sin \alpha - Y \cos \alpha)$$

$$\delta = \delta_0 + \mu_\delta t + \pi (X \cos \alpha \sin \delta - Y \sin \alpha \sin \delta - Z \cos \delta).$$

(13.41)

These equations are expressed in practical units – the star's parallax is in arc seconds.

(ii) **Planet.** The coordinates are first corrected for geocentric parallax, since this is a local effect. Let r be the geocentric distance of the planet, and let

(ρ, ϕ') be the geocentric distance and latitude of the observer. Then, using (4.59), we introduce corrected coordinates (α_1, δ_1) given by

$$\alpha_1 = \alpha + \frac{\rho}{r} \cos \phi' \sin H \sec \delta$$

$$\delta_1 = \delta - \frac{\rho}{r} (\cos \phi' \cos H \sin \delta - \sin \phi' \cos \delta). \qquad (13.42)$$

The coordinates (α_1, δ_1) give what is called the *astrometric place* of the planet.

The astrometric coordinates are an obvious standardization of the planet's position as determined from astrographic plate measurements. In the process of plate reduction, correction has been made for annual aberration but no account has been taken of the planet's own orbital motion. The simplest way to remedy this is to reintroduce the annual aberration and then to apply the correction for planetary aberration, which is particularly simple.

Suppose the observation was made at time t, and that τ is the estimated light-time. Planetary aberration may be allowed for simply by antedating the observation by a time τ. The correction for annual aberration is given by equations (8.17). We may, therefore, convert the astrometric place (α_1, δ_1) into a geometric place (α_2, δ_2) where

$$\alpha_2 = \alpha_1 + \sec \delta (\dot{Y} \cos \alpha - \dot{X} \sin \alpha)/c$$

$$\delta_2 = \delta_1 + (\dot{Z} \cos \delta - \dot{X} \cos \alpha \sin \delta - \dot{Y} \sin \alpha \sin \delta)/c. \qquad (13.43)$$

The geometric place is, of course, what will be required for any dynamical analysis and it is referred to the standard, and therefore fixed, equator and equinox. It must be remembered that this geometric place refers not to the time of observation, but to the earlier time $(t - \tau)$.

13.7   The method of dependences

Section 13.6 details what may be regarded as the traditional method of plate reduction. Although now superseded as an observational method, the use of standard coordinates does illustrate important principles. Moreover, the method described in the last section is perfectly feasible, and requires only modest computational aid. Modern methods for the routine reduction of photographic plates, however, involve iterative processes that necessitate the use of a high-speed computer. Principles of these methods are described in section 13.8. Here we develop the traditional method further by considering the method of dependences. This is intended to simplify the reduction of photographic plates when the same area of the sky has been repeatedly photographed.

The method of dependences was first developed for the determination of stellar parallax. While it has more general application, it is ideally suited to that problem. For parallax determination requires the same star field to be repeatedly photographed at different times of year. The parallax displacement of a target star is sought against the background of a star field that is considered fixed. If the same tangential point is adopted for all the plates, the reference stars have fixed standard coordinates and those of the target star will vary only by a small amount. The method calculates and makes use of a set of constants – the dependences – which quantify the extent to which the target star's position depends on that of each of the reference stars.

To simplify the discussion, let us assume that there are only three reference stars whose standard coordinates are (ξ_i, η_i), $i = 1, 2, 3$. While the measured coordinates (x_i, y_i) will vary from one plate to another, the standard coordinates are fixed and known *a priori*. Let (X_0, Y_0) be an *estimate* of the standard coordinates of the target star. This could be derived by reducing a single selected plate by the method of section 13.6. This is not essential, however, since the measured coordinates taken from a selected plate probably provide a sufficiently accurate estimate.

The following analysis is now applied to each plate. It is only necessary to consider the measurement of the x-coordinate in detail. Let x_i be the measured coordinate of the ith reference star and x that of the target star; further, let ξ be the standard coordinate of the target star – all on the plate under consideration. Equation (13.38) will give to sufficient accuracy

$$\xi_i - x_i = a\xi_i + b\eta_i + c, \quad i = 1, 2, 3 \tag{13.44}$$

$$\xi - x = aX_0 + bY_0 + c. \tag{13.45}$$

Consider the right-hand sides of these equations. They have been so written since the standard coordinates do not depend on the particular plate of the series. The plate constants, however, will vary from plate to plate, but they may be eliminated as follows.

Introduce constant multipliers D_i for the equations (13.44) and combine them and (13.45) to obtain, in the dummy suffix notation

$$D_i(\xi_i - x_i) - (\xi - x) = a(D_i\xi_i - X_0) + b(D_i\eta_i - Y_0) + c(D_1 + D_2 + D_3 - 1). \tag{13.46}$$

Now the multipliers D_i may be chosen so that the coefficients of a, b, c are zero. That is, writing it out in full, D_1, D_2 and D_3 are determined by the equations

$$D_1\xi_1 + D_2\xi_2 + D_3\xi_3 = X_0$$
$$D_1\eta_1 + D_2\eta_2 + D_3\eta_3 = Y_0 \tag{13.47}$$
$$D_1 + D_2 + D_3 = 1.$$

The solutions of equations (13.47) are called the *dependences*. Notice they are the same for all the plates of the series. The dependences are obtained by solving the three simultaneous equations (13.47), say, by determinants.

With this choice of multipliers equation (13.46) simplifies enormously. The right-hand side vanishes and the equation reduces to a simple expression for the target star's standard coordinate ξ. A similar equation holds for η. In fact, in dummy suffix notation,

$$\xi = x + X_0 - D_i x_i$$
$$\eta = y + Y_0 - D_i y_i. \tag{13.48}$$

These equations allow the standard coordinates of the target star to be obtained directly from its measured coordinates and those of the three reference stars. There is considerable saving of effort in that the plate constants do not need to be determined for each plate.

There is a geometrical interpretation of the three dependences. Let X_1, X_2, X_3 in Fig. 13.4 be the positions of the three stars plotted in a ξ–η plane, that is, using standard coordinates. Let X be the point (X_0, Y_0) representing the estimated position of the target star. Then it can be shown (cf. Problem 13.8) that the dependences are expressible as ratios of the areas of the triangles in the diagram. In fact

$$D_1 = \frac{\Delta X X_2 X_3}{\Delta X_1 X_2 X_3} \quad \text{etc.} \tag{13.49}$$

This allows a graphical determination of the dependences. They are, in fact, determined with sufficient accuracy if measured coordinates (from any of the plates) are used in place of standard coordinates in equation (13.47), or

Figure 13.4. The dependences for three reference stars.

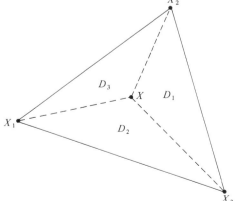

equivalently to plot the diagram of Fig. 13.4. The method works best when the target star is near the centroid of the triangle $X_1X_2X_3$. In any event, the three reference stars should be chosen so that the triangle they define encloses the target star – otherwise at least one of the dependences is negative.

The method of dependences may be extended to make use of additional reference stars. The analysis is then more complicated. Using standard rather than measured coordinates in (13.40), the plate constants are now solutions of the normal equations

$$a \sum \xi_i^2 + b \sum \xi_i \eta_i + c \sum \xi_i = \sum \xi_i (\xi_i - x_i)$$
$$a \sum \xi_i \eta_i + b \sum \eta_i^2 + c \sum \eta_i = \sum \eta_i (\xi_i - x_i) \qquad (13.50)$$
$$a \sum \xi_i + b \sum \eta_i + cN = \sum (\xi_i - x_i),$$

where N is the total number of stars. These equations are combined with (13.45) in a way that makes the coefficients of the plate constants zero. The matter may be investigated in Problems 13.9 and 13.10, where formulae for the resulting dependences D_i are given. The standard coordinates of the target star are again given by (13.48) but with the summation now extended over all of the N reference stars.

13.8   Direct use of rectangular coordinates

Standard coordinates have been discussed extensively in this chapter. They are without doubt an important theoretical concept and one that illustrates in an idealized form the imaging process on a photographic plate. In the past they were more than this, however. Before computers were available to allow complex iterative methods, standard coordinates were an essential intermediary between what was measured on the plate and the star's coordinates in the sky. They need no longer fulfil this function, however, since methods can be devised which reduce photographic measurements directly into rectangular coordinates.

After all, standard coordinates are equivalent to a special choice of coordinate axes, with one axis directed towards the tangential point. A reduction of the photographic plate based on vector methods will not depend on the choice of coordinate system and so it can be formulated to give rectangular coordinates directly in the form that they are eventually required. The traditional method of plate reduction, discussed in section 13.6, is, of course, much simpler, as indeed it had to be. It can be undertaken today with an efficient pocket calculator. The traditional method is only neat, however, if the second-order terms, in particular, the tilt terms, can be neglected. High precision reduction, even using standard coordinates, would still require an iterative method of computation. The advantages in the earlier method are, therefore, lost. Consequently, modern methods tend to dispense with standard coordinates altogether.

A brief outline is given below of a direct method of plate reduction. This is based on the idealization of central projection, already introduced in section 13.1. Since the method must go beyond first-order terms, it is probably most convenient to include all positional effects in the star's coordinates. Consequently, the coordinates of the reference stars that are used will be observed coordinates, including, in particular, the displacements for refraction and aberration and, if necessary, light deflection. The coordinates derived for any other objects will also include these effects. The positions of all stars are, however, still referred to the standard equator and equinox.

In Fig. 13.5, the point C represents the centre of projection. A star in the direction $\mathbf{s}$ is imaged at the point S on the photographic plate. The vector $\overrightarrow{SC}$ is in this direction. Suppose the star has measured coordinates (x, y) referred to an origin O. Ideally these coordinates should be expressed in units of the focal length. We shall define CO as the unit of length, so that $\overrightarrow{OC}$ is *strictly* a unit vector, $\mathbf{s}_0$, say. Now the vector $\overrightarrow{SO}$ on the plate may be written as

$$\overrightarrow{SO} = x\mathbf{i} + y\mathbf{j}. \tag{13.51}$$

Under ideal circumstances the vectors $\mathbf{i}$ and $\mathbf{j}$ would identify with the orthogonal unit vectors $\mathbf{I}$ and $\mathbf{J}$ introduced in section 13.2. Allowing for the instrumental effects listed in section 13.5, all that can be rigorously said is that they are near unit vectors. They are not necessarily perpendicular either to each other or to the unit vector $\mathbf{s}_0$.

Now let $\rho = CS/CO$. Then, with the above definitions, we have rigorously

$$\mathbf{s} = \rho^{-1}(\mathbf{s}_0 + x\mathbf{i} + y\mathbf{j}). \tag{13.52}$$

This equation must be applied to the reference stars to derive the components of the three vectors $\mathbf{s}_0$, $\mathbf{i}$ and $\mathbf{j}$. Once they are determined (13.52) may be used to derive the position vector of any object on the plate, since ρ is eliminated by the condition that $\mathbf{s}$ is a unit vector. Formally

$$\rho^2 = (\mathbf{s}_0 + x\mathbf{i} + y\mathbf{j})^2. \tag{13.53}$$

Suppose there are N reference stars with position vectors $\mathbf{s}_k$, and measured coordinates (x_k, y_k), $k = 1, \ldots, N$. For each reference star, define a residual vector given by

$$\mathbf{v}_k = \rho_k^{-1}(\mathbf{s}_0 + x_k\mathbf{i} + y_k\mathbf{j}) - \mathbf{s}_k. \tag{13.54}$$

Now it would appear at first sight that there are nine quantities to be derived, the components of the three unknown vectors. In fact, there are only eight, since s_0 is, by definition, a unit vector. The quantity to be minimized, therefore, is not the sum of the squares of the residuals, but rather the function ϕ, defined as

$$\phi = \sum_{k=1}^{N} v_k^2 + \lambda(s_0^2 - 1). \qquad (13.55)$$

Here λ is an undetermined Lagrangian multiplier. This function ϕ is treated as a function of all nine components of the three unknown vectors. The condition that ϕ should be a minimum leads to nine scalar normal equations, which, together with the condition $s_0^2 = 1$, allows a complete solution in principle.

The problem is clearly very complicated. Although certain simplifying approximations may be made, the normal equations are essentially non-linear and can only be solved iteratively. Difficulties arise because the values of the magnitudes ρ_k depend on the solution for the three vectors i, j and s_0, while they are required to establish these solutions. Satisfactory initial values for the ρ_k are provided by making the initial approximation that the three required vectors form an orthonormal set, that is, as a first approximation, take

$$\rho_k^2 = 1 + x_k^2 + y_k^2. \qquad (13.56)$$

The normal equations will then give the vectors i, j, s_0 explicitly, initiating the iteration between (13.53) and the normal equations. The process converges rapidly.

Figure 13.5

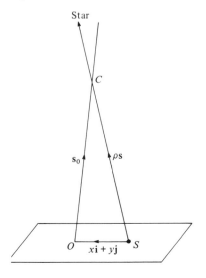

The above argument gives a brief outline of direct methods of plate reduction. A detailed discussion is beyond the scope of this book, and must be sought in a more advanced treatise, cf. Murray, 1983, chapter 8.

Problems

13.1 Show that, to a first approximation, the trail of a star of declination δ on a photographic plate is

$y = \text{const} + \frac{1}{2}x^2 \tan \delta$,

and that the projection of a meridian distant $\Delta\alpha$ from the central meridian makes with the latter the angle $\tan^{-1} (\tan \Delta\alpha \sin D)$, where D is the declination of the plate centre.

[Smart (1977)]

13.2 Prove that the equatorial coordinates (α, δ) of a star are expressed in terms of its standard coordinates (ξ, η) by

$$\alpha = A + \tan^{-1} \left(\frac{\xi}{\cos D - \eta \sin D} \right),$$

$$\delta = \sin^{-1} \left[\frac{\sin D + \eta \cos D}{(1 + \xi^2 + \eta^2)^{1/2}} \right],$$

where (A, D) are the equatorial coordinates of the tangential point.

13.3 Prove that the standard coordinates of a star (α, δ) are given by

$\xi = \cos q \sec (q - D) \tan (\alpha - A)$,

$\eta = \tan (q - D)$,

where (A, D) are equatorial coordinates of the tangential point, and

$q = \tan^{-1} [\tan \delta \sec (\alpha - A)]$.

13.4 The unit vectors $\mathbf{s}$ and $\mathbf{s}_A$ define the directions of a star and the tangential point in rectangular equatorial coordinates. Show that the standard coordinates of the star may be obtained by applying a rotation matrix $\mathbf{R}$ to the vector $(\mathbf{s} \cdot \mathbf{s}_A)^{-1}\mathbf{s}$, where

$$\mathbf{R} = \begin{pmatrix} -\sin A & \cos A & 0 \\ -\cos A \sin D & -\sin A \sin D & \cos D \\ \cos A \cos D & \sin A \cos D & \sin D \end{pmatrix}.$$

13.5 Three stars in the Pleiades have equatorial coordinates $(3^h 45^m 12^s.5, 24° 28' 03'')$, $(3^h 46^m 19^s.5, 23° 56' 55'')$ and $(3^h 49^m 09^s.7, 24° 03' 13'')$. Calculate their standard coordinates for a plate centred on the point $(3^h 47^m, 24° 00')$.

13.6 Using the results of the previous question calculate the dependences of a star with estimated standard coordinates (0.0002, 0.0018) on the three stars whose coordinates have been given.

13.7 To continue the investigation of the previous two questions, the measured coordinates of the three reference stars are $(-0.007\,12, 0.008\,34)$, $(-0.002\,71, -0.000\,72)$ and $(0.008\,56, 0.001\,12)$, while the target object has measured coordinates $(0.000\,19, 0.001\,96)$. Calculate the right ascension and declination of the target object.

13.8 Prove that the dependences may be expressed as in equation (13.49) as the ratios of the areas of triangles formed by the three reference stars and the target object.

13.9 If there are n comparison stars on a photographic plate with standard coordinates (ξ_i, η_i) and measured coordinates (x_i, y_i), $i = 1, \ldots, n$, show that the plate constants a, b, c are solutions of the normal equations

$$a \sum \xi_i^2 + b \sum \xi_i \eta_i + c \sum \xi_i = \sum \xi_i (\xi_i - x_i)$$
$$a \sum \xi_i \eta_i + b \sum \eta_i^2 + c \sum \eta_i = \sum \eta_i (\xi_i - x_i)$$
$$a \sum \xi_i + b \sum \eta_i + nc = \sum (\xi_i - x_i)$$

with similar equations for d, e and f. Here the summations extend over all n stars.

13.10 A target object has estimated standard coordinates (X_0, Y_0) on a plate with n comparison stars. Show that the true standard coordinates may be obtained from equation (13.48) using dependences D_i defined by

$$D_i = P\xi_i + Q\eta_i + R,$$

where P, Q and R are solutions of the simultaneous equations

$$P \sum \xi_i^2 + Q \sum \xi_i \eta_i + R \sum \xi_i = X_0$$
$$P \sum \xi_i \eta_i + Q \sum \eta_i^2 + R \sum \eta_i = Y_0$$
$$P \sum \xi_i + Q \sum \eta_i + nR = 1.$$

14

Stellar distances and movements

14.1 Some astrophysical background

Of all the displacements in a star's position considered so far, only proper motion and parallax have any direct bearing on the physical characteristics of the star itself. A trigonometric analysis of stellar parallax determines the star's distance. When this is known, the proper motion provides information about the star's motion perpendicular to the line of sight. This information couples neatly with that derived from the spectroscopic determination of radial velocity.

The application of positional methods in studying the stars and their overall distribution within the galaxy is important, but it is limited by the fact that positional displacements diminish with increasing distance. For most stars, parallax is simply unmeasurable. Trigonometric parallax becomes, therefore, only the first link in a chain of argument concerning the distances of the stars. It is an essential first link, however, and it is true to say that all distance determination is founded on the limited number of trigonometrical parallaxes that can be measured. Many inferences concerning the stars' distances are drawn from non-positional observations, using principally the techniques of spectroscopy and photometry. In this chapter it is intended to place the contribution made by positional astronomy in the wider context of the study of the structure and dynamics of the galaxy. Since proper motions play an important part and photographic techniques are generally used, the previous discussions of chapters 11 and 13 will be particularly relevant. Where we wish to discuss observational methods in more detail, the problems are sometimes conveniently formulated in terms of standard coordinates.

Methods of spectroscopy certainly provide radial velocities, the necessary complement of proper motions. The spectral analysis, however, provides much more information than this. A detailed spectral classifica-

Table 14.1. *The principal stellar spectral types*

Type	Prominent spectral lines	Temperature (K) (main sequence)
O	HeII, HeI	> 30 000
B	HeI, H	10 000–30 000
A	H lines at maximum strength	7500–10 000
F	H, metals (ionized and neutral)	6000–7500
G	CaII, H, neutral metals	4800–6000
K	Neutral lines dominate, some molecular lines	3500–4800
M	Extensive molecular bands (TiO)	2500–3500

Table 14.2. *Spectral luminosity classes*

I	Supergiants
II	Bright giants
III	Giants
IV	Sub-giants
V	Main sequence (dwarf)
VI	Sub-dwarf
VII	White dwarfs

tion of stars is possible. The principal spectral types are listed in Table 14.1. As the table indicates, spectral type is closely related to temperature. It is the surface temperature of the star which principally determines which spectral lines will be prominent. Subtler differences in the spectra allow the division of stars into the different luminosity classes, listed in Table 14.2. The majority of the stars are luminosity class V – the main sequence. Frequently, however, the intrinsically more luminous stars are particularly important, since they are effectively observable at greater distances. For example, all stars which can be studied individually in external galaxies are, of necessity, quite exceptionally luminous.

A star is assigned to its spectral type by using observational criteria based on the relative strengths of lines in its spectra. Each spectral type is divided into ten subdivisions, e.g. A0, A1, ..., A9, the last merging with F0. The temperatures quoted correspond to the range of *effective temperatures* of main-sequence stars in each of the principal spectral classes. Effective temperature is defined in terms of the star's continuum radiation, as the temperature of an equivalent perfect radiator – a black body of the same surface area.

The strength of a particular spectral line is most sensitive to temperature, but it is also affected by the abundance of the chemical element involved and by the ambient electron pressure. The latter is markedly different in giant and dwarf stars and produces the differences in the spectra referred to earlier. Further, there are slight temperature differences between giant and dwarf stars of the same spectral type.

The hot O- and B-type stars are often referred to as *early-type* stars, while those at the cool end of the spectral sequence are referred to as *late-type*. These are terms which have persisted from now largely abandoned ideas of stellar evolution. No temporal significance should be assigned to the spectral terms 'early' and 'late'.

A star's surface temperature influences not only the absorption line spectrum but also the spectral distribution of the underlying continuum emission. A star's colour indicates its surface temperature, the hottest stars being blue–white, the coolest orange or even red. The measurement of colour can be made quantitative by comparing the star's apparent brightness in different wavelength bands.

The *apparent magnitude* of a star is a logarithmic measure of the flux F of energy being received by the observer. The magnitude system is, in fact, of ancient Greek origin. Hipparchus divided the stars visible to the naked eye into six magnitude classes. Since the response of the eye is approximately logarithmic, the system has been made exact, so that the apparent magnitudes of two stars m_1 and m_2 are related to their fluxes F_1, F_2 by

$$m_2 - m_1 = 2.5 \log \frac{F_1}{F_2}. \tag{14.1}$$

The zero point of apparent magnitude is chosen so that the brightest stars have magnitudes near zero. An increase in five magnitudes corresponds to a decrease in flux of a factor of 100. Put technically, one magnitude is -4 decibels. The faintest stars that can be seen with the naked eye have magnitude of about 6. The faintest objects that can be seen with large powerful telescopes have magnitudes well beyond 20.

The fluxes occurring in (14.1) should be integrated fluxes extending over the entire electromagnetic spectrum. The corresponding magnitude is called *bolometric magnitude*. This cannot, of course, be measured directly, since allowance must be made for extinction in the earth's atmosphere and the limited spectral response of the photometer. Different magnitude systems have, therefore, been defined and standardized at different wavelengths. The most important of these is the U–B–V system of ultraviolet, blue and visual magnitudes. These and, in particular, the visual magnitude are more frequently used than the theoretical bolometric magnitude. The U, B and V magnitudes are all measured in the same way – by using a photomultiplier tube with the appropriate filter. The visual magnitude corresponds to the wavelength range to which the human eye is most sensitive. For this reason the term 'visual' has survived, although the measurements are made as impersonally as in any other system.

Equation (14.1) may be used to relate the magnitudes of two stars in any system, provided only that F_1 and F_2 are understood to be the (suitably weighted) fluxes in the correct wavelength range. This equation still allows an arbitrary choice of zero point. The U–B–V system is adjusted so that all three magnitudes are identical for a main-sequence star of spectral type A0.

The magnitude differences $(U–B)$ and $(B–V)$ are known as *colour indexes*. They provide information about the continuous spectrum of the star and, therefore, its temperature. The colour index $(B–V)$ is the more important. It correlates very well with spectral type and provides an observationally more accessible alternative.

The effect of interstellar absorption, by solid dust particles, is to redden the light from distinct stars. Detailed study of the star's spectra will still reveal the same spectral type, but the colour index will be altered since the dust particles absorb more strongly in the blue spectral region than in the red. The use of two-colour indexes allows the contribution of interstellar reddening to be identified and removed.

The discussion until now has been concerned with the apparent magnitude of a star, which is related to the flux of radiation from the star at the observer. If interstellar absorption is neglected, this flux is directly proportional to the star's intrinsic luminosity and is inversely proportional to the square of its distance. The *absolute magnitude* of a star is defined as the value its apparent magnitude would have if the star were situated at a standard distance of 10 parsecs from the sun. Absolute magnitude is, therefore, a measure of the star's luminosity, with the effect of distance and any intermediate absorption removed.

Suppose the star's distance is d parsecs, and its observed flux is F. Then the flux that the star would exhibit from the standard distance of 10 parsecs is F_s, where

$$F_s = \frac{d^2 F}{100}.$$ (14.2)

By equation (14.1), the star's absolute magnitude M is related to its apparent magnitude m by

$$M = m + 2.5 \log (F/F_s).$$

By (14.2) this reduces to

$$M = m + 5 - 5 \log d.$$ (14.3)

Alternatively if the star's parallax (in arc seconds) is used in place of d, the relationship is

$$M = m + 5 + 5 \log \pi. \tag{14.4}$$

An absolute magnitude may be defined to correspond to any apparent magnitude system, U, B, V, etc., or as an absolute bolometric magnitude. In each case an equation of the form of (14.4) will apply.

A traditional means of investigating and displaying physical characteristics of the stars, and their interrelation, has been the Hertzsprung–Russell (H–R) diagram. Each star is represented by a point in the diagram; the abscissa is determined by the star's spectral type, the ordinate by its luminosity. Points corresponding to actual stars do not occur randomly, but are restricted to certain definite regions. Much the most obvious concentration of stars in the H–R diagram is the main sequence extending from top-left to bottom-right. This is indicated in Fig. 14.1 together with some other luminosity classes.

The H–R diagram can be constructed from observations in two ways. Consider first a star in the immediate neighbourhood of the sun, sufficiently close for its parallax to be accurately measured. The apparent magnitude of

Figure 14.1. A schematic H–R diagram.

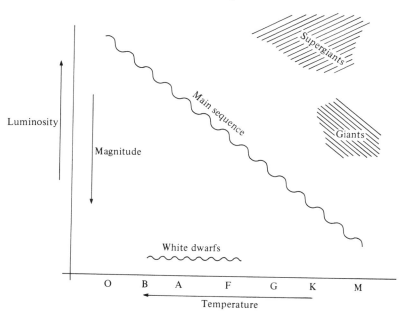

the star may then be converted into an absolute magnitude through (14.4). The spectral type of the star is ascertained, directly from observation. An H–R diagram may thus be constructed for all stars in the solar neighbourhood. This reveals only two luminosity classes, the main sequence and the intrinsically faint white dwarfs. The main sequence is incomplete, however, as there are no stars of spectral type earlier than A0.

H–R diagrams can also be constructed for star clusters. These are compact groups of stars, and so the working assumption is made that all stars in the cluster are at the same distance. This distance is unknown so the individual apparent magnitudes cannot be converted into absolute magnitudes. For every object in the cluster, however, the difference between its apparent and its absolute magnitude will be the same. This difference $(m - M)$ is known as the *distance modulus* of the cluster. If apparent magnitude is used as the ordinate, a satisfactory H–R diagram for the cluster may be constructed. Only the zero point of the luminosity scale is undetermined. Incidentally it is usually simpler to use the colour index $(B–V)$ in place of the detailed spectral classification.

Stars are certainly the principal constituent of the galaxy, but a considerable amount of matter is in the form of interstellar material. This is partly in the form of dust grains and partly in the form of interstellar gas. The two components usually occur together and are concentrated into clouds with typical diameters of a few parsecs. The distribution of interstellar material is more closely confined to the galactic plane than the general distribution of stars.

Absorption by interstellar dust grains can be severe close to the galactic plane. Consequently, effective observation at optical wavelengths is restricted to only part of the galaxy; the galactic centre is unobservable. Longer wavelengths are a definite advantage, since they penetrate the interstellar 'fog' more easily. Indeed, the absorption of radio emission by dust particles may be completely neglected. Moreover, the gas component of interstellar material may be studied in the 21 cm radio spectral line of atomic hydrogen. This provides an important means of mapping out the spiral structure of the galaxy.

The system of galactic coordinates was introduced in section 2.7. This is a formal definition, for, unlike an orbital plane, the plane of the galaxy does not have a precise and unambiguous definition. The galactic equator has been chosen to have a close correspondence to the distribution of interstellar hydrogen as indicated by 21 cm emission. The definition is slightly arbitrary, however, even subjective. The coordinate system is defined by the adopted coordinates of the galactic pole and the position angle of the galactic centre, namely

$$\alpha_G = 12^h\,49^m$$
$$\delta_G = 27°\,24' \tag{14.5}$$
$$\theta = 123°.$$

The equatorial coordinates of the point $b = 0$, $l = 0$, are determined by these definitions. This point corresponds closely to the galactic centre as indicated by the strong discrete radio source, Sagittarius A.

14.2   Measurement of parallax and proper motion

Plate reduction, as described in section 13.6, yields the right ascension and declination (α, δ) of a star corresponding to the time of observation. These coordinates contain the effects of annual parallax and proper motion. As an abbreviation, let us write

$$F_\alpha = \tfrac{1}{15} \sec \delta(X \sin \alpha - Y \cos \alpha)$$
$$F_\delta = X \cos \alpha \sin \delta - Y \sin \alpha \sin \delta - Z \cos \delta. \tag{14.6}$$

The quantities F_α and F_δ are referred to as the *parallax factors* in right ascension and declination respectively. Equations (13.41) are then more neatly expressed as

$$\alpha = \alpha_0 + \mu_\alpha t + \pi F_\alpha$$
$$\delta = \delta_0 + \mu_\delta t + \pi F_\delta \tag{14.7}$$

where (α_0, δ_0) is the standard mean place, and t is the interval in years from the standard epoch.

The determination of stellar parallax is linked to the determination of proper motion. The parallax factors are known oscillatory functions of time; parallax is determined as a small wobble in the star's position superimposed on its regular proper motion. Three photographic plates taken at different times are in principle sufficient to evaluate the unknown quantities on the right-hand side of equations (14.7). In fact, many more are required for a realistic determination of the parallax, which is derived, together with the mean place and the proper motion components, by the method of least squares.

The method described in section 13.6 and the last two paragraphs makes use of *a priori* information in the assumed coordinates of the comparison stars. If this information has been taken directly, or even deduced, from meridian observations, the derived proper motions may be described as *absolute proper motions*. The same term would be applied to proper motions obtained entirely from meridian observations. It is intended only to imply that the proper motions are consistent with a global coordinate system based on some adopted catalogue equinox, it is not necessarily claimed that this coordinate system is inertial.

Suppose, however, that there are insufficient stars on the photographic plate for which absolute positions are already known. It is still possible to make useful deductions about parallax and proper motion from relative measurements. It is convenient to work in terms of standard, rather than equatorial coordinates, so let us define parallax factors in the standard coordinates as

$$F_\xi = 15F_\alpha \cos \delta \sin 1''$$
$$F_\eta = F_\delta \sin 1''. \tag{14.8}$$

The conversion factors in (14.8) occur since the standard coordinates are expressed in terms of the radius of the celestial sphere. Similarly, proper motion components parallel to the standard coordinate axes may be defined as

$$\mu_\xi = 15\mu_\alpha \cos \delta \sin 1''$$
$$\mu_\eta = \mu_\delta \sin 1''. \tag{14.9}$$

The formulae in (14.7) may then be re-expressed in terms of standard coordinates as

$$\xi = \xi_0 + \mu_\xi t + \pi F_\xi$$
$$\eta = \eta_0 + \mu_\eta t + \pi F_\eta. \tag{14.10}$$

Suppose that the same region is photographed on a number of occasions, and that each of the resulting photographic plates has the same tangential point. Let us further assume that the same N reference stars will be used on each plate. These stars may be of unspecified position, but they must be chosen with care, since, in what follows, any displacement in their positions due to parallax or proper motion is ignored. They should, therefore, be distant stars, and photometric and spectroscopic evidence may be used in their selection.

(a)   Parallax measurement

This was discussed in section 13.7 as an illustration of the method of dependences. We shall, therefore, retain that analysis here, although other more sophisticated methods may be used. Since precise equatorial coordinates are not available for the reference stars, their standard coordinates are not known. Nevertheless, the dependences may be calculated by using measured coordinates (x_i, y_i) in their place in equations (13.50). A similar estimate (X_0, Y_0) is made of the standard coordinates of any target star.

A large number of plates is required spanning an interval of a whole year. Usually only ξ measurements are studied for the parallax determination of a target star, since the displacement is greater in right ascension than declination. For each plate that is measured an observational equation is established from (14.10) of the form

$$\xi_0 + \mu_\xi t + \pi F_\xi = M, \tag{14.11}$$

where M is a measured quantity. In fact, by (13.48), M is related to the target star's measured coordinate x, by

$$M = x + X_0 - \sum D_i x_i. \tag{14.12}$$

The method of least squares may be applied to the set of equations (14.11) to derive π and the two other unknowns ξ_0 and μ_ξ. The value of ξ_0 has no significance, but μ_ξ is the target star's proper motion in ξ relative to the reference stars. We are assuming that the derivation of the latter parameter is not the prime objective of the investigation. It can be determined more accurately by the method considered below. Small proper motions in the reference stars are unavoidable. Their effect will be reduced, to the extent that these proper motions are randomly orientated. Significant parallax displacements in the reference stars are quite unacceptable, however, because they would introduce systematic errors due to the fact that the parallax factor will be virtually the same over the whole plate.

(b)   Relative proper motions

These may be determined, in principle, from two plates taken many years apart. Once again, it is not necessary to have accurate knowledge of the positions of the reference stars. We shall briefly outline a method that uses plate constants rather than dependences.

Suppose the two plates are taken t years apart and at approximately the same time of year. Parallax displacements will then be the same on each plate and may be ignored. They will, in any case, be small compared with the accumulated proper motion displacements. In what follows, unprimed symbols refer to the first plate, and primed to the second; otherwise the notation is as before.

For each reference star, standard and measured coordinates are related by the equations

$$\begin{aligned} \xi_i - x_i &= a x_i + b y_i + c \\ \xi_i - x_i' &= a' x_i' + b' y_i' + c'. \end{aligned} \tag{14.13}$$

It may not be assumed that the plate constants are the same on each plate. Since they are small quantities, however, it will be sufficiently accurate to replace (x_i', y_i') on the right-hand side of the second equation by (x_i, y_i). The two equations are then combined to give

$$(a' - a)x_i + (b' - b)y_i + (c' - c) = x_i - x_i', \quad i = 1, \ldots, N. \tag{14.14}$$

These N equations are then solved by the method of least squares to yield the differences in the plate constants between the two plates.

In the above argument, any change in the standard coordinates of the reference stars has been neglected. This means that it has been assumed that each reference star has zero proper motion. An indication of the proper motion of each reference star is provided by the appropriate residual in equations (14.14). Any star, which has a large residual, almost certainly has a large proper motion and should be excluded.

Suppose now that measurements are made of a star for which the proper motion is required. Applying the above argument but allowing a change in the star's standard coordinate will yield

$$\mu_\xi t = \xi_i' - \xi_i = x_i' - x_i + (a'-a)x_i + (b'-b)y_i + (c'-c). \qquad (14.15)$$

Thus the proper motion component μ_ξ is determined. A similar argument will yield μ_η, and (μ_α, μ_δ) will follow from equation (14.9).

Proper motion components may be derived in this way for any star on the plate, even one of the reference stars. The proper motions obtained will be *relative proper motions*, relative that is to the mean proper motion of the reference stars. If this system of proper motions is to be significant, it is important that there should be no systematic errors depending on the star's position on the plate. This is most likely to be achieved if a large number of stars are used as reference stars so that the effect of individual proper motions in the reference stars is randomized. The set of reference stars will still have a small mean proper motion which ought to be added to the relative proper motions to make them absolute. This correction may be estimated by the method developed in section 14.4, case (a).

14.3   Distances of nearby stars

The astronomical distance scale is built up by a kind of leap-frogging argument. The distance of an object of class A is determined by some means. This allows its apparent magnitude m to be converted into an absolute magnitude, using (14.3). Once this absolute magnitude has been obtained the distance of any object of class A may be derived by reversing this equation. Fainter objects of this class are then studied which are in close proximity to some intrinsically more luminous object of class B. The difference in the apparent magnitudes of the two objects is equal to the difference in their absolute magnitudes, since they are at the same distance. So the absolute magnitude of objects of class B is then known. Objects of this class then provide a new standard which can be used at greater distances due to their greater luminosity. They may be used to calibrate the absolute magnitude and distances of objects which are still more luminous.

The argument may be extended indefinitely to ever-increasing distances. In this way the astronomical distance scale is constructed, starting with stars near the sun and reaching ultimately the extragalactic universe. While the subtle and intrinsic details of this chain of argument are beyond the scope of this book, we will discuss distance determination of the nearer stars which is essentially positional in character.

Direct measurements of stellar distances by trigonometric parallax are restricted to a useful range of about 20 parsecs from the sun. Even at this range, a 10 per cent error in the star's distance would ensue from errors of only 5 milliarcseconds in the star's position. Beyond 20 parsecs, parallax measurements are only used to give approximate distances of particular objects or as confirmatory evidence for distances derived by other means; they cannot be used as the basis of the calibration of the absolute magnitude of distance indicators.

Almost all the stars that are found within 20 parsecs of the sun are of luminosity class V, the main sequence. A few are white dwarfs but these stars are intrinsically so faint that they are of little use as distance indicators. The main sequence stars, however, can fulfil this function once their absolute magnitudes are derived from the parallax measurements. A clear relationship is established between spectral type and absolute magnitude in the H–R diagram of stars in the immediate solar neighbourhood. Since faint stars predominate, only the lower part of the main sequence is filled, but this may be matched to the main sequence of open clusters and enables the cluster distance to be established. The cluster will also contain stars that are intrinsically brighter than any in the solar neighbourhood. These will include both early main sequence stars and stars of other luminosity classes. As the distance of the cluster is known, the absolute magnitudes of these brighter stars may also be derived to provide new distance indicators.

An open or galactic cluster is a group of about 500 stars which have a common space motion. They are a homogeneous group, thought to have a common origin, often in the astronomically fairly recent past. The colour magnitude diagram exhibits a very narrow main sequence, so that absolute magnitude may be closely related to spectral type. By contrast, the stars in the immediate solar neighbourhood are a more heterogeneous group of different compositions and ages. The main sequence is decidedly broader. Stars of the same spectral type can differ by a factor of 2 in luminosity. It is desirable to refine the distance estimates of open clusters. In a few cases this can be done by utilizing the proper motions of the cluster members.

Stars in an open cluster are regarded as having a common proper motion. This provides a means of discriminating between cluster members and field stars which happen to lie in the same general direction. In a few cases, the clusters are sufficiently close to the sun that they cover an appreciable area of the sky. Slight differences in the star's proper motions within the cluster due to perspective effects may then be recognized. The proper motions appear either to converge or diverge according as the cluster is receding from or approaching the sun.

The assumption is made that the stars in the cluster have strictly parallel space motions, as illustrated in Fig. 14.2. Consider a particular star X of the cluster. Let OY be the direction from the observer that is parallel to the velocities of the stars in the cluster. This direction defines a point on the observer's celestial sphere with equatorial coordinates (α_0, δ_0), say, referred to the standard mean equator and equinox. There is a unique plane which contains the observer O, the star X and the star's velocity vector. It is, in fact, the plane OXY. This plane cuts the celestial sphere (Fig. 14.3) in a great circle through the point (α_0, δ_0). The great circle is, of course, that defined by the star's proper motion. Consequently, it is seen that the great circles defined by the star's proper motions will all converge on the point (α_0, δ_0). This point is referred to as the *convergent point* of the moving cluster. The determination of this point is difficult and requires a statistical investigation of the proper motions of as many stars in the cluster as possible.

Suppose that the cluster has been photographed on several, possibly wide-angle, plates. For each star, such as X, the standard coordinates (ξ, η) and the proper motion components (μ_ξ, μ_η) are determined. Now let (ξ_0, η_0) be the standard coordinates for the point Y. The great circle XY intersects the tangential plane in a straight line. Consequently, the proper motions of the cluster stars define a pencil of straight lines on the tangent plane. The

Figure 14.2. The common space motion of stars in a galactic cluster.

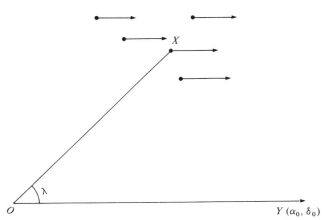

vertex of the pencil is unlikely to lie within the area covered by the photographic plates and its standard coordinates (ξ_0, η_0) will not, in general, be small.

For each star X in the cluster, there is a line of the pencil, given by the equation

$$\frac{\xi - \xi_0}{\mu_\xi} = \frac{\eta - \eta_0}{\mu_\eta}. \tag{14.16}$$

This may be rewritten as an equation of condition on the standard coordinates of the convergent point, viz.

$$\mu_\eta \xi_0 - \mu_\xi \eta_0 = \mu_\eta \xi - \mu_\xi \eta. \tag{14.17}$$

One equation of this form is derived for each star of the cluster, and the unknowns (ξ_0, η_0) may be determined by the method of least squares. The equatorial coordinates (α_0, δ_0) of the convergent point may then be derived by applying equation (13.13).

Let us denote the equatorial coordinates of X by (α, δ) and the angular distance of this star from the convergent point by λ. Then in the spherical triangle PXY of Fig. 14.3, one may identify the following parts: $XY = \lambda$, $PX = 90° - \delta$, $PY = 90° - \delta_0$, $X\hat{P}Y = \alpha_0 - \alpha$. So the cosine formula will yield the angle λ as

$$\cos \lambda = \sin \delta_0 \sin \delta + \cos \delta_0 \cos \delta \cos (\alpha_0 - \alpha). \tag{14.18}$$

The space motion of the star X may be resolved into a radial and

Figure 14.3. The convergent point of a galactic cluster.

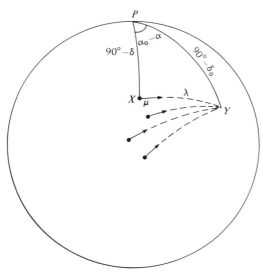

transverse velocity. The former, V_r say, is measured directly from the star's spectrum; the latter is related to its parallax and proper motion and is given in km s^{-1} by equation (11.8), namely

$$V_T = 4.74 \frac{\mu}{\pi}. \tag{14.19}$$

Referring to Fig. 14.2, it is immediately seen that

$$\begin{aligned} V_r &= V \cos \lambda \\ V_T &= V \sin \lambda, \end{aligned} \tag{14.20}$$

where V is the star's total velocity. Combining these results the parallax of the star X is given by

$$\pi = \frac{4.74\mu}{V_r \tan \lambda}. \tag{14.21}$$

The parallax of each star in the cluster may be determined in this way, once the convergent point of the cluster has been established. A mean parallax for the cluster may then be derived. A realistic determination of the convergent point, however, requires that the cluster should cover an appreciable area of sky. It is likely, therefore, that there will be a corresponding radial dispersion of stars within the cluster, so the individual parallaxes of cluster members will have more than statistical significance. There are very few clusters for which the convergent-point method is practicable, but it does provide an important link in the establishment of the astronomical distance scale.

14.4 The solar motion

The term *solar motion* refers to the motion of the sun with respect to the *local standard of rest* (LSR), not with respect to the centre of the galaxy, which might appear more natural. Optical astronomy suffers from observational difficulties in studying the galaxy as a whole, due to the effects of interstellar absorption. The LSR is, therefore, not only a convenient observational standard, it is, in fact, the only feasible one. It can, however, be defined in two distinct ways. The kinematical LSR is defined by the mean motion of stars in the solar neighbourhood. The solar motion with respect to this LSR produces systematic effects in proper motion and radial velocity which are discussed in this section. The alternative definition of the LSR is a dynamical one. It corresponds to a point in the sun's locality moving in a circular orbit about the centre of the galaxy. Dynamical considerations indicate that the kinematical LSR lags behind the dynamical one.

Suppose the solar motion has a magnitude U. It is directed towards the solar apex, a point on the celestial sphere with equatorial coordinates

(A, D), say. It will be convenient to use rectangular coordinates for the solar motion. So let us write

$$X = U \cos A \cos D$$
$$Y = U \sin A \cos D \qquad\qquad (14.22)$$
$$Z = U \sin D.$$

These components are expressed, like U, in km s^{-1}.

(a)   Proper motions

A series of photographic plates are taken of N regions of the sky centred on points with equatorial coordinates (α_i, δ_i), $i = 1, \ldots, N$. Within each region the proper motions of a homogeneous group of stars are measured. This restriction is necessary since it will be required that all the stars being studied should be approximately at the same distance from the sun. One would, therefore, select stars of a particular spectral type (or colour) within a narrow range of apparent magnitude. Let us suppose that there are n_i such stars within the ith region. With the restriction that has been made, it may be assumed that all the stars in the investigation have approximately the same secular parallax h, since from (11.57)

$$h = \frac{U\pi}{4.74}. \qquad\qquad (14.23)$$

The proper motion of each star may be thought of as being composed of two parts. There will be the proper motion components $(\mu_\alpha', \mu_\delta')$ which are due to the star's peculiar motion relative to the LSR, and the components of secular parallactic motion (P_α, P_δ) that were discussed in section 11.7. Let (μ_α, μ_δ) be the observed components for the star. Then using equations (11.61), with (14.22), it is found that

$$15(\mu_\alpha - \mu_\alpha') \cos \delta = \frac{h}{U}(X \sin \alpha - Y \cos \alpha)$$

$$\mu_\delta - \mu_\delta' = \frac{h}{U}(X \cos \alpha \sin \delta + Y \sin \alpha \sin \delta - Z \cos \delta).$$

$$\qquad\qquad (14.24)$$

These equations will hold for each star that is being studied. Let us now take the means of the equations obtained for the n_i stars that are within the ith region. Since it is a limited region, it will be sufficiently accurate to use (α_i, δ_i) for the coordinates of all n_i stars. Assuming that the space motions of the stars with respect to the LSR are randomly distributed the mean values of μ_α' and μ_δ' may be neglected. We, therefore, derive the following two equations

$$15\overline{\mu_{\alpha i}} \cos \delta_i = \frac{h}{U} X \sin \alpha_i - \frac{hY}{U} \cos \alpha_i \tag{14.25}$$

$$\overline{\mu_{\delta i}} = \frac{hX}{U} \cos \alpha_i \sin \delta_i + \frac{hY}{U} \sin \alpha_i \sin \delta_i - \frac{hZ}{U} \cos \delta_i, \tag{14.26}$$

where $\overline{\mu_{\alpha i}}$ and $\overline{\mu_{\delta i}}$ denote the mean values of the observed proper motions within the region.

When all N regions are treated in this way, $2N$ equations of condition are obtained involving three unknown quantities hX/U, hY/U, hZ/U, which may be derived by the method of least squares. It will be advisable, however, to weight the equations of condition according to the number of stars in each region. The normal equations are formed in the usual way, but the two forms of equation of condition make them rather complicated. Giving equal weight to proper motion measurements in right ascension and declination, they can be expressed as

$$a \frac{hX}{U} + H \frac{hY}{U} + G \frac{hZ}{U} = \sum_{i=1}^{N} n_i(\overline{\mu_{\delta i}} \cos \alpha_i \sin \delta_i + 15\overline{\mu_{\alpha i}} \sin \alpha_i \cos \delta_i)$$

$$H \frac{hX}{U} + b \frac{hY}{U} + F \frac{hZ}{U} = \sum_{i=1}^{N} N_i(\overline{\mu_{\delta i}} \sin \alpha_i \sin \delta_i - 15\overline{\mu_{\alpha i}} \cos \alpha_i \cos \delta_i)$$

$$G \frac{hX}{U} + F \frac{hY}{U} + c \frac{hZ}{U} = - \sum_{i=1}^{N} \overline{\mu_{\delta i}} \cos \delta_i. \tag{14.27}$$

The coefficients occurring on the left-hand sides of (14.27) are found, after a little trigonometric simplification, to be

$$a = \sum_{i=1}^{N} n_i(1 - \cos^2 \alpha_i \cos^2 \delta_i)$$

$$b = \sum_{i=1}^{N} n_i(1 - \sin^2 \alpha_i \cos^2 \delta_i)$$

$$c = \sum_{i=1}^{N} n_i \cos^2 \delta_i = 2 \sum_{i=1}^{N} n_i - (a+b)$$

$$F = - \sum_{i=1}^{N} n_i \sin \alpha_i \sin \delta_i \cos \delta_i \tag{14.28}$$

$$G = - \sum_{i-1}^{N} n_i \cos \alpha_i \sin \delta_i \cos \delta_i$$

$$H = - \sum_{i=1}^{N} n_i \sin \alpha_i \cos \alpha_i \cos^2 \delta_i.$$

The solution of the normal equations (14.27) provides values of hX/U,

hY/U and hZ/U. Using (14.22) we may then derive the coordinates of the solar apex and the secular parallax of the stars as

$$\tan A = \frac{hY}{U} \bigg/ \frac{hX}{U}$$

$$\tan D = \frac{hZ}{U} \bigg/ h,$$

(14.29)

where

$$h = \left[\left(\frac{hX}{U}\right)^2 + \left(\frac{hY}{U}\right)^2 + \left(\frac{hZ}{U}\right)^2 \right]^{1/2}.$$

(14.30)

This is as much information as the proper motions *per se* can provide. The solar apex is established but not the magnitude of the solar motion – only its direction. The secular parallax of the stars is also an observationally determined parameter, but by itself can provide no further information.

(b)   *Radial velocity*

The determination of the solar motion from radial velocity data is more straightforward as the analysis is uncomplicated by the presence of a secular parallax. The radial velocity of a star (α, δ) is composed of a radial velocity V_r', due to the star's motion with respect to the LSR, and a systematic effect V_p due to the solar motion. The latter is given by equation (11.62). On using equations (14.22), this may be rewritten as

$$V_p = -X \cos \alpha \cos \delta - Y \sin \alpha \cos \delta - Z \sin \delta.$$

(14.31)

The observed radial velocity of a star, V_r say, may, therefore, be written as

$$V_r = V_r' - X \cos \alpha \cos \delta - Y \sin \alpha \cos \delta - Z \sin \delta.$$

(14.32)

The radial velocities may be treated in the same way as the proper motions. It is assumed that the contributions V_r' are entirely random and produce no systematic effects in any part of the sky. The quantity V_r' is regarded as a random variable whose mean value, however weighted by the coordinates, may be neglected. Equations (14.32) may, therefore, be solved by the method of least squares. The normal equations are given by

$$X \sum \cos^2 \alpha \cos^2 \delta + Y \sum \cos \alpha \sin \alpha \cos^2 \delta + Z \sum \cos \alpha \sin \delta \cos \delta$$
$$= -\sum V_r \cos \alpha \cos \delta$$
$$X \sum \cos \alpha \sin \alpha \cos^2 \delta + Y \sum \sin^2 \alpha \cos^2 \delta + Z \sum \sin \alpha \sin \delta \cos \delta$$
$$= -\sum V_r \sin \alpha \cos \delta$$
$$X \sum \cos \alpha \sin \delta \cos \delta + Y \sum \sin \alpha \sin \delta \cos \delta + Z \sum \sin^2 \delta$$
$$= -\sum V_r \sin \delta,$$

(14.33)

where the summations extend over all stars whose radial velocities have been measured.

The solution of the normal equations will in this case give values for (X, Y, Z). The determination of the solar motion is, therefore, more complete, yielding both the solar apex and the magnitude of U. In fact, by (14.22)

$$U = (X^2 + Y^2 + Z^2)^{1/2}$$

$$\tan A = Y/X \tag{14.34}$$

$$\sin D = Z/U.$$

When using proper motions to investigate the solar motion, the method itself dictates that there should be some selection criteria for the stars used. While this is not prescribed by the radial velocity method it is nonetheless desirable. For it is found that there are slight systematic differences in the solar motion derived from different classes of stars. Some reasons for this will be considered in section 14.6. The two methods discussed separately here may be combined. Radial velocities will establish the magnitude of the sun's velocity, and proper motions will be used to improve the determination of its direction.

14.5   Statistical parallax

The statistical analysis of proper motions led to a determination of the secular parallax h of the class of stars being investigated. If the magnitude U of the sun's velocity has been derived from radial velocity measurements, h may be converted into the annual parallax to give the stars' mean distance.

By (14.23),

$$\pi = \frac{4.74h}{U}. \tag{14.35}$$

We shall now examine more generally the statistical determination of parallax from proper motions, assuming that the solar motion is already known.

In Fig. (14.4), the point A represents the apex of the solar motion on the celestial sphere, with equatorial coordinates (A, D). Let $X(\alpha, \delta)$ be a star at an angular distance λ from the apex. Then in spherical triangle PXA, $PX = 90° - \delta$, $PA = 90° - D$, $AX = \lambda$ and $A\hat{P}X = \alpha - A$. Further, let us denote the angle PXA by ψ. Applying the cosine formula to spherical triangle PXA will give λ as

$$\cos \lambda = \sin D \sin \delta + \cos D \cos \delta \cos (\alpha - A). \tag{14.36}$$

The angle ψ is obtained from the four-parts formula, namely,

$$\sin \delta \cos (\alpha - A) = \cos \delta \tan D - \sin (\alpha - A) \cot \psi.$$

Hence

$$\cot \psi = \cos \delta \tan D \operatorname{cosec} (\alpha - A) - \sin \delta \cot (\alpha - A). \qquad (14.37)$$

Now it is convenient to resolve the star's proper motion, not in the directions of increasing right ascension and declination, but rather along AX and perpendicular to this direction. Denote the star's proper motion components in these two directions by (v, τ), as shown in the diagram. Expressed in arc seconds yr^{-1}, these components are

$$v = 15\mu_\alpha \cos \delta \sin \psi - \mu_\delta \cos \psi$$
$$\tau = 15\mu_\alpha \cos \delta \cos \psi + \mu_\delta \sin \psi. \qquad (14.38)$$

The two new components of proper motion may be derived from (μ_α, μ_δ) once ψ has been calculated from equation (14.37).

Statistical parallaxes of a homogeneous group of stars, i.e. of the same spectral type and luminosity class, within a restricted magnitude range, may be derived by investigating either the v or the τ components of the proper motions. In each case it is assumed that the stars' velocities with respect to the LSR are purely random. If these velocities are larger than the sun's velocity, most satisfactory results are obtained by considering the τ-components. On the other hand, if the sun's velocity predominates, the statistical parallax of the group of stars is more readily derived from the v-components.

Figure 14.4. The τ- and v-components of proper motion. The point A is the solar apex.

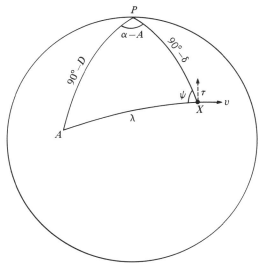

Let **U** and **V** denote the velocity of the sun and a star with respect to the LSR. Let **v** be the velocity of the star with respect to the sun, that is,

$$\mathbf{v} = \mathbf{V} - \mathbf{U}. \tag{14.39}$$

It is convenient to refer the velocities to coordinate axes in the υ- and τ-directions; an axis in the radial r-direction, along the line of sight to the star, completes a right-hand set. Then the two transverse components of the star's velocity relative to the sun are given by

$$\upsilon_\upsilon = \frac{4.74\upsilon}{\pi},$$
$$\upsilon_\tau = \frac{4.74\tau}{\pi}. \tag{14.40}$$

Since **U** is directed towards the point A, it is easily seen that, in the (υ, τ, r) coordinate system, its components are

$$\mathbf{U} = (-U \sin \lambda, 0, U \cos \lambda). \tag{14.41}$$

Consequently, equation (14.39) may be written as

$$\mathbf{v} = (V_\upsilon + U \sin \lambda, V_\tau, V_r - U \cos \lambda). \tag{14.42}$$

It is assumed that each of the components of **V** in this equation is a purely random variable. Two methods, based on this assumption, of determining the statistical parallax π of the group of stars will now be given in outline, but no more:

(i) υ-*Component.* Using (14.42), the first of equations (14.40) may be rewritten as

$$\upsilon - \frac{\pi U \sin \lambda}{4.74} = \frac{\pi V_\upsilon}{4.74}. \tag{14.43}$$

This equation is exact for any particular star. For the group of stars under consideration, the selection criteria should ensure that π is approximately constant. The right-hand side of (14.43) is regarded as a random variable. Then the principle of least squares will require that the function F be a minimum where

$$F(\pi) = \sum \left(\upsilon_i - \frac{\pi U}{4.74} \sin \lambda_i \right)^2. \tag{14.44}$$

This will give the mean parallax of the group of stars as

$$\pi = \frac{4.74}{U} \frac{\sum_i \upsilon_i \sin \lambda_i}{\sum_i \sin^2 \lambda_i}. \tag{14.45}$$

(ii) τ-*Component.* Using (14.42), the second of equations (14.40) will give

$$\tau = \frac{\pi V_\tau}{4.74} \tag{14.46}$$

for each star. The right-hand side of this equation will vary randomly for stars in the selected group, so we expect the mean value of τ to be zero. We therefore take the absolute value of (14.46) and derive the mean. Then, if N is total number of stars in the group, the formal equation is derived

$$\langle |V_\tau| \rangle \pi = \frac{4.74}{N} \sum_i |\tau_i|, \tag{14.47}$$

where $\langle |V_\tau| \rangle$ denotes the mean value of $|V_\tau|$. An estimate must now be made for this quantity.

If it is assumed that the velocities of the stars relative to the LSR are randomly oriented, we may make the approximation that

$$\langle |V_\tau| \rangle = \langle |V_r| \rangle. \tag{14.48}$$

But, from equation (14.42), it is seen that

$$V_r = v_r + U \cos \lambda. \tag{14.49}$$

Now v_r is the observed radial velocity. Consequently the components V_r of the stars' velocities with respect to LSR are individually known. The approximation of equation (14.48) will, therefore, give

$$\langle |V_\tau| \rangle = \frac{1}{N} \sum_i |v_r + U \cos \lambda|. \tag{14.50}$$

When this result is substituted into (14.47), we derive the statistical parallax for the group of stars as

$$\pi = 4.74 \frac{\sum_i |\tau_i|}{\sum_i |v_r + U \cos \lambda_i|}. \tag{14.51}$$

14.6 Galactic coordinates

The assumption that the star's velocities are purely random with respect to the LSR is something of an over-simplification. A more complicated model is required to match systematic effects in the proper motions and radial velocities. A comprehensive discussion of stellar dynamics is beyond the scope of this book. Such a discussion may be found in Mihalas and Binney (1981).

The solar motion has been considered mainly in the equatorial system of coordinates, in which the observations are made. A more physically significant system of coordinates is the system of galactic latitude and longitude, illustrated in Fig. 14.5. In this diagram, G represents the galactic pole (α_G, δ_G) and C the centre of the galaxy. The angle PGC is denoted by θ.

Let X be a general point on the celestial sphere with equatorial coordinates (α, δ) and galactic latitude and longitude (l, b). Let $\mathbf{s}$ be the unit vector giving the direction of X in the equatorial system and $\mathbf{s}_G$ the same vector expressed in the galactic system, that is,

$$\mathbf{s} = (\cos \alpha \cos \delta,\ \sin \alpha \cos \delta,\ \sin \delta)$$
$$\mathbf{s}_G = (\cos l \cos b,\ \sin l \cos b,\ \sin b).$$

(14.52)

Notice that, in the galactic system, the coordinates are measured respectively towards the galactic centre, in the direction $l = 90°$, $b = 0$, and finally perpendicular to the galactic plane. The directions of the equatorial axes (x, y, z) and the galactic axes (x_G, y_G, z_G) have been indicated in the diagram.

The relationship between equatorial and galactic angular coordinates has already been established in section 2.7 as equations (2.33) and (2.32). Making appropriate combinations, these equations may be written as

$$\cos l \cos b = -(\sin \theta \sin \alpha_G + \cos \theta \cos \alpha_G \sin \delta_G) \cos \alpha \cos \delta$$
$$+ (\sin \theta \cos \alpha_G - \cos \theta \sin \alpha_G \sin \delta_G) \sin \alpha \cos \delta$$
$$+ \cos \theta \cos \delta_G \sin \delta$$
$$\sin l \cos b = (\cos \theta \sin \alpha_G - \sin \theta \cos \alpha_G \sin \delta_G) \cos \alpha \cos \delta$$
$$- (\cos \theta \cos \alpha_G + \sin \theta \sin \alpha_G \sin \delta_G) \sin \alpha \cos \delta$$
$$+ \sin \theta \cos \delta_G \sin \delta$$

Figure 14.5. Galactic rectangular coordinates (x_G, y_G, z_G) and equatorial coordinates (x, y, z).

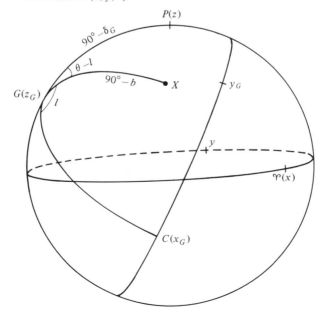

$$\sin b = + \cos \alpha_G \cos \delta_G \cos \alpha \cos \delta$$
$$+ \sin \alpha_G \cos \delta_G \sin \alpha \cos \delta + \sin \delta_G \sin \delta.$$

$$(14.53)$$

These equations are neatly expressed in vector form as

$$\mathbf{s}_G = \mathbf{R}_G \mathbf{s}, \tag{14.54}$$

where $\mathbf{R}_G$ is the rotation matrix, whose elements may be read off from (14.53). Using the value of (α_G, δ_G) and θ quoted in (14.5) the elements may be computed (cf. Problem 14.6). That form of the rotation matrix converts to galactic coordinates from equatorial coordinates based on the standard epoch B1950.0. A further application of a rotation matrix is needed to give the form of $\mathbf{R}_G$ for the epoch J2000.0. In numerical form it is

$$\mathbf{R}_G = \begin{pmatrix} -0.054\,876 & -0.873\,437 & -0.483\,835 \\ 0.494\,109 & -0.444\,830 & 0.746\,982 \\ -0.867\,666 & -0.198\,076 & 0.455\,984 \end{pmatrix}. \tag{14.55}$$

The rotation matrix may be applied to the solar velocity $\mathbf{U}$, which expressed in galactic coordinates will be

$$\mathbf{U}_G = \mathbf{R}_G \mathbf{U}. \tag{14.56}$$

Denote the components of $\mathbf{U}_G$ by (X_G, Y_G, Z_G). It has already been noted that different values of the solar motion are obtained from studies of different classes of stars. These variations are not entirely due to observational errors and can be understood in dynamical terms. The values derived for X_G and Z_G are fairly uniform and reflect the true components of the sun's velocity in these directions relative to the LSR. The values of Y_G – in the direction of galactic rotation – vary systematically, however, with the class of star. This means that the different classes of star are defining different kinematic LSRs. Stars moving in near circular orbits define a LSR that approximates to the dynamical LSR. Where the dispersion in the stars' velocities is greater, however, the kinematically defined LSR lags behind the dynamical one (cf. Problem 14.8).

Determinations of the solar motion for many stellar classes are given by Mihalas and Binney (1981). Two mean solar motions are defined. The *standard solar motion* is that derived essentially from all catalogue stars, omitting only some early type stars. This is found to be

$$X_G = +10.2 \text{ km s}^{-1}$$
$$Y_G = +15.1 \text{ km s}^{-1} \tag{14.57}$$
$$Z_G = + 7.4 \text{ km s}^{-1}.$$

On the other hand, the *basic solar motion* is defined less empirically. It is derived more selectively from stars thought to be moving in near circular orbits about the galactic centre. The components of the basic solar motion are

$$X_G = 9 \text{ km s}^{-1}$$
$$Y_G = 12 \text{ km s}^{-1}$$
$$Z_G = 7 \text{ km s}^{-1}.$$

(14.58)

The values in (14.57) and (14.58) are those given by Allen (1976).

14.7   Oort's constants

Previous discussion has been concerned with stellar motions either with respect to the sun or with respect to the LSR. It would obviously be more significant if the motions were referred to the galactic centre. This is a difficult observational problem, however, since the stars that can be effectively observed are confined to regions within about 1 kiloparsec (kpc) of the sun; the galactic centre itself is unseen. Even so, systematic effects in radial velocities and proper motions of the stars due to galactic rotation can be detected.

If peculiar stellar velocities are completely ignored, a star's velocity due solely to galactic rotation will be of the form

$$\mathbf{V} = \omega \times \mathbf{R},$$

(14.59)

where $\mathbf{R}$ is the star's position vector with respect to the centre of the galaxy. The system of galactic coordinates discussed in the last section is clearly the most appropriate one to use. Since the galaxy rotates clockwise as seen from the north galactic pole, it is convenient to write

$$\omega = [0, 0, -\Omega(R)].$$

(14.60)

Notice (cf. Fig. 14.6), that the sun's position vector with respect to galactic centre C is in the *negative* x-direction and the velocity of the LSR is in the positive y-direction, i.e. in the direction $l = 90°$. This simple view assumes that there is no overall expansion (or contraction) of the galaxy, a point that has been seriously questioned.

Let us use the subscript zero to indicate values of parameters corresponding to the LSR, the point O in Fig. 14.6. Then the local circular velocity of rotation is

$$V_0 = R_0 \Omega_0.$$

(14.61)

None of the three parameters occurring in (14.61) is determined by direct observation of nearby stars. Positional and spectroscopic data will reveal only the differential effects of galactic rotation. These are most neatly expressed in terms of Oort's constants A and B which are defined as

$$A = -\tfrac{1}{2}R_0\left(\frac{d\Omega}{dR}\right)_0,$$

$$B = A - \Omega_0. \tag{14.62}$$

Both A and B have the dimension of $(\text{time})^{-1}$; in practice, they are expressed in the units of km per second per kiloparsec.

Consider now a star S which, as indicated in Fig. 14.6, is at a distance ρ in the direction $\mathbf{s}$ from the LSR. Its position vector $\mathbf{R}$ with respect to the galactic centre is

$$\mathbf{R} = \mathbf{R}_0 + \rho\mathbf{s}. \tag{14.63}$$

If we assume that this star is moving only with the general galactic rotation its velocity may be derived from equation (14.59). Then the star's velocity relative to the LSR is $\mathbf{v} = \mathbf{V} - \mathbf{V}_0$, where

$$\mathbf{v} = (\boldsymbol{\omega} - \boldsymbol{\omega}_0) \times \mathbf{R}_0 + \rho\boldsymbol{\omega} \times \mathbf{s}. \tag{14.64}$$

The observed radial velocity v_r is immediately obtained from this equation as

$$v_r = \mathbf{v} \cdot \mathbf{s} = [(\boldsymbol{\omega} - \boldsymbol{\omega}_0), \mathbf{R}_0, \mathbf{s}]. \tag{14.65}$$

The components of the vector $\mathbf{s}$ are given in (14.52); so, using (14.60), the triple scalar product may be evaluated to give

$$v_r = R_0(\Omega - \Omega_0)\cos b \sin l. \tag{14.66}$$

Equation (14.66) is exact and is the formula that should be used in long-range radial velocity investigations, as, for example, in 21 cm hydrogen line observations. If, however, we assume that $\rho \ll R_0$, we may make the approximation

$$(\Omega - \Omega_0) = \left(\frac{d\Omega}{dR}\right)_0 (R - R_0). \tag{14.67}$$

Figure 14.6. O is the LSR and C the galactic centre.

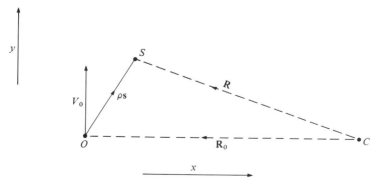

Now, from (14.63), the vector $\mathbf{R}$ has components.

$$\mathbf{R} = (-R_0 + \rho \cos b \cos l, \ \rho \cos b \sin l, \ \rho \sin b),$$

leading to

$$R^2 = R_0{}^2 - 2\rho R_0 \cos b \cos l + \rho^2.$$

This may be written with sufficient accuracy as

$$R - R_0 = -\rho \cos b \cos l,$$

which combines with (14.67) and (14.62) to yield

$$R_0(\Omega - \Omega_0) = 2A\rho \cos b \cos l. \tag{14.68}$$

The radial velocity is, therefore, derived from (14.66) as

$$v_r = A\rho \cos^2 b \sin 2l. \tag{14.69}$$

If A is in its conventional units, the distance ρ is then expressed in kiloparsecs to give the radial velocity in kilometres per second.

14.8   The effect of galactic rotation on proper motions

It will be convenient to express the star's velocity with respect to the LSR in the form

$$\mathbf{v} = \mathbf{v}_1 + \mathbf{v}_2, \tag{14.70}$$

where, by (14.64),

$$\mathbf{v}_1 = (\omega - \omega_0) \times \mathbf{R}_0$$
$$\mathbf{v}_2 = \rho\omega \times \mathbf{s}. \tag{14.71}$$

We then examine separately the proper motions produced by $\mathbf{v}_1$ and $\mathbf{v}_2$.

Figure 14.7 represents the celestial sphere centred on the observer. The star's direction is indicated by the point S, and X, Y and Z give the directions of the axes in the system of galactic coordinates. Then $SZ = 90° - b$ and $S\hat{Z}Y = 90° - l$. Let us denote the arc length SY by θ and the angle ZSY by ϕ.

The velocity $\mathbf{v}_1$ is in the positive y-direction and is, therefore, directed towards the point Y. The radial component of this velocity was investigated in the previous section; the transverse component will be $v_1 \sin \theta$. It will produce a proper motion μ_1 that is directed along the arc SY. Adapting equations (11.8), the components of this proper motion in galactic latitude and longitude are

$$4.74\mu_{1l} \cos b = \pi v_1 \sin \theta \sin \phi$$
$$4.74\mu_{1b} = \pi v_1 \sin \theta \cos \phi. \tag{14.72}$$

In these two equations v_1 is expressed in km s^{-1}, while the proper motion components and π must be in the same angular measure. As we shall see below, milliarcseconds are appropriate units, because then $\pi\rho = 1$.

Now applying the sine and analogue formulae to spherical triangle SZY yields the two results

$$\sin \theta \sin \phi = \cos l$$
$$\sin \theta \cos \phi = - \sin b \sin l. \tag{14.73}$$

The magnitude of $\mathbf{v}_1$ is, in fact, given implicitly by (14.68). Using this value and (14.73), it is found that

$$4.74\mu_{1l} = 2A \cos^2 l$$
$$4.74\mu_{1b} = -A \sin 2l \cos b \sin b. \tag{14.74}$$

Equation (14.71) indicates that the velocity $\mathbf{v}_2$ is wholly transverse and parallel to the galactic plane. It can only contribute to the proper motion component μ_l, in fact in the direction of decreasing galactic longitude, as shown by μ_2 in Fig. 14.7. Hence

$$4.74\mu_{2l} \cos b = -\pi v_2$$
$$\mu_{2b} = 0. \tag{14.75}$$

Now the magnitude $v_2 = \rho \Omega \cos b$ by (14.71).

To a first approximation, we may replace Ω by Ω_0; then using the definitions of Oort's constants (14.62) it is seen that

Figure 14.7

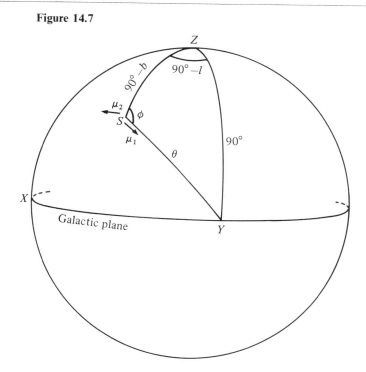

$$v_2 = (A - B)\rho \cos b. \tag{14.76}$$

Inserting this in (14.75) and combining that equation with (14.74) will lead to the final result

$$4.74\mu_l = A \cos 2l + B$$
$$4.74\mu_b = -A \sin 2l \cos b \sin b. \tag{14.77}$$

As mentioned above, Oort's constants are normally expressed in km s^{-1} per kiloparsec. Equations (14.77) then give the proper motion components in milliarcseconds per year. Notice that the proper motion effects indicated in these equations are independent of the star's distance. Equations (14.77), besides revealing systematic effects that depend on the star's galactic latitude and longitude, also uncover a rotation of the entire celestial sphere at angular rate of $B/4.74$ milliarcseconds per year. The value of B is very difficult to determine, but, if it is taken to be about 10 km s^{-1} kpc^{-1}, this rotation is 0″.21 per century. The implications of this rotation for the stellar reference system will be considered further in section 16.7 when a comparison with the reference system established by radio astrometry will be possible.

Problems

14.1 Calculate the right ascension and declination of the galactic centre for the epoch B1950.0.

14.2 A series of N astrographic plates are taken at times t_i, $i = 1, \ldots, N$. Measurement of these plates yields a set of standard coordinates (ξ_i, η_i) all referred to the same tangential point. Explain in detail how the star's parallax and proper motion components may be obtained by the method of least squares, stating the normal equations explicitly. (Note – there are $2N$ equations of condition, although the parallax is the only unknown that will appear in them all.)

14.3 Explain carefully how the solar apex may be established from a statistical investigation of stellar proper motions. In particular, justify the form of the normal equations quoted in equations (14.27).

14.4 Prove that for members of a star cluster with a common space motion the quantity

$$V_r \sin \delta + \frac{4.74\mu_\delta \cos \delta}{\pi} = \text{const},$$

where the symbols have their usual meanings.

14.5 Describe how the solar motion may be determined by investigating the radial velocities of stars, making clear the assumptions that are involved. Show that the components of the solar motion are the solutions of the normal equations (14.33).

14.6 Calculate the elements of the rotation matrix $\mathbf{R}_G$ which transforms a vector to galactic coordinates from its equatorial coordinates for the epoch B1950.0.

14.7 Apply the rotation matrix for precession to the matrix $\mathbf{R}_G$ derived in the previous question and verify the elements given in equation (14.55). Do you find small discrepancies in the last decimal place? How can these be accounted for?

14.8 Show that the Y-component of velocities of stars (i.e. in the direction $l=90°$, $b=0$) with respect to the LSR cannot be distributed symmetrically about zero. By drawing a comparison with planetary motion and Kepler's second law in particular, give qualitative reasons why the kinematically defined LSR should lag behind the dynamical one.

15

Elements of radio astronomy

15.1   Introduction

Up to this point, our discussion of positional astronomy has concentrated on optical sources with only occasional references to radio astronomy. It is true that most of the geometrical considerations are equally valid whatever the observational technique and whatever the wavelength of the radiation to which they are applied. When particular observational methods were examined, however, particularly in chapters 5, 13 and 14, they have involved observations of stars' positions and, therefore, methods of optical astronomy. The purpose of this chapter is to redress the balance.

To date, high-precision positional measurements have been restricted to the two traditional branches of ground-based astronomy, working at optical and radio wavelengths. Satellite observations, in other frequency bands, may provide a wealth of astrophysical information, but the positional measurements that are achieved can do little more than provide a general search area, for an optical or radio counterpart. Much the same was indeed true of radio astronomy during its formative era. It is only since the development of radio interferometers with transcontinental and intercontinental baselines that the precision of radio astrometry has matched and eventually surpassed that of optical astrometry. The two types of astrometry are, however, complementary. With its long history of development, optical positional astronomy has been far more systematically established and some measurements, like those of proper motions, of necessity require a long interval of time to be completed. Optical and radio astrometry involve different sources of radiation, in general, and, of course, the techniques and instrumentation are entirely different.

The most obvious difference is that of wavelength. The earth's atmosphere is opaque in most regions of the electromagnetic spectrum. The two most definite regions of transparency are the optical 'window' which covers the visible band of frequencies and a little bit more, and the vast radio window in which the atmosphere is almost completely transparent over several orders of magnitude in wavelength. It is difficult to give precise values for the limits of the radio window, since these vary with atmospheric conditions. As a rough guide, let us regard it as extending from wavelengths of 10 m down to 1 cm (in frequency terms from 33 MHz to 33 GHz). The short wavelength limit is imposed by molecular absorption in the atmosphere. The long wavelength limit is due to the ionosphere which reflects low-frequency cosmic radio waves. Ionospheric refraction, which is related to this, is considered in chapter 16.

The existence of the radio window means that cosmic radio emission may be detected from the ground over a range of frequencies in excess of 1000:1. The sources of emission are seldom stars, and are often associated with inconspicuous optical objects. In most cases the emission has a continuous spectrum arising from a non-thermal mechanism. Whereas the optical region of the spectrum is rich in spectral lines due to atomic transitions, there are, by comparison, few radio spectral lines. The lower-frequency molecular transitions tend to fall in the infra-red and millimetre bands, and it is this very paucity of molecular absorption bands that produces the atmospheric radio window.

Radio observational techniques differ from optical in two important respects. Straightforward optical observations tend to be made over a wide band of frequencies. Special spectroscopic techniques must be employed for spectral analysis. A radio observation, however, is normally made over a comparatively narrow band of frequencies, namely that to which the radio receiver is tuned. Moreover, the radio antenna is a polarized device, so that a radio measurement is made, of necessity, corresponding to a particular form of polarization. Whereas polarimetry is a comparatively recent refinement in optical astronomy, it is quite natural and basic to radio observations. A considerable percentage of polarization is detectable in some radio sources, and the measurements are comparatively simple.

The radio analogue of an optical reflecting telescope is the parabolic reflector, the radio dish. The surface of the dish is a paraboloid of revolution, as illustrated in Fig. 15.1. Rays from a distant point source will impinge on the reflecting surface as a parallel beam. After reflection they converge at the focus F. A small feed antenna is, therefore, placed at this point. The concentrated electric field at F drives a current in the feed antenna which is connected to the radio receiver, where the current is amplified and detected. The band of frequencies that is being accepted is determined partly by the design of the feed, but mainly by the receiving equipment. The great advantage of a radio dish is that it may be used over a wide band of frequencies, for the reflecting properties of the dish itself are independent of frequency. In order to change the radio frequency of observation, it is only necessary to change the feed antenna and retune the receiving equipment.

Parabolic reflectors are usually fully steerable. An equatorial mount would be ideal, but this is impracticable for large instruments. Mechanical considerations then prescribe an alt-azimuth mount. In these circumstances setting the radio telescope requires a transformation from equatorial to alt-azimuth coordinates. This problem has already been adequately covered in section 2.8. There are many circumstances in which the steering capability of a parabolic reflector is a great advantage. For example, when a radio spectral line is being studied, it may be necessary to sweep in frequency through the line with a narrow instantaneous bandwidth. This decreases the signal-to-noise ratio and a prolonged integration time is necessary to maintain the sensitivity of the equipment. Tracking of the source, to compensate for its diurnal motion, is therefore required.

We shall use the example of the parabolic reflector to illustrate a very important general principle that applies to all types of radio telescope. Figure 15.1 illustrates only the geometrical optics of a radio dish, it gives no indication of the diffraction pattern that is caused by edge effects, due to the finite size of the aperture. Let us denote the cross-sectional radius of the dish by D. The power from the incoming radio wave is not concentrated entirely at the geometrical point F but is spread over a small area of the focal plane. The resulting distribution of power P per m^2 against a distance x from the paraboloid axis is shown in Fig. 15.2. If the feed antenna were moved across the focal plane, this variation in power would be detected by the radio

Figure 15.1. Geometrical optics of a paraboloid radio antenna.

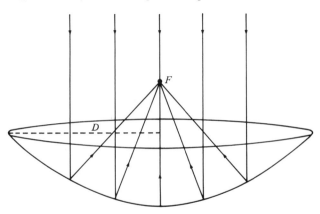

receiver. In fact, the feed antenna is normally fixed, but a similar variation is detected as the antenna beam traverses a point radio source.

These considerations are not peculiar to radio astronomy. A similar diffraction pattern (the Airy diffraction disc) is produced as the image of a star on a photographic plate. Usually this diffraction pattern is blurred by atmospheric scintillation caused by tropospheric irregularities. What is special to radio astronomy is that the diffraction pattern exhibited in Fig. 15.2 produces a practical limitation to the angular resolution of the radio telescope. The typical width of the diffraction pattern in the focal plane is of order $f\lambda/D$, where f is the focal length of the paraboloid. Expressed in angular terms, this means that the feed antenna is accepting significant radiation from a cone of directions within an angle λ/D radians of the paraboloid axis. We may express this by saying that the antenna system has an angular resolution R given by

$$R \sim \frac{\lambda}{D} \text{ radians.} \tag{15.1}$$

Equation (15.1) has purposely been kept imprecise. In order to make the equation exact, one must give a quantitative definition of R, introducing a numerical factor of order unity on the right-hand side. This factor will depend on the design of the particular antenna. As it stands, however, equation (15.1) is of quite general applicability. It gives the angular resolution of any radio instrument, provided that D is understood as the overall size of the antenna system.

Figure 15.2. Diffraction pattern for a paraboloid reflector.

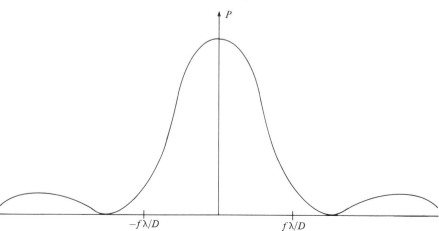

Table 15.1. *The antenna size necessary to achieve an angle of resolution*
R *at wavelength* λ

λ/R	1′	1″	0″.001
5 m	17 km	1000 km	—
50 cm	1.7 km	100 km	—
5 cm	170 m	10 km	10 000 km
	(Large array)	(SBI)	(VLBI)

15.2   Radio interferometry

The inherently low resolution of a radio telescope will be apparent
from equation (15.1). The largest fully-steerable dish in the world has a
diameter of 100 m. The angular resolution of even such a large instrument is
a matter of degrees at metre wavelengths and is still of the order of an arc
minute even at centimetre wavelengths. Table 15.1 gives approximate
numerical values for the size of instrument needed to provide angular
resolutions that will be useful in astrometry. Clearly this is more likely to be
achieved at short wavelengths, and indeed radio astrometric techniques
operate in practice at wavelengths of a few centimetres. The final row in
Table 15.1 indicates the observational technique that must be used to
achieve resolutions of these orders. They are described more fully below. It
should perhaps be noted that, barring systematic errors, which are certainly
very difficult to eliminate, positional accuracies using these techniques in
excess of the values of R quoted above are possible.

A large array will not normally be steerable, so tracking the source is not
possible. This is not necessarily a disadvantage in positional work. If the
source is allowed to move through the antenna beam with its diurnal
motion, timing of the maximum response provides information about the
source's position. Positional accuracies of the order of one arc minute are
sufficient to enable a realistic search to be made for an optical counterpart
of a radio source. For an independent positional determination, however,
interferometric methods must be used.

Short Baseline Interferometry (SBI), or connected interferometry, makes
use of two aerials separated by a considerable distance. The parameter D
used above now refers to the length of this baseline, not the size of the
individual aerials. The two component aerials are connected to the same
receiver and this requirement limits the length of the baseline to a few km.
There is a phase difference between the signals presented by the two aerials
due to the difference in path length for the radio wave as shown in Fig. 15.3.

The phase difference varies with the diurnal motion of the source to produce a modulated output from the receiver.

If the requirement that the two aerials should be connected to the same receiver can be circumvented, much longer baselines are possible. In connected interferometry the two radio-frequency signals from the aerials are mixed with the signal from a common local oscillator to produce signals at an intermediate frequency (IF). These are combined, amplified and detected by the main part of the receiver. In Very Long Baseline Interferometry (VLBI) independent local oscillators of extremely high stability are used at the two sites. The resulting IF signals are separately video-tape recorded. The tapes are later correlated to produce the interference pattern. This method places no restriction on the length of the baseline beyond the requirement that the radio source should be simultaneously observable at the two observing sites.

The basic principles of a two-element interferometer will now be discussed. Suppose that we have two aerials A_1 and A_2 which are separated by a baseline D, as shown in Fig. 15.3. The plane of this diagram is not necessarily a vertical plane, but is the plane defined by the baseline A_1A_2 and the direction of the source. The lines S_1A_1 and S_2A_2 represent parallel rays from the source to the two aerials. Let A_2B_1 be the perpendicular from the point A_2 to the ray S_1A_1. Then $A_1B_1 = D\cos\theta$, where θ is the angle that the direction of the source makes with the baseline. Now the line A_2B_1 is a wavefront, and so the radio emission is in phase along this line. It is not,

Figure 15.3. Two-element radio interferometer.

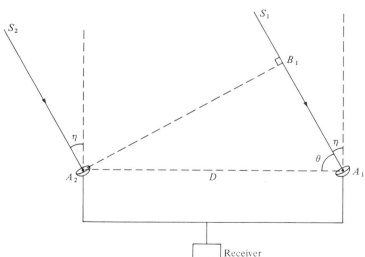

however, in phase at the two aerials due to the additional path-length to the aerial A_1. The phase difference between the signals presented to the receiver by the two aerials is given by

$$\phi = \frac{2\pi D}{\lambda} \cos \theta + \phi_c. \tag{15.2}$$

In this equation, ϕ_c is a constant phase difference of instrumental, rather than geometrical, origin. We have included this phase difference to make allowance for any difference in the lengths of connections of the two aerials to the receiver.

Let us assume, for definiteness, that the two aerials are individually fully steerable and are tracking the source. Then the mean voltages being supplied by each separately, V_1 and V_2, say, to the receiver will be constant. To simplify the discussion, let us assume that the source's emission may be treated as monochromatic. The source will have a continuous spectrum, but the receiving equipment will be rejecting all but a narrow band of frequencies. The effect of this finite bandwidth is – for the moment – being neglected. With these assumptions, the voltages $\mathscr{E}_1$, $\mathscr{E}_2$ being supplied by the two aerials to the receiver may be written as

$$\begin{aligned} \mathscr{E}_1 &= V_1 \cos \omega t \\ \mathscr{E}_2 &= V_2 \cos (\omega t + \phi). \end{aligned} \tag{15.3}$$

Now these two voltages may be combined in two different ways. In the *total power interferometer*, they are simply added together, and the output P from the receiver is proportional to $\langle (\mathscr{E}_1 + \mathscr{E}_2)^2 \rangle$. Here the symbol $\langle \; \rangle$ is being used to denote the long time average. In this type of interferometer, we measure

$$\begin{aligned} P \propto V_1{}^2 \langle \cos^2 \omega t \rangle + V_2{}^2 \langle \cos^2 (\omega t + \phi) \rangle \\ + V_1 V_2 \langle 2 \cos (\omega t + \phi) \cos \omega t \rangle. \end{aligned} \tag{15.4}$$

Now

$$\langle \cos^2 \omega t \rangle = \langle \cos^2 (\omega t + \phi) \rangle = \tfrac{1}{2}$$

while

$$\langle 2 \cos (\omega t + \phi) \cos \omega t \rangle = \langle \cos (2\omega t + \phi) \rangle + \langle \cos \phi \rangle = \cos \phi.$$

So the recorded output may be written in the form

$$P = P_0 + P_1 \cos \phi, \tag{15.5}$$

where P_0 and P_1 are constants.

The diurnal motion of the source produces a variation in the angle θ. The corresponding changes in the phase difference ϕ will be very rapid if $D \gg \lambda$. The output of the total power interferometer, therefore, exhibits a variation of amplitude P_1 imposed on a constant level P_0. If the two component aerials are identical, it is found that $P_1 = P_0$.

In the alternative form of interferometer the input voltages are multiplied rather than added. This is called a *correlation interferometer*. Since the output is proportional to $\langle \mathscr{E}_1 \mathscr{E}_2 \rangle$, it will be of the form

$$P = P_0' \cos \phi. \tag{15.6}$$

The advantage of this type of interferometer is that it suppresses any uncorrelated voltages from the two aerials and, therefore, greatly reduces background and receiver noise. Notice, however, that the two types of interferometer give the same interference pattern.

The pattern is seen most clearly when the source crosses the plane perpendicular to the baseline. Put $\theta = 90° - \eta$, and let us consider the situation when η is small. Then equation (15.2) yields

$$\phi - \phi_c = \frac{2\pi D}{\lambda} \eta. \tag{15.7}$$

So, near this normal plane, the interferometer output is

$$P = P_0' \cos\left(\frac{2\pi D}{\lambda} \eta + \phi_c\right). \tag{15.8}$$

This is graphed in Fig. 15.4, assuming for simplicity that ϕ_c is zero. Maximum output occurs when $\eta = 0$. Provided the correct fringe can be identified, the timing of this maximum locates the source on a great circle on the celestial sphere, namely the great circle that is instantaneously coincident with the plane $\theta = \pi/2$. From equation (15.8), it is seen that subsidiary maxima occur when

$$\eta = n \frac{\lambda}{D}. \tag{15.9}$$

Figure 15.4. Interferometer pattern with $\phi_c = 0$.

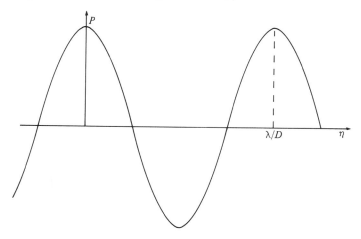

The separation between adjacent maxima is, as expected, λ/D radians. The temporal rate at which the fringe pattern is described also provides positional information, and may be used to locate the source's position on the plane $\theta = \pi/2$.

The above discussion refers strictly to a point source. If the source is greater in angular extent than λ/D radius, the fringe pattern will disappear. The interference pattern still contains useful information, though not in isolation. It may be regarded a Fourier component of period λ/D in the angular distribution of the source. We shall refer to this again when we consider aperture synthesis in section 15.8.

15.3   A radio transit instrument

A two-element interferometer operating along a horizontal east–west baseline provides the radio analogue of a meridian circle. The baseline defines two diametrically opposite points A_1, A_2 on the celestial sphere. Ideally, these points should coincide with the east and west points of the horizon. In order to allow for inevitable slight misalignment, however, let us suppose the point A_2 is displaced from W by amounts m and n in right ascension and declination respectively. Then, in Fig. 15.5

$$PA_2 = 90° - n \quad \text{and} \quad WPA_2 = m.$$

Figure 15.5

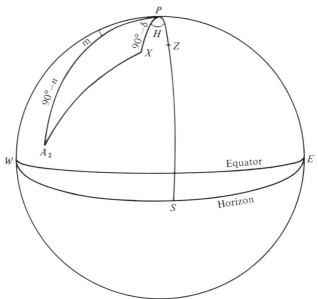

Now let X be the position of the source – of declination δ at a time when its hour angle is H. Then $PX = 90° - \delta$, and it is easily seen that

$$A_2PX = 90° - H - m.$$

Moreover the arc A_2X is equal to the angle θ introduced in the last section, cf. Fig. 15.3. Consequently applying the cosine formula to spherical triangle PXA_2 will yield the important result

$$\cos \theta = \sin n \sin \delta + \cos n \cos \delta \sin (H + m). \tag{15.10}$$

This last result allows the phase difference ϕ between the elements of the interferometer to be established in positional terms. For, by (15.2)

$$\phi - \phi_c = \frac{2\pi D}{\lambda} \left[\sin n \sin \delta + \cos n \cos \delta \sin (H + m)\right]. \tag{15.11}$$

This equation is exact and may be used for an interferometer of any orientation.

For the meridian type of instrument that we are considering here, m and n may, of course, be treated as small quantities, and the first-order approximations applied. Then near meridian transit (15.11) yields

$$H + m + n \tan \delta + \frac{\lambda}{2\pi D} (\phi_c - \phi) \sec \delta = 0. \tag{15.12}$$

Suppose now that the local sidereal time T is recorded corresponding to the central maximum of the fringe pattern. Allowing for the clock error ΔT, the value of the hour angle at this instant is

$$H = T + \Delta T - \alpha.$$

Inserting $\phi = 0$ in (15.12) then yields the source's right ascension as

$$\alpha = T + \Delta T + m + n \tan \delta + c \sec \delta, \tag{15.13}$$

where we have written

$$c = \frac{\lambda \phi_c}{2\pi D}. \tag{15.14}$$

Equation (15.13) is, by design, identical with the observational equation of a meridian circle. The misalignment of the interferometer baseline is exactly analogous to the misalignment of the rotation axis of a transit circle. Moreover, it is seen that the instrumental phase constant ϕ_c is equivalent to a collimation error. Taking the analogy still further, we may compare the fringe maxima to a series of cross-wires. Timing the passage of the source through a number of them distributed symmetrically about the central fringe and taking the mean will clearly give an improved value of the right ascension.

The source's declination may be derived from the observed fringe rate.

Since $dH/dT = 1$, differentiating equation (15.12) will yield

$$\cos \delta = \frac{\lambda}{2\pi D} \frac{d\phi}{dT}. \tag{15.15}$$

This equation does not allow us to discriminate between positive and negative declinations. This, in itself, is not a serious drawback, but the determination of δ becomes imprecise near the equator.

15.4   Phase ambiguity

The argument of the last section may only be applied if the central fringe ($\phi = 0$) can be correctly identified. But, as (15.5) or (15.6) indicate, the observed quantity is not the phase angle itself but its cosine. There is, therefore, an inherent ambiguity in the observed phase of $2n\pi$ which must be removed before interferometric measurements of position can be made definitive.

In deriving the observed fringe pattern for the two types of interferometer in section 15.2, it was assumed that each aerial was independently tracking the source and keeping it at the centre of its aerial beam. With this arrangement, which is not essential, the amplitude of the fringe pattern remains fixed, only the spacing of the fringes will vary as the source moves across the sky. Suppose, however, that as an alternative arrangement the two aerials are kept fixed pointing in a direction normal to the baseline. The fringe pattern will only be observed while the source is within the two aerial beams and the amplitude of the pattern will vary with the source's diurnal motion. It may be possible to identify the central fringe as the highest maximum in the pattern.

This procedure has one serious disadvantage, however. The identification depends on the pointing accuracy of the two aerials. This will be under imprecise control compared with the orientation of the baseline. The technique is, however, of necessity used when the component aerials are large and not fully steerable.

The analysis so far has treated the source's radiation as monochromatic, neglecting the fact that observation is being made over a finite bandwidth Δv. Since $\lambda v = c$, bandwidths in wavelength and frequency are related by

$$\frac{\Delta \lambda}{\lambda} = -\frac{\Delta v}{v}. \tag{15.16}$$

Now, as (15.2) or (15.7) indicates, the phase angle ϕ depends on wavelength. There is, therefore, a dispersion $\Delta \phi$ in the phase across the receiver bandwidth. Differentiating (15.7), this dispersion is found to be

$$\Delta \phi = 2\pi \frac{\Delta v}{v} \frac{D}{\lambda} \eta. \tag{15.17}$$

We conclude, therefore, that near the normal direction $\eta = 0$ all frequencies are in phase and the fringe pattern is clearly visible. Away from this normal direction, however, the phase dispersion builds up and reduces the visibility (or amplitude) of the fringes. The fringes will have entirely disappeared when $\Delta\phi \sim 2\pi$, corresponding to $\eta \sim \eta_{max}$, where

$$\eta_{max} = \frac{v}{\Delta v}\frac{\lambda}{D}. \tag{15.18}$$

An alternative way of expressing this is to say that the number of observable fringes is of order $v/\Delta v$.

The dependence of ϕ on wavelength may be regarded as both an advantage and a disadvantage. Considering positive aspects first, we note that the finite bandwidth imposes a variation on the fringe amplitude and this can aid the identification of the central maximum. The broader the bandwidth, the more restricted will be the fringe pattern and the central maximum will emerge more clearly. The same information may, however, be obtained without unduly degrading the fringe pattern, if simultaneous observations are made at two frequencies. For the two fringe patterns that are obtained will only match up at $\phi = 0$.

The effect of the finite bandwidth on fringe visibility implies, at first sight, that interferometric observations must be confined to within an angle η_{max} of the plane normal to the baseline. This will be quite a severe restriction with a long baseline. For example, if observations are being made at 6 cm wavelength ($v = 5000$ MHz) with a 1 km baseline, the fringe separation is about 12″. If the bandwidth is 10 MHz, this means that not more than 500 fringes can be observed, and so observation is restricted to 6000″ or 1° 40′ of the normal plane.

It is possible, however, to compensate artificially for the path-length difference between the two aerials by delaying the signal from one of them. One method of achieving this is to switch in an integral number of wavelengths into the connection between that aerial and the receiver. This will restore the visibility of the fringe pattern without affecting the phase difference or at least its cosine. Since the position of the source in the sky will be approximately known at all times, it is possible to calculate with sufficient accuracy the delay required to make the fringe pattern visible. Interferometer observations of the source may, by this technique, be extended to the whole of the visible hemisphere. In chapter 16 we shall consider how this observational technique permits the observation of a radio source's interference pattern over a 12-hour period. Since the entire pattern is observed, there is no ambiguity of phase in those circumstances. The geometrical considerations are similar to those discussed below for the simpler case of a north–south interferometer. As we shall see, it is then possible to use the disappearance of the fringe pattern as the source transits to remove the fringe ambiguity.

15.5   North–south interferometer

Consider an interferometer with a baseline of length D in a horizontal north–south direction. For definiteness, let us suppose that it is sited at north latitude Φ. To simplify the discussion we shall ignore any errors of alignment, and assume the orientation of the baseline is exact. The plane normal to the baseline is now that of the prime vertical WZE (Fig. 15.6). When a radio source crosses this great circle the phase difference ϕ between the two antennae will be zero. Such a crossing will only occur, however, if the source's declination lies within the range $0 < \delta < \Phi$.

Suppose the source is at the point X, with hour angle H. The phase difference corresponding to this point may be derived without further analysis from equation (15.11). The point A_2 on the celestial sphere of Fig. 15.5 corresponding to the baseline direction is now situated at S, the south point of the horizon. The displacement of this point in right ascension and declination from W are therefore given by

$$m = 90° \tag{15.19}$$
$$n = \Phi - 90°.$$

Inserting these values in the general result (15.11), we derive the phase difference as

$$\phi - \phi_c = \frac{2\pi D}{\lambda} (\sin \Phi \cos \delta \cos H - \cos \Phi \sin \delta). \tag{15.20}$$

Two alternative uses of the north–south interferometer will now be considered. In the first place, it may be used, as one might say conventionally, to observe the source near its transit over the prime vertical. The instrument is then equivalent to a prime vertical transit instrument. On the other hand, if a suitable line delay is introduced, the source may be observed near meridian transit. The interference pattern is then quite different.

In Fig. 15.6 the source's diurnal path is indicated as the small circle $X_1' X_2 X_1$. These three points are the source's position at transit of the prime vertical east, the observer's meridian and the prime vertical west respectively. Let $\mp H_0$ be the source's hour angle at its two transits of the prime vertical.

(i) Due to the bandwidth effects discussed in the last section, the fringe pattern is only observable when the source is close to the points X_1 and X_1'. For simplicity, let us take $\phi_c = 0$. Then the central maxima of the fringe pattern, $\phi = 0$, correspond to these points, and equation (15.20) yields

$$\cos H_0 = \tan \delta \cot \Phi. \tag{15.21}$$

Suppose that the two central maxima have been identified and their local sidereal times recorded. The mean of these two times determines the right ascension of the source, and half their difference yields the value of H_0. Consequently, equation (15.21) may be inverted to give the source's declination as

$$\tan \delta = \tan \Phi \cos H_0. \tag{15.22}$$

The above procedure clearly depends on the correct identification of the central fringes. Near the prime vertical the phase ϕ will be varying rapidly. The fringe rate may be derived from (15.20) which yields

$$\frac{d\phi}{dH} = -\frac{2\pi D}{\lambda} \sin \Phi \cos \delta \sin H. \tag{15.23}$$

Let us now suppose that an error of 2π has been made through misidentification of the fringes. The corresponding error in H_0 may be derived from (15.23), and substituted into the differential form of

Figure 15.6

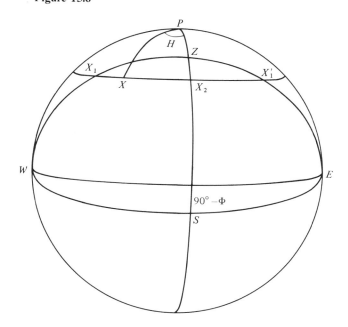

(15.22). This gives the resulting error in the derived value of the declination as

$$d\delta = \frac{\lambda}{D} \cos \delta \sec \Phi. \tag{15.24}$$

Let us now examine how this error will affect the fringe pattern near meridian transit.

(ii) In order to observe the fringe pattern near the meridian, it is necessary to compensate for the considerable path-length difference between the two aerials. Equation (15.23) indicates that the fringe rate goes to zero at $H = 0$. The phase difference has a maximum value at this point given, from (15.20), by

$$\phi_{max} = \frac{2\pi D}{\lambda} \sin (\Phi - \delta). \tag{15.25}$$

Suppose now that the compensation for path-length has been made by introducing an integral number of wavelength. The pattern near the meridian is then observable and, by (15.20) and (15.25), it will have the form

$$\phi = (\phi_{max} - 2n\pi) - \frac{4\pi D}{\lambda} \sin \Phi \cos \delta \sin^2 \tfrac{1}{2}H. \tag{15.26}$$

From this it is seen that the interferometer output is stationary at the meridian and the interference pattern does not build up until H is of the order of $(\lambda/D)^{1/2}$. The pattern is shown schematically in Fig. 15.7. The interferometer output near the meridian will, therefore, provide an unambiguous value of $\cos \phi_{max}$, and ϕ_{max} itself is determined apart from an integral multiple of 2π. Using (15.25) we relate the uncertainty in ϕ_{max} to that in the declination, obtaining

$$d\phi_{max} = -\frac{2\pi D}{\lambda} \cos (\Phi - \delta) \, d\delta.$$

Inserting the value of $d\delta$ from (15.24) then gives the error in the value of ϕ_{max} resulting from a misidentification of the fringe in (i). The result is

$$d\phi_{max} = -2\pi \cos \delta \sec \Phi \cos (\Phi - \delta). \tag{15.27}$$

It should be noted that this is of the order of 2π radians Consequently the misidentification of the central fringe on the prime vertical will provide a recognizable discrepancy between the observed value of $\cos \phi_{max}$ and its values as computed from (15.25). The phase ambiguity may, therefore, in principle, be removed.

The above argument makes the tacit assumption that the latitude of the instrument is accurately known. If this is not the case, even with correct phase identification on the prime vertical, there will still be a discrepancy between the calculated and observed values of $\cos \phi_{max}$. We may proceed as follows. Differentiating (15.22) relates the errors in Φ and δ, yielding

$$\sec^2 \delta \, d\delta = \sec^2 \Phi \cos H_0 \, d\Phi. \tag{15.28}$$

On the other hand, differentiating (15.25) gives a further relationship, namely

$$d\phi_{max} = \frac{2\pi D}{\lambda} \cos (\Phi - \delta)(d\Phi - d\delta). \tag{15.29}$$

Thus we have two simultaneous equations for $d\Phi$ and $d\delta$ – in terms of the discrepancy in ϕ_{max}. Assuming that there are no undetermined multiples of 2π in $d\phi_{max}$, we may solve for the required corrections to declination and latitude obtaining:

$$\begin{aligned} d\Phi &= \frac{\lambda}{D} \frac{d\phi_{max}}{2\pi} \frac{\cos^2 \Phi \sec (\Phi - \delta)}{(\cos^2 \Phi - \cos^2 \delta \cos H_0)} \\ d\delta &= \frac{\lambda}{D} \frac{d\phi_{max}}{2\pi} \frac{\cos^2 \delta \sec (\Phi - \delta) \cos H_0}{(\cos^2 \Phi - \cos^2 \delta \cos H_0)}. \end{aligned} \tag{15.30}$$

Figure 15.7. Interferometer pattern at meridian transit with a north–south interferometer.

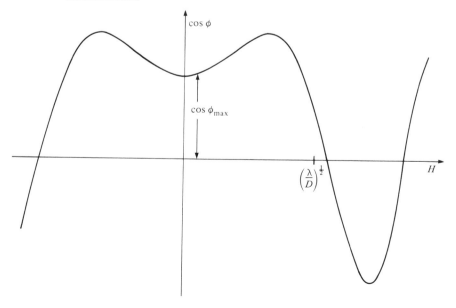

This provides, in principle at least, a method of investigating polar motion.

15.6   The interferometer of general orientation

We shall conclude our discussion of interferometers by considering one of general orientation. First of all, let us define its parameters as follows: the baseline is of length D, the operating wavelength is λ and the instrument is sited at latitude Φ and east longitude Λ.

Since the baseline may be several kilometres long, one must be quite specific about which point of the baseline these geographical coordinates refer to. Several conventions are possible; it might be the midpoint or one of the aerials. It does not really matter which convention is used as long as it is followed consistently. Let us state that (Λ, Φ) are the geographical coordinates of the aerial A_1, which is the more easterly of the aerials. Suppose further that the other aerial A_2 is in a direction at azimuth A (west of north) from A_1. Then $0 < A < 180°$. Finally let us assume that the baseline $A_1 A_2$ is at an altitude a referred to A_1's horizontal plane, meaning that this altitude is positive if A_2 is higher than A_1.

The celestial sphere in Fig. 15.8 is centred on the aerial A_1. Z is the zenith point for that aerial and N, W, S the corresponding cardinal points of the

Figure 15.8

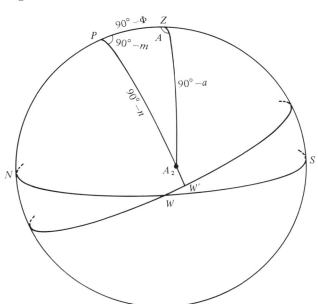

horizon. The direction of the baseline defines a point A_2 on the celestial sphere. This point has been drawn near the west point of the horizon, but this is not essential. It is presumably close to the horizon, but it may be at any azimuth. From our earlier definitions it is seen that $PZA_2 = A$ and $ZA_2 = 90° - a$, and, of course, $PZ = 90° - \Phi$.

The treatment given below is exact and compatible with the general expression for the phase given in equation (15.11). The two parameters m and n occurring in that equation are given by

$$Z\hat{P}A_2 = 90° - m$$
$$PA_2 = 90° - n. \tag{15.31}$$

These two parameters may be derived from spherical triangle PZA_2; the cosine and four-parts formulae respectively yield

$$\sin n = \sin a \sin \Phi + \cos a \cos \Phi \cos A$$
$$\tan m = \cos \Phi \tan a \operatorname{cosec} A - \sin \Phi \cot A. \tag{15.32}$$

The values of m and n derived from these equations may then be inserted into (15.11) to give the phase angle ϕ. Let us modify this very slightly, however, writing it as

$$\phi - \phi_c = \frac{2\pi D}{\lambda} [\sin n \sin \delta + \cos n \cos \delta \sin H'], \tag{15.33}$$

where H' is a modified hour angle, simply given by

$$H' = H + m.$$

H' is, in fact, the hour angle for an observer at east longitude Λ', where

$$\Lambda' = \Lambda + m. \tag{15.34}$$

This longitude is referred to as the *effective longitude* of the interferometer. It is determined by the direction of the baseline. It is obviously not the same as the longitude of either aerial or even their midpoint. For the case illustrated in Fig. 15.8, the effective longitude is east of both the aerials. If the orientation of the baseline is markedly not east–west, the effective longitude will differ appreciably from the longitudes of the two aerials.

Now join PA_2 on the celestial sphere and produce it (if necessary) to intersect the equator in the point W'. Since $WW' = m$, the point W' is the west point of the equator for any observer at east longitude Λ'. As such, it provides a useful reference direction.

The baseline may be represented by a vector **D**, of magnitude D, in the direction of A_2. Let us refer this vector to rectangular equatorial axes fixed with respect to the earth with the z-axis directed towards P. It is convenient to choose the x-axis in the direction of W', and the y-axis will complete a right-handed set. The y-axis, therefore, lies in the meridian plane of longitude Λ'. Referred to these same axes, the baseline vector $\mathbf{D}$ has components

$$\mathbf{D} = D(\cos n, 0, \sin n). \tag{15.35}$$

Suppose next that $\mathbf{s}$ is a unit vector in the direction of the source. Referred to these same axes, this vector has components

$$\mathbf{s} = (\cos \delta \sin H', \cos \delta \cos H', \sin \delta). \tag{15.36}$$

Consequently equation (15.33) may be expressed in vector form as

$$\phi - \phi_c = \frac{2\pi}{\lambda} \mathbf{D} \cdot \mathbf{s}. \tag{15.37}$$

This amounts to no more than a verification of the fundamental expression for the phase (15.2). The vector form is helpful in that it permits the use of any coordinate system. Up until now, we have used an earth-based coordinate system, so that $\mathbf{D}$ is regarded as a fixed vector and $\mathbf{s}$ is subject to a diurnal variation. It is perfectly possible, however, to use a coordinate system fixed with respect to the stars. Then $\mathbf{s}$ is regarded as a fixed vector and the diurnal variation must be applied to $\mathbf{D}$. This approach is convenient in VLBI, but for the present we shall continue with earth-based coordinate systems.

The baseline vector $\mathbf{D}$ will be subject to small systematic changes arising from the pointing mechanism of the two aerials. The resulting change in the phase is given by

$$d\phi = \frac{2\pi}{\lambda} d\mathbf{D} \cdot \mathbf{s}. \tag{15.38}$$

It is appropriate to express $d\mathbf{D}$ in the coordinate system used in (15.35), particularly if the component antennas have equatorial mounts. Let us write $d\mathbf{D} = (dx, dy, dz)$ in this coordinate system. Then, from (15.36) and (15.38), it follows that

$$d\phi = \frac{2\pi}{\lambda} (dx \cos \delta \sin H' + dy \cos \delta \cos H' + dz \sin \delta). \tag{15.39}$$

It should be noted, in particular, that the effect of dz is independent of hour angle. It has the effect, therefore, of a declination dependent change in the value of the instrumental phase ϕ_c.

15.7   Polar motion

The sidereal direction of the baseline vector is affected by polar motion, whereas the direction of the radio source is fixed. With this understanding the change in the phase is given by (15.38). The vector $d\mathbf{D}$ may not, however, be derived simply by differentiating (15.35), since the coordinate system used in that equation is itself affected by the polar motion. Differentiate instead the phase itself in (15.33) to obtain

$$d\phi = \frac{2\pi D}{\lambda} [dn \cos n \sin \delta - dn \sin n \cos \delta \sin H'$$

$$+ \cos n \cos \delta \cos H' dH']. \quad (15.40)$$

Comparing this with (15.39), and noting that $dH' = d\Lambda'$, we identify the components of $d\mathbf{D}$ as

$$dx = -D \sin n \, dn$$
$$dy = D \cos n \, d\Lambda' \quad (15.41)$$
$$dz = D \cos n \, dn.$$

The vector $\mathbf{D}$ is inclined at an angle n to the equatorial plane and is perpendicular to the meridian plane of longitude Λ'. It is, therefore, parallel to the geocentric position vector of an observer at latitude Φ^* and longitude Λ^*, where

$$\Phi^* = n$$
$$\Lambda^* = \Lambda' - 90°. \quad (15.42)$$

Polar motion affects the vector $\mathbf{D}$ and this observer's position vector in the same way. Suppose that (x_p, y_p) represent the displacement of the pole along the meridians of longitude 0 and 270° E, respectively, then using (5.29), the changes in the geographical coordinates (Λ^*, Φ^*) are given by

$$d\Phi^* = x_p \cos \Lambda^* - y_p \sin \Lambda^*$$
$$d\Lambda^* = (x_p \sin \Lambda^* + y_p \cos \Lambda^*) \tan \Phi^*.$$

From this equation and (15.42), we find

$$dn = x_p \sin \Lambda' + y_p \cos \Lambda'$$
$$d\Lambda' = (y_p \sin \Lambda' - x_p \cos \Lambda') \tan n. \quad (15.43)$$

These values may now be substituted into (15.41) to give the components of $d\mathbf{D}$. The change in the interferometer phase may then be calculated from (15.39). The effect is complicated for an interferometer of general orientation, but simplifies considerably for an east–west baseline. In that case n is a small quantity, and (15.43) indicates that $d\Lambda'$ may be regarded as second order. By (15.41), the same is true of dx, and so only the z-component of $d\mathbf{D}$ is significant. Applying (15.39), with these approximations, yields

$$d\phi = \frac{2\pi D}{\lambda} (x_p \sin \Lambda' + y_p \cos \Lambda') \sin \delta. \quad (15.44)$$

As this is independent of hour angle, it is equivalent to a change in the instrumental phase ϕ_c.

15.8   Aperture synthesis

It is intended to complete the chapter with a brief discussion of aperture synthesis. No more than a skeleton outline is attempted of this important but complex facility in radio astronomy. Many results are quoted below with little justification. The reader interested in further details is recommended to consult the text by W. N. Christiansen and J. A. Högbom (1984).

Only point radio sources have been discussed so far. This is, of course, an idealization. A radio source must have a finite angular size, but it has been assumed that this is below the resolution of the radio telescope or interferometer. The intensity of an unresolved source is specified by its flux density S, which will be a function of radio frequency. The flux density is the power per hertz (Hz) crossing unit surface area of the wavefront. It is measured in Janskys (Jy) – $1 \text{ Jy} = 10^{-26}$ watts $\text{m}^{-2} \text{ Hz}^{-1}$. Now, suppose that P is the power per Hz accepted by a radio antenna from a source of flux density S. Then one may write

$$P = SA(H, \delta)$$

where A is an aerial parameter called the *effective area*. Now A is a function of direction, which will be effectively zero outside the aerial beam. It should perhaps be mentioned that S is strictly that part of the flux density which is in the form of polarization accepted by the aerial.

The flux density is the only measurable source parameter for an unresolved radio source. When the source more than fills the aerial beam, however, its structure may be investigated. Its intensity is then given in terms of its radio *brightness b* which will vary over the source. The SI unit of brightness is watts $\text{m}^{-2} \text{ Hz}^{-1}$ steradian^{-1}. The power P accepted by the antenna from a distributed source may be written as

$$P = \iint b(H, \delta) A(H, \delta) \, d\Omega \tag{15.45}$$

where $d\Omega$ is the element of solid angle.

Let us now introduce a rectangular coordinate system (ξ, η, ζ) on the celestial sphere. Normally the ξ–η plane is chosen as the plane of the antenna system. Since the element of solid angle is

$$d\Omega = \cos \delta \, dH \, d\delta = \frac{d\xi \, d\eta}{(1 - \xi^2 - \eta^2)^{1/2}}, \tag{15.46}$$

equation (15.45) may now be written as

$$P = \int_{-\infty}^{\infty} \int_{-\infty}^{\infty} \frac{b(\xi, \eta) A(\xi, \eta)}{(1 - \xi^2 - \eta^2)^{1/2}} \, d\xi \, d\eta. \tag{15.47}$$

The limits in ξ and η are, for convenience, extended to $\pm \infty$, although they are coordinates on the unit sphere, by taking A or b zero when $\xi^2 + \eta^2 > 1$.

For a total power telescope, P, and therefore A, is necessarily positive in all directions. Suppose, however, that two aerials of effective area A_1 and A_2 are connected as a correlation interferometer. Since the output may now be positive or negative, the same is true of the effective area A_c of the interferometer combination. In fact,

$$A_c = 2(A_1 A_1)^{1/2} \cos \phi, \tag{15.48}$$

where ϕ the geometrical phase difference between the two aerials. Supposing the interferometer baseline lies in the ξ–η plane, we may write its vector as

$$\mathbf{D} = \lambda(u, v, 0). \tag{15.49}$$

It then follows that the phase is given by

$$\phi = 2\pi(u\xi + v\eta). \tag{15.50}$$

Consequently the measured correlated power P_c with this interferometer arrangement is

$$P_c = \int_{-\infty}^{\infty} \int_{-\infty}^{\infty} \frac{2[A_1(\xi, \eta) A_2(\xi, \eta)]^{1/2}}{(1 - \xi^2 - \eta^2)^{1/2}}$$
$$\times b(\xi, \eta) \cos[2\pi(u\xi + v\eta)] \, d\xi \, d\eta. \tag{15.51}$$

Equation (15.51) gives the power detected by a single instantaneous observation by the interferometer arrangement. An independent measurement can be made at the same time by inserting a quarter of a wavelength in one arm of the interferometer and thereby increasing the phase difference by $\pi/2$. The correlated power detected by this arrangement, P_s say, is given by the same integral equation (15.51), except that a sine replaces the cosine. Now a complex correlated power for the interferometer spacing may be introduced as $P(u, v) = P_c + iP_s$. Notice that this is a quantity determined from observation and given by

$$P(u, v) = \int_{-\infty}^{\infty} \int_{-\infty}^{\infty} 2\left[\frac{A_1(\xi, \eta) A_2(\xi, \eta)}{1 - \xi^2 - \eta^2}\right]^{1/2}$$
$$\times b(\xi, \eta) e^{2\pi i(u\xi + v\eta)} \, d\xi \, d\eta. \tag{15.52}$$

Now this equation is a Fourier integral equation. It implies that $P(u, v)$ is a Fourier component of the brightness distribution weighted by the factor indicated. The factor is a known function of position. Consequently if the interferometer spacing is varied to provide the components $P(u, v)$ over a wide range of values of u and v, a Fourier inversion becomes possible to yield the required brightness distribution as

$$b(\xi, \eta) = \frac{1}{2} \left[\frac{1 - \xi^2 - \eta^2}{A_1(\xi, \eta) A_2(\xi, \eta)} \right]^{1/2}$$

$$\times \int_{-\infty}^{\infty} \int_{-\infty}^{\infty} P(u, v) e^{-2\pi i(u\xi + v\eta)} \, du \, dv. \tag{15.53}$$

This equation is the mathematical basis of aperture synthesis. Let us examine its implications by considering a particular example.

Suppose that the ξ–η plane is the horizontal plane, with the ξ-axis aligned east–west. Let us consider two identical aerials A_1 and A_2, which may be regarded as square with linear dimensions $n\lambda$. By themselves, either would provide a resolving power of n^{-1} radians. They are, however, connected as a correlation interferometer with variable spacing as indicated in Fig. 15.9. The aerial A_1 is moved through $2N$ positions along the ξ-axis, while A_2 is moved independently through N positions along the η-axis. There are $2N^2$ combinations in all, and the complex correlated power $P(u, v)$ may be obtained for each of them.

The values of u and v that are sampled by this process lie within the ranges $-Nn \leqslant u \leqslant Nn$, $0 \leqslant v \leqslant Nn$. It is easily seen, however, that $P(-u, -v) = P^*(u, v)$, the complex conjugate of $P(u, v)$. The region of u–v plane that is effectively covered is $|u| \leqslant Nn$, $|v| \leqslant Nn$.

When the brightness distribution is derived using the discrete analogue of (15.53), all interferometer spacings are included up to a maximum separation of $Nn\lambda$ in each direction. These are exactly the spacings that are

---

Figure 15.9. Aperture synthesis. Moving the aerials A_1 and A_2 along their tracks synthesizes the area A.

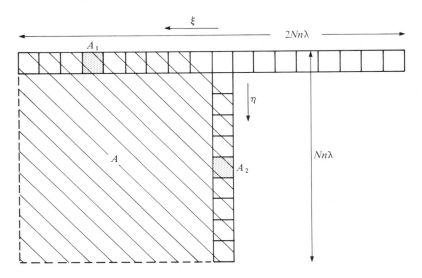

present between elements of the filled aperture A indicated in Fig. 15.9. So this aperture has been effectively synthesized. The derived brightness distribution is a map of resolution $(Nn)^{-1}$ radians, the same as that for the filled aperture.

The economic advantages of aperture synthesis are obvious. The filled aperture, however, achieves its high resolution with a single observation, while the method of aperture synthesis requires $2N^2$ sequential observations to be combined. This disadvantage is at least partly offset by the fact that these $2N^2$ observations provide a radio map of angular extent n^{-1} radians, and the filled aperture must scan the region to provide the equivalent information.

In order to reduce the problems associated with frequent movement of the two aerials, one of them may be replaced by an extended line array. Suppose, for example, the aerial A_1 is replaced by a complete line array, then only a one-dimensional synthesis is required, moving the aerial A_2. There will be a considerable saving of time, but the sky area covered by the synthesis is reduced in the east–west direction.

In the example cited above, the considerations of angular resolution are only strictly valid for a source at zenith, that is, in a direction normal to the antenna plane. If the source is at zenith distance z, however, it is still possible to direct the beam from a fixed horizontal array towards the source by introducing suitable phase differences between different elements within the array. The effective area is, however, reduced by a factor $\cos z$ due to foreshortening and the angular resolution of the instrument is correspondingly reduced.

A similar effect is found in the map obtained by aperture synthesis. Here the ξ–η plane has special significance as the plane in which the variable baseline lies. The brightness distribution (15.53) contains Fourier components up to frequency Nn and we, therefore, expect the radio map to have a resolution of $(Nn)^{-1}$ radians. This resolution, however, corresponds to a map on the ξ–η plane, not on the celestial sphere itself. The angular resolution on the celestial sphere is maintained in the azimuthal direction $\zeta = \text{const}$, but in the polar direction the angular resolution becomes $(Nn)^{-1}\zeta$ radians. Since $\zeta^2 = 1 - \xi^2 - \eta^2$, it is seen that the aperture synthesis is most effective in the direction $\xi = 0$, $\eta = 0$.

15.9   Rotational synthesis

In some circumstances the need for moving the two antennas may be removed by making use of the rotation of the earth to alter the baseline. Again we consider a particular example, an east–west correlation interferometer of baseline D. Suppose that the radio source to be investigated is centred on a point (α_0, δ_0) on the celestial sphere. Let us choose an equatorial coordinate system in which the η–ζ plane is the meridian of right ascension α_0. So in Fig. 15.10, the ξ-, η- and ζ-axes are directed towards the points X, Y, P respectively, at a time when the source's hour angle is H. Then the coordinates of a general point (α, δ) are given by

$$\xi = \cos \delta \sin (\alpha_0 - \alpha)$$
$$\eta = \cos \delta \cos (\alpha_0 - \alpha) \tag{15.54}$$
$$\zeta = \sin \delta.$$

Now the baseline is directed towards the point W. Since $WY = 90° - H$, the baseline vector is

$$\mathbf{D} = D(\cos H, \sin H, 0). \tag{15.55}$$

Consequently, by (15.49),

$$u = \frac{D}{\lambda} \cos H$$
$$\tag{15.56}$$
$$v = \frac{D}{\lambda} \sin H.$$

The method of rotational synthesis makes use of the variation in u, v as the source's hour angle changes. If the observation is continued for 12^h, a semicircle is completed on the u–v plane. This may be extended to a full circle, since $P(-u, -v) = P^*(u, v)$. A one-dimensional synthesis is thereby completed.

Figure 15.10

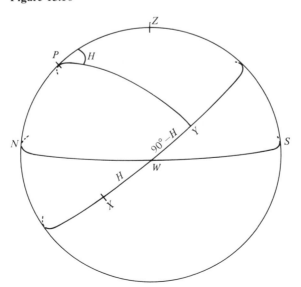

A two-dimensional synthesis will require repetition of the 12^h observation with varying spacing. Alternatively, if a line antenna is connected as a series of interferometers of different spacings the whole aperture may be synthesized in one 12^h period. The elements in the interferometers must be fully steerable so that $A_1(\xi, \eta)$ and $A_2(\xi, \eta)$ can be kept constant. In practice this means that they are usually equatorially mounted paraboloid antennae.

Phase compensation, as discussed in sections 15.4 and 15.5, is necessary with rotational synthesis in order that the interferometry may be performed with a reasonable bandwidth. The problems with foreshortening of the aperture, which were considered earlier, also arise. With the coordinate system that has been adopted, the source centre is in the direction $(0, \cos \delta_0, \sin \delta_0)$. The synthesis map of radio brightness, therefore, has a reduced resolution in declination; features smaller in angular extent than $(\lambda/D) \csc \delta_0$ cannot be investigated. Rotational synthesis, at least along the lines described here, becomes ineffective at low declinations, and it is most successful for sources near the celestial pole.

In conclusion, one other point should be noted. The requirement of a 12^h period of continuous observations means that the source must lie in the same hemisphere as the instrument. Otherwise the source will not remain above the horizon for a sufficient length of time to cover all necessary orientations in the u–v plane (cf. Problem 15.7).

Problems

15.1 Two identical aerials are connected as a total power interferometer, with each aerial tracking a radio source independently and accepting the same mode of polarization. Prove that the output from the receiver is proportional to $\cos^2 \frac{1}{2}\phi$, where ϕ is the phase difference between the two aerials.

15.2 Explain how a simple two-element interferometer with an east–west baseline may be used as a radio transit instrument to measure the right ascension and declination of radio sources. If the interferometer phase can be measured with a precision of $\Delta\phi$, show that the measurement of n fringes will lead to uncertainties in right ascension and declination approximately given by

$$\Delta\alpha = \frac{\lambda \, \Delta\phi \sec \delta}{2\pi\sqrt{n} \, D}$$

$$\Delta\delta = \frac{\Delta\phi \cot \delta}{2\pi n}.$$

15.3 A radio transit instrument is operating at a radio frequency of 5000 MHz with a receiver bandwidth of 10 MHz. No phase compensation is being used. Estimate the extent of the equatorial region within which no discrimination in declination is possible, if the interferometer phase is determined to a precision of 5°.

15.4 A radio source at declination 15° is being observed at meridian transit with a north–south interferometer of baseline 3 km. The instrument is sited at latitude 65° and a constant phase compensation is being used. The radio frequency of observation is 6000 MHz and a bandwidth of 15 MHz is being used. Calculate the approximate interval of time over which the fringes will be observable.

15.5 Explain how phase measurements and timings with a north–south interferometer allow the determination of a source's right ascension and declination together with the latitude of the observing site. Examine the errors of such determinations and, in particular, prove equations (15.30).

15.6 In section 15.6 the interferometer characteristics are developed from the location of A_1, the more easterly of the two aerials. Give the corresponding argument that takes A_2 as its starting point, and show that it leads to the same value for the effective longitude of the instrument.

15.7 An east–west line antenna of total length $n\lambda$ is used for rotational synthesis at north latitude ϕ. Show that if a source declination $-\delta(\delta < 0)$ were observed, the region of u–v plane that is covered would be restricted. Determine the shape of this region and show that its area is $2n^2 \cos^{-1}(\tan \phi \tan \delta)$.

16

Radio astrometry

16.1 Introduction

A reference system may be constructed based upon the observed positions of radio sources, instead of stars. The two reference systems will be independent since they are based on observations of different objects. The stellar reference system uses observations of galactic stars. Stellar motions are subject to systematic effects caused by the solar motion, already discussed in chapter 14, and by the overall galactic rotation. The radio reference system, on the other hand, is based primarily on compact radio sources most of which are extragalactic. Mach's principle would, therefore, suggest that the radio system is the more fundamental, and a closer approximation to an inertial frame of reference.

In optical astrometry stars are treated as point sources. There is a firm theoretical foundation for this assumption in that their physical sizes are known from the theory of stellar structure and their distances are known with reasonable, if not high, accuracy. One can calculate, therefore, how far a particular star is below the resolution of an optical telescope, at least to order of magnitude, and this calculation carries a fair degree of assurance. No such universal consensus exists about the nature of compact radio sources. Particularly important are the radio QSOs. They exhibit short time-scale variations that set upper limits to their physical dimensions. Their distances are derived from the assumption that the large optical red-shifts that are observed are of cosmological origin, being due to the universal expansion. While most astronomers today accept this interpretation, there remains a small minority that is unconvinced of the cosmological nature of the QSO red-shifts. They doubt, therefore, the extreme extragalactic nature of the QSOs. The discussion of the radio reference frame later in this chapter rests in part on an acceptance of the orthodox view that quasars and similar objects are, in fact, very distant galactic nuclei.

Aperture and rotational synthesis methods when applied to very compact sources reveal a brightness distribution that is virtually identical to the antenna pattern of the synthesized antenna. This merely indicates that the angular extent of the source is below the resolution of the radio telescope. For a mathematical point source the brightness distribution is formally a Dirac delta function, infinite at the source and zero elsewhere. From (15.51) one easily deduces that, for a point source of flux density S in the direction (ξ_0, η_0, ζ_0), the complex correlated power is

$$P(u, v) = 2 \left[\frac{A_1(\xi_0, \eta_0) A_2(\xi_0, \eta_0)}{1 - \xi_0^2 - \eta_0^2} \right]^{1/2} S e^{2\pi i (\xi_0 u + \eta_0 v)}. \qquad (16.1)$$

Consequently $|P(u, v)|$ is a constant. The Fourier transform of (16.1) gives the mathematical singularity in the desired position.

As explained in the last chapter, a one-dimensional rotational synthesis is achieved by a 12-hour observation with a fixed baseline east–west interferometer. When the source is unresolved by the synthesis, this is indicated by the constancy of $|P(u, v)|$. It is clearly much simpler to interpret the data as observations of a point source being made with a fixed spacing interferometer. Positional information of this point source is then extracted from the time variation of the interferometer phase over the 12^h period (Ryle and Elsmore, 1973). The instrument is then an extension of the radio transit instrument considered in section 15.3.

16.2   Twelve-hour connected interferometry

Consider an east–west interferometer with baseline D. We shall not include discussion of the errors of alignment since they have already been adequately covered in sections 15.3 and 15.6. The equation for the interferometer phase (15.11) then simplifies to

$$\phi - \phi_c = \frac{2\pi D}{\lambda} \cos \delta \sin H. \qquad (16.2)$$

If phase observations are made over the 12^h period it is possible to locate the points at which the phase has a turning point. From (16.2) it is easily seen that ϕ has a maximum at $H = 6^h$ and a minimum at $H = 18^h$. The maximum and minimum values are

$$\phi_{max} = \phi_c + \frac{2\pi D}{\lambda} \cos \delta$$

$$\phi_{min} = \phi_c - \frac{2\pi D}{\lambda} \cos \delta. \qquad (16.3)$$

These two equations may readily be solved for $\cos \delta$ and ϕ_c, to yield

$$\cos \delta = \frac{\lambda(\phi_{max} - \phi_{min})}{4\pi D}, \qquad (16.4)$$

$$\phi_c = \frac{(\phi_{max} + \phi_{min})}{2}. \tag{16.5}$$

Now the difference $(\phi_{max} - \phi_{min})$ is simply the total phase change over the 12^h period. Since this is directly observed, the declination is derived from (16.4). The actual values of the phase at any instant, however, may contain an undetermined multiple of 2π. The same will be true of ϕ_c. This is of no consequence, however; since this undetermined multiple is present throughout the phase pattern, it may be conveniently ignored. We may assume, for example, that $-\pi < \phi_c \leqslant \pi$, in order to make (16.5) definitive.

The value of ϕ at meridian transit is simply ϕ_c. This point in the interferometer pattern may be identified. The corresponding local sidereal time gives the right ascension of the source.

Although both coordinates (α, δ) are determined by the above argument, it is inadequate in this respect. The argument only makes use of the observed phase at three points corresponding to $H = 0, \pm 6^h$, but the interferometer phase is available over the entire 12^h period, so only a tiny fraction of the data has been used. The position derived may, however, be used as a first approximation. Suppose that it contains errors $(\Delta\alpha, \Delta\delta)$ and that instrumental phase is in error by $\Delta\phi_c$. The phase may be then calculated for any time from the derived values (α, δ) and ϕ_c, as

$$\phi_{calc} = \phi_c + \frac{2\pi D}{\lambda} \cos \delta \cos (T - \alpha), \tag{16.6}$$

where T is the local sidereal time. Let ϕ_{obs} be the observed value of the phase. Then, since the error in the hour angle $\Delta H = -\Delta\alpha$, equation (16.2) predicts that

$$\phi_{obs} - \phi_{calc} = \Delta\phi_c - \frac{2\pi D \, \Delta\delta}{\lambda} \sin \delta \sin H$$

$$- \frac{2\pi D}{\lambda} \Delta\alpha \cos \delta \cos H, \tag{16.7}$$

where H is the calculated hour angle $T - \alpha$.

Equation (16.7) provides an equation of condition that may be applied over the 12^h period during which H varies. The method of least squares will yield values of $\Delta\phi_c$ and of the coefficients of $\sin H$ and $\cos H$. Consequently, the corrections $(\Delta\alpha, \Delta\delta)$ to the source's position are determined.

For simplicity, the deviations (m, n) of the baseline from the exact east–west direction have been omitted. Their effects must, of course, be included in the derivation of ϕ_{calc} (cf. Problem 16.3). The local sidereal time should be that corresponding to the effective longitude Λ' of the baseline. This was introduced in section 15.6. There will, in addition, be a clock error

ΔT to be included in the complete analysis and all right ascensions will be displaced by this amount. A similar problem arose with the optical transit circle, which could only be resolved by locating the equinox from observations of the sun. This is not a feasible solution in radio interferometry since the sun is a large, irregular and erratic radio source. Consequently, right ascensions derived by radio interferometry are relative right ascensions. The zero point must be located by other means. The declination measurements, on the other hand, are absolute.

16.3   Very long baseline interferometry

The technique known as *Very Long Baseline Interferometry* (VLBI) differs from connected interferometry in that the intermediate frequency (IF) signals are tape recorded independently at the two sites of observation. They are later cross-correlated to produce the interference pattern. A schematic diagram of the equipment at one observing site is shown in Fig. 16.1.

The two observing sites may be at widely separated geographical locations. It makes no sense, therefore, to speak of the latitude and longitude of the interferometer or to work in terms of hour angles and local sidereal times. The basic geometry of the interferometer is unchanged but its formulation will be different.

Let us consider VLBI observations of a point radio source in the direction **s**, where

$$\mathbf{s} = (\cos \delta \cos \alpha, \cos \delta \sin \alpha, \sin \delta). \tag{16.8}$$

Now let $\mathbf{r}_1$ and $\mathbf{r}_2$ be the geocentric position vectors of the two antennas of

Figure 16.1. Very long baseline interferometry.

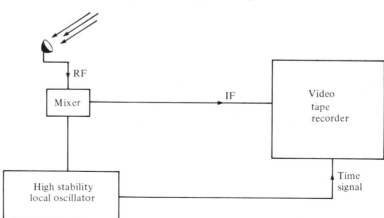

the interferometer as shown in Fig. 16.2. The components of these vectors depend on the geographical locations of the two observing sites. Suppose we denote their geocentric latitudes and longitudes by (ϕ_1, λ_1) and (ϕ_2, λ_2). The geocentric distances of the two sites ρ_1 and ρ_2 may be calculated using the methods of section 4.5. The vector $\mathbf{r}_1$ is directed towards the point on the celestial sphere with declination ϕ_1 and right ascension equal to the local sidereal time. Hence, if T is the Greenwich sidereal time, we may write the two vectors as

$$\mathbf{r}_1 = \rho_1[\cos \phi_1 \cos (T+\lambda_1), \cos \phi_1 \sin (T+\lambda_1), \sin \phi_1]$$
$$\mathbf{r}_2 = \rho_2[\cos \phi_2 \cos (T+\lambda_2), \cos \phi_2 \sin (T+\lambda_2), \sin \phi_2].$$

$$(16.9)$$

The baseline vector $\mathbf{D}$ is derived from these vectors as

$$\mathbf{D} = \mathbf{r}_2 - \mathbf{r}_1. \tag{16.10}$$

The *geometrical* phase difference between the two aerials may then be derived as

$$\phi = \frac{2\pi}{\lambda} \mathbf{D} \cdot \mathbf{s} = \frac{\omega}{c} \mathbf{D} \cdot \mathbf{s}, \tag{16.11}$$

where ω is the angular radio frequency, i.e. $\omega = 2\pi\nu$.

Figure 16.2. The geometry of VLBI.

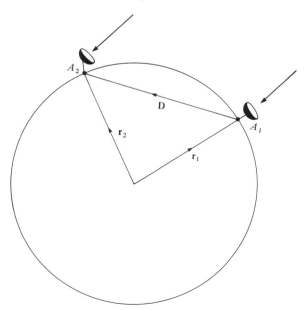

The observed interferometric phase, ϕ_0 say, will differ from the geometric phase for a number of reasons. Following Counselman (1976) one may write

$$\phi_0 = \frac{\omega}{c} \mathbf{D} \cdot \mathbf{s} + \phi_m + \phi_i + 2n\pi. \tag{16.12}$$

The notation is explained below.

The quantity ϕ_m represents the phase difference introduced in the propagation medium. After all, the radio wave must pass through different parts of the earth's atmosphere at quite different angles to the two antennae. Phase changes which are closely analogous to refraction effects, arise in both the troposphere and the ionosphere.

The phase difference of instrumental origin is represented by ϕ_i. In connected interferometry, it is possible to treat the instrumental phase as an unknown constant. In VLBI, however, ϕ_i will depend on the two independent frequency standards. Any slight instability in these will produce a variation in ϕ_i. This is an important limiting factor in the precision of VLBI.

In order that the interferometric phase be discernible in spite of bandwidth effects, it is necessary that the phase difference at cross-correlation should be quite small. If we identify this with ϕ_0, then there must be a large instrumental phase to compensate for the geometric phase $\omega \mathbf{D} \cdot \mathbf{s}/c$. This is achieved by introducing an offset between the two IF signals, so that the signal received at time $t + \Delta t$ at the first aerial is cross-correlated with signal received at time t at the second. Considered classically, Δt is simply

$$\Delta t = (\mathbf{D} \cdot \mathbf{s})/c. \tag{16.13}$$

In fact, relativistic effects are included in the reduction of VLBI observations. They are considered in section 16.6, but for the present we shall be content with the classical treatment.

Equation (16.12) includes a term $2n\pi$ representing the inherent phase ambiguity. The integer n is not normally precisely determinable in VLBI. The geometric phase, the first term on the right-hand side of (16.12), cannot be isolated. In order to make any deductions from that equation it is necessary to use a differential form. This can be realized in three ways; the differentiation may be made with respect to (i) angle, (ii) time, or (iii) frequency. Any of these procedures will remove the phase ambiguity. To simplify the discussion, we shall treat both ϕ_i and ϕ_m as constants, in spite of what has been said above. In this way, the geometrical principles of VLBI will emerge most clearly.

It is convenient to use a coordinate system aligned with the inter-ferometer baseline. Let $\mathbf{I}$ be the unit vector in the direction of $\mathbf{D}$, and $\mathbf{J}$ the unit vector perpendicular to the baseline which lies in the earth's equatorial plane. The unit vector $\mathbf{K}$ completes the right-handed set. If (α_0, δ_0) gives the direction of the baseline, then the three vectors just defined have components in the equatorial system given by

$$\mathbf{I} = (\cos \alpha_0 \cos \delta_0, \sin \alpha_0 \cos \delta_0, \sin \delta_0)$$
$$\mathbf{J} = (-\sin \alpha_0, \cos \alpha_0, 0) \qquad (16.14)$$
$$\mathbf{K} = (-\cos \alpha_0 \sin \delta_0, -\sin \alpha_0 \sin \delta_0, \cos \delta_0).$$

To transform any vector from the equatorial system to the baseline system, it is only necessary to apply the appropriate rotation matrix $\mathbf{R}$. This has components given by

$$\mathbf{R} = \begin{pmatrix} \mathbf{I} \\ \mathbf{J} \\ \mathbf{K} \end{pmatrix}. \qquad (16.15)$$

Remember the inverse of $\mathbf{R}$ is simply its transpose.

(i) *Angle*

Suppose there are two sources close together in the sky, in directions $\mathbf{s}$ and $\mathbf{s} + \mathrm{d}\mathbf{s}$. Taking the differential of (16.12) will yield

$$\mathrm{d}\phi_0 = \frac{\omega}{c} \mathbf{D} \cdot \mathrm{d}\mathbf{s}. \qquad (16.16)$$

The component of the source separation parallel to the baseline may, therefore, be determined from the observed phase difference.

When the separation is very small, the two sources will be simultaneously within the beam width of each of the component antennae. This need not, however, lead to confusion between the interference patterns of the two sources, for bandwidth effects will limit the visibility of the fringes to only one source at a time. To switch from one source to the other it is necessary to make a change in the time offset Δt.

When the two sources are farther apart, two antennae must be used at each observing site. Such observations are known as connected phase interferometry. The connection applies at each end of the baseline, since the two antennae there share a common frequency standard.

(ii) *Time*

Taking the time derivative of (16.12) will yield

$$\frac{\mathrm{d}\phi_0}{\mathrm{d}t} = \frac{\omega}{c} \frac{\mathrm{d}\mathbf{D}}{\mathrm{d}t} \cdot \mathbf{s}. \qquad (16.17)$$

The baseline vector is varying due to the rotation of the earth. Since $\mathbf{D}$ is fixed with respect to the earth, it follows that

$$\frac{d\mathbf{D}}{dt} = \mathbf{\Omega} \times \mathbf{D},\tag{16.18}$$

where $\mathbf{\Omega}$ is the earth's angular velocity vector. Consequently, if time is being measured in sidereal seconds

$$\frac{d\mathbf{D}}{dt} = \frac{2\pi D}{86400}\,(\mathbf{k} \times \mathbf{I}),$$

where $\mathbf{k}$ is a unit vector directed to the north celestial pole. From (16.14) it follows that $\mathbf{k} \times \mathbf{I} = \cos \delta_0\,\mathbf{J}$. Equation (16.17), therefore, yields

$$\mathbf{J} \cdot \mathbf{s} = \frac{86400}{2\pi D}\,\frac{c}{\omega}\,\frac{d\phi_0}{dt}\,\sec \delta_0.\tag{16.19}$$

Determination of fringe rate, therefore, allows one component of a single source's direction to be obtained. To complete the derivation of the source's position, a further VLBI measurement is required over a different baseline. Notice that the component that is obtained of necessity lies in the equatorial plane.

Any differentiation process applied to observational data tends to degrade those observations. This is particularly true of the fringe rate method. Time variations in the instrumental phase will be amplified. If the geometric phase could be determined without any ambiguity, one would expect VLBI to provide, like connected interferometry, an astrometric precision of the order of λ/D radians. This can indeed be achieved by method (i) for the relative positions of closely spaced sources. The fringe-rate method reduces this precision substantially.

(iii)   Frequency

The third method of VLBI involves deriving the group delay τ. This is simply the derivative of the phase with respect to the angular radio frequency, i.e.

$$\tau = \frac{d\phi_0}{d\omega}.\tag{16.20}$$

It is easily seen from (16.12) that determination of the group delay will yield positional information in the form

$$\mathbf{I} \cdot \mathbf{s} = \frac{c\tau}{D}.\tag{16.21}$$

The group delay is the observable that is normally used in VLBI. It is found to be far more satisfactory than the fringe rate. As equation (16.21) indicates, it gives the component of **s** parallel to the baseline. Exactly the same information would be obtained if the geometric phase angle itself were accessible. The group delay cannot, of course, be as precisely determined as an unambiguous phase. Over a bandwidth Δv, there will be a reduction of order $v/\Delta v$ in precision. Consequently as large a bandwidth as possible is used.

The time rate of change of the group delay provides the same information as the fringe rate itself, namely $\mathbf{J} \cdot \mathbf{s}$. In fact, by an argument similar to (ii), one derives

$$\mathbf{J} \cdot \mathbf{s} = \frac{86400}{2\pi D} c \frac{d\tau}{dt} \sec \delta_0. \tag{16.22}$$

The third component, $\mathbf{K} \cdot \mathbf{s}$, can then be obtained from the consideration that **s** is a unit vector. To derive **s** in equatorial coordinates it is only necessary to apply the inverse rotation matrix.

The above argument shows that the source's position may be derived from observations of the group delay over a period of time. The differentiation implied by (16.22) need not be explicitly performed. Instead, writing $\mathbf{s} = (l, m, n)$ and $\mathbf{D} = (x, y, z)$ in the equatorial system, we may use (16.21) in the form

$$xl + ym + zm = c\tau.$$

The right-hand side is determined from observation and the variation of the components of **D** are known. At each instant, this equation provides an equation of condition on (l, m, n). The components of **s** may then be derived by the method of least squares, using a Lagrangian multiplier to ensure that the requirement $l^2 + m^2 + n^2 = 1$ is fulfilled.

16.4 Source directions and baseline vectors

The applications of VLBI considered in the last section have been entirely astrometric. It was assumed that the baseline vector **D** was known with absolute precision and the direction of the source, or at least a component of it, was derived from the condition that

$$\mathbf{D} \cdot \mathbf{s} = c\tau. \tag{16.23}$$

This argument may be inverted, however, so that an unknown baseline is determined from VLBI observations of sources of known position. Consequently VLBI measurements have important applications to geodesy as well as astrometry, in particular, measurements may be made of the tidal distortion in solid earth. Moreover, baseline measurements provide a powerful means of investigating polar motion.

Our prime concern is, of course, the astrometric application of VLBI. The determination of the baseline, however, cannot be divorced from the measurement of the source's position. It is equivalent to the determination of the instrumental errors (of azimuth and level) in optical meridian measurements. Let us, therefore, examine the requirements for the simultaneous determination of radio source positions and baseline vectors.

Suppose that observations are made simultaneously over two baselines with vectors $\mathbf{D}_1$ and $\mathbf{D}_2$. These may be provided by VLBI recordings being made at only three separate locations, but it is immaterial whether the two baselines being considered share a common end-point or not. Three locations would provide three baselines, but only two of them are independent. We shall consider the information that may be extracted from the simultaneous observation of three sources.

Let $\mathbf{s}_1, \mathbf{s}_2, \mathbf{s}_3$ be the directions of the three sources, and let τ_{ij} denote group delay of the jth source measured over the ith baseline. Then

$$\mathbf{D}_i \cdot \mathbf{s}_j = c\tau_{ij}. \tag{16.24}$$

Now approximate *a priori* positions are assumed for the three directions $\mathbf{s}_j$. They will, of course, be unit vectors. Using these, equation (16.24) will yield the components of each baseline in the directions $\mathbf{s}_1, \mathbf{s}_2, \mathbf{s}_3$. This is clearly sufficient information to determine each baseline vector uniquely, provided the three directions $\mathbf{s}_j$ are not coplanar. Since the three directions are not, in general, mutually orthogonal, the derivation of the vectors $\mathbf{D}_i$ in equatorial coordinates is slightly involved, cf. Problem 16.4.

Any errors in the assumed source positions will, of course, produce corresponding errors in the baseline vectors. To examine these we must use the time derivatives of (16.24). By (16.18), it follows that

$$(\mathbf{\Omega} \times \mathbf{D}_i) \cdot \mathbf{s}_j = c \frac{d\tau_{ij}}{dt}, \tag{16.25}$$

where $\mathbf{\Omega}$ is the earth's angular velocity vector. Using the derived values of $\mathbf{D}_i$, these equations may now be employed to obtain improved values of the source directions. For each vector $\mathbf{s}_j$, two components are derived in the directions $(\mathbf{\Omega} \times \mathbf{D}_1)$ and $(\mathbf{\Omega} \times \mathbf{D}_2)$. Each of these lies in the equatorial plane, so no information is acquired about the polar components of the $\mathbf{s}_j$. This must be obtained from the condition that they are unit vectors.

It is possible to iterate in this way, using (16.24) to obtain the baselines and (16.25) with the unit vector condition to derive the source directions. Some care is required, however, to ensure convergence, since the solution is not unique. Equation (16.24) is a vector equation which will remain valid under any rotation of the coordinate system. The use of the equatorial coordinate system is only imposed through the components of the vector $\mathbf{\Omega}$ in (16.25). This establishes the direction of the pole and the rate of the earth's rotation, but no more. The zero point of right ascension is arbitrary.

It is clear from this argument that if three or more sources are observed over at least two baselines, sufficient data is accumulated to determine all source and baseline vectors. As far as the source positions are concerned, the declinations are absolute, but the right ascensions are relative. No increase in the number of sources or baselines being used can alter the indeterminacy in right ascension.

When considering polar motion an earth-based coordinate system is helpful. In Fig. 16.3, the point P_0 represents the reference pole, fixed with respect to the earth, that is, the pole of figure. Let X_0 and Y_0 be points on the equator of east longitude $0°$ and $90°$ respectively. Then the directions of X_0, Y_0, P_0 define a set of right-handed orthogonal axes. Let $\mathbf{D}_0$ be the baseline vector referred to these axes.

Suppose now that, due to polar motion, the celestial pole is displaced to a point P. The effect of this displacement of the pole may be represented by a small rotation ω of the coordinate system. Let us write

$$\omega = (\omega_1, \omega_2, \omega_3). \tag{16.26}$$

The inclusion of the component ω_3 may appear surprising. It does not affect the position of the pole, but it does allow variations in the earth's rotation rate, to be included. All three components of ω are functions of time.

Figure 16.3

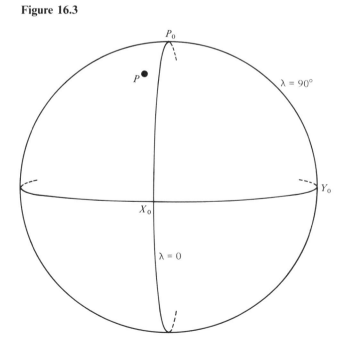

Whereas ω_1 and ω_2 are purely oscillatory, ω_3 contains a secular term in addition to periodic seasonal variations.

Let $\mathbf{k}_0$ be the unit vector in the direction of P_0. The displacement of the celestial pole $d\mathbf{k}_0$ is given by

$$d\mathbf{k}_0 = \omega \times \mathbf{k}_0 = (\omega_2, -\omega_1, 0). \tag{16.27}$$

The displacement of the pole is normally resolved into displacements (x, y) along the meridians of longitude $0°$ and $270°$ E. It follows, from (16.27), that

$$x = \omega_2$$
$$y = \omega_1. \tag{16.28}$$

The vector formulation of polar motion used here allows the change in the baseline vector to be very simply expressed. It is given by

$$d\mathbf{D}_0 = \omega \times \mathbf{D}_0. \tag{16.29}$$

Once this correction has been applied, the baseline vector $\mathbf{D}$ in the space-fixed coordinate system is obtained by a simple rotation about the pole P through an angle Ωt. Since variations in the earth's rotation are included in the vector ω, the earth's angular velocity Ω is treated as a constant, and the time t is measured from an epoch of zero ephemeris sidereal time.

16.5   Media effects

The discussion of groups delay in the last two sections has been drastically simplified in order to consider purely geometrical effects. Indeed, the use of the term group delay itself requires some justification. It arises from the derivative of the radio phase difference between the two interferometer elements. The observed phase difference was given in equation (16.12) and contains instrumental and media effects. These effects will still be present in the differentiated form. It is, in fact, the presence of a derivative of ϕ_m that requires us to identify the derivative of ϕ_0 as a group phenomenon. The definition of the group delay τ in (16.20) remains valid, but let us replace (16.23) with the more general equation

$$c\tau = \mathbf{D} \cdot \mathbf{s} + c\tau_m, \tag{16.30}$$

where τ_m is the delay due to media effects. It follows from (16.20) that

$$\tau_m = \frac{d\phi_m}{d\omega}. \tag{16.31}$$

In optical astronomy the effect of the atmosphere is treated as a refraction is the lower atmosphere. The situation is rather different in radio interferometry, however, and refraction is not the main concern. Certainly it affects source directions and, therefore, the pointing of the two component antennas. The precise astrometric measurements, however, involve phase differences or group delays and require a slightly different investigation of atmospheric effects. Moreover, these effects are not confined to the lower atmosphere; the ionosphere also makes an important contribution.

The refractive properties of the ionosphere are related to the plasma frequency v_p. This is determined by the local free electron density N and is given in SI units by

$$v_p = 8.978 N^{1/2} \text{ Hz.} \tag{16.32}$$

The electron density varies in a complicated way within the earth's atmosphere, reaching a maximum value at a height normally between 250 km and 400 km above the earth's surface. Moreover, the electron density distribution varies, often rather erratically, with time, being dependent on the time of day, the season and the level of solar activity. The value of N at any point within the ionosphere may vary by at least an order of magnitude. Particularly significant is the maximum electron density in the ionosphere, the corresponding plasma frequency is known as the critical frequency of the ionosphere. Its value is normally between 5 and 10 MHz.

The plasma frequency determines the refractive index n of the plasma through the equation

$$n = \left(1 - \frac{v_p^2}{v^2} \right)^{1/2} = \left(1 - \frac{\omega_p^2}{\omega^2} \right)^{1/2}. \tag{16.33}$$

Clearly n becomes imaginary at radio frequencies less than the plasma frequency and normal radio propagation is not possible. This sets the long wavelength limit to the radio window, since no radiation at a frequency lower than the critical frequency can penetrate the ionosphere.

Ionospheric refraction is severe at metre wavelengths, but falls off rapidly at the higher frequencies used in radio astrometry. Accurate modelling of the ionosphere is very difficult. It extends over a height of about 1000 km; so a plane-parallel approximation, and even the assumption of spherical symmetry, is rather crude. Even so, the unrealistic plane parallel situation, illustrated in Fig. 16.4, may still be used to elucidate certain features that are at least valid to order of magnitude. In the analysis that follows we shall neglect the contribution of the troposphere which must be investigated separately. We shall, therefore, take $n = 1$ at ground level.

Consider a ray that is observed at the point O after entering the ionosphere at S at a height h above the ground. As Fig. 16.4 indicates the radio wave is first refracted towards the horizontal and then back towards the zenith. If z denotes the inclination of the ray to the vertical at any level, then, as shown in section 4.2, the plane-stratification assumption leads to the result

$$n \sin z = \sin z_0. \tag{16.34}$$

Here z_0 represents the initial direction of the ray above the ionosphere and also its zenith distance at ground level. For within a plane stratified atmosphere the refraction is clearly self-cancelling. The greatest deflection from the vertical will occur at the point T where the electron density is greatest and the refractive index has its minimum value, n_c say.

At cm wavelengths $1 - n_c$ is very small. For example, a radio frequency of 5000 MHz and a maximum ionospheric electron density of 10^{12} m^{-3} yields $1 - n_c = 1.6 \times 10^{-6}$. The maximum deflection at the point T then amounts to no more than $0''.3$. This still produces important phase effects, but we are certainly justified in neglecting terms of order $(1 - n)^2$. Equation (16.34) then reduces to

$$z = z_0 + (1 - n) \tan z_0. \tag{16.35}$$

Now let ds be the element path-length along the ray. We small measure s from the point S and let s_0 be the path-length of an undeflected ray through the ionosphere, i.e. $s_0 = h \sec z_0$. Then the actual path-length SO is given by

$$s = \int_0^h \sec z \, \mathrm{d}x$$

where x is the coordinate in the vertical direction. Using (16.35), this is easily evaluated as

$$s = s_0 + \sec z_0 \tan^2 z_0 \int_0^h (1 - n) \, \mathrm{d}x, \tag{16.36}$$

Figure 16.4. Ionospheric refraction.

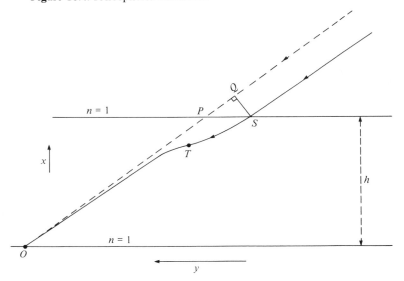

correct to first order in $(1-n)$. Very roughly, the additional term in (16.36) will be of order $h(1-n_c)$, about 160 cm, which is far from negligible.

Now, in the absence of any ionospheric refraction, the ray observed at O would pass through the point P rather than S. Let y_0 be the horizontal displacement of an undeflected ray through the ionosphere, i.e. $y_0 = h \tan z_0$. Further, let δy be the additional horizontal displacement caused by the refraction. Then

$$y_0 + \delta y = \int_0^h \tan z \, dx.$$

Applying once again the approximation of (16.35) we find

$$\delta y = \tan z_0 \sec^2 z_0 \int_0^h (1-n) \, dx. \tag{16.37}$$

Now we can make the following identifications in Fig. 16.4: $s_0 = OP$, $\delta y = PS$ and so $PQ = \delta y \sin z_0$. It then follows from (16.36) and (16.37) that, to first order in $(1-n_c)$, $OQ = s$.

The radio phase at the point Q is the same as at the point S. The above argument does not prove that the 'observed' phase at O is unaffected by ionospheric refraction, but only that the path length increase as such has negligible influence. Therefore, in deriving the phase at the point O, we are entitled to treat the ray-path as a straight line.

At a general point a distance s along the ray-path the phase is given by

$$\phi = \omega \left(t - \frac{s}{v} + \text{constant} \right)$$

where v is the phase velocity. In fact

$$v = \frac{c}{n} = c \left(1 - \frac{\omega_p^2}{\omega^2} \right)^{-1/2}. \tag{16.38}$$

Consequently, the phase difference $\Delta \phi$ between the points S and O will be

$$\Delta \phi = - \int \frac{\omega \, ds}{v} = - \int \frac{\omega n \, ds}{c}. \tag{16.39}$$

In the plane stratified case this reduces to

$$\Delta \phi = - \frac{\sec z_0}{c} \int_0^h \omega \left(1 - \frac{v_p^2}{v^2} \right)^{1/2} dx.$$

The integrand may be expanded by the binomial theorem. Then, setting $\omega/c = 2\pi/\lambda$ and using (16.32), we find

$$\Delta \phi = - \frac{2\pi h \sec z_0}{\lambda} + \frac{80.6\pi \sec z_0}{\lambda v^2} \int_0^h N \, dx. \tag{16.40}$$

The first term in (16.40) is merely the geometrical phase difference to be expected between the points S and O. The second term gives the effect of ionospheric refraction on the phase. It is far from insignificant, and, from the values quoted earlier, one can estimate the medium's contribution as about 150 radians, or 25 complete cycles. In this over-simplified model, the ionospheric phase depends only on the total electron content of the ionosphere and the source's zenith distance. The model is obviously inadequate, and for a more realistic model the phase difference must be computed from the more general equation (16.39). However, the path-length considerations remain valid, and so the integration may be performed along a straight ray-path. At high frequencies, at least, it is not necessary to compute the ray-path before performing the integration in (16.39).

Extracting the geometrical phase from (16.39), it is seen that the effect of the ionosphere is to advance the radio phase by $\delta\phi$ where

$$\delta\phi = \int \frac{\omega(1-n)}{c} \, ds. \tag{16.41}$$

This same formula may be used to derive the effect of the lower atmosphere. At these levels radio refraction is similar to optical refraction and n is greater than unity. The refractive index depends on atmospheric pressure and temperature but varies with the water vapour content. Following Allen (1976) we may use the formula

$$n - 1 = \frac{3.787 \times 10^{-6}}{(1 + T/273)} \left[P + 0.87f + \frac{17.69f}{(1 + T/273)} \right], \tag{16.42}$$

where T is the temperature in degrees centigrade, and P and f are respectively the total atmospheric pressure and the water vapour both in mm Hg.

Although $(n-1)$ at cm wavelengths is considerably larger in the troposphere than in the ionosphere, the integration is carried out over a much shorter distance. The contribution of the lower atmosphere to $\delta\phi$ is, therefore, of the same order of magnitude; it is about twice as big at 5000 MHz, but it has the opposite sign. The total effect of the medium is given by (16.41) with the integration extended over both the troposphere and the ionosphere. The values of the refractive index in the two regions are derived from equations (16.42) and (16.33).

The contribution of the media to the interferometric phase is simply the difference between the values of $\delta\phi$ obtained at each antenna. Formally, therefore, we may write

$$\phi_m = \delta\phi_2 - \delta\phi_1, \tag{16.43}$$

where the subscripts 1 and 2 are used to distinguish between the two antennae.

The group delay is obtained by differentiating the phase difference with respect to ω. It will be convenient to split the media delay into the tropospheric and ionospheric components, writing

$$\tau_m = \frac{\mathrm{d}\phi_m}{\mathrm{d}\omega} = \tau_{\mathrm{trop}} + \tau_{\mathrm{ion}}. \tag{16.44}$$

Since the refractive index in the lower atmosphere is independent of frequency, we may write

$$c\tau_{\mathrm{trop}} = \int (1 - n_2)\,\mathrm{d}s_2 - \int (1 - n_1)\,\mathrm{d}s_1. \tag{16.45}$$

This is formally a constant. In fact, two integrals in (16.45) are subject to random fluctuations which over the wide baselines used in VLBI will be uncorrelated. Indeed, uncertainties in the value of τ_{trop}, caused mainly by fluctuations in the water vapour content, are one of the most serious limitations to positional accuracy in VLBI.

The analysis of the ionospheric group delay is rather different. The refractive index now depends on frequency, and (16.33) yields with sufficient accuracy

$$\omega(1 - n) = \frac{1}{2} \frac{\omega_p^2}{\omega}.$$

Inserting this in (16.41), we derive τ_{ion} from (16.43) and (16.44) as

$$c\tau_{\mathrm{ion}} = \frac{1}{2\omega^2} \left(\int \omega_{p_1}^2\,\mathrm{d}s_1 - \int \omega_{p_2}^2\,\mathrm{d}s_2 \right),$$

where the integrations are carried out over the two path lengths through the ionosphere. Finally, by (16.32), it is found that

$$c\tau_{\mathrm{ion}} = \frac{40.30}{\nu^2} \left(\int N_1\,\mathrm{d}s_1 - \int N_2\,\mathrm{d}s_2 \right). \tag{16.46}$$

Although varying fairly rapidly, the ionospheric delay is not subject to quite such severe random fluctuations as the tropospheric equivalent. Moreover, the dependence on frequency is useful, since it will allow the bracket in (16.46) to be determined from multi-frequency observations.

16.6 Relativistic effects

The group delay is the time interval between the arrival of a signal wave front at the two antennae of the interferometer. The definition needs to be made with a little care, however, when relativistic effects are allowed for, and the definition of τ in (16.23) is correspondingly modified. Let us suppose that the antennae sites have geocentric position vectors $\mathbf{r}_1$ and $\mathbf{r}_2$, and that each site is equipped with an ideal atomic clock. If the signal reaches the first antenna at time T_1 on its clock and reaches the second site at time T_2 on its clock, then the group delay is defined as

$$\tau = T_2 - T_1. \tag{16.47}$$

The delay τ may be regarded as a function of T_1.

Let $\mathbf{R}$ be the barycentric position vector of the geocentric point. Now the vectors $\mathbf{R}, \mathbf{r}_1, \mathbf{r}_2$ will all be functions of time. It is necessary to use the coordinate time of general relativity, and then the components of the vectors may be identified in terms of the isotropic coordinate system. Some simplification is possible, however, if appropriate scaling factors are introduced. The isotropic form of the Schwarzschild metric (cf. equations (3.53) and (8.50)) may be written, with sufficient accuracy, as

$$ds^2 = \left(1 - \frac{2m}{\tilde{r}}\right) d\tilde{t}^2 - \frac{1}{c^2}\left(1 + \frac{2m}{\tilde{r}}\right)(d\tilde{r}^2 + \tilde{r}^2\, d\theta^2 + \tilde{r}^2 \sin^2\theta\, d\phi^2). \tag{16.48}$$

Rather than use the coordinate time $\tilde{t}$ itself, it is convenient to use TDB as the time coordinate t. Then, if a is the semimajor axis of the earth's orbit, we may use equation (10.41) to write

$$dt = \left(1 - \frac{3m}{2a}\right) d\tilde{t}. \tag{16.49}$$

The time coordinate t that has been introduced has only long-term variations with respect to TDT. This choice will most nearly match with the time kept by atomic clocks. We will now make an *ad hoc* adjustment to the radial coordinate $\tilde{r}$. Let us, therefore, select a new radial coordinate r so that the coordinate velocity of light is exactly c, taking

$$r = \left(1 + \frac{3m}{2a} - \frac{2m}{R}\right)\tilde{r}, \tag{16.50}$$

where R is the value of the earth's radial coordinate at the instant of observation. Then introducing Cartesian coordinates (x, y, z) with the usual transformation (3.48), we find that the metric in the immediate vicinity of the earth takes the form

$$ds^2 = \left(1 + \frac{3m}{a} - \frac{2m}{R}\right)\left[dt^2 - \frac{1}{c^2}(dx^2 + dy^2 + dz^2)\right]. \tag{16.51}$$

It is to be understood that the three vectors introduced earlier and illustrated below in Fig. 16.5 have components in the Cartesian coordinate system introduced in (16.51) and are regarded as functions of the time coordinate t.

Suppose now that interferometric observations are made of a source in the direction of unit vector $\mathbf{s}$. This is the direction that would be determined by an observer at rest in the barycentric reference frame. Let the signal reach the first antenna at time t. The signal wave front at this time is shown as the dotted line in Fig. 16.5 in which G is the geocentric point and A_1 and A_2 are the two antennae. It is a plane through the point with barycentric position vector $\mathbf{R}(t) + \mathbf{r}_1(t)$ perpendicular to the vector $\mathbf{s}$. Consequently, any point on this wave front has a position vector $\mathbf{r}$ that satisfies the equation

$$\mathbf{r} \cdot \mathbf{s} = (\mathbf{R} + \mathbf{r}_1) \cdot \mathbf{s}.$$

At the subsequent time $t + \Delta t$ the same wave front will be given by the equation

$$\mathbf{r} \cdot \mathbf{s} = (\mathbf{R} + \mathbf{r}_1) \cdot \mathbf{s} - c\,\Delta t. \tag{16.52}$$

If $t + \Delta t$ is the time at which the signal reaches A_2, then we may substitute, with sufficient accuracy,

$$\mathbf{r} = \mathbf{R} + \mathbf{r}_2 + \Delta t (\dot{\mathbf{R}} + \dot{\mathbf{r}}_2).$$

This leads to the result that

$$\Delta t = \frac{(\mathbf{r}_1 - \mathbf{r}_2) \cdot \mathbf{s}}{c} \left[1 + \frac{(\dot{\mathbf{R}} + \dot{\mathbf{r}}_2) \cdot \mathbf{s}}{c} \right]^{-1}. \tag{16.53}$$

It should be noted that all the vectors are to be evaluated at the barycentric time t, and equation (16.53) gives the delay in barycentric time as a function of t. Of course, the dot denotes differentiation with respect to barycentric time.

The quantity Δt must next be related to the clock times, T_i, $i = 1, 2$. The time being kept by each clock is the interval along its world line. This is obtained by inserting that clock's coordinates $\mathbf{R} + \mathbf{r}_i$ in the metric (16.51).

Figure 16.5

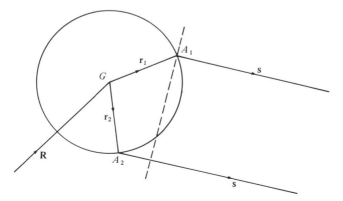

One then derives with sufficient accuracy

$$dT_i^2 = dt^2 \left[1 + \frac{3m}{a} - \frac{2m}{R} - \frac{\dot{\mathbf{R}}^2 + 2\dot{\mathbf{R}} \cdot \dot{\mathbf{r}}_i + \dot{\mathbf{r}}_i^2}{c^2} \right]. \tag{16.54}$$

Now substitute for $\dot{\mathbf{R}}^2$ and R in terms of two-body orbital theory. Using (6.15) and (6.24), we may write

$$\frac{\dot{\mathbf{R}}^2}{c^2} = \frac{2m}{R} - \frac{m}{a},$$

and   (16.55)

$$R = a(1 - e \cos E).$$

Then equation (16.54) will reduce to

$$dT_i = dt \left[1 - \frac{1}{2} \frac{\dot{\mathbf{r}}_i^2}{c^2} - \frac{\dot{\mathbf{R}} \cdot \dot{\mathbf{r}}_i}{c^2} - \frac{2m}{a} \frac{e \cos E}{(1 - e \cos E)} \right]. \tag{16.56}$$

The expression on the right-hand side will integrate – almost completely – with the help of Kepler's equation (6.26). Moreover, $\dot{\mathbf{r}}_i^2$ is constant, and so (16.56) will reduce to

$$T_i = t \left(1 - \frac{1}{2} \frac{\dot{\mathbf{r}}_i^2}{c^2} \right) - \frac{2m}{an} e \sin E - \frac{\dot{\mathbf{R}} \cdot \mathbf{r}_i}{c^2} + \int \frac{\ddot{\mathbf{R}} \cdot \mathbf{r}_i}{c^2} \, dt. \tag{16.57}$$

The first term represents time-dilation due to the clock's movement with respect to the centre of the earth; the second term is the periodic differences between TDT and TDB discussed in section 10.7; the third term is perhaps the most important, being of order of 2 µs and having a diurnal variation. The final term is rather smaller, but also has a diurnal variation and is not negligible. Although it cannot be integrated analytically it is not difficult to model. For analytic convenience let us write

$$T_i = t - \frac{\dot{\mathbf{R}} \cdot \mathbf{r}_i}{c^2} + \tau_i, \tag{16.58}$$

where τ_i contains the other terms in (16.57) and in addition any departures of the clock from the ideal.

Now, including the effect of the atmospheric media, the group delay is given by

$$\tau = T_2(t + \Delta t) - T_1(t) + \tau_m. \tag{16.59}$$

Using a Taylor expansion, this gives

$$\tau = \frac{\dot{\mathbf{R}} \cdot (\mathbf{r}_1 - \mathbf{r}_2)}{c^2} + \tau_2 - \tau_1 + \tau_m + \Delta t \frac{dT_2}{dt}.$$

The derivative in the final term is obtained from equation (16.56). Then writing

$$\tau_c = \tau_2 - \tau_1, \tag{16.60}$$

we obtain the final result as

$$\tau = \Delta t \left[1 - \frac{\dot{\mathbf{R}} \cdot \dot{\mathbf{r}}_2}{c^2} - \frac{2m}{a} \frac{e \cos E}{(1 - e \cos E)} - \frac{1}{2} \frac{\dot{\mathbf{r}}_2{}^2}{c^2} \right]$$
$$+ \frac{\dot{\mathbf{R}} \cdot (\mathbf{r}_1 - \mathbf{r}_2)}{c^2} + \tau_c + \tau_m. \tag{16.61}$$

The terms in the first bracket are arranged in order of decreasing size; the last two are not generally significant. All functions in equation (16.61) are evaluated at time t, the coordinate time of arrival of the signal at the first site.

Complicated algorithms are necessary to reduce VLBI observations, since the source's coordinates, the baseline coordinates and the parameters determining media and instrumental delays are all derived simultaneously from the best fit to the data. Even so, we may note the following features of principle. The observed quantity is the group delay τ; so Δt may be derived from (16.61) if the baseline vector is known and the media-propagation and instrumental effects are satisfactorily modelled. The source's direction, or at least its component parallel to the baseline, may then be obtained iteratively from (16.53).

The source direction $\mathbf{s}$ derived in this way will automatically be corrected for all aberrational effects both annual and diurnal. Parallax effects, however, and light deflection are not included. The corrections for annual parallax and light deflection may be applied by the methods considered in chapter 8. Thus if $\mathbf{s}_0$ is the barycentric direction of the source, modifying equation (8.73) yields

$$\mathbf{s}_0 = \mathbf{s} - \left[\pi + \frac{2m}{R(R + \mathbf{R} \cdot \mathbf{s})} \right] \mathbf{s} \times (\mathbf{s} \times \mathbf{R}), \tag{16.62}$$

where π is the source's parallax. For many radio sources, parallax will be negligible, but light deflection should always be included.

Geocentric parallax effects, which can only be significant for bodies within the solar system, will require a slightly different treatment. When this effect is included, it implies that the direction of the source is not the same at the two antenna sites. The spherical shape of the wavefront must then be allowed for – a problem considered in Problems 16.7 and 16.8.

16.7   The radio and stellar reference frames

The majority of radio sources that are observed for astrometric purposes are very distant extragalactic objects. If Mach's principle is invoked, it follows that the radio reference system defined by these objects should be an inertial system. It is hardly essential to invoke this principle which denies the possibility of the universe as a whole having a rotation. Some unorthodox cosmologies do, indeed, envisage a universal rotation, but its time-scale must be of the order of the Hubble time and this would imply proper motion of about 10^{-5} arc seconds per year, which is wholly negligible.

In the previous two sections we saw how the effects of aberration, of light deflection and of the atmospheric media may be removed from VLBI observations. Let (α_m, δ_m) be the right ascension and declination of a radio source subsequently to all these corrections. If the source is extragalactic, parallax and proper motion may be ignored; and so the measured coordinates (α_m, δ_m) give the barycentric place referred to the true equator and equinox of date. This last statement needs a slight qualification – since the determination of right ascension is relative, the value of α_m will require an equinox correction, Δ say, due to the discrepancy between the dynamical equinox and its adopted position. The radio source's true place is, therefore, $(\alpha_m + \Delta, \delta_m)$.

Radio astrometry provides a powerful direct means of investigating precession. Since the complication of a proper motion peculiar to the source is absent for an extragalactic object, any change in the source's measured coordinates may, apart from a variation in Δ, be ascribed entirely to a change in the coordinate system, that is, to precession and nutation. The nutational effects are periodic changes in (α_m, δ_m). Once these are removed there remain only the secular variations due to precession which merit further consideration.

The changes in right ascension and declination due to precession over a short time interval τ were given by equations (9.13). These give, in terms of (α_m, δ_m),

$$d\alpha_m = \left(m - \frac{d\Delta}{dt} \right)\tau + n\tau \sin \alpha_m \tan \delta_m$$

$$d\delta_m = n\tau \cos \alpha_m. \qquad (16.63)$$

Clearly the precessional constant n may be determined from the measured change in δ_m. The change in α_m will then yield the quantity $(m - d\Delta/dt)$, but the value of m itself is not known until the equinox correction Δ is determined. To achieve this, some recourse to solar system dynamics is necessary.

Two methods of investigation have been applied to this problem. An indirect method involves using timing of pulsars. It was explained in section 11.5 how the ecliptic coordinates (λ, β) of a pulsar may be obtained from the annual variation in the arrival time of the pulses. This method assumes a detailed knowledge of the earth's orbit round the sun. It is at this point that dynamical principles are invoked. The absolute declination of the pulsar is derived as before from interferometric measurements. This, together with the ecliptic coordinates, will then give an absolute value of the right ascension, for example, from the formula

$$\cos \alpha \cos \delta = \cos \lambda \cos \beta. \tag{16.64}$$

The equinox correction is then derived as $\alpha - \alpha_m$.

Alternatively, interferometric measurements may be made of radio sources within the solar system itself. Unfortunately, the sun, moon and planets are not really suitable, as their radio emission is either too weak or not sufficiently localized. Use may, however, be made of the artificial radio sources that are provided by transmitters placed either on the surface of natural bodies or in well-established heliocentric orbits.

The stellar reference frame is in many ways more comprehensive than the radio reference frame, but is probably subject to greater systematic errors. Much the same problem presents itself: fundamental observations of stars alone with meridian instruments provide absolute declinations, but only relative right ascensions. Consequently, an equinox correction has to be derived.

The investigation of precession is far more complicated with star observations. First of all, individual stellar positions are complicated by the presence of peculiar proper motions, so only a statistical approach is valid. Secondly, the effect of the solar motion must be removed. Only when these problems are overcome, can the question of the rotation of the stellar reference frame be examined.

Suppose the reference frame has an unknown rotation ω within a certain time interval. Then the change in a star's direction $\mathbf{s}$ is given by

$$d\mathbf{s} = \omega \times \mathbf{s}. \tag{16.65}$$

Only the change in the star's declination can be measured with certainty, that is, the component of $d\mathbf{s}$ in the polar direction $\mathbf{k}$, viz.

$$\mathbf{k} \cdot d\mathbf{s} = (\omega \times \mathbf{s}) \cdot \mathbf{k} = \omega \cdot (\mathbf{s} \times \mathbf{k}). \tag{16.66}$$

This equation implies that observations of different stars may reveal the component ω in different equatorial directions, but $\omega \cdot \mathbf{k}$ cannot be derived from observations of stars alone.

The same argument applies to the radio reference frame. Then, however, we can be certain that the rotation ω is due solely to precession. In the case of the stellar reference frame, there is a further complication in that ω is the sum of two independent rotations, namely that due to precession and that due to galactic rotation and represented by the Oort parameter B, cf. section 14.8.

The stellar reference system is normally related to an inertial frame through observations of solar system objects, the inertial frame itself being established by the dynamics of the solar system. The natural bodies of the solar system in this case provide suitable sources for direct measurements with meridian instruments. A comparison with extragalactic objects has also been undertaken at two observatories, Lick and Pulkova. The programs are incomplete since they involve long-term studies of apparent extragalactic proper motions in the stellar reference frame. The comparison is of a secondary nature, since the external galaxies provide rather faint sources which cannot be observed with meridian instruments. Their positions are, therefore, derived photographically relative to the stellar background. An additional complication arises since most optical extragalactic sources provide rather diffuse photographic images.

A direct relationship between the stellar and radio reference frames is difficult to establish. Most compact radio sources have rather faint optical counterparts, and the radio emission from normal stars is generally weak or undetectable. Consequently, the two reference systems are based on quite different sources. Opportunities for direct comparison are very few. One object that can be studied both by fundamental optical techniques and by radio interferometry is the variable star Algol. Most other compact radio sources can only be investigated optically, if at all, by photographic means.

The radio reference system is based on sources which are virtually fixed in an inertial frame. So long as observations are performed by earth-based instruments, the measurements produce absolute declinations and relative right ascensions. Interferometric positions therefore contain an uncertain equinox correction. This cannot be removed without appeal to other methods. The reason for this is essentially that the observations cover only a tiny fraction of a precessional period. If the observations were continued over a significantly longer period, the equinox correction itself could be determined from interferometric positions alone – see Problem 16.10.

Problems

16.1 An east–west interferometer is operating at radio frequency 5000 MHz over a baseline of 3.6 km. Calculate the total phase change $\Delta\phi$ occurring during 12^h observations of a source at declination $15°$. Calculate further the change in $\Delta\phi$ corresponding to a difference in declination of one arc second.

16.2 Show that diurnal aberration has no effect on the value of $\Delta\phi$ defined in the previous question, but will introduce a spurious instrumental phase given by

$$d\phi_c = 2''.011 \frac{D}{\lambda} \cos \Phi \sin^2 \delta,$$

where Φ is the latitude of the interferometer.

16.3 The baseline of the radio interferometer is approximately east–west and is directed to the point on the celestial sphere with hour angle and declination $(6^h - m, n)$. Show that 12^h observation of a radio source will provide the source's declination and the instrumental phase through the equations

$$\cos \delta = \frac{\lambda(\phi_{max} - \phi_{min}) \sec n}{4\pi D}$$

$$\phi_c = \tfrac{1}{2}(\phi_{max} + \phi_{min}) - \frac{2\pi D}{\lambda} \sin n \sin \delta.$$

Hence show that observation of a number of sources at different declinations will in principle allow the determination of n, but that the determination of m is not possible without assuming, *a priori*, the right ascension of at least one source. Deduce that an absolute measurement of declination is possible, but only a relative measurement of right ascension.

16.4 Consider the solution of equation (16.24) for three sources in known directions $\mathbf{s}_1, \mathbf{s}_2, \mathbf{s}_3$. Supposing that $\mathbf{i}_1, \mathbf{i}_2, \mathbf{i}_3$ are unit vectors in the directions of the standard rectangular equatorial axes, show that the baseline vectors $\mathbf{D}_i$ are given by

$$\mathbf{D}_i = c \, \mathbf{S}^{-1} \, \mathbf{T}_i,$$

where $\mathbf{T}_i$ is the vector whose components are τ_{ij} and $\mathbf{S}$ is the matrix whose elements are defined as

$$\mathbf{S}_{jk} = \mathbf{s}_j \cdot \mathbf{i}_k.$$

16.5 Solutions $\mathbf{D}_j$, $j = 1, 2$, and $\mathbf{s}_i$, $i = 1, 2, 3$, have been found for equations (16.24) and (16.25). Prove that $\mathbf{D}_j + \varepsilon(\mathbf{\Omega} \times \mathbf{D}_j)$ and $\mathbf{s}_i + \varepsilon(\mathbf{\Omega} \times \mathbf{s}_i)$ are also approximate solutions, provided that ε is a small quantity. What are the implications of this result?

16.6 A VLBI baseline is directed towards the point on the celestial sphere with equatorial coordinates (α_0, δ_0). Prove that polar motion produces a change in the group delay $\delta\tau$ for a source with equatorial coordinates (α, δ), where

$$\delta\tau = \frac{Dx}{c}(\cos\alpha\cos\delta\sin\delta_0 - \cos\alpha_0\cos\delta_0\sin\delta)$$

$$-\frac{Dy}{c}(\sin\alpha\cos\delta\sin\delta_0 - \sin\alpha_0\cos\delta_0\sin\delta).$$

$((x, y)$ are the displacement of the pole along the meridians of longitude 0 and $90°$ W.)

16.7 Incorporate the effect of geocentric parallax into the argument of section 16.6. In particular, show that the coordinate time interval Δt becomes the solution of a quadratic equation which may be written as

$$(\Delta t(\dot{\mathbf{R}} + \dot{\mathbf{r}}_2 + c\mathbf{s}) - \mathbf{r}_1 + \mathbf{r}_2) \cdot [\Delta t(\dot{\mathbf{R}} + \dot{\mathbf{r}}_2 - c\mathbf{s}) + 2(\mathbf{R} - \mathbf{R}_0) + \mathbf{r}_1 + \mathbf{r}_2] = 0.$$

Here $\mathbf{R}_0$ is the barycentric position vector of the source at the time of emission, but all other quantities are evaluated at the arrival time of the radiation at the first antenna. Further, the unit vector $\mathbf{s}$ is in the direction of $(\mathbf{R}_0 - \mathbf{R} - \mathbf{r}_1)$, but otherwise the notation is exactly that of section 16.6.

16.8 Deduce from the equation in the previous question the result (16.53) when geocentric parallax is neglected, and derive its modified form when the geocentric parallax is treated correct only to first order.

16.9 The transverse velocity of an extragalactic source is a fraction η of its radial velocity and the latter is assumed to be cosmological in origin. Show that the proper motion of the source is

$$\mu = 2''.11 \times 10^{-7}\eta H$$

where H is Hubble's constant (in km s^{-1} Mpc^{-1}).

16.10 Show that, if interferometric measurements of right ascension and declination of extragalactic sources were continued for a sufficiently long period of time, it should be possible to determine from the rate of change of declination, not just the precessional constant n but also the quantity $n\Delta$, where Δ is the equinox correction. Show that for this determination to be effective it is necessary to assume that Δ is a constant, or at least varies in some prescribed way.

17

Planetary phenomena and surface coordinates

17.1   The apparent motion of an inferior planet

The two planets Mercury and Venus are referred to as inferior planets. The term is not intended in any derogatory sense, but merely implies that they are always closer to the sun than the earth. It is convenient to discuss the two inferior planets separately from the rest, since the phenomena associated with them are qualitatively different. The following basic definitions, however, apply to any planet.

Let $\mathbf{r}$ and $\mathbf{R}$ be the heliocentric position vectors of a planet and the earth respectively. Then the planet's geocentric direction $\mathbf{s}$ and its geocentric distance ρ are given by the equation

$$\rho\mathbf{s} = \mathbf{r} - \mathbf{R}. \tag{17.1}$$

In Fig. 17.1, the points S, E and P represent the sun, the earth and the planet respectively, and the vectors of equation (17.1) are shown. The angles in the triangle SEP are also important. The angle at E is called the *elongation*, while that at P is called the *phase angle*, ϕ say. The angles are easily derived as

$$\cos E = -\frac{1}{R}\mathbf{R}\cdot\mathbf{s}, \tag{17.2}$$

$$\cos\phi = \frac{1}{r}\mathbf{r}\cdot\mathbf{s}. \tag{17.3}$$

Denote the third angle in Fig. 17.1, i.e. $P\hat{S}E$, by θ. This angle increases approximately uniformly, which is not true of either E or ϕ. The apparent brightness of the planet, of course, depends upon its distance from the earth; it is inversely proportional to ρ^2. The brightness will also depend on the phase angle, since this angle determines the inclination of the illuminated hemisphere of the planet to the direction of the earth. A third cause of variation in the planet's brightness is its changing distance from the sun, due to the ellipticity of its orbit.

The elongation E affects the visibility of the planet, since it is the planet's angular distance from the sun in the sky. For inferior planets, E is restricted to a finite range. This means that these planets can generally only be observed shortly before sunrise or after sunset.

In order to examine qualitatively the phenomena associated with the two inferior planets, we shall make the very rough approximation that their orbits, and that of the earth, are circular and coplanar. Figure 17.2 illustrates the most significant of the configurations that an inferior planet may assume with respect to the earth. The point P_1 corresponds to inferior conjunction, P_3 to superior conjunction, while P_2 and P_4 represents the planet's position at its greatest elongation west and east of the sun respectively.

Let n_E and n_P denote the mean motions of the earth and the planet. Then the sun–earth line has an angular velocity n_E with respect to an inertial frame. We are imagining that this rotation has been subtracted in Fig. 17.2 and that the line SE is fixed. It is then clear that the sun–planet line SP will advance in this diagram with an angular velocity $(n_P - n_E)$. The period in which this line rotates through $360°$ with respect to SE is known as the *synodic period* of the planet. It is the period between any two identical relative configurations of the planet, the sun and the earth. The synodic period S is given by

$$S = 360°/(n_P - n_E),\qquad(17.4)$$

or in terms of the orbital periods T_P, T_E of the planet and the earth

$$S^{-1} = T_P^{-1} - T_E^{-1}.\qquad(17.5)$$

Figure 17.1

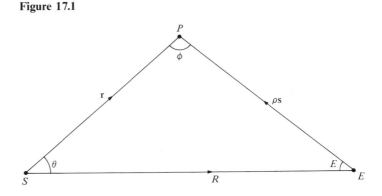

Let us consider the cycle of events during one synodic period. At inferior conjunction, we have the formal results that the elongation is zero and the phase angle is 180°. The planet is unobservable at this time for two reasons – its unilluminated hemisphere is presented towards the earth, and it is too close to the sun for effective observation. The planet's elongation then increases as it moves west of the sun. It may then be observed as a 'morning star'. The planet reaches maximum elongation, at the point P_2, where the earth–planet line is tangential to the planet's orbit. Hence $SP_2E = 90°$. The solution of triangle SP_2E is, therefore, particularly simple. The phase angle, the elongation, and time interval from inferior conjunction, all corresponding to P_2, are

$$\phi = 90°$$
$$E = \sin^{-1} a \tag{17.6}$$
$$t = \frac{S}{2\pi} \cos^{-1} a,$$

where a is the radius of the planet's orbit in astronomical units. Following maximum elongation west the planet appears to approach the sun again and reaches superior conjunction after half of the synodic period. After that the planet reappears at an easterly elongation which allows it to be seen in the evening sky following sunset. Maximum elongation east is reached at P_4, after which the planet moves back towards the sun more rapidly to the new inferior conjunction.

Suppose P is the planet's position at a general time t after inferior

Figure 17.2. Configurations for an inferior planet.

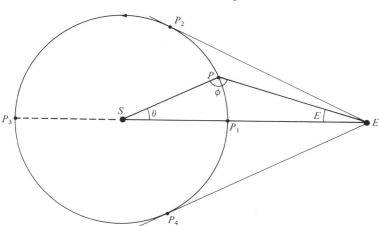

conjunction. Then in the circular coplanar approximation $\theta = (n_P - n_E)t$. Furthermore, the sine formula applied to triangle SPE yields

$$\sin E = a \sin \phi. \tag{17.7}$$

Now it is possible to eliminate either the elongation or the phase angle from this equation by using the fact that

$$\theta + \phi + E = 180°,$$

i.e. $\tag{17.8}$

$$E + \phi = 180° - n_s t,$$

where we have written n_s for the mean synodic motion ($n_s = n_P - n_E$). For example, on eliminating E, (17.7) yields

$$\sin (n_s t + \phi) = a \sin \phi.$$

Expanding the left-hand side allows this to be rewritten as

$$(\cos n_s t - a) \sin \phi + \sin n_s t \cos \phi = 0$$

or

$$\tan \phi = \frac{\sin n_s t}{a - \cos n_s t}. \tag{17.9}$$

On the other hand, eliminating ϕ instead of E leads in the same way to

$$\tan E = \frac{a \sin n_s t}{1 - a \cos n_s t}. \tag{17.10}$$

Equations (17.9) and (17.10) indicate the time variation of the phase angle and the planet's elongation, but, of course, only approximately. For precise calculations the inclination of the planet's orbit to the ecliptic, its orbital eccentricity, and that of the earth must all be included. These effects are particularly important for the planet Mercury. Exact values of ϕ and E must be derived from equations (17.2) and (17.3).

17.2   The apparent motion of a superior planet

Rather different phenomena are associated with a superior planet. The possible alignments that the planet can assume with respect to the earth are illustrated in Fig. 17.3. In this diagram we are again treating the sun–earth line as fixed. Since the earth's mean motion, which has been suppressed, exceeds that of a superior planet, the planet will move in the direction shown with respect to the earth–sun line. Under the approximation of circular coplanar orbits, the relative motion of P will be described with constant angular velocity ($n_E - n_P$). The synodic period is, therefore, given by

$$S^{-1} = T_E^{-1} - T_P^{-1}. \tag{17.11}$$

During the synodic period, the elongation of the planet from the sun will assume all values between $0°$ and $180°$, but the phase angle has a restricted range.

When the planet is at the point P_1, it is said to be at *opposition*. It is then observable for most of the hours of darkness. By contrast, the planet cannot be observed when it is close to the point P_3, which is again called (superior) conjunction. When the planet is at either of the two points P_2 or P_4, at which its elongation is $90°$, it is said to be in *quadrature*. It is at these two points that the phase angle reaches its maximum value.

Although the geocentric configurations of a superior planet are quite different from those of an inferior planet, there is a close mathematical relationship between them. After all, essentially the same problem is being considered, only the roles of the two planets are interchanged. From the point of view of the planet, the phase angle ϕ is just the earth's elongation from the sun, and E would be the earth's phase angle. In fact, the formulae which refer to quadrature are

$$\phi = \sin^{-1}(1/a)$$
$$E = 90°$$
$$t = \frac{S}{2\pi}\cos^{-1}(1/a).$$

(17.12)

Figure 17.3. Configurations for a superior planet.

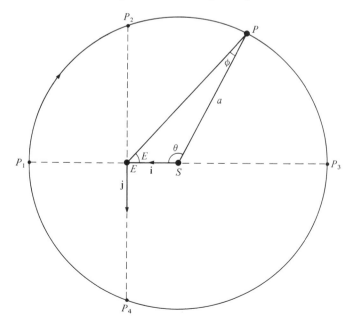

As before, these formulae are very approximate, and precise values of the phase angle and the planet's elongation should be calculated from equations (17.2) and (17.3).

17.3   Stationary points

All the planets have direct orbital motions about the sun; their heliocentric ecliptic longitudes are monotonically increasing. This is not true of their apparent geocentric longitudes, however. For a certain part of the synodic period the combination of the planet's and the earth's orbital motion produces an apparent retrograde motion. The overall motion is still direct, but near opposition a superior planet describes a 'loop' against the stellar background, as shown in Fig. 17.4. The two points Q_1 and Q_2 at which the planet's apparent motion in longitude reverses are known as *stationary points*. There are similar stationary points in the motion of an inferior planet.

Let **s** be the unit vector defined in equation (17.1) giving the geocentric direction of the planet. Furthermore, let **k** be the unit vector normal to the ecliptic plane. Now the sense of revolution of the planet's motion on the celestial sphere is given by the vector $\mathbf{s} \times \dot{\mathbf{s}}$. This will be direct so long as

$$(\mathbf{s} \times \dot{\mathbf{s}}) \cdot \mathbf{k} = [\mathbf{s}, \dot{\mathbf{s}}, \mathbf{k}] > 0. \tag{17.13}$$

Differentiating (17.1) yields

$$\dot{\rho}\mathbf{s} + \rho\dot{\mathbf{s}} = \dot{\mathbf{r}} - \dot{\mathbf{R}}.$$

Taking the vector product of this with (17.1) gives the result that

$$\mathbf{s} \times \dot{\mathbf{s}} = \frac{1}{\rho^2}(\mathbf{r} - \mathbf{R}) \times (\dot{\mathbf{r}} - \dot{\mathbf{R}}). \tag{17.14}$$

The condition for a stationary point is, therefore, seen from (17.13) and (17.14) to be that

$$[\mathbf{r} - \mathbf{R}, \dot{\mathbf{r}} - \dot{\mathbf{R}}, \mathbf{k}] = 0. \tag{17.15}$$

Figure 17.4. Stationary points in a planet's apparent motion.

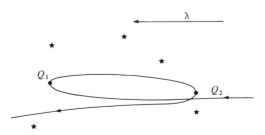

We wish to investigate the circumstances connected with the stationary points of a superior planet, using the approximation of coplanar circular orbits developed in the last two sections. In particular, the analysis of section 17.2 will be useful. Let **i** and **j** be, respectively, unit vectors in the sun–earth direction and in the perpendicular direction within the ecliptic. These vectors are respectively parallel to SE and EP_4 in Fig. 17.3, so that with **k** they form a right-handed set. Then it is easily seen from that diagram that

$$\mathbf{r} - \mathbf{R} = (a \cos \theta - 1)\mathbf{i} - a \sin \theta \mathbf{j}. \tag{17.16}$$

The two vectors **i** and **j** are not constant vectors, and so we must write

$$\dot{\mathbf{r}} - \dot{\mathbf{R}} = (a \cos \theta - 1)\frac{d\mathbf{i}}{dt} - a \sin \theta \frac{d\mathbf{j}}{dt}$$

$$- a \frac{d\theta}{dt}(\sin \theta \mathbf{i} + \cos \theta \mathbf{j}). \tag{17.17}$$

Now the angle θ in Fig. 17.3 is $(n_E - n_P)t$ where t is the time measured from opposition. Moreover, this whole diagram is subject to a rotation with angular velocity n_E. Consequently

$$\frac{d\mathbf{i}}{dt} = n_E \mathbf{k} \times \mathbf{i} = n_E \mathbf{j}$$

$$\frac{d\mathbf{j}}{dt} = n_E \mathbf{k} \times \mathbf{j} = -n_E \mathbf{i}. \tag{17.18}$$

Equation (17.17), therefore, reduces to

$$\dot{\mathbf{r}} - \dot{\mathbf{R}} = an_P \sin \theta \mathbf{i} + (an_P \cos \theta - n_e)\mathbf{j}. \tag{17.19}$$

Combining this equation with (17.16) will then yield

$$[\mathbf{r} - \mathbf{R}, \dot{\mathbf{r}} - \dot{\mathbf{R}}, \mathbf{k}] = (a^2 n_P + n_E) - a(n_E + n_P) \cos \theta. \tag{17.20}$$

The condition for a stationary point, equation (17.15), may, therefore, be written as

$$\cos \theta = \frac{a^2 n_P + n_E}{a(n_E + n_P)}. \tag{17.21}$$

Since the angle θ is directly proportional to time, this condition also determines the times of the stationary points. The condition for a stationary point may also be expressed in terms of the planet's semimajor axis. For, neglecting the mass of the planet and the earth in comparison with the sun's mass $M_\odot$,

$$n_P^2 a^3 = GM_\odot = n_E^2.$$

Equation (17.21), therefore, reduces to the condition that

$$\cos\theta = \frac{a^{1/2}+1}{a+a^{-1/2}}.$$ (17.22)

The corresponding elongation of the planet is found, after some reduction, to be

$$\tan E = \pm\frac{a}{(a+1)^{1/2}}.$$ (17.23)

The appropriate values of E are in the second and third quadrants.

The details of the solution derived above depend on the approximations that have been made to the planetary orbits. A precise determination of the stationary points must involve an iterative solution of equation (17.15), which is exact.

17.4 Planetary phases

A planet presents a disc to the earth which will, in general, be only partly illuminated. Treating the planet as a sphere will simplify the discussion, although some planets, in particular Jupiter, Saturn and Mars, are noticeably oblate and some refinement of the analysis may be necessary for them. It is convenient to work in units of the planet's radius so that the methods of spherical geometry may be used without complication.

Let C be the centre of the planet (Fig. 17.5), and suppose that the directions from C to the sun and the earth respectively intersect the planetary surface in the points S and E. Then S is called the subsolar point and E the subearth point. The arc SE is the phase angle ϕ. The point S is the central point of the illuminated hemisphere which is bounded by the great circle $PQRT$ which has S as pole. Let $AQBT$ be the great circle which has E as its pole. Since the plane of this great circle is perpendicular to the line of sight, it may be referred to as the planet's disc. Moreover, the illuminated part of the disc is bounded by the curve $QP'T$ which is the projection of the semi-great circle QPT. The shaded area in Fig. 17.5 is not visible. The great circle QPT, or sometimes its projection on the disc, is called the *geometric terminator*.

Suppose now that X is a general point on the geometric terminator which projects into the point X' on the disc. Denote the spherical arc EX by θ, and the spherical angle QEX by ψ. Now $EQX = 90° - \phi$, and $QE = 90°$. So applying the four-parts formula will yield

$$\cot\theta - \sin\psi\tan\phi = 0$$

i.e.

$$\tan\phi\sin\theta\sin\psi = \cos\theta.$$ (17.24)

Let us introduce a Cartesian coordinate system (x, y, z) centred on the point C with axes in the directions CQ, CA and CE. Then the coordinates of the point X are

$$x = \sin \theta \cos \psi$$
$$y = \sin \theta \sin \psi \qquad\qquad (17.25)$$
$$z = \cos \theta.$$

Equation (17.24) may, therefore, be written as

$$y \tan \phi = z = (1 - x^2 - y^2)^{1/2}.$$

So any point on the geometric terminator will satisfy the equation

$$x^2 + y^2 \sec^2 \phi = 1. \qquad\qquad (17.26)$$

Let us now consider the appearance of the planet's disc. This is shown in Fig. 17.6 using the same notation as in the previous diagram. The phase Φ of the planet is defined as the fraction of the disc that is illuminated. The area of the whole disc, still working in terms of the planet's radius, is π. The illuminated area consists of the semicircle QTB, of area $\frac{1}{2}\pi$, and the area QTP. By (17.26) this is half of an ellipse of semimajor axis unity and

Figure 17.5

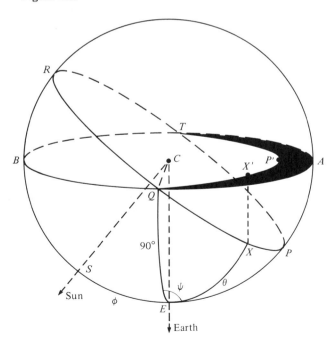

semiminor axis cos ϕ. Its area is, therefore, $\frac{1}{2}\pi\cos\phi$, and so the planetary phase is

$$\Phi = \tfrac{1}{2}(1 + \cos\phi). \tag{17.27}$$

Now let D be the angular diameter QT of the planet, expressed in arc seconds. The angular distance PA is termed the *defect of illumination*. This quantity, q, say, is given by

$$q = \tfrac{1}{2}D(1 - \cos\phi). \tag{17.28}$$

Let EN be the direction of north in the plane perpendicular to the line of sight. We shall denote the angle NES by χ. This is the position angle of the subsolar point, sometimes called the *position angle of the bright limb*. Denote by d the angular distance ES. Then the parameters (d, χ) provide a convenient way of specifying the position of the subsolar point on the disc and consequently the planet appearance. As indicated in Fig. 17.6, the position angle χ is measured from north in an anticlockwise direction as seen by the observer.

The point S has $x = 0$, $y = -\sin\phi$. The angular separation ES is, therefore, given by

$$d = \tfrac{1}{2}D\sin\phi. \tag{17.29}$$

The convention has been adopted in the *Astronomical Almanac* of tabulating d as negative whenever $\phi > 90°$. This is intended to indicate that the subsolar point is on the hidden hemisphere.

Figure 17.6

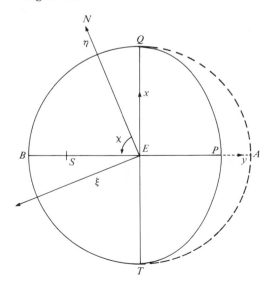

Now let (α, δ) be the planet's geocentric right ascension and declination. These, of course, refer strictly to the point E. Suppose that (ξ, η) represent standard coordinates using the point E as tangential point. Then EN is the η-axis. If $(\alpha_\odot, \delta_\odot)$ are the right ascension and declination of the sun, then, by equation (13.12), its standard coordinates are

$$\xi_\odot = \frac{\cos \delta_\odot \sin (\alpha_\odot - \alpha)}{\sin \delta \sin \delta_\odot + \cos \delta \cos \delta_\odot \cos (\alpha_\odot - \alpha)}$$

$$\eta_\odot = \frac{\cos \delta \sin \delta_\odot - \sin \delta \cos \delta_\odot \cos (\alpha_\odot - \alpha)}{\sin \delta \sin \delta_\odot + \cos \delta \cos \delta_\odot \cos (\alpha_\odot - \alpha)}. \tag{17.30}$$

The sun itself is not, of course, close to the point E, but the formulae (17.30) are rigorous. Furthermore, the straight line ESB is the projection of the planet's radius vector on to the tangent plane at E, and the ratio $\xi:\eta$ is maintained along this line. Consequently, the position angle of the subpolar point is given by

$$\tan \chi = \frac{\xi_\odot}{\eta_\odot} = \frac{\cos \delta_\odot \sin (\alpha_\odot - \alpha)}{\cos \delta \sin \delta_\odot - \sin \delta \cos \delta_\odot \cos (\alpha_\odot - \alpha)}. \tag{17.31}$$

The ambiguity in (17.31) is easily removed; χ is in the first two quadrants when the numerator is positive.

The phase of a planet is one important contributory factor to its apparent brightness. Another is the geocentric distance. For an inferior planet, in particular, both are subject to very considerable variation. It may be correctly, but incompletely, stated that the planet's brightness will be directly proportional to the phase and inversely proportional to the geocentric distance squared. The illumination of the planet by the sun will vary inversely as the square of its heliocentric distance, and, what is most difficult to model, the reflectivity of the planetary surface varies with direction. Simple formulae which ignore this last fact are apt to be in considerable error, cf. Harris (1961).

For a superior planet, maximum brightness will occur close to opposition, when the phase is greatest and the geocentric distance least. The situation is more complicated for an inferior planet, since its distance is least at inferior conjunction when its phase is zero. It is found that Mercury has its greatest brilliance close to superior conjunction. Venus, on the other hand, has two clear maxima which occur between inferior conjunction and the planet's greatest elongations. A simplified representation of Venus' varying brightness may be investigated in Problem 17.4.

17.5 Planetographic coordinates

The ecliptic is an observationally convenient reference plane for planetary motion, but it is particular to the earth and is not fundamental to the planetary system as a whole. Moreover, it is subject to slow change due to planetary precession. A dynamically more significant reference plane is the *invariable plane* of the system. This is defined as the plane through the barycentre that is normal to the angular momentum vector of the entire system. Since this vector is a constant of the motion, the invariable plane justifies its name. The ecliptic is within a few degrees of the invariable plane, and planetary precession is due to a very long period oscillation about it.

Unlike its orbital motion, a planet's axial rotation may be direct or retrograde. This is taken into account when defining a coordinate system on its surface. Since the majority of planets have direct rotations – Venus and Uranus are the exceptions – this case will be studied in more detail and is illustrated in Fig. 17.7.

Suppose P_0 is the north pole of the planet. This is chosen as the pole which is north of the invariable plane. The planet's rotation is direct if it is anticlockwise when seen from this pole. The planet's rotational axis may, to a first approximation, be regarded as defining a fixed direction in space. Like the earth's it will be subject to precessional effects, but, where these have been determined, they are found to be considerably smaller than those relating to the earth. Let (α_0, δ_0) be the right ascension and declination of P_0 on the celestial sphere referred to mean equator and equinox of date. Then if P is the celestial pole, $PP_0 = 90° - \delta_0$.

Let UQV in Fig. 17.7 represent the planet's equator. The point Q is this equator's ascending node on the celestial equator. A system of planetographic latitude and longitude (β, λ) may be defined using P_0 as the pole of the coordinate system, but first a prime meridian of longitude must be selected. Ideally, this is chosen through a recognizable surface feature, but this is not always possible, and some degree of arbitrariness may be involved. Let P_0B in Fig. 17.7 represent this prime meridian, B being its intersection with the planet's equator. We shall denote the arc QB by W. It is called the *argument of the prime meridian*.

The system of planetographic coordinates is defined by the values of (α_0, δ_0) and W. It is assumed that the prime meridian rotates uniformly. For direct rotation W increases with time; it decreases if the planet's rotation is retrograde. The IAU makes recommendations for the values of (α_0, δ_0) and W, giving the latter at the standard epoch J2000.0 and its rate of change. These are naturally referred to the equator and equinox of the standard epoch. The *Astronomical Almanac* provides values referred to the mean equator and equinox of date in the form

$$W = W_0 + \dot{W}d, \tag{17.32}$$

where W_0 is the value at the beginning of the year in question and d is the number of days from that epoch.

Suppose now that E is any point on the planet's surface. Then its planetographic latitude and longitude are given by

$$\beta = 90° - P_0 E$$
$$\lambda = B P_0 E. \tag{17.33}$$

The convention for longitude is that it is measured from $0°$ to $360°$ in the direction opposite to that of the planet's rotation.

Let us now identify the point E with the subearth point. Its planetocentric right ascension and declination (α_E, δ_E) will be given by

$$\alpha_E = \alpha + 12^h$$
$$\delta_E = -\delta, \tag{17.34}$$

where (α, δ) are the geocentric coordinates of the planet. Consequently, we may make the following identifications in spherical triangle $P_0 P E$: $P P_0 = 90° - \delta_0$, $P E = 90° + \delta$, $P_0 E = 90° - \beta$, $P_0 \hat{P} E = 180° + (\alpha - \alpha_0)$, and $P \hat{P}_0 E = 90° - W + \lambda$. The last of these equations assumes that the planetary rotation is direct. When the rotation is retrograde the sign of λ must be changed.

The planetographic coordinates of the point E may now be calculated from the solution of spherical triangle $P_0 P E$. By the cosine, sine and analogue formulae, we obtain the following results.

Figure 17.7. Planetographic longitude and latitude (λ, β).

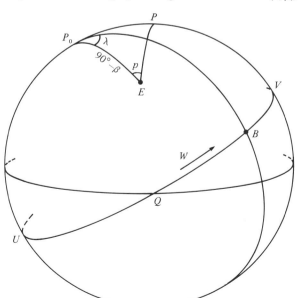

$$\sin \beta = -\sin \delta_0 \sin \delta - \cos \delta_0 \cos \delta \cos (\alpha - \alpha_0)$$

$$\cos \beta \cos (W - \lambda) = -\cos \delta \sin (\alpha - \alpha_0) \qquad (17.35)$$

$$\cos \beta \sin (W - \lambda) = \cos \delta \sin \delta_0 \cos (\alpha - \alpha_0) - \sin \delta \cos \delta_0.$$

The first of these equations will give the latitude of the subearth point without any ambiguity. The other two equations are both needed to determine $(W - \lambda)$ in the correct quadrant. Since $\cos \beta > 0$, we may rewrite them as

$$\lambda = W \mp \cos^{-1} [\cos \delta \sin (\alpha_0 - \alpha) \sec \beta], \qquad (17.36)$$

the principal value being taken, and the minus or plus sign according as

$$\cos \delta \sin \delta_0 \cos (\alpha - \alpha_0) - \sin \delta \cos \delta_0 \gtrless 0. \qquad (17.37)$$

When the rotation is retrograde, this procedure yields $-\lambda$ rather than λ.

The planetographic latitude and longitude of the subsolar point may be found in the same way. It is only necessary to replace the planet's geocentric right ascension and declination with the equivalent heliocentric coordinates.

The subearth point E is at the centre of the planet's disc and is, therefore, the natural observational reference point for planetary features. The planet's north pole is displaced from E by an arc $90° - \beta$ on its surface. This projects into a distance d on the disc given by

$$d = \tfrac{1}{2}D \cos \beta, \qquad (17.38)$$

where D is again the planet's angular diameter. It is convenient to treat d as negative when the pole lies on the hidden hemisphere.

To complete the specification of the position of the planet's pole on the disc, let us consider its position angle, p say. This is the unused part of spherical triangle P_0PE, namely spherical angle P_0EP. The four-parts formula applied to this triangle yields

$$\sin \delta \cos (\alpha - \alpha_0) = \cos \delta \tan \delta_0 + \sin (\alpha - \alpha_0) \cot p,$$

i.e.

$$\cot p = \cos \delta \tan \delta_0 \operatorname{cosec} (\alpha_0 - \alpha) - \sin \delta \cot (\alpha_0 - \alpha). \qquad (17.39)$$

The position angle is in the first two quadrants when $0 < \alpha_0 - \alpha < 180°$.

The concept of planetographic coordinates may also be applied to satellites. The determination of the rotation axis is usually less precise, and an *ad hoc* assumption is often made that the axis of rotation is perpendicular to the orbital plane. The same assumption is made for the planet Mercury, so that its subsolar point will automatically have planetographic latitude zero. The major planets, Jupiter and Saturn, do not rotate even approximately as solid bodies, and several different longitude systems may be defined for them. System I applies to visible features in the equatorial regions of each planet. System II, which is only defined for Jupiter, corresponds again to optical features but at higher latitudes. On the other hand, system III is determined by periodicities in the radio emission from the two planets and corresponds to the rotation of their magnetic fields. A separate expression of the form of equation (17.32) is necessary to give the value of W in each longitude system. The detailed formulae may be found in the *Astronomical Almanac*.

17.6   Planetographic coordinates of a point on the disc

Figure 17.8 represents the planet as seen from the earth. The subearth point is E, and EN represents the direction of zero position angle. The position of a feature X on the planet's disc may be expressed in terms of measured quantities (d, θ), respectively its angular distance from the centre of the disc and the angle $N\hat{E}X$. We will now consider how these directly measured quantities may be converted into planetographic latitude and longitude.

The location of the planet's north pole P_0 on the apparent disc must be determined and with it the longitude of the central meridian. The first step is to derive the value of W – the argument of the prime meridian. This computation, however, must be carried out not for the actual time t of

Figure 17.8

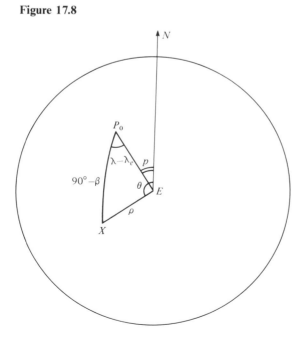

observation, but for the earlier time $t - \tau$, where τ is the light travel time. This correction is essential, particularly for the distant but rapidly rotating outer planets. Once the appropriate value of W is known, the method of the previous section is employed to derive the planetographic latitude and longitude of the subearth point, (β_e, λ_e), say, and the position angle p of the planet's north pole.

Let (λ, β) be the planetographic longitude and latitude of X. Then in the spherical triangle P_0XE on the planet's surface $P_0X = 90° - \beta$, and $X P_0 E = \lambda - \lambda_e$. Moreover, $P_0 E = 90° - \beta_e$ and $P_0 \hat{E} X = \theta - p$, both of which will be known. Furthermore, the angular separation d is easily converted into an angle ρ on the planetary surface. In fact

$$\rho = \sin^{-1} (2d/D), \tag{17.40}$$

where D is the planet's apparent angular diameter.

Now apply the cosine formula to spherical triangle PXE to obtain

$$\sin \beta = \sin \beta_e \cos \rho + \cos \beta_e \sin \rho \cos (\theta - p). \tag{17.41}$$

The formula allows the latitude of X to be derived. The longitude may be derived from the analogue formula, which gives

$$\cos \beta \cos (\lambda - \lambda_e) = \cos \rho \cos \beta_e - \sin \rho \sin \beta_e \cos (\theta - p), \tag{17.42}$$

in all circumstances. There is still an ambiguity in the inverse cosine required to give λ. This may be removed by the sine formula which yields

$$\cos \beta \sin (\lambda - \lambda_e) = \pm \sin \rho \sin (\theta - p). \tag{17.43}$$

With the conventions of longitude measurement, the plus sign is taken when the rotation is direct, the minus sign when the rotation is retrograde.

17.7   Heliographic coordinates

The basic ideas of the last two sections may also be applied to define coordinate systems for the sun and the moon. There are, however, some differences in the conventions that are followed, so a separate discussion will be helpful.

The rotation of the sun is evident from the apparent movement of sunspots across the disc. From such observations a rotation axis and a solar equator may be determined. It is also found, however, that the sun's angular velocity decreases markedly with latitude. The longitude system adopted is rather arbitrary, and corresponds to the average rotation of equatorial regions.

Figure 17.9 represents the heliocentric celestial sphere; the point K is the pole of the ecliptic and P_0 is the sun's rotational pole. The great circle UNV

is the solar equator, N being its ascending node on the ecliptic – the sun's rotation is direct.

The rotation axis, or equivalently the solar equator, is specified by the two parameters I and Ω. These are respectively the inclination of the equator to the ecliptic and the longitude of the node N. The adopted values are

$$I = 7°\,15'$$
$$\Omega = 73°\,40' + 50''.25(t - 1850.0), \tag{17.44}$$

where t is the time expressed, in years. As the second of these equations suggests, the system has been in use for more than 100 years. The variation in Ω is due to precession, and the secular variation in the ecliptic is ignored. The prime meridian is shown as P_0O; the point O where it cuts the equator is assumed to rotate with a sidereal period of 25.38 days. This reference point was originally chosen to coincide with the node at 1854 January 1 12^h UT (JD2 398 220.0). The position of the point O is given by the arc NO, denoted by W, which is

$$W = \frac{360°}{25.38}\,(JD - 2\,398\,220.0). \tag{17.45}$$

Figure 17.9. Heliographic coordinates.

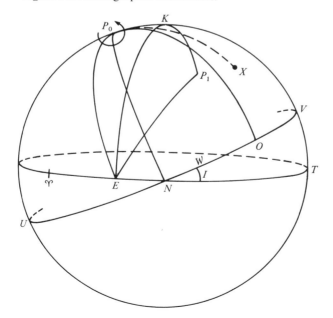

Now suppose that X is any point on the solar surface. Its heliographic latitude and longitude are defined as

$$B = 90° - P_0 X$$
$$L = OP_0 X. \tag{17.46}$$

It should be noted that heliographic longitude increases in the direction of the solar rotation – the opposite convention to that used for planetographic longitude. The heliographic longitude of the node is $360° - W$.

Let E be the centre of the apparent solar disc, or in earlier parlance the subearth point. This point lies on the ecliptic at longitude $\odot - 180°$, where $\odot$ is the true geocentric longitude of the sun. Hence $EN = \Omega - \odot + 180°$. Suppose that (L_0, B_0) are the heliographic coordinates of the centre of the disc. Then $P_0 E = 90° - B_0$ and $E\hat{P}_0 N = 360° - W - L_0$. We may identify two other parts in spherical triangle $P_0 NE$, namely $P_0\hat{N}E = 90° - I$, and $P_0 N = 90°$. Applying in turn the cosine, sine and analogue formulae to this triangle yields the result

$$\sin B_0 = \sin (\odot - \Omega) \sin I$$
$$\cos B_0 \sin (L_0 + W) = \cos I \sin (\Omega - \odot) \tag{17.47}$$
$$\cos B_0 \cos (L_0 + W) = -\cos (\Omega - \odot).$$

These equations are sufficient to determine (L_0, B_0) for any time, once W is calculated from (17.45).

The values of (L_0, B_0) are tabulated at daily intervals in the *Astronomical Almanac*, together with the position angle P of the rotation axis. If the point P_1 in Fig. 17.9 represents the celestial pole, then the position angle may be computed as the sum of two angles, namely

$$P = P_1 EK + KEP_0. \tag{17.48}$$

These two angles must be derived separately.

Consider first, spherical triangle $KP_1 E$, in which $KE = 90°$ and $KP_1 = \varepsilon$, the obliquity of the ecliptic. The spherical angle $E\hat{K}P_1$ is the ecliptic longitude of P_1 minus the ecliptic longitude of E. So $E\hat{K}P_1 = 90° - (\odot - 180°) = 270° - \odot$. Now apply the four-parts formula to this spherical triangle to obtain

$$\tan P_1 EK = -\cos \odot \tan \varepsilon. \tag{17.49}$$

Consider next the spherical triangle KEP_0. Again $KE = 90°$, while $P_0 K = I$. The point P_0 has ecliptic longitude $\Omega - 90°$, and so we find that $P_0\hat{K}E = \odot - 180° - (\Omega - 90°) = \odot - \Omega - 90°$. Applying the four-parts formula to spherical triangle KEP_0 yields the result

$$\tan KEP_0 = -\tan I \cos (\Omega - \odot). \tag{17.50}$$

Hence the position angle of the axis is given by

$$P = -\tan^{-1}[\tan I \cos(\Omega - \odot)] - \tan^{-1}(\cos\odot \tan\varepsilon). \qquad (17.51)$$

No ambiguity is encountered in practice when this formula is used, since the two inverse tangents are, respectively, restricted to the ranges $(-I, I)$ and $(-\varepsilon, \varepsilon)$. Principal values may be taken throughout.

17.8   Heliographic coordinates of a sunspot

The angular semidiameter S of the sun varies slightly in the course of the year due to its changing distance. The percentage variation is smaller than for a planet, but, since the solar disc is so much larger, some modification is required to the treatment of angular measurements. It is not quite sufficient to treat the disc as a plane.

In Fig. 17.10, the point C represents the heliocentric point and E the centre of the earth. The plane of the diagram is defined by the points E, C and X, where X is any surface marking, for example, a sunspot. Let R be the sun's radius and denote EC by r. Now let one of the tangents – in the plane of the diagram – from the point E to the sun touch the solar surface at T. The angle TEC is the sun's angular semidiameter, and, since $E\hat{T}C = 90°$, it is rigorously given by

$$\sin S = \frac{R}{r}. \qquad (17.52)$$

Suppose now that the sunspot X is observed at an angular distance ρ_1 from the centre of the disc. Let ρ be its heliocentric angular displacement from the subearth point. Then in the triangle ECX, $EC = r$, $CX = R$, $C\hat{E}X = \rho_1$, $E\hat{C}X = \rho$ and consequently $C\hat{X}E = 180° - (\rho + \rho_1)$. Using the sine formula, we derive

$$\sin(\rho + \rho_1) = \frac{r \sin \rho_1}{R} \simeq \frac{\rho_1}{S}. \qquad (17.53)$$

The value of ρ may be computed from this formula either in its precise or approximate form.

Figure 17.10

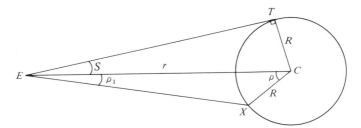

Suppose θ is the position angle of the point X. The heliographic coordinates (L, B) of this point may be derived in terms of the three parameters (L_0, B_0) and P discussed in the last section. The method is essentially identical to that used to derive planetographic coordinates in section 17.6, and need not be repeated. With the appropriate modification in notation, equations (17.41) to (17.43) are replaced by

$$\sin B = \sin B_0 \cos \rho + \cos B_0 \sin \rho \cos (P - \theta)$$

$$\cos B \sin (L - L_0) = \sin \rho \sin (P - \theta) \tag{17.54}$$

$$\cos B \cos (L - L_0) = \cos \rho \cos B_0 - \sin \rho \sin B_0 \cos (P - \theta).$$

Due to the convention that heliographic longitude increases towards the *apparent* west limb of the sun, the value of L_0 decreases with time. A *synodic* rotation period – with a mean value of 27.2753 days – corresponds to a decrease of L_0 of 360°. Such synodic rotations are designated by their *Carrington rotation number*. The synodic period begins when $L_0 = 0$, and the first rotation of the series commenced on 1853 November 9. The rotation number N corresponding to a particular Julian date may be derived from

$$N = 1750 + \text{int} \left[\frac{\text{JD} - 2\,445\,871.91}{27.2753} \right], \tag{17.55}$$

where int denotes the integral part. Rotation number 1750 began on 1984 June 20.41.

17.9   Selenographic coordinates

It has been possible to treat the rotations of the sun and the planets as constant, regarding their rotational axes as having fixed sidereal directions. This approximation cannot be made for the moon since its rotation is tied to the changes that occur in its orbit about the earth. In this section we shall not attempt to do more than define the selenographic coordinate system and indicate the main periodic phenomena – the librations.

The moon's mean rotation is about an axis directed to the point P_0 on the selenocentric celestial sphere as indicated in Fig. 17.11. We speak of the mean rotation since the axis is subject to oscillations about this mean position, which may be compared to the nutation in the earth's axis. This effect which is small is known as the *physical libration*.

The mean lunar equator is shown in the diagram as UQV where Q is the ascending node on the ecliptic. The point P_0 is not a fixed sidereal direction but precesses about the pole of the ecliptic in an 18.6 year period. This precession is linked to the regression of the node of the moon's orbit, in that the ascending node of the lunar equator Q is also the descending node of the mean lunar orbit. The ecliptic longitude of Q is, therefore, $180° + \Omega$, where Ω is the longitude of the ascending node of the moon's mean orbit. The variation in Ω was given in equation (7.17); the value of the mean equator's inclination I is fixed. In fact

$$I = 1° 32' 32''.7$$
$$\Omega = 55°.204\,723 - 0°.052\,953\,78\,d, \qquad (17.56)$$

where d is the number of days that have elapsed from 1985 January 0 0^h TDT.

It is almost a truism to say that the moon keeps the same face towards the earth. It is necessary to invest this statement with some exactness, however. First of all, we should note that the moon's mean rotation is direct and its period is equal to the mean sidereal period of its orbit. Suppose that L is the mean geocentric longitude of the moon. Then on the selenocentric celestial sphere we may define a 'mean earth' at the point E_0 with a mean longitude of $L + 180°$. Consequently, the arc $QE_0 = L - \Omega$. Now let us define a *mean centre of the disc* as the point M which moves round the mean lunar equator with its rotational period and which coincides with E_0 at the two nodes. It then follows that

$$MQ = L - \Omega. \qquad (17.57)$$

This point M is a fixed point on the lunar surface and P_0M is adopted as the prime meridian of selenographic longitude. The longitude, like heliographic longitude, increases in the sense of the moon's rotation. So the

Figure 17.11. Selenographic coordinates.

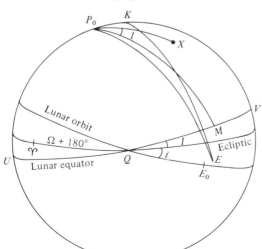

selenographic longitude and latitude (l, b) of a general point X are formally defined as

$$l = MP_0X$$
$$b = 90° - P_0X. \tag{17.58}$$

It is sometimes found convenient to use the colongitude, i.e. $90° - l$, of a point rather than the longitude. This practice is followed for the subsolar point, leading to the following approximate relationships.

longitude of morning terminator $= 360°$ − colongitude of sun

longitude of evening terminator $= 180°$ − colongitude of sun.

$$\tag{17.59}$$

Let E be the true subearth point in Fig. 17.11. This point will be close to E_0; the departure of E from E_0 is due to the eccentricity of the moon's orbit and to perturbations of it. Moreover, the angle $MQE_0 = I + i$ where i is the orbital inclination. This angle is fairly small (about $6°$), so the departure of E from M is also small. The true centre of the disc oscillates about M and its departure is conveniently expressed as the selenographic coordinates (l_e, b_e) of the point E.

Suppose (λ, β) are the geocentric ecliptic coordinates of the moon. The ecliptic coordinates of the point E are then $(\lambda + 180°, -\beta)$. Let us consider spherical triangle P_0KE. We may identify, without too much difficulty, the following parts: $P_0K = I$, $P_0E = 90° - b_e$, $KE = 90° + \beta$. While $KP_0E = 90° - L' + \Omega - l_e$, and $P_0KE = \lambda + 180° - (\Omega + 90°) = 90° - (\Omega - \lambda)$. Now apply the cosine, sine and analogue formulae to this triangle to derive

$$\sin b_e = -\cos I \sin \beta + \sin I \cos \beta \sin (\Omega - \lambda)$$
$$\cos b_e \cos (L' + l_e - \Omega) = +\cos \beta \cos (\Omega - \lambda) \tag{17.60}$$
$$\cos b_e \sin (L' + l_e - \Omega) = -\sin \beta \sin I - \cos \beta \cos I \sin (\Omega - \lambda).$$

The values of (l_e, b_e) may be deduced from these equations.

What has been established here is the *geocentric optical libration*. The *Astronomical Almanac* gives daily values of (l_e, b_e) which incorporate the physical libration as well. The latter is a much smaller effect and is tabulated separately so that the optical libration may be isolated if desired. When the libration in longitude is positive, i.e. $l_e > 0$, a normally unseen part of the moon near the apparent west limb is exposed. Similarly, if the libration in latitude is positive $(b_e > 0)$, a region near the north limb is revealed. Due to libration, therefore, as much as 59 per cent of the entire lunar surface may be seen from earth at one time or another.

A further cause of optical libration is the observer's displacement from the geocentric point. This is just the effect of geocentric parallax; it can be allowed for by using the topocentric coordinates of the moon in equations (17.60). The *Astronomical Almanac* gives a differential procedure for modifying the tabulated geocentric librations to derive the full topocentric librations.

The position angle of the moon's rotation axis may be obtained by the method that was used in section 17.7. Since the point E is not on the ecliptic the formulae that replace (17.51) are rather more complicated. It is left to the reader to prove that the position angle C of the axis is given by

$$C = \cot^{-1}\left[\sin \beta \tan (\Omega - \lambda) + \cos \beta \cot I \sec (\Omega - \lambda)\right]$$
$$+ \cot^{-1}\left[\sin \beta \tan \lambda - \cos \beta \cot \varepsilon \sec \lambda\right]. \tag{17.61}$$

Problems

(It is to be assumed in Problems 17.1–17.5 that the earth's orbit is circular of radius 1 AU. The planets' orbits are also treated as circular and are assumed to lie in the ecliptic.)

17.1 Show that the phase angle ϕ for a superior planet has its maximum value at quadrature. Calculate the maximum values of ϕ for the planets Mars and Jupiter and the corresponding minimum phases, assuming that their mean distances from the sun are 1.52 AU and 5.20 AU respectively.

17.2 Prove that the apparent motion of a superior planet is retrograde at opposition, but direct at quadrature. Can any equivalent statements be made about the apparent motion of an inferior planet?

17.3 Show that the value of $\cos \theta$ given by equation (17.22) is necessarily less than unity. Prove that the same formula gives the angle θ corresponding to the stationary points for an inferior planet. Show further that the formula for the elongation of a planet at its stationary point is given in either case by (17.23), namely

$$\tan E = \pm \frac{a}{(a+1)^{1/2}},$$

but that, for an inferior planet, the solutions are in the first and fourth quadrants, while for a superior planet they are in the second and third quadrants.

17.4 Assuming that the brightness of an inferior planet is directly proportional to its phase and inversely proportional to its geocentric distance r, show that it attains maximum brilliance when r is a solution of the equation

$$r^2 + 4ar - 3(1 - a^2) = 0.$$

Apply this to the planet Venus, assuming that $a = 0.7233$ AU, and derive its geocentric distance and elongation at maximum brilliance.

17.5 A minor planet is at a stationary point when its elongation is $125°$. Estimate its heliocentric distance.

17.6 Give reasons why it would be expected that precessional effects on the planets' rotational axes should, with the exception of Pluto, be considerably less than luni–solar precession on the earth.

17.7 Prove that the position angle of the solar axis is zero on two occasions during the year which are approximately six months apart, and that the sun's true longitude is then given by

$$\odot = -\tan^{-1}\left(\frac{\tan \varepsilon + \tan I \cos \Omega}{\tan I \sin \Omega}\right).$$

17.8 Calculate the maximum value of the position angle of the solar axis during 1985. (Assume $\varepsilon = 23° \, 26'.5$.)

17.9 Prove that the altitude of sun above the horizon at a point of the moon's surface with selenographic longitude and latitude (l, b) is given by

$$\sin a = \sin b_0 \sin b + \cos b_0 \cos b \sin (c_0 + l),$$

where b_0 and c_0 are respectively the selenographic latitude and colongitude of the subsolar point.

17.10 Prove that the inclination i' of the moon's mean equator to the earth's equator is given by

$$\cos i' = \cos \varepsilon \cos I + \sin \varepsilon \sin I \cos \Omega$$

where the symbols on the right-hand side have their usual meanings.

17.11 Prove that the position angle of the moon's axis of rotation is given by equation (17.61).

18

Eclipses and occultations

18.1   Classification of eclipses

An eclipse of the sun or the moon will occur whenever the sun, the earth and the moon are nearly in a straight line. This can only happen when the moon and the sun are in conjunction or at opposition. If the moon's orbit lay precisely in the ecliptic, a solar eclipse would occur at each new moon and a lunar eclipse at each full moon. Owing to the inclination of the moon's orbit, however, an exact alignment does not normally occur. The moon lies in the ecliptic only on two occasions each month – when it passes through the nodes of its orbit. Consequently, it is only when the full or the new moon is close to one of these points that an eclipse is possible. A solar eclipse is due to the moon's shadow falling on to part of the surface of the earth so that the sun is at least partially obscured. In a lunar eclipse it is the earth's shadow that is falling on the surface of the moon.

In Fig. 18.1, the points S and E represent the centres of the sun and the earth respectively. Let us treat these two bodies as spheres with radii $R_\odot$ and $R_\oplus$. The external common tangents such as XA generate the *umbral cone* with vertex V. For an observer within the umbral cone the sun is totally obscured. On the other hand, the internal common tangents, such as YB, generate the *penumbral cone* with vertex W. Clearly, within the penumbra, part of the solar disc is hidden, but the obscuration is not complete.

The solar radius SA, and the radius of the earth EX are each perpendicular to the common tangent AXV. Consequently, the triangles EXV and SAV are similar, and

$$EV/R_\oplus = SV/R_\odot. \qquad (18.1)$$

Let us now denote the earth's distance from the sun by r. Then $SV = EV + r$, and equation (18.1) may be reorganized to give

$$EV = \frac{rR_\oplus}{R_\odot - R_\oplus}.$$ (18.2)

This equation gives the length of earth's umbral cone. It varies owing to the annual variation in r, but only slightly and its mean length is 1.384×10^6 km. This is several times the moon's distance from the earth. In fact, not only does the umbra of the earth's shadow reach the moon, its extent is such that it can completely contain it.

We must distinguish, therefore, between three types of lunar eclipse. They are

   (i) A total eclipse of the moon, meaning that the moon is wholly within the earth's umbral cone. The moon is then very dim, but it does not completely disappear, since a small amount of light is refracted by the earth's atmosphere on to its surface.

  (ii) A partial eclipse of the moon occurs when the moon is partly within the earth's umbra, and partly within the penumbra. The boundary between the two appears as a false terminator.

 (iii) A penumbral eclipse of the moon implies that the moon has only entered the earth's penumbra. There is then no obvious change in the moon's appearance. This is the most common and least remarkable type of lunar eclipse. Naturally a penumbral phase must precede (and follow) any umbral eclipse.

The general circumstances associated with solar eclipses may be examined by considering the umbral and penumbral cones of the moon's shadow. The length of the umbral cone may be derived from (18.2) simply by replacing the earth's radius with that of the moon. The mean length is found to be 3.735×10^5 km. This is slightly less than the average distance of the earth from the moon. The umbra of the moon's shadow does not necessarily reach the earth, but the eccentricity of the moon's orbit is

Figure 18.1. The umbral and penumbral cones.

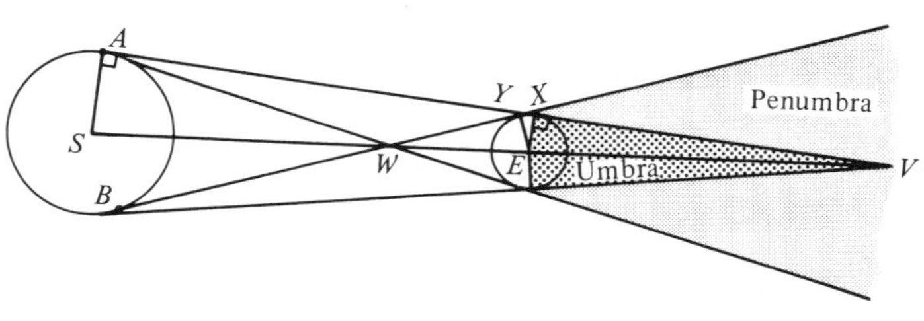

sufficient to ensure that it can on occasion do so. When this does occur, however, the umbral cone falls on a very small portion of the earth's surface.

Solar eclipses are classified into three kinds depending on the nature of maximum eclipse for the most favoured observer:

(i) A total eclipse of the sun implies that the moon's umbral cone reaches the earth. Observers within that very small region will see the sun totally obscured by the moon's disc.

(ii) An annular eclipse occurs when the umbral cone falls short of the earth, but its axis intersects the earth's surface. A centrally placed observer sees the sun wholly obscured except for a narrow annulus.

(iii) A partial eclipse of the sun results when the axis of the umbral cone does not intersect the earth's surface at all, but only the penumbra of the moon's shadow is involved.

Owing to the smallness of the region of intersection of the moon's umbral cone with the earth, the occurrence of a total or annular solar eclipse at a particular locality is a very rare event. On the other hand, most solar eclipses may be classified as total or annular in the sense that these terms apply over a limited region. The partial phase is observable over a considerable part of the earth since the penumbra can cover a large area of the earth's surface, equivalent to nearly 30° in latitude at normal incidence.

18.2 Conditions for a lunar eclipse

For most of this chapter, it will be convenient to use the following notation:

P – equatorial horizontal parallax of the sun

P_1 – equatorial horizontal parallax of the moon

S – sun's angular semidiameter

S_1 – moon's angular semidiameter

r – geocentric distance of the sun

r_1 – geocentric distance of the moon.

(18.3)

Consider Fig. 18.2 in which AXV and $A'X'V$ are two external common tangents to the sun and the earth. The point V is the vertex of the umbral

Figure 18.2

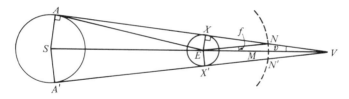

cone; let $v = X\hat{V}E$ be the half-angle of this cone. The dashed curve NMN' represents the moon's orbit, and $f = NEM$ is the geocentric angular radius of the umbral cone which is required.

Consider the angle $X\hat{N}E$. It is the angle subtended by the earth's radius at the moon's distance and is, therefore, equal to P_1. Moreover, since $X\hat{N}E$ is an exterior angle to the triangle ENV, it follows that

$$f = P_1 - v. \tag{18.4}$$

A similar argument may be applied to triangle EAV. The exterior angle SEA is the sun's semidiameter S, while the angle EAV is the angle subtended by the earth's radius at the sun, the sun's geocentric parallax P. Hence

$$v = S - P. \tag{18.5}$$

Now eliminating the vertex angle from equations (18.4) and (18.5) will yield the angular radius of the umbral cone as

$$f = P + P_1 - S. \tag{18.6}$$

A parallel argument may be applied using an internal tangent and the vertex of the penumbral cone. The angular radius of the penumbral cone is then obtained as

$$f' = P + P_1 + S. \tag{18.7}$$

Equations (18.6) and (18.7) give the geometrical formulae for the angular radii of the two shadow cones. The effect of the earth's atmosphere, however, is found to give an increase of about 2 per cent. The formulae used for predicting eclipses are, therefore,

$$f = 1.02(P + P_1 - S)$$
$$f' = 1.02(P + P_1 + S). \tag{18.8}$$

Figure 18.3 represents a geocentric celestial sphere; K is the pole of the ecliptic and N a node of the moon's orbit. Suppose that M represents the centre of the moon and S' the antisolar point. It is, therefore, the direction of the axis of the umbral cone. Let us denote by η the separation of the points S' and M on the celestial sphere. The shaded area in the diagram represents the umbral cone at the moon's distance. For an umbral eclipse to take place, it is clearly necessary that some part of the moon fall within this shaded area, i.e. $\eta < f + S_1$.

Using (18.8), we may deduce that conditions for this and the other types of lunar eclipse formally as

Penumbral eclipse	$\eta < 1.02(P + P_1 + S) + S_1$
Partial eclipse	$\eta < 1.02(P + P_1 - S) + S_1$
Total eclipse	$\eta < 1.02(P + P_1 - S) - S_1.$

$$\tag{18.9}$$

18.3   Conditions for a solar eclipse

The conditions for the occurrence of a solar eclipse are formulated in terms of the angular separation of the sun and the moon as seen from the centre of the earth. We shall again use the symbol η to denote the relevant separation, which is the angle SEM in Fig. 18.4. That diagram illustrates the relative configuration of the sun, the moon and the earth when a solar eclipse is just beginning. The earth is about to enter the penumbra of the moon's shadow, and the diagram shows the limiting configuration for a partial eclipse of the sun.

For simplicity it is assumed that all three bodies (sun, moon and earth) are spherical. The line $AWXO$ is tangential to them all touching the earth at O. The line OZ drawn through O perpendicular to this common tangent will give the zenith direction for an observer at O. This observer will see both the sun and the moon very close to the horizon which is given by the common tangential line. The zenith distance of the moon will be $90° - S_1$,

Figure 18.3

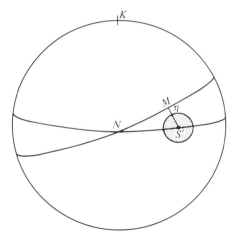

Figure 18.4. Limiting configuration for a partial solar eclipse.

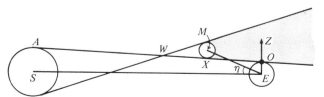

and that of the sun will be $90° + S$. The observed separation of the two bodies is just the difference of the two zenith distances.

These are, however, topocentric zenith distances; refraction has been ignored. The angle η is the difference of the geocentric zenith distances referred to the same zenith direction; the effect of geocentric parallax must be included. Using equation (4.52), for a spherical model of the earth, we find that the geocentric zenith distances of the sun and the moon are z and z_1, where

$$
\begin{aligned}
z &= 90° + S - \sin^{-1}(\sin P \cos S) \\
z_1 &= 90° - S_1 - \sin^{-1}(\sin P_1 \cos S_1).
\end{aligned}
\tag{18.10}
$$

The required separation $\eta = z - z_1$.

The limiting requirement for a total (or annular) eclipse is derived similarly by considering an external common tangent. Only the sign of S in (18.10) is changed. With sufficient accuracy we may write the conditions for solar eclipses as:

Partial eclipse $\eta < S_1 + P_1 + S - P$

Total (or annular) eclipse $\eta < S_1 + P_1 - S - P.$
\qquad (18.11)

The second condition gives the requirement for a total eclipse somewhere on the earth's surface. The area experiencing totality is extremely limited.

18.4   Ecliptic limits

The conditions for a solar and lunar eclipse have been stated in terms of the angular separation of the two bodies. It is more convenient, however, to express these conditions in terms either of the sun's position with respect to the moon's orbit, in particular its node, or alternatively in terms of the moon's position with reference to the ecliptic, that is, its latitude. We shall give the argument pertinent to solar eclipses; that for lunar eclipses is easily deduced by replacing the sun with the antisolar point.

Figure 18.5 represents a geocentric celestial sphere. The points S and M correspond to the positions of the centres of the sun and the moon respectively at their conjunction in ecliptic longitude. Then $M\hat{S}N = 90°$. We shall denote MS by β and NS by ξ; these are the two parameters that will be used to specify the limiting requirements for a solar eclipse. Let I be the inclination of the moon's orbit to the ecliptic. This angle is about $5°$, but is subject to some variation. Using the four-parts formula in spherical triangle MNS yields a relationship between the two parameters ξ and β, namely

$$\sin \xi = \cot I \tan \beta. \tag{18.12}$$

If an eclipse is to take place, the latitude β will be a small quantity. We may, therefore, treat the triangle MNS as a plane triangle in our investigation. The moon is, of course, moving along the side NM of this triangle, while the sun is moving more slowly along the side NS. In considering the relative position of the two bodies it is convenient, however, to treat the sun rather than the node as fixed. This is what is envisaged in Fig. 18.6, the points M, S and N are defined as before, while $N'M$ gives the motion of the moon relative to the sun. The least angular separation η of the two bodies is SM', where M' is the foot of the perpendicular from S to MN'.

Let I' be the angle $MN'S$. Then

$$\eta = \beta \cos I'. \tag{18.13}$$

The angle I' is clearly a modified inclination of the moon's orbit to the ecliptic. It is related to the true inclination by the equation

Figure 18.5. The eclipse parameters ξ and β.

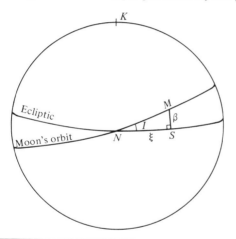

Figure 18.6

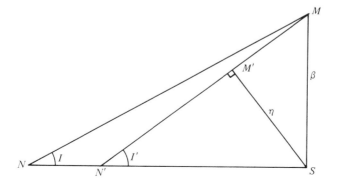

$$\tan I' = \frac{SN}{SN'} \tan I.$$
(18.14)

Consider next the time interval between the moon passing through the node and its conjunction in longitude with the sun. The moon's longitude increases by the amount SN in this time, while the sun's longitude increases by NN'. If, therefore, q denotes the ratio of the moon's motion in longitude to that of the sun, then equation (18.14) implies that

$$\tan I' = \frac{q}{q-1} \tan I.$$
(18.15)

The requirements for an eclipse may now be expressed as a condition on β or ξ. Full details are given in Tables 18.2 and 18.3; here we shall only examine the requirement for a partial solar eclipse in detail.

Combining equations (18.13) and (18.11) allows us to express the requirement for a partial eclipse of the sun as

$$\beta < (S_1 + P_1 + S - P) \sec I'.$$
(18.16)

This is, of course, a condition on the absolute value of the moon's latitude when it is in conjunction in longitude with the sun. The corresponding condition for the sun's angular distance from the node may be derived from (8.12) as

$$\xi < \sin^{-1} (\cot I \tan \beta).$$
(18.17)

In making use of these conditions it must be recognized that the parameters involved are not strictly constant. The two semidiameters and horizontal parallaxes vary due to the eccentricities of the earth's and the moon's orbit. The parameter q will vary due to the same cause. Moreover, the inclination I also varies due to the solar perturbation of the moon's orbit. The extreme variations of all of these quantities is given in Table 18.1 which is based on values taken from the *Explanatory Supplement*. The entries in the final row of this table are derived from equation (18.15). For example, the maximum value of I' is calculated from the maximum value of I and the minimum value of q.

The variations in the angular diameters and the horizontal parallaxes of the sun and moon are not independent. Moreover, the value of q is linked to these quantities. The extreme values given in Table 18.1 may, however, be used to provide upper and lower bounds on β and ξ for the occurrence of eclipses. For example, suppose that the right-hand side of the inequality (18.16) is computed, using maximum values of S_1, P_1, S and I' and, formally at least, the minimum value of P, although, in fact, the last quantity does not vary significantly. Call the computed quantity β_{max}. Then we may be certain that no solar eclipse is possible for $\beta > \beta_{max}$. Similarly if β_{min} corresponds to

Table 18.1. *Extreme and mean values of solar eclipse parameters*

	Maximum	Minimum	Mean
S_1	16′ 45″	14′ 41″	15′ 33″
P_1	61′ 27″	53′ 53″	57′ 03″
S	16′ 18″	15′ 46″	16′ 00″
P	0′ 09″	0′ 09″	0′ 09″
I	5° 18′	4° 59′	5° 08′
q	16.2	10.9	13.5
I'	5° 50′	5° 18′	5° 32′

Table 18.2. *Some conditions for the occurrence of solar eclipses*

No eclipse possible	$\beta > 1° 34′ 50″$	$\xi > 18°.45$
Partial eclipse certain	$\beta < 1° 24′ 33″$	$\xi < 15°.38$
No central eclipse	$\beta > 1° 02′ 36″$	$\xi > 12°.06$
Central eclipse certain	$\beta < 0° 52′ 20″$	$\xi < 9°.45$

the right-hand side of (18.16), using minimum values of S_1, P_1, S and I' and maximum P, then a partial solar eclipse is certain in $\beta < \beta_{min}$. Equation (8.12) may then be used to derive corresponding limits on ξ. Playing very safe, we may take

$$\sin \xi_{max} = \cot I_{min} \tan \beta_{max}$$
$$\sin \xi_{min} = \cos I_{max} \tan \beta_{min}. \tag{18.18}$$

A partial eclipse of the sun is certain for $\xi < \xi_{min}$ and impossible for $\xi > \xi_{max}$.

Using the second requirement in (18.11) the corresponding ecliptic limits for a central solar eclipse may be derived. The results are collected in Table 18.2.

The ecliptic limits obtained in this way are too severe, particularly for ξ, and refinement is, of course, possible. The statements contained in Table 18.2 are perfectly accurate; they should be interpreted literally. For cases which are not included in the table, e.g. $1° 34′ 50″ > \beta > 1° 24′ 33″$, it will be necessary to test for an eclipse by inserting the actual values of all parameters in (18.16).

The occurrence of eclipses of the moon may be investigated similarly. The conditions on η are given in (18.9), and corresponding conditions on β and ξ may be deduced from (18.13) and (18.17). The quantity ξ is the departure of the antisolar point from the node when it is in conjunction in longitude with the moon. It may equally well be regarded as the sun's separation from the other node when it is in opposition to the moon. The conditions for the occurrence of lunar eclipses are given in Table 18.3.

Table 18.3. *Some conditions for the occurrence of lunar eclipses*

No eclipse possible	$\beta > 1° 36' 43''$	$\xi > 18°.83$
Penumbral eclipse certain	$\beta < 1° 26' 15''$	$\xi < 15°.69$
No umbral eclipse	$\beta > 1° 03' 50''$	$\xi > 12°.30$
Umbral eclipse certain	$\beta < 0° 53' 24''$	$\xi < 9°.64$
No total eclipse	$\beta > 0° 32' 14''$	$\xi > 6°.17$
Total eclipse certain	$\beta < 0° 21' 50''$	$\xi < 3°.93$

18.5   Frequency and recurrence of eclipses

The moon passes through a particular node once a month. The sun, however, moves much more slowly, and the mean interval between its passages through a node is the eclipse year of $346^{d}.62$ (cf. equation (10.6)). The sun's distance ξ from the node is, therefore, the most useful indicator of when eclipses will be possible. Since ξ increases by 360° in one eclipse year, it follows that the rate of change of this parameter is

$$\frac{d\xi}{dt} = 1°.0386 \text{ per day.} \tag{18.19}$$

Now the mean interval from one new moon to the next is the synodic month – cf. equation (7.20) – of length $29^{d}.5306$. Within this interval of time the parameter ξ will increase by an amount $\Delta\xi$ which is

$$\Delta\xi = 30°.67. \tag{18.20}$$

Let S_1 and S_2 in Fig. 18.7 represent the sun's positions at two consecutive new moons, strictly at two consecutive conjunctions with the moon in

Figure 18.7

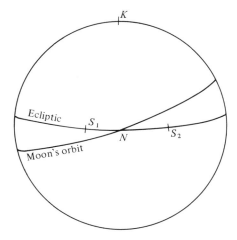

longitude. In the diagram the node N is regarded as fixed, and it is assumed that S_1 and S_2 are situated on either side of this point. Since $S_1 S_2 = \Delta\xi$, it follows that one of the intervals $S_1 N$ or $S_2 N$ is less than $\frac{1}{2}\Delta\xi = 15°.34$. This should be compared with the value of ξ ($15°.38$) which is stated in Table 18.2 as the sufficient condition for a partial eclipse to occur. We, therefore, conclude that each passage of the sun through a node will be accompanied by at least one solar eclipse. It may even produce two solar eclipses one month apart if S_1 and S_2 are nearly symmetrically placed with respect to the node.

Similar considerations apply to the occurrence of lunar eclipses. The requirements for penumbral eclipses are slightly less stringent than for a partial eclipse of the sun. Each passage of the sun through a node is, therefore, accompanied by at least one penumbral eclipse of the moon.

There must then be at least four eclipses every year, two solar eclipses and two lunar eclipses. The latter are quite likely to be penumbral, and sometimes these are not counted as eclipses, since there is nothing remarkable to be observed. Exceptionally the passage of the sun through a node may produce three eclipses. These will either have the sequence of partial solar–total lunar–partial solar, or a central eclipse of the sun straddled by two penumbral lunar eclipses.

Since the eclipse year is rather shorter than the calendar year, it is possible to have three passages of the sun through a node in one calendar year. The maximum number of eclipses in any one year is seven. For example, in 1982 there were three total eclipses of the moon and four partial eclipses of the sun.

An approximate commensurability may be constructed between the eclipse year and the synodic month. It is known as the *Saros*. This is the period of approximately 18 years 11 days which corresponds closely to 19 eclipse years and 223 synodic months. In fact

$$19 \text{ eclipse years} = 6585^d.78$$
$$223 \text{ synodic months} = 6585^d.32. \tag{18.21}$$

Let us consider the change in the parameter ξ over 223 synodic months. Using (18.19) and (18.21), it is found that, when the multiples of $360°$ are extracted, the change in ξ over this interval is

$$\Delta\xi_S = -0°.48. \tag{18.22}$$

If, therefore, an eclipse has occurred at a certain date, after an interval of $6585^d.32$ the moon will have the same phase and ξ almost the same value. Favourable conditions for the same kind of eclipse are consequently reproduced. Eclipses may, therefore, be arranged in series, similar eclipses recurring at intervals of about 18 years 11 days. The solar eclipses are likely to occur at quite different points on the earth's surface, so the series is more obvious for lunar eclipses. A complete series will last over a thousand years. It consists of a sequence of umbral eclipses preceded and followed by a series of penumbral eclipses.

The conclusions of this section are based on the assumptions that the sun and the moon's node proceed uniformly round the ecliptic, neither of which is exactly true. Eclipses depend on the positions of the true sun and moon, and the departures of these bodies from their mean positions may affect the circumstances of particular eclipses. The general conclusions, however, remain valid, cf. Problem 18.3.

18.6   Besselian elements of a solar eclipse

The detailed discussion will concentrate on eclipses of the sun. Eclipses of the moon are generally regarded as less-important phenomena. Moreover, the circumstances and timing of a lunar eclipse are the same for all observers, and such data may be given definitively in the annual almanacs. The methods for deriving these circumstances are very similar to those discussed below in connection with solar eclipses. In the latter case, however, local circumstances are also very important and these cannot be tabulated in the almanacs. Besselian elements provide the data from which an observer may compute the local circumstances of the solar eclipse.

Let $\mathbf{r}$ and $\mathbf{r}_1$ be the geocentric position vectors of the sun and the moon, which are indicated as $\overrightarrow{ES}$ and $\overrightarrow{EM}$ in Fig. 18.8. It will be convenient to use the earth's equatorial radius as our unit of distance. Consequently the magnitudes of these vectors are given by

$$r = \operatorname{cosec} P$$
$$r_1 = \operatorname{cosec} P_1. \tag{18.23}$$

Further, let us formally write the two vectors as

$$\mathbf{r} = \operatorname{cosec} P(\cos \alpha \cos \delta, \sin \alpha \cos \delta, \sin \delta)$$
$$\mathbf{r}_1 = \operatorname{cosec} P_1(\cos \alpha_1 \cos \delta_1, \sin \alpha_1 \cos \delta_1, \sin \delta_1),$$

$$\tag{18.24}$$

where (α, δ) and (α_1, δ_1) are respectively the apparent coordinates of the sun and the moon.

It is desirable to introduce a new system of coordinates (x, y, z) with origin E. The z-axis is chosen parallel to the vector $\overrightarrow{MS}$ and the plane $z = 0$ is called the *fundamental plane*. The x-axis is the intersection of the fundamental plane with the equator; its positive direction being east. The y-axis completes the right-handed set; it, therefore, points north.

Let **I**, **J**, **K** be unit vectors in the x-, y- and z-directions respectively. Then

$$\mathbf{K} = \frac{\mathbf{r} - \mathbf{r}_1}{|\mathbf{r} - \mathbf{r}_1|}. \tag{18.25}$$

The equatorial components of this vector may, therefore, be computed from the ephemerides for the sun and the moon. If the z-axis points in the direction of right ascension a and declination d, then these equatorial components are

$$\mathbf{K} = (\cos d \cos a, \cos d \sin a, \sin d). \tag{18.26}$$

Consequently (a, d) are determined by equation (18.25). Their values may then be used to obtain the equatorial components of the vectors **I** and **J**, which, by their definitions, are given as

$$\begin{aligned} \mathbf{I} &= (-\sin a, \cos a, 0) \\ \mathbf{J} &= (-\cos a \sin d, -\sin a \sin d, \cos d). \end{aligned} \tag{18.27}$$

Thus all three unit vectors are now determined.

Now let (x, y, z) and (x_1, y_1, z_1) be the coordinates of the sun and the moon. From the choice of axes it follows that $x = x_1$ and $y = y_1$. Moreover, by (18.24), (18.27) and (18.26), it follows that

$$\begin{aligned} x_1 &= \mathbf{r}_1 \cdot \mathbf{I} = \operatorname{cosec} P_1 \cos \delta_1 \sin(\alpha_1 - a) \\ y_1 &= \mathbf{r}_1 \cdot \mathbf{J} = \operatorname{cosec} P_1 [\sin \delta_1 \cos d - \cos \delta_1 \sin d \cos(\alpha_1 - a)] \\ z_1 &= \mathbf{r}_1 \cdot \mathbf{K} = \operatorname{cosec} P_1 [\sin \delta_1 \sin d + \cos \delta_1 \cos d \cos(\alpha_1 - a)]. \end{aligned} \tag{18.28}$$

The line SM when produced is the axis of the moon's shadow. The coordinates of the point F, the intersection of the axis with the fundamental

Figure 18.8. The geometry of a solar eclipse.

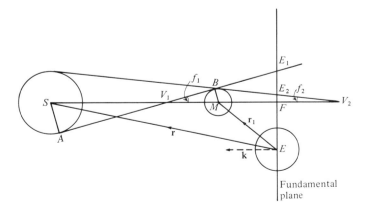

plane, are clearly $(x_1, y_1, 0)$. So x_1 and y_1 specify the position of the axis with respect to the centre of the earth. The coordinate z_1 gives the moon's distance from the fundamental plane. The spherical coordinates (a, d) determine the direction of the axis in space. In fact, the right ascension of the axis is not normally used as an eclipse element, as it is more convenient to use the Greenwich hour angle. This is given by

$$\mu = T_s - a \tag{18.29}$$

where T_s is the apparent Greenwich sidereal time.

In Fig. 18.8 an internal tangent AV_1BE_1 has been indicated. Only a single tangent has been drawn to avoid unnecessary complications in the diagram, but this tangent indicates the extent of the penumbral cone. The tangent touches the sun at A and the moon at B and intersects the fundamental plane at E_1. The point V_1 where it intersects the axis of the shadow is, of course, the vertex of the penumbral cone. Let us denote the half-angle of the cone by f_1, that is, the angle BV_1M. Similarly, a single external tangent has been drawn which cuts the fundamental plane in E_2 and the axis in V_2. The half-angle of the umbral cone (E_2V_2F) is denoted by f_2.

Let R and k be respectively the radii of the sun and the moon, each expressed in units of the earth's equatorial radius. From the similar triangles BMV_1 and ASV_1, it follows that

$$\sin f_1 = \frac{k}{MV_1} = \frac{R}{SV_1} = \frac{R+k}{MS}$$

i.e.

$$\sin f_1 = \frac{R+k}{|\mathbf{r} - \mathbf{r}_1|}. \tag{18.30}$$

The angle f_1 may, therefore, be computed from assumed values of R and k. A similar argument provides the half angle of the umbral cone as

$$\sin f_2 = \frac{R-k}{|\mathbf{r} - \mathbf{r}_1|}. \tag{18.31}$$

Let c_1 and c_2 be the z-coordinates of the vertices V_1 and V_2 of the two shadow cones. Then, as shown in Fig. 18.8, $c_1 = FV_1$ and $c_2 = -FV_2$. Now, $V_1M = c_1 - z_1$, so consideration of triangle V_1MB indicates that

$$c_1 = z_1 + k \operatorname{cosec} f_1. \tag{18.32}$$

The vertex of the penumbral cone has coordinates (x_1, y_1, c_1), its axis is parallel to the z-axis and its half angle is f_1. It is easy to see that the equation of the penumbral cone is, therefore,

$$(x - x_1)^2 + (y - y_1)^2 = (c_1 - z)^2 \tan^2 f_1.$$

On using (18.32), this reduces to

$$(x-x_1)^2 + (y-y_1)^2 = ((z_1-z)\tan f_1 + k \sec f_1)^2. \qquad (18.33)$$

A similar argument may be applied to the umbral cone to give its equation as

$$(x-x_1)^2 + (y-y_1)^2 = ((z_1-z)\tan f_2 - k \sec f_2)^2. \qquad (18.34)$$

Let us now introduce two new parameters l_1 and l_2 defined as

$$\begin{aligned} l_1 &= z_1 \tan f_1 + k \sec f_1 \\ l_2 &= z_1 \tan f_2 - k \sec f_2. \end{aligned} \qquad (18.35)$$

The equations of the two cones will then become

$$\begin{aligned} (x-x_1)^2 + (y-y_1)^2 &= (l_1 - z \tan f_1)^2 \\ (x-x_1)^2 + (y-y_1)^2 &= (l_2 - z \tan f_2)^2. \end{aligned} \qquad (18.36)$$

Setting $z=0$ in this equation indicates that l_1 and l_2 are the radii of the intersections of the penumbral and umbral cones with the fundamental plane. From the definitions (18.35), the radius of the penumbral cone is always positive, but that the umbral cone may be negative. The convention being applied is that l_2 is positive when $c_2 > 0$, that is, when the umbral cone's vertex falls short of the fundamental plane. In general, l_2 is positive for annular eclipses but negative for total eclipses.

The *Astronomical Almanac* tabulates the following quantities at 10-minute intervals during a solar eclipse: $x_1, y_1, \sin d, \cos d, \mu, l_1$ and l_2. They are known as the *Besselian elements of the solar eclipse*. In addition, single values are given for $\tan f_1, \tan f_2, \mu'$ and d' – the prime indicating the *hourly* rates of change. The variations in these four quantities are sufficiently small that they may be regarded as constants for the duration of the eclipse.

18.7 Solar eclipse calculations

Together with the numerical data just described, the *Astronomical Almanac* provides an eclipse map indicating approximate circumstances for the eclipse in question. Superimposed on the outline world map are contours over which the eclipse starts or terminates at the same time. The path of the total (or annular) phase is also indicated. Using this map the beginning (or end) of the eclipse for any observing site may be determined, certainly to within an accuracy of 10 minutes. Suppose T_0 is the approximate universal time derived in this way. We shall now examine how an improved value may be calculated.

The geocentric position vector of the observer O may be written in equatorial coordinates as

$$\mathbf{r}_0 = \rho[\cos\phi \cos(T_s + \lambda), \cos\phi \sin(T_s + \lambda), \sin\phi]. \qquad (18.37)$$

Here T_s is the Greenwich sidereal time, and ϕ is the *geocentric* latitude, λ the east longitude and ρ the geocentric distance of O. Let (ξ, η, ζ) be the observer's coordinates in the system that was introduced in the last section. Then, by (18.26), (18.27) and (18.29),

$$\xi = \mathbf{r}_0 \cdot \mathbf{I} = \rho \cos \phi \sin (\mu + \lambda)$$
$$\eta = \mathbf{r}_0 \cdot \mathbf{J} = \rho[\sin \phi \cos d - \cos \phi \sin d \cos (\mu + \lambda)] \qquad (18.38)$$
$$\zeta = \mathbf{r}_0 \cdot \mathbf{K} = \rho[\sin \phi \sin d + \cos \phi \cos d \cos (\mu + \lambda)].$$

Now at the instant $(UT = T_1)$ at which the eclipse begins the point O lies on the penumbral cone. By (18.36) the condition for this is that

$$(\xi - x_1)^2 + (\eta - y_1)^2 = (l_1 - \zeta \tan f_1)^2. \qquad (18.39)$$

The problem is then to find the universal time when this condition is satisfied.

Let us write the condition (18.39) in the form $F(T) = 0$, where

$$F(T) = (\xi - x_1)^2 + (\eta - y_1)^2 - (l_1 - \zeta \tan f_1)^2. \qquad (18.40)$$

All the parameters on the right-hand side are functions of the universal time T, although f_1 is virtually constant and l_1 varies slowly. Now $F(T)$ is evaluated for $T = T_0$, the estimated beginning of the eclipse, using the Besselian elements that are tabulated in the *Astronomical Almanac* both directly in (18.40) and to compute (ξ, η, ζ) from (18.38). The value $F(T_0)$ obtained will presumably be small, indicating that T_0 is close to the required time. An improved time $T_0 + \Delta T$ may be obtained by the Newton–Raphson method which gives

$$\Delta T = -\frac{F(T_0)}{F'(T_0)}. \qquad (18.41)$$

In obtaining the derivative $F'(T_0)$ from (18.40), the elements l_1 and f_1 may be treated as constants. We, therefore, write

$$\tfrac{1}{2} F'(T) = (\xi - x_1)(\xi' - x_1') + (\eta - y_1)(\eta' - y_1') + (l_1 - \zeta \tan f_1)\zeta' \tan f_1. \qquad (18.42)$$

The derivatives (x_1', y_1') can only be obtained by differencing these tabulated elements, but (ξ', η', ζ') are derived by differentiating equations (18.38). It is found, after a little reduction, that

$$\xi' = \mu'(\zeta \cos d - \eta \sin d)$$
$$\eta' = \mu' \xi \sin d - d' \zeta \qquad (18.43)$$
$$\zeta' = -\mu' \xi \cos d + d' \eta.$$

The Newton–Raphson method is analytically the cleanest method but may require several iterations to yield a sufficiently precise result. A single application will only determine the time of the beginning of the eclipse to the nearest minute, unless the choice of T_0 happens to be rather close. The iteration process is fairly straightforward. The necessary new values of the Besselian elements are derived by writing

$$x_1 = x_1 + \Delta T x_1'$$
$$y_1 = y_1 + \Delta T y_1'$$
$$\mu = \mu + \Delta T \mu' \tag{18.44}$$
$$\sin d = \sin d + d' \cos d$$
$$\cos d = \cos d - d' \sin d.$$

Then (ξ, η, ζ) and hence F may be recomputed. In deriving the new values of $F'(T)$, it is sufficiently accurate to treat x_1' and y_1' as constants and so use the previous values. New values of (ξ', η', ζ') should, however, be obtained from (18.43).

We shall now consider an alternative method which in practice avoids the necessity of iteration. This will be applied instead to the problem of determining the times associated with the central phase of an eclipse, either total or annular. For definiteness, we shall refer to a total eclipse, and suppose now that T_0 is an estimated time at which the total phase begins for the observer O. Detailed tabulations of the central line and the limits of totality in the *Astronomical Almanac* allow T_0 to be estimated to within a few minutes.

From the second of equations (18.36) the condition that the observer lie on the umbral cone is

$$(\xi - x_1)^2 + (\eta - y_1)^2 = (l_2 - \zeta \tan f_2)^2. \tag{18.45}$$

This, of course, applies at the beginning and end of totality. For brevity this equation may be written as

$$u^2 + v^2 = L^2 \tag{18.46}$$

where

$$u = x_1 - \xi$$
$$v = y_1 - \eta \tag{18.47}$$
$$L = l_2 - \zeta \tan f_2.$$

The condition (18.46) is approximately satisfied at time T_0; suppose that it applies exactly at time $T_0 + \Delta T$. Let us, therefore, write

$$u = u_0 + \Delta T u'$$
$$v = v_0 + \Delta T v', \tag{18.48}$$

where the subscript zero indicates the value at time $T = T_0$.

The derivatives (u', v'), i.e. $(x_1' - \xi', y_1' - \eta')$, are derived exactly as before. In solving (18.46) second derivatives will be neglected and L is treated as a constant. The latter, in particular, is a stronger assumption than that made previously. The method is more powerful, however, since some second-order terms in ΔT are retained. In fact, substituting from (18.48) into (18.46) yields the quadratic equation

$$n^2 \, \Delta T^2 + 2D \, \Delta t + F(T_0) = 0, \tag{18.49}$$

where n^2 and D are a shorthand notation, viz.

$$n^2 = u'^2 + v'^2$$
$$D = u_0 u' + v_0 v'. \tag{18.50}$$

The quadratic equation is solved by the usual methods to give the two roots

$$\Delta T = -\frac{D}{n^2} \pm \frac{(D^2 - n^2 F(T_0))^{1/2}}{n^2}. \tag{18.51}$$

The two values of ΔT derived in this way correspond to the beginning and the end of totality. Since these two times are within a few minutes of each other, both roots are significant. The time of mid-eclipse is

$$T = T_0 - \frac{D}{n^2}, \tag{18.52}$$

and the duration of totality is δT where

$$\delta T = \frac{2(D^2 - n^2 F(T_0))^{1/2}}{n^2}. \tag{18.53}$$

The times calculated by this method will have an accuracy of about one second if the first estimate is within five minutes of the true time.

The method just described may also be used to derive the times of commencement and termination of the partial phase of an eclipse. The only change required is that L is now defined as

$$L = l_1 - \zeta \tan f_1. \tag{18.54}$$

Otherwise the formalism is identical. In practice T_0 will be chosen near one of the times of contact. Two roots ΔT may then be derived from (18.51). One of these, that with the plus sign, will be small and gives an accurate value for the time of contact. The other will give only a very approximate value for the time of the other contact. It should be recomputed from a new value of T_0 close to that contact.

The *Astronomical Almanac* provides eclipse data with universal time as the argument. This data is prepared with an assumed value of $TDT - UT$. If, in the event, the true value exceeds the estimated one by δT, then some slight adjustment is required. The tabulated universal times are reduced by δT, μ by $1.0027 \, \delta T$, and λ is increased by the latter amount. All other quantities in the tabulation are unaffected.

18.8   Eclipse magnitudes

The magnitude of a solar eclipse is defined as the fraction of the solar diameter that is obscured. Let us consider the magnitude of an eclipse for an observer O situated within the penumbral, but not the umbral, cone. In Fig. 18.9 the lines SS' and MM' represent the solar and lunar discs, and the approximation is made that the internal and external tangents both pass through the points M and M'. The points V_1 and V_2 represent the vertices of the penumbral and umbral cones. The diagram corresponds to the circumstances where the central eclipse, if possible, will be annular. The line PP' represents the plane through the observer parallel to the fundamental plane, that is, the plane $z = \zeta$.

Suppose now that the axis of the shadow intersects this plane in the point A, which is, therefore, the point (x_1, y_1, ζ). Let $OA = m$; it then follows from (18.47) that

$$m = (u^2 + v^2)^{1/2}. \tag{18.55}$$

As we shall wish to consider simultaneously the penumbral and umbral cones, we must distinguish between the two values of L used previously, writing

$$L_1 = l_1 - \zeta \tan f_1$$
$$L_2 = l_2 - \zeta \tan f_2. \tag{18.56}$$

The condition (18.46) investigated in the last section implies that $AP = L_1$ and $AU = L_2$.

Let the line OM produced intersect the solar disc in the point D. Then, by considering the similar triangles with a common vertex M, the magnitude of

---

Figure 18.9. Eclipse magnitudes. SS' and MM' represent the sun's and the moon's discs. The observer is at O and POP' represents the fundamental plane.

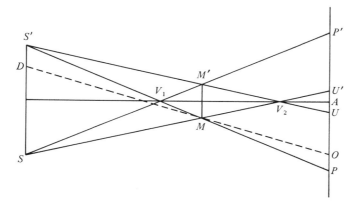

the eclipse for the observer O is seen to be M_1, where

$$M_1 = \frac{DS'}{SS'} = \frac{OP}{U'P}.$$

Now

$$OP = AP - AO = L_1 - m,$$

and

$$U'P = U'A + AP = L_2 + L_1.$$

Hence

$$M_1 = \frac{L_1 - m}{L_1 + L_2}. \tag{18.57}$$

This is the general formula for the magnitude of the eclipse during the partial phase. If, however, the observer were situated on the umbral cone at U or U', the magnitude would be M_2 where

$$M_2 = \frac{L_1 - L_2}{L_1 + L_2}, \tag{18.58}$$

for at these points $m = L_2$. In fact, this equation applies throughout the annular phase (cf. Problem 18.8). If the central phase is total, rather than annular, equation (18.57) still applies during the partial phase, but L_2 is then negative. When the observer lies on the umbral cone, $m = -L_2$ and, consequently, the magnitude is unity, as it remains throughout the total phase. On the other hand, the formula for M_2 is sometimes formally used to describe the magnitude of a total eclipse as it indicates how much larger the moon appears than the sun.

Only a privileged minority of observers will lie along the path of the central eclipse; other observers will only experience a partial phase. Such an observer may wish to determine the time of maximum phase at his position and the eclipse magnitude at that time. An estimated time T_0 is selected, probably that of mid-eclipse. Let $T_0 + \Delta T$ be the required time. It is necessary to maximize M_1 in equation (18.57). Since L_1 and L_2 vary slowly compared with m, it is usually sufficient to minimize m. From (18.55), the condition for this is that

$$uu' + vv' = 0. \tag{18.59}$$

The procedure is similar to that in the last section, i.e. put

$$u = u_0 + u'\,\Delta T$$
$$v = v_0 + v'\,\Delta T. \tag{18.60}$$

Then in the notation of (18.50), ΔT is given by

$$\Delta T = -\frac{D}{n^2}. \tag{18.61}$$

The minimum value of m may then be calculated from (18.55), using (18.60). This value is usually denoted by Δ and is found, after some reduction, to be

$$\Delta = |u_0 v' - v_0 u'|/n. \tag{18.62}$$

The greatest magnitude is then given as

$$M_1 = \frac{L_1 - \Delta}{L_1 + L_2}. \tag{18.63}$$

One final point should be noted. Suppose the partial eclipse is just beginning so that the observer is coincident with the point P. This point then has coordinates (ξ, η, ζ), while A has coordinates (x_1, y_1, ζ). The angle θ defined by

$$\tan \theta = \frac{u}{v}, \tag{18.64}$$

therefore, gives the inclination of PA to the y-axis, which points north. It follows from the geometry of the figure that $\theta + 180°$ is the position angle of the point S' just being obscured from the centre of the solar disc. The positional angles of the contacts may thus be derived from (18.64). The quadrant is simply decided by noting that, when the eclipse is starting, the moon appears west of the sun in the sky.

18.9   Occultations by the moon

An eclipse of the sun may be regarded as representative of a whole set of phenomena that are governed by similar geometrical principles. These include lunar occultations of stars, radio and X-ray sources, transits of the inferior planets across the solar disc, solar occultations of radio sources, and occultations of stars and radio sources by planets and even asteroids. The solar eclipse presents the most complicated problem since both the occulted body (the sun) and the occulting body (the moon) have large angular diameters, and the motion of the occulted body must be taken into account. The chapter will be concluded with a discussion of the much simpler problem of a lunar occultation of a star. Such phenomena are far more frequent than eclipses, even for naked-eye stars. As in a solar eclipse, local circumstances vary over the surface of the earth.

As before, a coordinate system (x, y, z) is adopted with the geocentric point E, as origin. The x–y plane is called the *fundamental* plane. The z-axis is now chosen in the apparent direction of the star (α, δ) which may be regarded as fixed for the duration of the occultation, cf. Fig. 18.10. The x- and y-axes are chosen in the same way as in a solar eclipse. Unit vectors in these directions are (cf. (18.27)), therefore,

$$\mathbf{I} = (-\sin \alpha, \cos \alpha, 0)$$
$$\mathbf{J} = (-\cos \alpha \sin \delta, -\sin \alpha \sin \delta, \cos \delta). \tag{18.65}$$

Now let the moon's geocentric place be (α_1, δ_1). Then the rectangular coordinates of the moon's centre M are (x_1, y_1, z_1) where, adapting (18.28),

$$x_1 = \operatorname{cosec} P_1 \cos \delta_1 \sin (\alpha_1 - \alpha)$$
$$y_1 = \operatorname{cosec} P_1 [\sin \delta_1 \cos \delta - \cos \delta_1 \sin \delta \cos (\alpha_1 - \alpha)].$$

(18.66)

A similar adaptation may be made to (18.38) to give the x- and y-coordinates of the observer O as

$$\xi = \rho \cos \phi \sin (h + \lambda)$$
$$\eta = \rho(\sin \phi \cos \delta - \cos \phi \sin \delta \cos (h + \lambda)),$$

(18.67)

where h is the Greenwich hour angle of the star. The rest of the notation is the same as before.

In this problem z-components are not needed. Since it is assumed that the star is a point source at an effectively infinite distance, there are no umbral and penumbral cones to be considered. The shadow may instead be represented by a cylinder whose axis is along the line $x = x_1$, $y = y_1$. The cross-section of this cylinder is a circle of radius k. The disappearance of the star behind the moon's disc is referred to as its *immersion*, and its reappearance is called its *emersion*. At either of these instants the observer lies on the surface of the shadow cylinder, and the condition for this is

$$(x_1 - \xi)^2 + (y_1 - \eta)^2 = k^2.$$

(18.68)

Let us now consider how a prediction may be made of the times of

---

Figure 18.10. The shadow cylinder at the lunar occultation of a star.

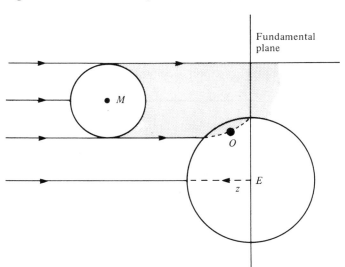

immersion and emersion for a particular observer. Besselian elements of the occultation are defined as follows:

T_c the universal time of conjunction of the star and the moon in right ascension

H the Greenwich hour angle of the star at this instant

Y the value of y_1 at time T_c

x', y' the hourly rates of change of x_1 and y_1

α, δ the apparent right ascension and declination of the star.

Notice that at the instant of time T_c, $x = 0$. For prediction purposes it is sufficiently accurate to treat x' and y' as constants during the occultation.

Suppose that T_0 is an estimated universal time of immersion, where $T_0 = T_c + \Delta T_c$. Then the Besselian elements allow (x_1, y_1) to be computed for time T_0. To derive (ξ, η) for this time it is necessary to know the value of h. This is given by

$$h = H + \Delta T_c', \tag{18.69}$$

where $\Delta T_c'$ is the interval ΔT_c expressed in sidereal rather than solar time.

Setting $u = x_1 - \xi$, $v = y_1 - \eta$, the condition (18.68) has exactly the form of (18.46) and is solved by the method discussed in section 18.7. The derivatives of x_1 and y_1 are Besselian elements, while differentiating (18.67) yields

$$\begin{aligned} \xi' &= \rho h' \cos \phi \cos (h + \lambda) \\ \eta' &= \rho h' \cos \phi \sin \delta \sin (h + \lambda), \end{aligned} \tag{18.70}$$

since δ is now a constant. The value of h' is also a constant, being 2π radians per sidereal day. Expressed in the required units of radians per mean solar hour, it is

$$h = 0.262\,516. \tag{18.71}$$

Predictions for stellar occultations are not published in the *Astronomical Almanac* as they are regarded as comparatively local phenomena. The *Explanatory Supplement* gives a list of publications which publish occultation predictions appertaining to different countries. The predictions take the form of the approximate times of immersion and emersion for selected sites together with coefficients which allow the approximate times to be derived in the geographical neighbourhood of these sites. The predictions for North America are published in *Sky and Telescope*, while the *Handbook of British Astronomical Association* provides details for the United Kingdom, New Zealand and parts of Australia.

The times obtained from these sources provide the value of T_0 necessary to derive a precise timing. Exact Besselian elements may be calculated from the lunar data in the *Astronomical Almanac*.

An observation of an occultation yields the universal time, usually of immersion, together with the position (ρ, λ, ϕ) of the observing site. The reduction of such observations is a rather specialized task which is the responsibility of the International Lunar Occultation Centre, Tokyo. The reduction of such observations provides the classical method of improving the lunar orbit, although this is now largely superseded by new methods involving radar and laser ranging.

Problems

18.1 Show that atmospheric refraction is sufficiently large to allow some part of the sun and one-half of the totally eclipsed moon to be simultaneously visible on the horizon.

18.2 Discuss the limits that permit solar and lunar eclipses in terms of the parameter ξ. In particular answer the following questions:
  (i) Can a central eclipse of the sun follow a fortnight after a total eclipse of the moon?
  (ii) If two solar eclipses occur a month apart, must they both be partial?
  (iii) Is it possible to have two total (as opposed to annular) eclipses of the sun within an interval of six months?

18.3 Describe what is meant by the Saros. Show that, in addition to its definition in equation (18.21), it may be equivalently described as the near equality of 223 synodic months and 242 draconic months. Show further that the Saros is also approximately 239 anomalistic months and comment on the significance of this additional commensurability.

18.4 Prove that the direction (a, d) of the axis of the moon's shadow at the time of a solar eclipse is approximately given by the equations

$$a = \alpha - \frac{b \sec \delta \cos \delta_1}{1 - b} (\alpha_1 - \alpha),$$

$$d = \delta - \frac{b}{1 - b} (\delta_1 - \delta),$$

where (α_1, δ_1) and (α, δ) are respectively the right ascension and declination of the moon and sun and b is the ratio of their geocentric distances.

18.5 The times of a partial eclipse of the sun are to be worked out below for a site at sea level with geodetic latitude $55° 54' 13''$ and longitude $4° 17' 28'' W$. As a preliminary, calculate the geocentric latitude and the geocentric distance of the observing site.

18.6 The eclipse referred to in the previous question occurred on 1984 May 30, and the *Astronomical Almanac* provides the following data for $18^h 50^m$ on that day:

$\sin d = 0.372\,654$

$\cos d = 0.927\,970$

$\mu = 103°.123\,86$.

Calculate the coordinates (ξ, η, ζ) of the observing site at that time.
    Given further that $\mu' = 0.261\,797$ and $d' = 0.000\,099$, both in radians per hour, calculate the derivatives (ξ', η', ζ').

18.7 The *Astronomical Almanac* gives the following additional information for the eclipse considered in the previous two questions

UT	x_1	y_1	l_1	$\tan f_1$
18.50	1.018\,929	0.544\,001	0.550\,819	0.004\,612
19.00	1.105\,765	0.566\,070	0.550\,791	0.004\,612

Select the necessary data from this table to compute the times at which the partial eclipse begins and ends at the observing site whose coordinates were given in Problem 18.5.
    (The time for the end of the eclipse will be accurate to about one second, but the time for the beginning of the eclipse is nearly 5 minutes in error. It may be used, however, to indicate a first approximation for a new calculation.)

18.8 Show that throughout the annular phase of a solar eclipse, the eclipse magnitude is given by

$$M_2 = \frac{L_1 - L_2}{L_1 + L_2}.$$

In what sense can this formula also be applied to total eclipses of the sun.

18.9 Prove that the coordinates (x_1, y_1) of the moon during the occultation of a star with equatorial coordinates (α, δ) are approximately given by

$$x_1 = \frac{15}{P_1} (\alpha_1 - \alpha) \cos \delta_1,$$

$$y_1 = \frac{\delta_1 - \delta}{P_1},$$

where (α_1, δ_1) and P_1 are respectively the right ascension, declination and horizontal parallax of the moon.

18.10 Estimate the approximate dates at which transits of Mercury and Venus over the solar disc can occur.

19

Binary stars

19.1   Introduction

The terms *'binary star'* or *'binary system'* are used exclusively to describe two stars which are sufficiently close together in space so that their motion is dominated by their mutual gravitational attraction. The apparent proximity of two stars in the sky is in itself no guarantee that the stars form a binary system. The near alignment of the two stars may be fortuitous, the stars being at quite different distances from the sun. In that case the pair are referred to as an *optical double*. Many such cases may be excluded if the proper motions of the two stars show no similarity. In a genuine binary system the two stars will have a common proper motion and in addition one may be observed to describe an orbit with respect to the other over a long period of time. Binary stars evidently provide examples of the two-body gravitational problem investigated in chapter 6.

What has just been described is, however, only one kind of binary system – the *visual binary*. In this case both stars are visible separately through the telescope, and the orbital motion may be studied from relative positional changes of the two stars. In many cases, however, the angular separation of the two stars is too small to allow them to be seen separately. The binary nature is then detected and studied by other means.

Spectroscopic binaries reveal their nature through a variable radial velocity determined from the Doppler shift in the spectrum of an apparently single star. In some cases two sets of spectral lines are observed and the Doppler shift in each is variable. Such a system is known as a *double line spectroscopic binary*. When, however, one star is considerably brighter than the other, the spectrum of the fainter star is lost in the continuum of the brighter one. This type of system is known as a *single-line spectroscopic binary*. The binary nature is still detectable from the variable Doppler shift in the observed spectrum.

Photometric (or eclipsing) binaries result from periodic eclipses taking place in close binary systems. The stars are too close to be resolved telescopically, but, due to the eclipses, the magnitude of the apparently single star varies in a regular manner. A fourth class of binary might be categorized as *polarimetric binaries*, where the binary nature of the system is indicated from periodically varying polarization parameters.

The classification of binary systems given above is basically one of observational techniques. There is no reason, in principle, however, why a particular system should not be a binary of several, or indeed all, of the above classes. In practice, a natural dichotomy exists, and the following comments are generally true, although there are some exceptions. Visual binaries are widely separated stars which are, therefore, moving comparatively slowly and are not easily studied spectroscopically. By contrast spectroscopic binaries are close systems which are moving rapidly and are, therefore, subject to large variable Doppler shifts. There are few stars that can be adequately studied as both visual and spectroscopic binaries. On the other hand, if the alignment of the orbital plane of a close binary is particularly favourable, the system will be an eclipsing binary as well. Both eclipsing and polarimetric binaries are best regarded as special cases of spectroscopic binaries, although polarimetric variation is possible in the absence of detectable spectroscopic variation when the true binary's orbit is nearly perpendicular to the line of sight. The interpretation of the polarimetry, however, requires detailed astrophysical modelling.

The study of binary stars provides the *only* means of measuring stellar masses. Suppose that the two stars have masses M_1 and M_2, and that their relative orbit is an ellipse of semimajor axis a and period T. Then the amended form of Kepler's third law cf. (6.31) yields

$$M_1 + M_2 = \frac{4\pi^2 a^3}{GT^2},$$

(19.1)

where G is the gravitational constant. A value for $4\pi^2/G$ may be inserted depending on the units that are employed.

In a visual binary the semimajor axis a is determined as an angle – in arc seconds. It is expressed in AU as a/π^*, where π^* is the system's parallax. The period T is conveniently expressed in years, and masses in terms of the mass of the sun. In these units (19.1) reduces to

$$M_1 + M_2 = \frac{a^3}{\pi^{*3}T^2}.$$

(19.2)

In a spectroscopic binary, however, the measurements involve velocities which are normally expressed in km s^{-1}. The information that is derived about the semimajor axis is, therefore, in kilometres. Moreover, the shorter orbital periods are usually expressed in days. In such units, the masses of the stars, in solar units, are given by

$$M_1 + M_2 = 3.985 \times 10^{-20} \frac{a^3}{T^2}.$$
(19.3)

The use of equation (19.1) only allows the sum of the masses to be determined. In order to establish how this mass is divided between the two stars it is necessary to locate the centre of mass of the system. It is not sufficient to study the relative motion of the two stars.

In this chapter we shall examine how the orbital elements of a binary system are derived from observations in the two main classes of binary – the visual binary and the spectroscopic binary. A detailed study of the photometric and polarimetric data is beyond the scope of the book.

19.2   Orbital elements of a binary system

Let A and B be two stars that form a visual binary system; the brighter star A is called the *primary*, the fainter star B is called the *secondary* or *companion*. Observation of the visual binary involves making measurements of the separation ρ of the two stars and the position angle $\theta = N\hat{A}B$. Figure 19.1 shows the binary system as it appears to the observer. Notice that the position angle is measured from the initial line AN which is in the direction of north, i.e. along the meridian of right ascension towards the north celestial pole. The position angle θ is measured in an anticlockwise direction as seen by the observer.

Due to the orbital motion of the binary, both the separation and the position angle will change perceptably over a long period of time. The change in the position angle will be monotonic. If the position angle

---

Figure 19.1. The appearance of a visual binary system with separation ρ and position angle θ.

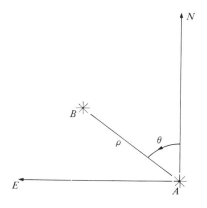

increases with time the orbital motion is said to be *direct*. If θ decreases with time the orbit is *retrograde*.

These observational conventions which should be strictly adhered to present us with a dilemma. The reference plane for the orbit is the plane through A perpendicular to the line of sight. This is taken as the x–y plane. If only visual binaries were to be considered, it would be natural to choose the initial line AN as the x-axis and AE as the y-axis – a convention that is usually followed in reducing observations of visual binaries. In that case the z-axis, completing the right-handed set, must be directed from the star to the observer. This is incompatible with what is required for spectroscopic binaries, where the outward radial direction must be regarded as the positive direction, since recessional radial velocities are always treated as positive. We, therefore, make the following definitions: the x-axis is chosen in the direction AE and the y-axis in the direction AN. This is compatible with the conventions followed for standard coordinates on astrographic plates, and the choice of directions in the fundamental plane for eclipses and occultations. Then the observed position (x, y) of the companion with respect to the primary is

$$x = \rho \sin \theta$$
$$y = \rho \cos \theta. \tag{19.4}$$

Figure 19.2 represents a celestial sphere centred on the primary star A. The great circle SLT is the plane perpendicular to the line of sight. The

---

Figure 19.2. The orbital elements of a binary system $\Omega = NL, i = VLS, \omega = LP'$.

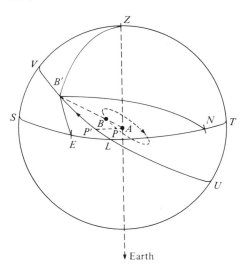

directions of north and east are indicated by the points N and E. Let Z be the pole of this great circle which marks the direction of the z-axis. The great circle ULV is the orbital plane. It intersects the reference plane in the two nodes. With the direction of rotation as indicated in the diagram, L is the ascending node. The arc NL is the *position angle of the ascending node* and will be denoted by Ω. Spherical angle VLS is the orbital inclination i. As usual, i is restricted to the first two quadrants; when $0 \leqslant i \leqslant 90°$, the orbit is direct.

Let P be the position of the companion at the time when it is closest to the primary. This point is called *periastron*. Furthermore let B be the position of the companion at a general time t. Now join AP and AB and produce these lines to cut the celestial sphere at P' and B' respectively. Let us denote the arc LP' by ω; this is the *argument of periastron*. Furthermore, if v is the true anomaly at time t, then $P'B' = v$, that is, $LB' = (v + \omega)$.

The remaining orbital elements of a binary system are the semimajor axis a, the orbital eccentricity e, the time of periastron τ, and the orbital period T. The last of these, or alternatively the mean motion, must be considered as an independent orbital element; it is required for insertion in (19.2) to determine the sum of the masses. There are in all, therefore, seven orbital elements which are $(a, e, i, \omega, \Omega, \tau, T)$.

The orbital elements have been defined for the orbit of the companion relative to the primary. This is what is normally studied for a visual binary. In any binary the centre of mass of the system will have a constant velocity with respect to the solar system barycentre. It will, therefore, exhibit a constant proper motion and a constant radial velocity. If, therefore, the positional variations of each star in a visual binary are measured with respect to the stellar background, the elements of each star's orbit about the centre of mass may be determined. Let a_1 and a_2 be the semimajor axes of these orbits, then

$$a_1 + a_2 = a$$
$$M_1 a_1 = M_2 a_2. \tag{19.5}$$

The rest of the orbital elements are essentially the same, as in the relative orbit.

In a spectroscopic binary the variations in the star's radial velocities are evidence of their motion with respect to the centre of mass. The orbital elements should, therefore, be defined on a celestial sphere centred on the system's barycentre rather than the primary. Apart from that there is no change in the definitions of orbital elements. The discussion of the relative orbit is easily adapted to the space-orbits of the two stars.

Let us refer again to Fig. 19.2. A system of rectangular Cartesian coordinates is adopted with the centre of the celestial sphere as origin and the axes in the directions of the points E, N and Z. As pointed out earlier, this provides a right-handed set. Let r be the radius vector at the time t, when the companion is at the point B. Then the coordinates of this star with respect to the primary are given by

$$x = r \cos EB'$$
$$y = r \cos NB' \qquad (19.6)$$
$$z = r \cos ZB'.$$

The three direction cosines in (19.6) may be derived by applying the cosine formula to spherical triangles $EB'L$, $NB'L$ and $ZB'L$. The side $B'L = v + \omega$ is common to all three triangles, and the remaining parts that are needed are easily read off as: $EL = 90° - \Omega$, $NL = \Omega$, $ZL = 90°$, $B'LE = i$, $B'LN = 180° - i$, $B'LZ = 90° - i$. Equations (19.6), therefore, yield the results

$$x = r[\cos (v + \omega) \sin \Omega + \sin (v + \omega) \cos \Omega \cos i]$$
$$y = r[\cos (v + \omega) \cos \Omega - \sin (v + \omega) \sin \Omega \cos i] \qquad (19.7)$$
$$z = r \sin (v + \omega) \sin i.$$

The radius vector r may itself be expressed in terms of the true anomaly, or if preferred the eccentric anomaly, using the dynamical equations

$$r = \frac{a(1 - e^2)}{1 + e \cos v} = a(1 - e \cos E). \qquad (19.8)$$

In a visual binary the variations of the coordinates x and y are investigated for the relative orbit. Purely positional observations provide no information about z. On the other hand, in a spectroscopic binary, the radial velocity measurements provide information of the rate of change of the z-coordinate of the star's position vector with respect to the mass centre. With such different, indeed complementary, observational data, quite different methods of investigation are required for the two kinds of binary. In each case, however, the orbital elements of the system are determined from the reduction of the observations.

19.3   The Thiele–Innes method

Observations of a visual binary must normally extend over a very long interval of time to allow the orbit to be derived. Orbital periods in excess of 100 years are quite common. Each observation provides the angular separation of the two stars, the position angle and, of course, the time to which these apply. Thus a series of values (ρ, θ, t), or equivalently (x, y, t) are obtained. Plotting the series of points (x, y) gives the *apparent orbit* in the x–y plane as shown in Fig. 19.3.

The true orbit of the companion about the primary is an ellipse with the primary A situated at one focus. The apparent orbit is the projection of this ellipse on to the x–y plane and will itself be an ellipse. The semimajor axis of the apparent orbit does not correspond to the semimajor axis of the true orbit, nor will A appear at a focus. The centre C of the apparent orbit, however, is the projection of the centre of the true ellipse.

The nature of the projection must be analysed to deduce the orbital elements of the true orbit. This can be done entirely from geometrical considerations. It is much more satisfactory, however, if dynamical principles are used to supplement such considerations. This is a feature of the Thiele–Innes method, the method most commonly used to reduce observations of a visual binary. The considerations given below are based on this method.

We shall assume that at least one complete orbit has been observed and that the period T has thereby been measured. Let us further suppose that the centre C of the apparent ellipse has been successfully identified. Let CA produced intersect the apparent ellipse in points P and Q as shown in Fig. 19.3. Now one salient feature of the projection process is that straight lines project into straight lines and length ratios along the line are conserved. So the points Q, C, A and P are respectively the projections of apastron, the centre, the focus and periastron. The time that the companion is at the point

Figure 19.3. The apparent relative orbit of a visual binary.

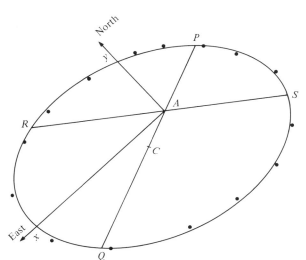

P is the time of periastron τ. Moreover, since length ratios are preserved, the eccentricity e is also derived as

$$e = \frac{CA}{CP}. \tag{19.9}$$

Since the time of periastron and the period are now known, the mean anomaly for any time t may be computed as

$$M = \frac{2\pi}{T}(t - \tau). \tag{19.10}$$

The eccentric anomaly may then be derived from the solution of Kepler's equation

$$M = E - e \sin E, \tag{19.11}$$

and indeed the true anomaly is then available from the condition that

$$\tan \frac{v}{2} = \left(\frac{1+e}{1-e}\right)^{1/2} \tan \frac{E}{2}. \tag{19.12}$$

Let us consider the point on the orbit when $v = 90°$. Then

$$\tan^2 \frac{E}{2} = \frac{1-e}{1+e}$$

which allows us to derive the eccentric anomaly. In fact,

$$\cos E = \frac{1 - \tan^2 \frac{1}{2}E}{1 + \tan^2 \frac{1}{2}E} = e. \tag{19.13}$$

The corresponding time may then be derived from equations (19.10) and (19.11) as

$$t = \tau + \frac{T}{2\pi}[\cos^{-1} e - e(1 - e^2)^{1/2}]. \tag{19.14}$$

Using this value of the time we may locate the point R on the apparent ellipse which corresponds to $v = 90°$. In Fig. 19.3, the line RS represents the projection of the *latus rectum* of the true ellipse.

Now let (x_1, y_1) and (x_2, y_2) be the coordinates of the points P and R respectively. These will, by the above argument, be known measured quantities. But the points P and R correspond to $v = 0$ and $v = 90°$. Moreover, the values of the radius vector corresponding to these two points are, from (19.8), respectively $a(1 - e)$ and $a(1 - e^2)$.

Therefore inserting these values into the first two equations of (19.7) yields the four results

$$X_1 \equiv \frac{x_1}{1-e} = a(\cos \omega \sin \Omega + \sin \omega \cos \Omega \cos i)$$

$$Y_1 \equiv \frac{y_1}{1-e} = a(\cos \omega \cos \Omega - \sin \omega \sin \Omega \cos i)$$

$$X_2 \equiv \frac{x_2}{1-e^2} = a(-\sin \omega \sin \Omega + \cos \omega \cos \Omega \cos i) \tag{19.15}$$

$$Y_2 \equiv \frac{y_2}{1-e^2} = a(-\sin \omega \cos \Omega - \cos \omega \sin \Omega \cos i).$$

The quantities X_1, Y_1, X_2, Y_2 will be referred to as the Thiele–Innes constants. They are now known quantities and the orbital elements on the right-hand side of (19.15) may be expressed in terms of them.

Combining equations (19.15) gives the following results:

$$\begin{aligned}
X_1 - Y_2 &= a \sin (\omega + \Omega)(1 + \cos i) \\
X_1 + Y_2 &= -a \sin (\omega - \Omega)(1 - \cos i) \\
X_2 - Y_1 &= -a \cos (\omega - \Omega)(1 - \cos i) \\
X_2 + Y_1 &= a \cos (\omega + \Omega)(1 + \cos i).
\end{aligned} \tag{19.16}$$

Consequently we find that

$$\tan (\omega + \Omega) = \frac{X_1 - Y_2}{X_2 + Y_1}$$

$$\tan (\omega - \Omega) = \frac{X_1 + Y_2}{X_2 - Y_1}. \tag{19.17}$$

Hence $(\omega + \Omega)$ and $(\omega - \Omega)$ may be determined. In fact this determination may be made unambiguously, for the signs of the cosines and sines of both angles may be derived from (19.16) by noting that the factors $(1 + \cos i)$ and $(1 - \cos i)$ are necessarily positive. The values of ω and Ω themselves can then be found, but, although their sum and difference are uniquely determined, ω and Ω themselves share an ambiguity. It is evident from (19.16) that adding 180° to both ω and Ω will not alter the right-hand sides. This ambiguity is removed artificially by restricting the position angle of the node to the range $0 \leqslant \Omega \leqslant 180°$. With this convention ω is uniquely defined.

Next, combining the first two equations of (19.16) gives

$$\tan^2 \frac{i}{2} = \frac{1 - \cos i}{1 + \cos i} = -\frac{(X_1 + Y_2) \sin (\omega + \Omega)}{(X_1 - Y_2) \sin (\omega - \Omega)}, \tag{19.18}$$

which yields i. Finally a may be determined from any of the original equations (19.16).

The determination of the orbital elements is now complete, except for the single ambiguity in Ω. This means that, although the two nodes of the orbit may be identified, it is not known which is the ascending node from purely positional measurements. This uncertainty does not affect the determination of the stellar masses and may be removed if spectroscopic data is also available.

The essence of the Thiele–Innes method is that the times corresponding to the different points in the orbit are used in the derivation of orbital elements. Once T, τ and e are derived, the true anomaly may be computed for any point. In fact we only used two points, P and R, in determining the Thiele–Innes constants, but any points in the apparent orbit may be used. Expanding $\cos (v + \omega)$ and $\sin (v + \omega)$ in (19.7) leads by (19.15) to the results

$$x = \frac{r}{a} \cos v X_1 + \frac{r}{a} \sin v X_2$$

$$y = \frac{r}{a} \cos v Y_1 + \frac{r}{a} \sin v Y_2. \tag{19.19}$$

Using equation (19.8), this may be rewritten as

$$\cos v X_1 + \sin v X_2 = \frac{(1 - e \cos v)}{(1 - e^2)} x$$

$$\cos v Y_1 + \sin v Y_2 = \frac{(1 + e \cos v)}{(1 - e^2)} y. \tag{19.20}$$

So for any point (x, y) on the apparent orbit, the value of v may be computed from its time t using equations (19.10) to (19.12). Equations (19.20) then provide equations of condition on the Thiele–Innes constants which may then be derived by the method of least squares. The orbital elements are calculated as before.

19.4   The Lehmann–Filhés method

We must now examine a method for determining orbital elements in a spectroscopic binary. For definiteness a single-line spectroscopic binary will be considered; the principles for double-line stars are the same. The observations now yield the radial velocity V_r of this star as a function of time t. We shall assume that corrections have already been made to V_r so that this is the barycentric radial velocity of section 11.4, and that the time t has also been corrected by applying (10.23) so that this, too, refers to the solar system barycentre.

Spectroscopic binaries generally have fairly short periods, and consequently the measurements will extend over many orbital periods. This has the advantage that the period T will be determined with high accuracy. Suppose next that V_0 is the long-term average value of the radial velocity or, what is the same thing, the average value of V_r over exactly one orbital period. This quantity V_0 is the radial velocity of the centre of mass of the binary system. The oscillation of V_r about this mean value is due to the radial component of the star's orbital motion about the centre of mass. Since the z-axis is the line of sight, we may write

$$V_r = V_0 + \frac{dz}{dt}. \tag{19.21}$$

Note that in (19.21), and in the analysis that follows, the origin of the coordinate system is the binary's centre of mass. We are examining the orbit of one star with respect to this point, rather than the relative orbit. Consequently, the third equation of (19.7) is modified to give

$$z = \frac{a_1}{a} r \sin (v + \omega) \sin i, \tag{19.22}$$

where a_1 is the semimajor axis of the star's orbit about the centre of mass, but r and a refer to the relative orbit as before.

Differentiating (19.22) will yield

$$\frac{dz}{dt} = \frac{a_1}{a} \sin i \left[\frac{dr}{dt} \sin (v + \omega) + r \frac{dv}{dt} \cos (v + \omega) \right]. \tag{19.23}$$

The square brackets may be simplified by using results from two-body orbital theory. For example, dr/dt and $r\,dv/dt$ are just the radial and transverse orbital velocities considered in section 6.6. So using equations (6.41) to (6.43) yields

$$\frac{dz}{dt} = \frac{ha_1 \sin i}{a^2(1 - e^2)} (\cos (v + \omega) + e \cos \omega). \tag{19.24}$$

The constant h is expressed in terms of the mean motion n by the means of the equations

$$h^2 = \mu a(1 - e^2)$$
$$n^2 a^3 = \mu.$$

This gives the final result as

$$V_r - V_0 = \frac{dz}{dt} = K_1 [\cos (v + \omega) + e \cos \omega], \tag{19.25}$$

where K_1 is the constant defined as

$$K_1 = \frac{na_1 \sin i}{(1 - e^2)^{1/2}}. \tag{19.26}$$

Let us suppose that the observed radial velocities have been corrected as described above and that the mean value V_0 has been extracted. The data may then be displayed as a *velocity curve* as shown in Fig. 19.4. The form of this curve is predicted by equation (19.25). If the abscissa were the true anomaly instead of time, it would be a simple sine curve. Since the true anomaly does not increase uniformly with time, there may, however, be considerable distortion. The choice of t-axis as $V_r = V_0$ will mean that the area under the segment BCD will equal the area above the segment DEB'.

Let α and β denote the maximum and minimum values of dz/dt in absolute value, so that $\alpha = CY$ and $\beta = ZE$. The maximum at C corresponds to $\cos(v + \omega)$ having its maximum value of unity in (19.25). Consequently $v = -\omega$ at the point C. Similarly, $v = 180° - \omega$ at the point E. Applying (19.25) at these two points provides the results that

$$\alpha = K_1(1 + e \cos \omega)$$
$$\beta = K_1(1 - e \cos \omega). \tag{19.27}$$

We may, therefore, solve these two simultaneous equations to derive

$$K_1 = \tfrac{1}{2}(\alpha + \beta), \tag{19.28}$$

$$e \cos \omega = \frac{1}{2K_1}(\alpha - \beta) = \frac{\alpha - \beta}{\alpha + \beta}. \tag{19.29}$$

The constants K_1 and $e \cos \omega$ are then satisfactorily determined. If a determination can be made of $e \sin \omega$ as well, it would be possible to derive e and ω separately. In order to achieve this, it is necessary to make use of other features of the velocity curve besides its maximum and minimum values. The Lehmann–Filhés method utilizes the two areas BCY and CDY shown shaded in Fig. 19.4.

Figure 19.4. The velocity curve of a single-line spectroscopic binary.

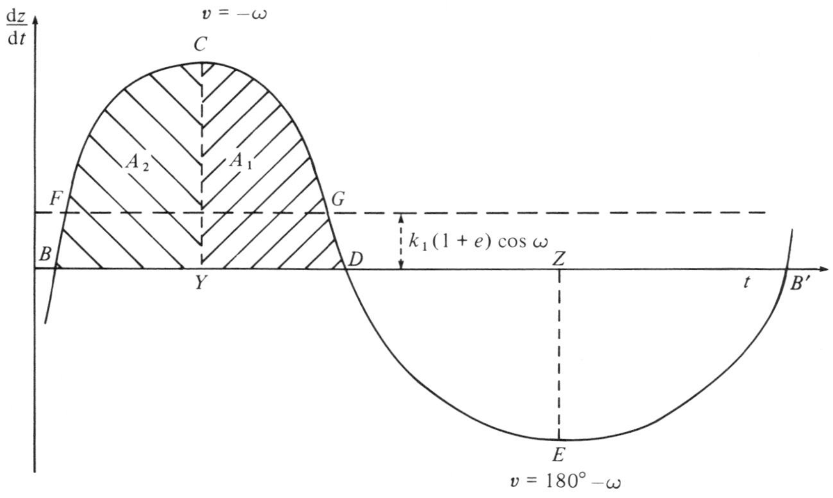

Let us examine the area CDY which we denote by A_1. Clearly,

$$A_1 = \int_C^D \frac{dz}{dt} \, dz = z_D - z_C.$$

Now at the point C, $(v+\omega) = 0$ which implies by (19.22) that $z_C = 0$. A similar argument is applied to the area BCY, denoted by A_2. We conclude that

$$A_1 = z_D$$
$$A_2 = -z_B. \tag{19.30}$$

Since $dz/dt = 0$ at the points D and B, it is easy to see that the values of z in (19.30) are the maximum and minimum values in the orbit.

By (19.22) and (19.8) it follows that for any point in the orbit

$$z = \frac{a_1(1-e^2)\sin i \sin (v+\omega)}{(1+e\cos v)}.$$

Remembering that we seek a value of $e \sin \omega$ it is convenient to replace $\cos v$ with $\cos \overline{(v+\omega - \omega)}$ in the denominator to derive

$$z = \frac{a_1(1-e^2)\sin i \sin (v+\omega)}{[1+e\cos \omega \cos (v+\omega)+e\sin \omega \sin (v+\omega)]}. \tag{19.31}$$

Now at the two points in question, $dz/dt = 0$ and so, by (19.25),

$$\cos (v+\omega) = -e\cos \omega$$
$$\sin (v+\omega) = \pm (1-e^2 \cos^2 \omega)^{1/2}. \tag{19.32}$$

Referring to the original formula for z (19.22), it is clear that the plus sign corresponds to the maximum and the minus sign to the minimum. So combining equations (19.30), (19.31) and (19.32) leads to the result that

$$\frac{A_1}{A_2} = \frac{1-e^2\cos^2\omega - e\sin\omega(1-e^2\cos^2\omega)^{1/2}}{1-e^2\cos^2\omega + e\sin\omega(1-e^2\cos^2\omega)^{1/2}}. \tag{19.33}$$

Following cross-multiplication and some rearrangement, this gives

$$e\sin\omega = +\left(\frac{A_2-A_1}{A_2+A_1}\right)(1-e^2\cos^2\omega)^{1/2}. \tag{19.34}$$

Now the areas A_1 and A_2 may be measured from the velocity curve. Moreover, the value of $e\cos\omega$ is already determined in (19.29). Consequently, the value of $e\sin\omega$ may be obtained. In fact,

$$e\sin\omega = +\left(\frac{A_2-A_1}{A_2+A_1}\right)\frac{2\sqrt{\alpha\beta}}{\alpha+\beta}. \tag{19.35}$$

As both $e\cos\omega$ and $e\sin\omega$ are now known, a unique determination is possible for both the orbital eccentricity and the argument of periastron.

Once the value of e is known, it is a simple matter to derive the time of periastron τ. At this time $v=0$, and so by (19.25),

$$\frac{dz}{dt}=K_1(1+e)\cos\omega.$$

This value of the ordinate will define two points on the velocity curve, say F and G in Fig. 19.4. Now the true anomaly, by definition, increases monotonically with time. Since v is already known at the two points C and E, it is easy to decide which of F and G corresponds to $v=0$ and to periastron. The element τ is the abscissa of this point.

The quantity K_1 is often quoted as an orbital element of a spectroscopic binary. By (19.26), however, we have

$$a_1\sin i=\frac{K_1(1-e^2)^{1/2}}{n}.$$

Now the mean motion $n=2\pi/T$. A little care is required with the units since K_1 will usually be in km s^{-1}, while T is conveniently expressed in days. So

$$a_1\sin i=\frac{86400\,TK_1(1-e^2)^{1/2}}{2\pi}.$$

Using (19.28) this reduces to

$$a_1\sin i=\frac{21600\,T}{\pi}(\alpha+\beta)(1-e^2)^{1/2}. \tag{19.36}$$

This allows the quantity $a_1\sin i$ to be derived in kilometres.

To sum up, the analysis of the velocity curve of a single-line spectroscopic binary enables the following orbital elements to be derived ($a_1\sin i, e, \tau, \omega, T$). The elements a_1 and i cannot be decoupled from purely spectroscopic measurements, and, hardly surprisingly, we obtain no knowledge at all of Ω, which does not appear in the formula for z.

19.5   The masses of visual binaries

The last two sections have been concerned with methods of determining the orbital elements for a visual and spectroscopic binary. Although the methods may be varied the end product is always the same. The indeterminacies encountered are inherent in the two types of system. In a visual binary the only indeterminacy is in Ω and this does not affect the determination of the stars' masses.

The Thiele–Innes method, described in section 19.5, will yield the semimajor axis of the relative orbit in the units in which the original measurements of the separation were made. The method, therefore, normally yields T in years and a in arc seconds. The total mass of the system is expressed in solar units by equation (19.2). Writing π for π^*, this is

$$M_1 + M_2 = \frac{a^3}{\pi^3 T^2}.$$

The use of this formula presupposes that the star's trigonometric parallax can be and has been measured. While there are nearby binaries for which this is the case, it is far from being invariably true.

If the binary is at too great a distance to allow an accurate determination of π, all need not be lost. The argument may be inverted as follows. The masses of the two stars are estimated from their spectral classification. These estimates are themselves founded on more definitive mass determinations of stars of similar spectral type in more favourable examples of binaries. Equation (19.2) can then be rewritten to give the binary's parallax as

$$\pi = \frac{a}{T^{2/3}(M_1 + M_2)^{1/3}}. \tag{19.37}$$

The parallax determined in this way is known as the *dynamical parallax*. Since only the cube root of the total mass is involved in (19.37), the dynamical parallax is rather insensitive to the mass estimate that has been made. The dynamical parallax is likely, therefore, to give a useful determination of the star's distance. This, in turn, may be used to calibrate the luminosities of stars of the spectral type involved.

Equation (19.2) can only yield the total mass of the binary system. No more can be expected of an analysis of the relative orbit. As explained in section 19.2, it is only possible to determine how the total is divided between the two components if the centre of mass of the system can be identified. This requires that the stars' motion with respect to the stellar background be investigated.

In Fig. 19.5, the points A and B represent the primary and companion in a visual binary system, and G is their centre of mass. The position angle and separation of B with respect to A are measured in the usual way – ρ being expressed in arc seconds. The separation of G from A is easily found to be

$$GA = \frac{M_2}{M_1} \, GB = \frac{M_2 \rho}{M_1 + M_2}. \tag{19.38}$$

Suppose now that the standard coordinates (ξ, η) of the primary are measured from a series of photographic plates taken at different times. Then if (ξ_G, η_G) represent the standard coordinates of the point G

$$\xi = \xi_G - \frac{M_2}{M_1 + M_2} \, \rho \, \sin \theta \sin 1''$$

$$\eta = \eta_G - \frac{M_2}{M_1 + M_2} \, \rho \, \cos \theta \sin 1''. \tag{19.39}$$

The coordinates (ξ_G, η_G) are, of course, initially unknown, but the point G will have a uniform proper motion. The standard coordinates of A will, therefore, have the form

$$\xi = \xi_0 + \mu_\xi t - \frac{M_2}{M_1 + M_2} \rho \sin\theta \sin 1''$$

$$\eta = \eta_0 + \mu_\eta t - \frac{M_2}{M_1 + M_2} \rho \cos\theta \sin 1''. \tag{19.40}$$

Astrographic observations of the primary over a long period of time are fitted to the predicted form of (19.40) by the method of least squares. This leads to a determination of the unknowns (ξ_0, η_0), (μ_ξ, μ_η) and $M_2/(M_1 + M_2)$. As the sum of the masses is already known from (19.2), the two stellar masses are individually determined.

Equation (19.40) may be applied to an apparently single star, whose binary nature is indicated solely by periodic variations in its proper motion. Such objects are known as *astrometric binaries*. The analysis of the proper motion will lead to a determination of $M_2\rho/(M_1 + M_2)$ and θ as functions of time. The Thiele–Innes method may then be applied to this data to yield the period T and the quantity $M_2a/(M_1 + M_2)$ which is simply a_1. Now from equations (19.2) and (19.5) we have the following two relationships

$$M_1 a_1 = M_2 a_2$$

$$M_1 + M_2 = \frac{(a_1 + a_2)^3}{\pi^3 T^2}. \tag{19.41}$$

Assuming that the parallax of the star is measurable, there are three unknown quantities in equations (19.41), the two masses and a_2. The masses

Figure 19.5

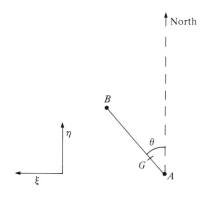

cannot, therefore, be derived with certainty. What we can determine is the *mass function* $f(M)$ defined, by eliminating a_2 from equations (19.41), as

$$f(M) = \frac{M_2{}^3}{(M_1 + M_2)^2} = \frac{a_1{}^3}{\pi^3 T^2}. \tag{19.42}$$

A similar mass function is defined for single-line spectroscopic binaries.

An astrometric binary establishes the existence of an unseen companion. The method of investigation was pioneered by Bessel in the mid 19th century. He showed that Sirius had a companion which could not (at that time) be observed, but which was of stellar mass. The companion, Sirius *B*, was discovered telescopically some decades later; its apparent faintness is a consequence of its very small size as a white dwarf.

Several nearby stars are astrometric binaries. In a number of cases, most notably Barnard's star, the masses of the unseen companion are so small that they are probably planets rather than companion stars. It is easily seen from (19.42) that, if $M_2 \ll M_1$,

$$M_2 \simeq \left(\frac{a_1{}^3 M_1{}^2}{\pi^3 T^2} \right)^{1/3}. \tag{19.43}$$

If an estimate is made for M_1 from spectroscopic evidence, the mass of the unseen companion may be derived from this equation, or more accurately from (19.42) itself.

Astrometric binaries provide a means of establishing the existence of planetary systems about other stars. The periodic variations in the proper motion of the primary star when perturbed by a planetary mass are so minute that some astronomers have suspended judgement on this matter. It is expected that the high-precision astrometry of the space telescope will settle the question. It must be stressed that equation (19.40) is an oversimplification. The extremely high precision needed in investigating astrometric binaries requires the inclusion of second- and, indeed, third-order terms in the proper motion. The proper motion components cannot be treated as constants, but the intrinsic variations considered in section 11.2 must be included.

19.6   Spectroscopic binary masses

When both spectra are visible in a spectroscopic binary, the two velocity curves may be analysed separately. The orbital elements of the two stars about their centre of mass are identical with the exceptions of a and ω, for which

$$M_1 a_1 = M_2 a_2$$
$$\omega_1 = \omega_2 + 180°. \tag{19.44}$$

It follows from equations (19.25) and (19.26) that the two velocity curves have the form

$$\frac{dz_1}{dt} = K_1[\cos(v+\omega) + e \cos\omega]$$

$$\frac{dz_2}{dt} = -K_2[\cos(v+\omega) + e \cos\omega].$$

(19.45)

When the two curves are analysed by the methods of section 19.4, $a_1 \sin i$ and $a_2 \sin i$ will be obtained.

In this case there is no difficulty in deriving the ratio of the stars' masses, since

$$\frac{M_1}{M_2} = \frac{a_2 \sin i}{a_1 \sin i} = \frac{K_2}{K_1}.$$

(19.46)

The total mass of the system cannot, however, be determined from spectroscopic data alone. The best that can be achieved, using (19.3) is

$$(M_1 + M_2)\sin^3 i = 3.985 \times 10^{-20}\frac{(a_1 \sin i + a_2 \sin i)^3}{T^2},$$

(19.47)

where a_1 and a_2 are expressed in km and T is in days.

The presence of $\sin^3 i$ in (19.47) means the stellar masses cannot be determined for an individual spectroscopic binary, even when both spectra are available. The determination of $(M_1 + M_2)\sin^3 i$ is still useful, however, for statistical investigations. Suppose that a large number N of spectroscopic binaries of similar spectral type are studied and the quantity $(M_1 + M_2)\sin^3 i$ is determined in each case. The assumption is made that the masses are in fact identical and that the orbital planes are distributed at random. The mean value of $\sin^3 i$ is then taken to be $\frac{2}{3}$ and one derives the binary mass M as

$$M = \frac{3}{2N}\sum(M_1 + M_2)\sin^3 i.$$

(19.48)

With single-line spectroscopic binaries, not even this much information is available. As with an astrometric binary, one has to be content with the determination of a mass function which, in this case will involve the orbital inclination as well. The mass function $f(M)$ is now defined as

$$f(M) = \frac{M_2{}^3 \sin^3 i}{(M_1 + M_2)^2}.$$

(19.49)

By a similar argument to that given in the previous section, one finds that the mass function may be evaluated as

$$f(M) = 3.985 \times 10^{-20}\frac{(a_1 \sin i)^3}{T^2}.$$

(19.50)

The mass function is, of course, in solar units.

The velocity curve will give no indication at all of the orbital inclination, and this quantity is generally unknown for a spectroscopic binary. The situation is rather different, however, if the binary undergoes eclipses. There is then additional data in the form of a light curve, a plot of the star's magnitude against time. The rather artificial form of this graph (Fig. 19.6) usually allows such eclipsing binaries to be distinguished from genuinely variable stars. Figure 19.6 shows an idealization of such a light curve. There are generally two eclipses per orbit, the two components, interchanging eclipsing roles. In Fig. 19.6 the segment of the light curve BC coincides with the total eclipse of a small hot star by a larger, cooler one. The segment FG is the corresponding annular eclipse when the small star is nearer the earth. The sloping segments are the accompanying partial phases, and between eclipses the light level is at a steady maximum. If the stars are not greatly dissimilar in size the occurrence of such total and annular eclipses requires a near perfect alignment in that the line of sight lies very close to the plane of the orbit. Let us examine the requirement for eclipses in slightly more detail.

The projected separation q between the centres of the two stars is simply

$$q = (x^2 + y^2)^{1/2}. \tag{19.51}$$

Using equation (19.7), this may be expressed as

$$q = r[\cos^2 (v+\omega) + \sin^2 (v+\omega) \cos^2 i]^{1/2}$$

i.e. $\hspace{8cm}$ (19.52)

$$q = r[1 - \sin^2 (v+\omega) \sin^2 i]^{1/2}.$$

Figure 19.6

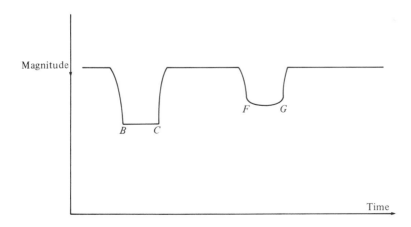

The variation of q is awkward to treat analytically due to its dependence on the eccentricity. It is evident from (19.52), however, that if eclipses are to take place it will be near the points $v + \omega = \pm 90°$. The conditions for partial and central eclipses to take place are respectively

$$q < R_1 + R_2$$
$$q < |R_1 - R_2|,$$

(19.53)

where R_1 and R_2 are the radii of the two stars. For a circular orbit, it follows from (19.52) that these conditions reduce to

$$|\cos i| < \frac{R_1 + R_2}{a}$$

and

(19.54)

$$|\cos i| < \frac{|R_1 - R_2|}{a}.$$

The analysis of the light curve is rather complex and will not be considered here. To a first approximation, the diminution in the light received is proportional to the eclipsed area of the more distant star; even the expression for this is not simple. The variation in brightness, however, over a stellar disc (limb darkening) needs to be taken into account. Furthermore, as equation (19.54) indicates, eclipses are more likely to occur if the stellar radii are an appreciable fraction of the semimajor axis of the orbit. For such close binary systems, tidal, and indeed rotational effects, distort the stars' shapes so that they are decidedly non-spherical. Moreover, reflection effects can be important. All in all there is a very complicated astrophysical situation to be disentangled from the observations.

The light curve gives no information about the absolute size of the orbit of the stars; it is sensitive to their relative dimensions. The quantities that can be determined from an analysis of the light curve are R_1/a, R_2/a and i. This can produce a valuable supplement to spectroscopic data. Let us consider the optimum case of an eclipsing binary which is also a double-line spectroscopic binary. The spectroscopic data provides, among other parameters, the quantities $a_1 \sin i$, $a_2 \sin i$. The ratio of the masses is, therefore, known. The light curve yields the inclination i which may be inserted in (19.47) to give the total mass. The individual masses may then be calculated. Moreover, the semimajor axis is also available as $a = a_1 + a_2$ and the ratios R_1/a, R_2/a obtained from the light curve may be converted into values for the star's radii. Thus the masses and sizes of both stars are obtained.

19.7   The binary pulsar

Visual, spectroscopic and eclipsing binaries are the three classical types of binary star that have been studied for many years and have provided information about stellar masses. More recently other more exotic binary systems have been discovered and studied, such as X-ray binaries and the famous black-hole candidates. We shall end the chapter with a discussion of one such exotic object, the binary pulsar, but first some general comments.

The end-points of stellar evolution are compact objects – white dwarfs, neutron stars or black holes. White dwarfs appear to be the most common, but they are intrinsically faint. They are only observable within the solar neighbourhood and certainly undetectable beyond a kiloparsec. Neutron stars are intrinsically even fainter and, in general, they are not detected optically at all. They are, however, recognizable as pulsars and are found throughout the galaxy and even in the closest external galaxies. There must be many neutron stars for which the pulsar emission has died out and which are, therefore, quite unobservable. They could only be detected if they were members of binary systems. The same is true of black holes. By definition, no radiation can escape from the black hole itself. Certainly there may be evidence of violent activity in its vicinity caused by its intense gravitational field, but the quantification of that field is only possible if the black hole is in a binary system.

The strongest black hole candidate is the X-ray source Cygnus X1. This is associated with the 9th magnitude star HDE 226868 which is a single-line spectroscopic binary. The X-ray emission clearly arises from the immediate vicinity of the unseen companion, and short time-scale fluctuations in this emission indicate that the companion must be a compact object. Analysis of the single-line spectroscopic binary leads to a mass function which is rather small, about one-quarter of a solar mass. Equation (19.49) then provides the condition.

$$\frac{M_2{}^3}{(M_1+M_2)^2} \simeq 0.25 \, \mathrm{cosec}^3 \, i. \tag{19.55}$$

Now the mass function always gives an underestimate of the mass of the unseen companion, often, as in this case, a gross underestimate. For a realistic estimate of M_2 we must insert a value for M_1 in (19.55). The star that is observed has the spectral characteristics of a massive supergiant. Estimates of M_1 are, therefore, made in the range of 20–30 $M_\odot$. A lower bound for M_2 may be deduced from (19.55) by solving the algebraic equation obtained by setting $M_1 = 20$ and $i = 90°$. It is then found that

$$M_2 \geqslant 5.45 \, M_\odot. \tag{19.56}$$

Now the theory of stellar structure predicts maximum masses for both white dwarfs and neutron stars. While the neutron star maximum is not precisely known, it is generally agreed that both maxima are less than the

Table 19.1. *Orbital elements of*
PSR 1913+16

$a_1 \sin i$	7.02×10^5 km
e	0.617
ω	179°
T	$0^d.323$

right-hand side of (19.56). It is then deduced that the unseen companion of HDE 226868 is a black hole.

Some astronomers would argue that there are uncertainties in the foregoing argument concerning Cygnus XI. A much more clear-cut example of a binary containing compact objects is provided by the binary pulsar PSR 1913+16. The orbital elements of this object are measured with extraordinarily high precision.

The binary pulsar was discovered in 1974. It was almost immediately possible to determine the orbital elements by using the pulse repetition frequency as a spectral line to measure Doppler shifts. This can be done in the first instance following purely classical procedures. The resulting orbital elements are given to low precision in Table 19.1. A mass function may then be computed by using equation (19.50). It is found that

$$f(M) = 0.13. \tag{19.57}$$

What makes the binary pulsar particularly interesting is the fact that relativistic effects become apparent. Most important of these is the advance of the apse line, the increase in the argument of periastron. Adapting equation (6.81), one finds that the advance of periastron in radians per revolution is given by

$$\Delta\omega = \frac{6\pi\mu}{c^2 a(1-e^2)}, \tag{19.58}$$

where $\mu = G(M_1 + M_2)$.

The unknown a may be eliminated from this equation by using the condition that $n^2 a^3 = \mu$. Using the known values of the period and the eccentricity, it is then found that

$$\dot{\omega} = 2°.1 \, (M_1 + M_2)^{2/3} \, \text{yr}^{-1} \tag{19.59}$$

where M_1, M_2 are in solar units.

Equation (19.59) gives the predicted relativistic advance of periastron, which is certainly large enough to be measurable. Before comparing this with the observed rate ($4°.226 \, \text{yr}^{-1}$), it is necessary to rule out any advance due to purely Newtonian causes. The nature of the companion to the pulsar is unknown, but it cannot be a normal star. Notice that the value of $a_1 \sin i$

that is derived is about a solar radius. Consequently, any normal star would suffer severe tidal distortion due to its proximity to the pulsar, and this would produce a periastron shift considerably greater than that observed. The companion must, therefore, be a compact object, whether a white dwarf, neutron star or black hole. For all of these it is found that the periastron shift due to modification of the Keplerian orbit arising from tidal distortion is negligible. We make the assumption therefore that the observed periastron shift is entirely relativistic, although it is just possible that there could be a small Newtonian contribution if the companion is a rapidly rotating white dwarf.

Giving $\dot{\omega}$ the observed value of $4°.226$ yr^{-1} in equation (19.59) allows the total mass to be determined as

$$M_1 + M_2 = 2.85 \, M_{\odot}. \tag{19.60}$$

How this mass is divided between the pulsar and its companion depends on the orbital inclination. Equations (19.57) and (19.60) may be combined to give

$$M_2 \sin i = 1.02 \, M_{\odot}. \tag{19.61}$$

The observations that are made of the pulsar signals are not in practice measurements of a Doppler shift but consist of recording the arrival times of the pulses over an extended period of time. Assumed values for the orbital parameters and the pulsar period are inserted in an elaborate timing formula to predict the arrival times of particular pulses. The differences between the predicted and observed arrival times are then used to improve the parameters by the method of least squares. This type of analysis derived not only the orbital elements and the pulsar period but their rates of change as well. A useful account of these methods, and of the binary pulsar in general, will be found in Will (1981). Here we shall be content with establishing the general form of the timing formula.

Suppose that a pulse is emitted at coordinate time t and observed at time t_0. Then using a quasi-Cartesian coordinate system with origin at the centre of mass, equation (11.52) yields in this application

$$t_0 = t_e + \frac{\rho}{c} + \frac{2m}{c} \ln \left(\frac{r + r_0 + \rho}{r + r_0 - \rho} \right). \tag{19.62}$$

Here r_0 is the distance of the centre of mass and ρ the distance of the pulsar from the observer at the time of emission, while r is, as before, the separation of the two bodies. The parameter m is half the Schwarzschild radius for the companion. The effect of the pulsar's own gravitational field on the pulse timing will be constant and can be ignored. Since $r_0 \gg r$, we have, to sufficient accuracy $\rho = r_0 + z$, and so (19.62) reduces to

$$c(t_0 - t_e) = r_0 + z + \frac{2GM_2}{c^2} \ln\left(\frac{2r_0}{r-z}\right).$$ (19.63)

This formula gives the pulse's arrival time for a barycentric observer, for we are assuming that the observed arrival times have already been reduced to that point. Consequently t_0 is the proper time for the barycentric observer, but t_e is not the proper time s_e for the pulsar. Using (11.47) one derives with sufficient accuracy

$$\frac{dt_e}{ds_e} = 1 + \frac{m}{r} + \frac{1}{2}\frac{V_1^2}{c^2}.$$ (19.64)

A little care is required in interpreting this equation; the parameter $m = GM_2/c^2$ as before, but V_1 is the pulsar's velocity with respect to the centre of mass.

In fact,

$$V_1^2 = \frac{a_1^2}{a^2} V^2 = \frac{GM_2^2}{M_1 + M_2}\left(\frac{2}{r} - \frac{1}{a}\right).$$ (19.65)

Substituting this into (19.64) yields the result

$$\frac{dt_e}{ds_e} = 1 - \frac{M_2}{2(M_1 + M_2)}\frac{m}{a} + \frac{GM_2}{c^2}\left(\frac{M_1 + 2M_2}{M_1 + M_2}\right)\frac{1}{r}.$$ (19.66)

In integrating (19.66) we need not be too concerned about the constant term. This may be absorbed by redefining the units of s. The problem is similar to the scaling of TDB and TDT considered in section 10.7. Again, to perform the integration, one uses the eccentric anomaly and derives the parallel equation to (10.42), namely

$$t_e = s + \frac{GM_2}{c^2 a}\left(\frac{M_1 + 2M_2}{M_1 + M_2}\right)\frac{e \sin E}{n}.$$ (19.67)

Substituting this result into (19.63) gives

$$c(t_0 - s) = r_0 + z + cT_1 \sin E + \frac{2GM_2}{c^2} \ln\left(\frac{2r_0}{r-z}\right),$$ (19.68)

where the parameter T_1 has the dimensions of time and is given by

$$T_1 = \frac{GM_2}{c^2 a}\left(\frac{M_1 + 2M_2}{M_1 + M_2}\right)\frac{eT}{2\pi}.$$ (19.69)

The last term in (19.68) is included in the analysis, but it is of marginal importance. The parameter T_1, however, can over a long series of timing be determined with an accuracy of slightly better than 10 per cent. The only unknown in (19.69) apart from the masses is the semimajor axis a. This is again eliminated by using $n^2 a^3 = \mu$. It is then found that

$$T_1 = 2.94 \times 10^{-3} \, M_2 \, \frac{(M_1 + 2M_2)}{(M_1 + M_2)} \text{ seconds,} \tag{19.70}$$

where the masses are in solar units.

The value determined for T_1 is 4.4 milliseconds. In terms of radial velocity, this parameter gives the orbital variation due to the second-order Doppler effect and the gravitational red-shift. Using this value, another condition is imposed on the masses. Since it is already known that the total mass is 2.85 $M_\odot$, the two masses can, in principle, be determined separately. Using this argument it is deduced the two masses are approximately equal within the accuracy of the determination (about 5 per cent). It can then be deduced from (19.61) that $\sin i \simeq 0.7$.

In view of these conclusions the most favoured explanation of the binary pulsar is that the system comprises two neutron stars, only one of which is active as a pulsar.

Problems

19.1 Prove that, in spite of projection effects, Kepler's second law still holds for the apparent orbit of a visual binary, in the sense that the radius vector from the primary to the companion sweeps out equal areas in equal times.

19.2 In an ellipse, any chord through the centre is termed a diameter. The locus of the midpoints of chords parallel to this diameter lie on a straight line which is called the conjugate diameter. Prove that the property of conjugate diameters is preserved under projection. Hence show that the projection of the minor axis may be identified in the apparent orbit of a visual binary from geometrical principles alone without recourse to any dynamical considerations.

19.3 The apparent orbit of a (fictitious) visual binary of period 125 years is a circle of diameter 10″. The binary has a parallax of 0″.25. The centre of the apparent orbit is 3″ from the primary star. Find the total mass of the binary system.

Explain why the data is insufficient to apply the Thiele–Innes method.

19.4 Prove that, at any point of the orbit, the position angle θ of a binary is related to the true anomaly v by the equation

$$\tan (\theta - \Omega) = \cos i \tan (v + \omega).$$

19.5 A star is observed to describe an orbit of semimajor axis 2″.6 (after correction for projection) about an invisible companion. The period is 50 years. From spectroscopic observations it is estimated that the star's mass is 2.2 solar masses. Calculate the mass of the invisible companion and the size in AU of the relative orbit, if the parallax of the star is 0″.375.

19.6 Show that the assumption that the orbital planes of double-line spectroscopic binaries are randomly oriented leads to a formal mean value of $\sin^3 i$ equal to $3\pi/16$. Explain why the larger (empirical) value of 2/3 is used in equation (19.48).

19.7 It is found that the velocity curve of a single-line spectroscopic binary can be adequately represented as

$R = 90 \cos (1.5t + 0.25) + 3 \cos (3t - 0.5)$

where R is the radial velocity in km s^{-1} and t is the time measured in days. Derive values for the orbital elements e, ω and τ and calculate the mass function.

APPENDIX A

Tensor methods

A.1   Introduction

The general theory of relativity requires us to study curved spacetimes. This is best undertaken by using the methods of the tensor calculus which allow statements to be made which are valid in any coordinate system. This is possible because the transformation laws for tensors may be precisely formulated in a quite general manner. Consequently, if a tensor's components are known in one coordinate system they may also be derived in any other using only the transformation relations between the two sets of coordinates. A tensor equation once derived is, by its nature, a relationship that is in no way dependent on a particular choice of coordinate system.

It is only occasionally necessary in the main text of this book to have recourse to tensor methods. More usually it is sufficient to make use of 'ordinary vectors'. Strictly speaking, such vectors are a special case of tensors; they are tensors (of rank one) being applied in three-dimensional Euclidean space in Cartesian coordinates. In the next section we shall discuss tensors without any of these simplifying restrictions, but first the rules of ordinary vector analysis, which result from these restrictions, will be discussed.

A vector **a** has three components and may, therefore, be written as

$$\mathbf{a} = (a_x, a_y, a_z). \tag{A.1}$$

The components depend on the particular choice of Cartesian coordinates (x, y, z). If, for example, a rotation is applied to the coordinate axes, the components of the vector are modified in exactly the same way as the coordinates themselves. Suppose that the new coordinates (x', y', z') are related to the original ones by the set of equations

$$x' = R_{11}x + R_{12}y + R_{13}z$$
$$y' = R_{21}x + R_{22}y + R_{23}z \tag{A.2}$$
$$z' = R_{31}x + R_{32}y + R_{33}z.$$

This transformation may be represented by a rotation matrix $\mathbf{R}$ whose elements appear on the right-hand side of equation (A.2). The vector's components in the new coordinate system are often written in matrix form as

$$\mathbf{a}' = \mathbf{R}\,\mathbf{a} \tag{A.3}$$

but notice that both $\mathbf{a}$ and $\mathbf{a}'$ are to be understood in this equation as column vectors, not row vectors as (A.1) might suggest.

Vectors may be added, subtracted and multiplied by a scalar quantity in an immediately obvious way. Thus the vector $K\mathbf{a} + \mathbf{b}$ has components $(Ka_x + b_x,\ Ka_y + b_y,\ Ka_z + b_z)$. The product of two vectors is defined, however, in two essentially different ways.

Definition. The *scalar product* (or inner product) of two vectors $\mathbf{a}$ and $\mathbf{b}$ is written as $\mathbf{a} \cdot \mathbf{b}$ and is the single scalar quantity

$$\mathbf{a} \cdot \mathbf{b} = a_x b_x + a_y b_y + a_z b_z. \tag{A.4}$$

Definition. The *vector product* of two vectors $\mathbf{a}$ and $\mathbf{b}$ is the vector $\mathbf{a} \times \mathbf{b}$ whose components are

$$\mathbf{a} \times \mathbf{b} = (a_y b_z - a_z b_y,\ a_z b_x - a_x b_z,\ a_x b_y - a_y b_x). \tag{A.5}$$

It is easily seen that

$$\mathbf{a} \times \mathbf{b} = -\mathbf{b} \times \mathbf{a} \tag{A.6}$$

and so the order in a vector product is crucial.

The square of the *magnitude* of a vector is derived by taking the scalar product of that vector with itself, that is, in the usual notation

$$a^2 = \mathbf{a} \cdot \mathbf{a} = a_x{}^2 + a_y{}^2 + a_z{}^2. \tag{A.7}$$

A vector is a directed quantity, however, and the angle θ between two vectors $\mathbf{a}$ and $\mathbf{b}$ is given in terms of their scalar product as

$$\mathbf{a} \cdot \mathbf{b} = ab \cos\theta. \tag{A.8}$$

Consequently the scalar product of two orthogonal vectors is zero.

The vector product $\mathbf{a} \times \mathbf{b}$, on the other hand, is a vector that is perpendicular to both $\mathbf{a}$ and $\mathbf{b}$. Its magnitude is found to be $ab \sin\theta$.

The *triple scalar product* of three vectors is written as $[\mathbf{a}, \mathbf{b}, \mathbf{c}]$. It is the scalar quantity which is defined as

$$[\mathbf{a}, \mathbf{b}, \mathbf{c}] = \mathbf{a} \cdot (\mathbf{b} \times \mathbf{c}). \tag{A.9}$$

This quantity may also by the earlier definitions be written as a third-order determinant, namely

$$[\mathbf{a}, \mathbf{b}, \mathbf{c}] = \begin{vmatrix} a_x & a_y & a_z \\ b_x & b_y & b_z \\ c_x & c_y & c_z \end{vmatrix}. \tag{A.10}$$

It will then be readily acknowledged that only the cyclical order of the vectors in the square brackets is significant.

Finally, we note that considerable use has been made in the text of triple vector products and, in particular, of the important identity

$$\mathbf{a} \times (\mathbf{b} \times \mathbf{c}) = (\mathbf{a} \cdot \mathbf{c})\mathbf{b} - (\mathbf{a} \cdot \mathbf{b})\mathbf{c}. \tag{A.11}$$

The above is a cursory listing of the main formulae of ordinary vector analysis. As mentioned earlier, they are special cases of the more general results of tensor calculus which are developed in the remainder of this appendix.

The vectors that have been discussed in this section are more properly described as Cartesian 3-vectors. While all the results discussed here may be generalized, it will not in fact be profitable to generalize vector products, which are, therefore, discussed no further.

A.2   Tensors

When vectors are introduced in a non-Euclidean space it is found that it is necessary to distinguish between two kinds of vector, a *contravariant vector* and a *covariant vector*. These will be formally defined below, but it should be noted that the former is indicated by a superscript and the latter by a subscript. It will be found that there are certain symmetries in the use of subscripts and superscripts that must be rigidly followed. The analysis is enormously simplified by the use of the dummy suffix notation. Let us begin, therefore, by stating two conventions.

Convention 1. In n-dimensional space the single appearance of a suffix (whether superscript or subscript) in each term of an equation implies n such equations, the suffix taking all values from 1 to n.

This just means in practice that a set of equations like

$$a_\lambda = b_\lambda + c_\lambda \quad \lambda = 1, \ldots, n$$

will be written simply as

$$a_\lambda = b_\lambda + c_\lambda \tag{A.12}$$

without further comment.

Convention 2. In n-dimensional space the double appearance of a suffix in any term of an equation implies the summation of that term over all n values of the suffix.

This implies that, for example

$$a^{\lambda}b_{\lambda} \equiv \sum_{\lambda=1}^{n} a^{\lambda}b_{\lambda}. \tag{A.13}$$

It will be found that in tensor equations the following rules apply and may be used as checks:

(i) If a suffix appears once in any term of an equation, it must appear in the same position (i.e. either as a subscript or superscript) in every term of that equation.

(ii) If a suffix appears twice in a term, it is a dummy suffix and should appear symmetrically in an upper and lower position [as in (A.13)].

(iii) If a suffix appears three times in any term, the notation is being used wrongly!

In n-dimensional space, a point is labelled by n real coordinates $(x^1, x^2, \ldots, x^n)$. The coordinate system may be quite general. Cartesian coordinates are a very special case; they do even exist in a non-Euclidean space. Suppose now that a transformation is made to a new set of coordinates $(\tilde{x}^1, \tilde{x}^2, \ldots, \tilde{x}^n)$. These coordinates will each be general functions of the n original coordinates. They are not restricted to linear combinations as in the transformation equations (A.2). The only requirement is that there should exist a unique inverse transformation from $\tilde{x}^{\lambda}$ to x^{λ}, so that we may write

$$\tilde{x}^{\lambda} = \tilde{x}^{\lambda}(x^1, \ldots, x^n)$$
$$x^{\lambda} = x^{\lambda}(\tilde{x}^1, \ldots, \tilde{x}^n). \tag{A.14}$$

We shall now proceed to give definitions of tensors of several different types. What defines a tensor of a particular type is the transformation rules which apply when a coordinate transformation is made. For a tensor of rank zero, this is particularly simple.

Definition. A tensor of rank zero, or a *scalar*, is a single function of the n coordinates whose *value* at any point is unchanged under a coordinate transformation.

The definitions of tensors of rank one, that is, of vectors, is a little more complicated. They are of two kinds defined as follows.

Definition. A contravariant vector, U^{λ} say, and a covariant vector, V_{λ} say, each have n components which are functions of the coordinates which transform according to the following rules

$$\tilde{U}^{\lambda} = \frac{\partial \tilde{x}^{\lambda}}{\partial x^{\alpha}} U^{\alpha} \tag{A.15}$$

$$\tilde{V}_{\lambda} = \frac{\partial x^{\alpha}}{\partial \tilde{x}^{\lambda}} V_{\alpha}. \tag{A.16}$$

It is not necessary to make the distinction between contravariant and covariant vectors in Euclidean space when only Cartesian coordinates are being used, since they are then the same. Otherwise the distinction is essential. Notice that the right-hand side of each of (A.15) and (A.16) is the sum of n terms, since the suffix α is dummy.

An important example of a contravariant vector is provided by the set of n differentials $(\mathrm{d}x^{\alpha})$. For, under a coordinate transformation the new set of differentials $(\mathrm{d}\tilde{x}^{\lambda})$ is given by

$$\mathrm{d}\tilde{x}^{\lambda} = \frac{\partial \tilde{x}^{\lambda}}{\partial x^{\alpha}} \mathrm{d}x^{\alpha} \tag{A.17}$$

which has exactly the same form as equation (A.15).

Covariant and contravariant tensors, and indeed mixed tensors, of higher rank may be defined by generalizations and combinations of the transformation rules (A.15), (A.16). There is only one higher-rank tensor which has been used in the text of this book, namely the covariant tensor of rank 2 which is defined as follows:

Definition. A covariant tensor of rank 2, $W_{\lambda\mu}$ say, has n^2 components which are functions of the coordinates which transform according to the following rule

$$\tilde{W}_{\lambda\mu} = \frac{\partial x^{\alpha}}{\partial \tilde{x}^{\lambda}} \frac{\partial x^{\beta}}{\partial \tilde{x}^{\mu}} W_{\alpha\beta}. \tag{A.18}$$

Observe that the right-hand side of this equation is summed over both α and β.

Tensors of the same kind may be combined by addition, subtraction or scalar multiplication. The result will evidently be a tensor of the same kind. A number of other operations may be performed on tensors, but only one is of immediate concern, the inner product.

Definition. The inner product of two tensors is defined formally as their product in which a contravariant suffix in one tensor and a covariant suffix in the other are suppressed by summation.

If the inner product is taken of two vectors, one must be a contravariant vector and the other a covariant vector. The inner product of two covariant or two contravariant vectors is not defined. The result of taking the inner product of two tensors is always a tensor whose rank is two less than the sum of the ranks of the two tensors.

Example. Prove that the inner product of a contravariant vector U^λ and a covariant vector V_λ is a scalar.

It is necessary to show that the value of the inner product $U^\lambda V_\lambda$ is independent of the coordinate system. Its value may be derived in one coordinate system (x'). When the transformation is made to another coordinate system $(\tilde{x}')$, the scalar product is $\tilde{U}^\lambda \tilde{V}_\lambda$, which by (A.15) and (A.16) may be written as

$$\tilde{U}^\lambda \tilde{V}_\lambda = \frac{\partial \tilde{x}^\lambda}{\partial x^\alpha} \frac{\partial x^\beta}{\partial \tilde{x}^\lambda} U^\alpha V_\beta. \tag{A.19}$$

Now let us introduce the Kronecker delta δ_v^μ with the definition that

$$\begin{aligned}
\delta_v^\mu &= 1 \quad \text{when } \mu = v \\
\delta_v^\mu &= 0 \quad \text{when } \mu \neq v.
\end{aligned} \tag{A.20}$$

Then it is easily seen that equation (A.19) becomes

$$\tilde{U}^\lambda \tilde{V}_\lambda = \frac{\partial x^\beta}{\partial x_\alpha} U^\alpha V_\beta = \delta_\alpha^{\ \beta} U^\alpha V_\beta$$

i.e.

$$\tilde{U}^\lambda \tilde{V}_\lambda = U^\alpha V_\alpha = U^\lambda V_\lambda.$$

The result is proved.

Similar methods only slightly more complicated may be used to derive the important results stated in the problems below.

Problem A.1. Show that the Kronecker delta is a mixed tensor of rank 2, whose components have the same values in every coordinate system. This means that it is necessary to prove that

$$\tilde{\delta}_v^\mu = \frac{\partial \tilde{x}^\mu}{\partial x^\alpha} \frac{\partial x^\beta}{\partial \tilde{x}^v} \delta_\beta^{\ \alpha}. \tag{A.21}$$

Problem A.2. Prove that the inner product of a covariant tensor of rank 2 and a contravariant vector is a covariant vector.

A.3   The metrical tensor

So far nothing has been said and nothing has been inferred about the nature of the geometry of the space under discussion. This is contained in a fundamental tensor $g_{\mu v}$ which is known as the *metrical tensor*. As the notation indicates, it is a covariant tensor of rank 2. It is also symmetric, meaning that

$$g_{\mu v} = g_{v \mu}. \tag{A.22}$$

The metrical tensor determines the geometry of the space through the metric which may, with sufficient generality for our purposes, be written as

$$ds^2 = g_{\mu\nu}\, dx^{\mu}\, dx^{\nu}. \tag{A.23}$$

A space with this form of the metric, that is, a quadratic form in the differentials, is said to be *Riemannian*.

The right-hand side of equation (A.23) is formed by taking the inner product of the metrical tensor twice with the differentials, which are components of a contravariant vector. The quantity ds^2, and therefore ds, is thus seen to be a scalar. The latter is known as the *element of interval*; in a pure space it is equivalent to an element of length. If the components of $g_{\mu\nu}$ are evaluated at the point with coordinates (x^i), then the metric allows the element of interval between two neighbouring points (x^i) and $(x^i + dx^i)$ to be derived.

The metric of three-dimensional Euclidean space may be written as

$$ds^2 = dx^2 + dy^2 + dz^2 \tag{A.24}$$

where (x, y, z) have replaced the formal coordinates (x^1, x^2, x^3). Two points should be noted. Firstly, this simple form of the metric is a consequence of using Cartesian coordinates and, secondly, such coordinates do not exist in a more general curved space. Consider, for example, the following two problems:

Problem A.3. Prove that the metric of three-dimensional Euclidean space in spherical polar coordinates (r, θ, ϕ) is

$$ds^2 = dr^2 + r^2\, d\theta^2 + r^2 \sin^2 \theta\, d\phi^2. \tag{A.25}$$

Problem A.4. By considering the subspace $r = 1$ of three-dimensional Euclidean space, show the metric on the unit sphere is

$$ds^2 = d\theta^2 + \sin^2 \theta\, d\phi^2. \tag{A.26}$$

Consider now any contravariant vector V^{λ}. A covariant vector may be formed by taking the inner product with the metrical tensor. The resulting new vector and the original one are called *associated vectors*, and they are usually denoted by the same symbol, i.e.

$$V_{\lambda} = g_{\lambda\mu} V^{\mu} \tag{A.27}$$

because they are regarded as different forms of the same vector. This is taken one step further by the following definition.

Definition. The magnitude squared of a vector is the inner product of its contravariant and covariant forms. Hence

$$V^2 = V^{\lambda} V_{\lambda} = g_{\lambda\mu} V^{\lambda} V^{\mu}. \tag{A.28}$$

The magnitude of the vector – the positive square root of V^2 – is thus seen to be a scalar.

Suppose now that we have two vectors a^λ and b^λ defined at the same point of the space. The magnitudes a and b of these vectors may be derived from the definition just given, and the angle θ between the two vectors is defined by the equation

$$ab \cos \theta = a^\lambda b_\lambda. \tag{A.29}$$

Problem A.5. Consider the contravariant vector defined as

$$V^\lambda = (\dot{x}, \dot{y}, \dot{z})$$

in Cartesian coordinates. Derive the components of the associated vector V_λ in spherical polar coordinates and show that the magnitude of this vector is given by

$$V^2 = \dot{r}^2 + r^2 \, \dot{\theta}^2 + r^2 \sin^2 \theta \, \dot{\phi}^2.$$

Problem A.6. Prove that the proper motion components of a star in right ascension and declination (μ_α, μ_δ) may be regarded as the components of a contravariant vector and that the magnitude of this vector is given by

$$\mu^2 = \cos^2 \delta \, \mu_\alpha{}^2 + \mu_\delta{}^2.$$

A.4   Geodesics

A curve in n-dimensional space is a continuous one-parameter family of points, which may be written as

$$x^\lambda = f_\lambda(p), \tag{A.30}$$

indicating that each of the coordinates of points on the curve is a function of the curve parameter p. Now let A and B be two points of a curve Γ at which the curve parameter has values P_A and P_B respectively. Then the length of the curve from A to B, strictly the interval, may be calculated by integrating ds along the curve. We may, therefore, write this interval as

$$s = \int_{\Gamma A}^{B} \mathrm{d}s = \int_{P_A}^{P_B} \left(g_{\mu\nu} \frac{\mathrm{d}x^\mu}{\mathrm{d}p} \frac{\mathrm{d}x^\nu}{\mathrm{d}p} \right)^{1/2} \mathrm{d}p. \tag{A.31}$$

The curve parameter p may be chosen in many ways, and it is generally possible, and most convenient, to adopt s itself as the curve parameter. The interval from A to B that is derived will not, of course, depend on the choice of curve parameter, but it will depend on the curve Γ that links the two points. The curve that yields the minimum interval is particularly significant and is termed the *geodesic*.

It is possible to show, e.g. McVittie (1965), that the geodesics of the space must satisfy the n differential equations

$$\frac{\mathrm{d}}{\mathrm{d}p} \left(g_{\lambda\mu} \frac{\mathrm{d}x^\mu}{\mathrm{d}p} \right) - \frac{1}{2} \frac{\partial g_{\mu\nu}}{\partial x^\lambda} \frac{\mathrm{d}x^\mu}{\mathrm{d}p} \frac{\mathrm{d}x^\nu}{\mathrm{d}p} = 0. \tag{A.32}$$

This curve parameter p cannot, however, be chosen in an arbitrary way; it must be an *affine parameter*, which means that it is proportional to the interval along the geodesic measured from an arbitrarily chosen point.

Each of the n equations (A.32) is a second-order differential equation. Integration of the geodesic differential equations will, therefore, introduce $2n$ constants of integration. This indicates that it is possible to require the geodesic to go through two arbitrarily selected points, whose coordinates will determine the constants of integration. Finally the adopted curve parameter may be related to the interval s through the metric, which may be written as

$$\left(\frac{ds}{dp}\right)^2 = g_{\mu\nu} \frac{dx^\mu}{dp} \frac{dx^\nu}{dp}. \tag{A.33}$$

Problem A.7. Write down the geodesic differential equations for the unit sphere using the metric given in equation (A.26). Show that the equator $\theta = \frac{1}{2}\pi$ and the meridians $\phi = $ constant are geodesics.

A.5   Curved spacetimes

Up to this point, it has been tacitly assumed that the quadratic form of the metric

$$ds^2 = g_{\mu\nu} \, dx^\mu \, dx^\nu \tag{A.23}$$

is positive definite at all points. This corresponds to a pure space. As a consequence the element of interval is of necessity real, as are the magnitudes of all vectors. In general relativity, however, four-dimensional spacetimes are investigated and this is no longer the case. The points of the spacetime represent events with three spatial coordinates and one time coordinate. It is now necessary to distinguish between intervals of length and of time. Each may be expressed in terms of the interval s, but in one case this interval is real and in the other it is imaginary.

The situation is best examined by using the simple metric of special relativity in which space is Euclidean. The most natural extension of the metric (A.24) to a spacetime would be to write

$$ds^2 = dx^2 + dy^2 + dz^2 - c^2 \, dt^2.$$

Then space-like intervals would be real and time-like intervals imaginary. While this is a perfectly valid formulation, the opposite convention has been preferred in this book for this reason. Time intervals may be measured directly by the astronomer, but astronomical distances are always inferred either from timings or from angular measurements. The metric that has been adopted for special relativity, therefore, has the form

$$ds^2 = dt^2 - \frac{1}{c}(dx^2 + dy^2 + dz^2). \tag{A.34}$$

It is then seen that ds has the dimensions of time and is real for time-like intervals.

The same considerations pertain to the curved spacetimes of general relativity. The interval between two distinct neighbouring events may be real, imaginary or zero. The interval is then described as being time-like, space-like or null respectively. The same terms may be applied to vectors according as their magnitudes are real, imaginary or zero.

All the analysis of the previous three sections is still valid in a spacetime as opposed to a space. In particular, the geodesic differential equations (A.32) are unchanged. There are now, however, three types of geodesic; they may be time-like, space-like or null, depending on the nature of the interval along them. A curve in spacetime, and in particular a geodesic, is a continuous sequence of events. If these events correspond exactly to those experienced by an observer, then the curve is referred to as the observer's *world line*.

Time-like and null geodesics are very important in general relativity. They correspond respectively to the world lines of material particles (or observers) and of photons moving freely in the gravitational field. We shall have no use for space-like geodesics as such, but space-like unit vectors are used to represent directions. The following definitions will make this clear.

Definition. Two vectors are said to be orthogonal if their inner product is zero.

Definition. A vector will be described as a unit vector if its magnitude squared is equal to either $+1$ or -1. For a space-like vector the magnitude will be interpreted as i times a positive number.

Now for any curve we may define a tangent vector q^i at any point as

$$q^i = \frac{dx^i}{dp} \tag{A.35}$$

where p is a curve parameter. If this curve is the world line of a real observer, it is a time-like curve and the tangent vector q^i is also time-like. We now make the following important statement for the interpretation of general relativity in relation to positional astronomy.

Statement. A direction, e.g. a coordinate axis, for an observer may be defined as a space-like unit vector which is orthogonal to the observer's world line.

The angle between two vectors was defined in equation (A.29), and this definition, which is still valid, may also be written as

$$ab \cos \theta = g_{\lambda\mu}a^{\lambda}b^{\mu}. \tag{A.36}$$

Since general relativity can always be matched locally with special relativity, the result of the following problem has general significance.

Problem A.8. Prove that, in special relativity, the angle between two space-like vectors may be real, but that between two time-like vectors is imaginary.

This appendix has, of necessity, given a somewhat cursory and selective description of tensors methods. The reader desiring a fuller account of these, and indeed of general relativity, should consult one of the more specialized texts on the subject, e.g. Schutz (1984).

Astronomical constants

B.1 Fundamental constants

The most recent revision of the system of astronomical constants was approved by the IAU in 1976. This system (IAU, 1977) has been used in the compilation of annual almanacs from 1984 onwards. The *Astronomical Almanac* for that year included a special supplement which explained the new system of constants very fully and detailed the changes that were necessary from earlier volumes of the almanac (*Astronomical Almanac Supplement*, 1984).

The transition had several distinct features. In addition to the use of new fundamental constants, the basis of the computation of planetary ephemerides has been altered, and a new time-scale has been introduced. This is TDT which is defined as

$$\text{TDT} = \text{TAI} + 32^s.184. \tag{B.1}$$

Moreover, the relationship between UT and sidereal time has been restated as

$$\text{GMST at } 0^h \text{ UT1} = 6^h 41^m 50^s.548\,41$$
$$+ 8\,640\,184^s.812\,866\ T_U \tag{B.2}$$
$$+ 0^s.093\,104\ T_U{}^2 - 6^s.2 \times 10^{-6}\ T_U{}^3$$

where T_U is the time in Julian centuries of 36 525 days that has elapsed from 2000 January 1, 12^h UT1 (JD 2 451 545.0). The corresponding date in TDB is the new standard epoch J2000.0, and in general a Julian epoch is to be used, defined as

$$\text{J2000.0} + (\text{JD} - 2\,451\,545.0)/365.25. \tag{B.3}$$

The constants in the IAU system are of three kinds. The distinction is made between defining constants – there are only two, c and k – and primary and derived constants. Table B.1 is based on the IAU system, but

Table B.1. *Fundamental astronomical data (J2000.0)*

Velocity of light	$c = 2.99792458 \times 10^5$ km s^{-1}
Gaussian gravitational constant	$k = 0.01720209895$
Constant of gravitation	$G = 6.672 \times 10^{-11}$ m^3 kg^{-1} s^{-2}
Mass of the sun	$M_\odot = 1.9891 \times 10^{30}$ kg
Heliocentric gravitational constant	$GM_\odot = 1.32712438 \times 10^{20}$ m^3 s^{-2}
Astronomical unit	1 AU $= 1.49597870 \times 10^8$ km
Solar parallax	$\pi_\odot = 8''.794148$
Solar radius	$R_\odot = 6.96 \times 10^5$ km
Mass of the earth	$M_\oplus = 5.974 \times 10^{24}$ kg
Geocentric gravitational constant	$GM_\oplus = 3.986005 \times 10^{14}$ m^3 s^{-2}
Equatorial radius of the earth	$R_\oplus = 6.378140 \times 10^3$ km
Flattening factor for the earth	$f = 0.00335281$
Mass of the moon	$M_{\mathrm{C}} = 0.01230002\, M_\oplus$
Radius of the moon	$R_{\mathrm{C}} = 1.738 \times 10^3$ km
Obliquity of the ecliptic	$\varepsilon = 23° 26' 21''.448$
Annual general precession in longitude	$p = 50''.290966$
Annual precession in right ascension	$m = 3^s.0749575$
Annual precession in declination	$n = 20''.043109$
Constant of nutation	$N = 9''.2025$
Constant of aberration	$\kappa = 20''.49552$
Equatorial coordinates of the galactic pole	$(12^h 51^m.3, 27° 07')$
Equatorial coordinates of the galactic centre	$(17^h 45^m.4, -28° 56')$
Equatorial coordinates of the solar apex	$(18^h 08^m, 30°.5)$

the distinction between primary and derived constants has been omitted. Moreover, some further derived values have been included together with some additional material connected with galactic coordinates and the solar apex. Almost all the values quoted have already appeared within the text of this book, but they are collected here for easy reference.

All the data in Table B.1 is standardized to the epoch J2000.0. The equatorial coordinates quoted are referred to this standard equator and equinox, and, where constants have a secular variation, the values quoted correspond to those at J2000.0. It will be noticed that the gravitational constants for the sun and the earth are quoted to many more significant figures than their masses. The reason for the comparatively low precision in the sun and the earth's mass is the uncertainty in the constant of gravitation G in SI units.

B.2 Planetary data

The IAU system of constants contains some recommendations concerning the planets' dimensions and physical characteristics. These have

504

Table B.2. Physical and rotational elements of the planets

Planet	Equatorial radius (km)	Flattening	Reciprocal mass ($M_\odot^{-1}$)	Mean density (kg m^{-3})	Period of axial rotation	Orbital period in Julian years	Inclination of equator to orbit
Mercury	2439	0	6 023 600	5433	$58^{\mathrm{d}}.65$	0.240 85	$0°.01$
Venus	6051	0	408 523.5	5246	$243^{\mathrm{d}}.01$	0.615 19	$177°.34$
Earth	6378.14	0.003 352 81	332 946.0	5515	$23^{\mathrm{h}} 56^{\mathrm{m}}$	1.0000	$23°.44$
Mars	3393.4	0.005 186 5	3 098 710	3942	$24^{\mathrm{h}} 37^{\mathrm{m}}$	1.8807	$25°.19$
Jupiter	71 398	0.064 808 8	1 047.355	1332	$9^{\mathrm{h}} 55^{\mathrm{m}}$	11.861	$3°.13$
Saturn	60 000	0.107 620 9	3 498.5	704	$10^{\mathrm{h}} 39^{\mathrm{m}}$	29.570	$26°.75$
Uranus	25 400	0.030	22 869	1310	$15^{\mathrm{h}} 34^{\mathrm{m}}$	84.746	$97°.86$
Neptune	24 300	0.0259	19 314	1760	$18^{\mathrm{h}} 26^{\mathrm{m}}$	166.57	$29°.55$
Pluto	1500	0	130 000 000	1100	$6^{\mathrm{d}}.39$	249.98	$117°.57$

Table B.3. Osculating elements of planets at the epoch 1985 January 15.0 TDT, referred to the ecliptic and mean equinox of J2000.0

Planet	Semimajor axis in AU a	Mean motion degrees per day n	Eccentricity e	Inclination i	Longitude of the ascending node Ω	Longitude of perihelion ϖ	Mean longitude at epoch L_0
Mercury	0.387 099	4.092 34	0.205 629	$7°.0059$	$48°.350$	$77°.434$	$209°.6643$
Venus	0.723 326	1.602 15	0.006 772	$3°.3947$	$76°.723$	$131°.70$	$67°.1413$
Earth	1.000 018	0.985 582	0.016 773	$0°.0020$	$354°.9$	$102°.972$	$114°.7122$
Mars	1.523 638	0.524 061	0.093 298	$1°.8509$	$49°.604$	$335°.942$	$11°.8733$
Jupiter	5.202 48	0.083 099 1	0.048 058	$1°.3047$	$100°.467$	$15°.662$	$300°.3429$
Saturn	9.563 29	0.033 331 5	0.050 916	$2°.4846$	$113°.713$	$93°.444$	$227°.2512$
Uranus	19.293 7	0.011 630 3	0.047 285	$0°.7746$	$74°.056$	$177°.048$	$249°.2488$
Neptune	30.274 3	0.005 917 02	0.006 851	$1°.7693$	$131°.800$	$357°.07$	$272°.2648$
Pluto	39.682 3	0.003 942 83	0.252 786	$17°.1313$	$110°.421$	$224°.499$	$217°.4574$

been slightly modified since 1976 (IAU, 1983a), and Table B.2 is based on these modified recommendations. In particular, the equatorial radii and flattening correspond to the reference spheroids which are used in current mapping programmes. Some derived quantities such as the mean densities have been added.

The last two columns in Table B.2 depend on the planet's orbital elements, which are given in Table B.3. These are osculating elements for the date 1985 January 15.0 TDB[δ]. They are referred to the ecliptic and equinox of J2000.0. Such osculating elements will only allow accurate positions of the planets to be determined over comparatively short intervals of time. The *Astronomical Almanac* provides such osculating elements at 40-day intervals and, of course, the current values should be used for any precise work. If the values in Table B.3 are used for times far from the stated epoch, say a few years, then the derived positions of the planets may, in unfavourable cases, be in error by as much as a degree. Using the data of Table B.3 over a prolonged period would allow a general search area for the planet to be established, but little more.

The orbital elements are, in fact, secondary quantities. The planetary and lunar ephemerides in the *Astronomical Almanac* are based on a numerical integration (denoted by DE 200/LE 200) of the entire solar system which covers the years 1800 to 2050. By this means the position and velocity vectors are computed in rectangular coordinates referred to the equator and equinox of J2000.0. The osculating elements are then computed from the instantaneous position and velocity vectors by the methods considered in section 7.6.

Answers to the numerical problems

Chapter 1

1.5   5869 nautical miles, 63° 35′ W.

1.10   4460 nautical miles,
       56° 34′ west of north (48° 39′ N, 169° 38′ W).

Chapter 2

2.9   60° 11′, 6° 23′ and 186° 23′.

2.10 $-1°.4$.

Chapter 3

3.4   79°.

Chapter 4

4.2 30^{d}.

4.5 $\ln n_0$, $r_0(n_0 - 1)$.

4.8   6363.5 km, 55° 41′, 10′ 44″.

4.9   6° 51′.

4.11   Astronomical and geocentric.

Chapter 5

5.3 $-5^{\mathrm{s}}.21$.

Chapter 6

6.3   1984 November 7.7.

6.4 3.778×10^5 km.

6.5   79° 51′.

6.10   23.08 km.

Chapter 7

7.1 $19\,322\, M_\odot^{-1}$.

7.2 $408\,540\, M_\odot^{-1}$.

7.3 $(55° 13' 38'', -4° 36' 26'')$.
7.4 $(2^h 47^m 32^s.9, 9° 15' 29'')$.
7.8   2.7668 AU, $80°.718$, $10°.606$,
       0.0784, $72°.841$, 1986 February 5.2.

Chapter 8
8.1 $0''.4$, June 21, March 21, $5\sqrt{2}/2$ pc.

Chapter 9
9.2 $(2^h 16^m 22^s.7, 89° 11' 47''.3)$.
9.3 $-345''.926$, $-345''.908$, $-300''.656$.
9.6 $(2^h 16^m 26^s.5, 89° 11' 46''.9)$.
9.7 $(2^h 16^m 22^s.7, 89° 11' 47''.3)$.
9.10 2102.4, $89° 32' 27''.1$.

Chapter 10
10.2 $p = +50''.4$, $\Delta\varpi = +61''.7$.
10.3 1985 March 2 $11^h 15^m$, 1985 March 2 $19^h 40^m$.

Chapter 11
11.2 11830 AD $(17^h 34^m 11^s, 54° 52')$,
       $25''.35$, $0''.859$.
11.3 $V_T \gg V_r$.
11.5 0.0125 Å.
11.6 $(3^h 32^m.0, 54° 32')$.

Chapter 12
12.4 $(5^h 58^m 43^s.156, -0° 02' 31''.40)$.
12.5 $-0.003\,18$, $-0.130\,52$, $-0.137\,32$, $-0.164\,73$,
       $-0.640\,29$, $0.768\,13$, $0.864\,86$, $-0.608\,57$.

Chapter 13
13.5 $(-0.007\,116, 0.008\,171)$, $(-0.002\,692, -0.000\,895)$,
       $(0.008\,613, 0.000\,952)$.
13.6 $0.227\,05$, $0.428\,29$, $0.344\,67$.
13.7 $(3^h 47^m 03^s.3, 24° 06' 09'')$.

Chapter 14
14.1 $(17^h 42^m 26^s.6, -28° 55' 00'')$.

14.6 $\begin{pmatrix} -0.066\,989 & -0.872\,756 & -0.483\,539 \\ 0.492\,728 & -0.450\,347 & 0.744\,585 \\ -0.867\,601 & -0.188\,375 & 0.460\,200 \end{pmatrix}$.

14.7 Julian and Besselian epochs of the same date use slightly different
       equinoxes, cf. equation (12.56).

Chapter 15

15.3 36′.

15.4 57^m.

Chapter 16

16.1 364 145.46 radians. 27°.1.

16.5 Only relative right ascension can be derived.

Chapter 17

17.1 41°.1, 0.876, 11°.1, 0.991.

17.2 Retrograde at inferior conjunction, direct at maximum elongation.

17.4 0.4306 AU, 39°.72.

17.5 2.77 AU.

17.8 26°.32.

Chapter 18

18.2 Yes, Yes, No.

18.5 55° 43′ 29″, 0.997 707.

18.6 (0.555 215, 0.797 213, 0.227 171),
($-0.022\,587$, 0.054 144, $-0.134\,805$).

18.7 17^h 14^m 42^s, 18^h 52^m 54^s.

18.10 May 9, November 8, June 7, December 6.

Chapter 19

19.3 1 $M_\odot$.

19.5 1.14 $M_\odot$, 20.29 AU.

19.7 1/30, 1 radian, 0^d.5, 0.316 $M_\odot$.

REFERENCES

Allen, C. W. (1976). *Astrophysical Quantities*, 3rd edn. London: The Athlone Press.

APFS. *Apparent Places of the Fundamental Stars*. Annual publication issued by Astronomisches Rechen-Institut in Heidelberg. Karlsruhe: Verlag G. Braun.

Astronomical Almanac. Annual publication. Washington: US Government Printing Office, and London: Her Majesty's Stationery Office.

Astronomical Almanac Supplement (1984). The Improved IAU System. In the *Astronomical Almanac (1984)*, Section S. Washington: US Government Printing Office, and London: Her Majesty's Stationery Office.

Christiansen, W. M. and Högbom, J. A. (1984). *Radio Telescopes*, 2nd edn. Cambridge University Press.

Counselman, C. C. (1976). Radio Astrometry. *Annual Reviews of Astronomy and Astrophysics*, **14**, 197.

Danjon, A. (1960). The Impersonal Astrolabe. In *Stars and Stellar Systems*, Vol. 1, *Telescopes*, ed. G. P. Kuiper and B. M. Middlehurst, pp. 115–137. University of Chicago Press.

Duffett-Smith, P. (1981). *Practical Astronomy with Your Calculator*, 2nd edn. Cambridge University Press.

Explanatory Supplement (1974). *Explanatory Supplement to the Astronomical Ephemeris and the American Ephemeris and Nautical Almanac*. London: Her Majesty's Stationery Office.

FK4 – Fricke, W. and Kopff, A. (1963). Fourth Fundamental Catalogue (FK4). *Veröffentlichungen Astronomisches Rechen-Institut:* Heidelberg.

Fricke, W. (1980). On the Determination of the Equator and Equinox of the New Fundamental Reference Coordinate System, the FK5. *Celestial Mechanics*, **22**, 113.

Handbook of the British Astronomical Association. Annual publication. London: British Astronomical Association.

Harris, D. L. (1961). Photometry and Colorimetry of Planets and Satellites. In *Planets and Satellites*, ed. G. P. Kuiper and B. M. Middlehurst, pp. 272–342. University of Chicago Press.

IAU (1977). *Transactions of the International Astronomical Union*, **XVI** B, 58.

IAU (1982). *Transactions of the International Astronomical Union*, **XVIII** A, 8.

IAU (1983). *Transactions of the International Astronomical Union*, **XVIII** B, 67.
IAU (1983a). *Transactions of the International Astronomical Union*, **XVIII** B, 147.
McVittie, G. C. (1965). *General Relativity and Cosmology*, 2nd edn. London: Chapman and Hall.
Markowitz, W. (1960). The Photographic Zenith Tube and Dual-Rate Moon Position Camera. In *Stars and Stellar Systems*, Vol. 1, *Telescopes*, ed. G. P. Kuiper and B. M. Middlehurst, pp. 88–114. University of Chicago Press.
Mihalas, D. M. and Binney, J. J. (1981). *Galactic Astronomy – Structure and Kinematics of Galaxies*. Oxford: W. H. Freeman and Company.
Murray, C. A. (1983). *Vectorial Astrometry*. Bristol: Adam Hilger Ltd.
Newcomb, S. (1895). Tables of the Sun. *Astronomical Papers of the American Ephemeris*, VI. Part 1. Washington: US Government Printing Office.
Orlov, B. A. (1956). *Refraction Tables of the Pulkova Observatory*, 4th edn. Moscow, Leningrad: Academy of Sciences Press.
Pound, R. V. and Snider, J. L. (1965). Effect of Gravity on Gamma Radiation. *Physical Review*, **140B**, 788.
Roy, A. E. (1982). *Orbital Motion*, 2nd edn. Bristol: Adam Hilger Ltd.
Ryle, M. and Elsmore, B. (1973). Astrometry with the 5-km Radio Telescope. *Monthly Notices of the Royal Astronomical Society*, **164**, 223.
Sadler, D. H. and Clemence, G. M. (1954). *The Improved Lunar Ephemeris 1952–1959*. Joint Supplement to the American Ephemeris and the (British) Nautical Almanac. Washington: US Government Printing Office.
Schutz, B. (1984). *A First Course in General Relativity*. Cambridge University Press.
Sky and Telescope. Monthly publication. Cambridge, Mass.: Sky Publishing Corporation.
Smart, W. M. (1977). *Textbook on Spherical Astronomy*, 6th edn. Cambridge University Press.
Will, C. M. (1981). *Theory and Experiment in Gravitational Physics*. Cambridge University Press.
Woolard, E. W. and Clemence, G. M. (1966). *Spherical Astronomy*. New York: Academic Press.

INDEX

aberration, 61–2, 67–9, 75–9, 206, 315, 329, 409; constant of, 192, 270; diurnal, 82, 109–11, 122–3, 179, 317; ellipse, 192; planetary, 167, 179, 190, 193–5, 300, 325; relativistic, 298, 302; second order, 69, 204, 317; secular, 190

absolute: magnitude, 336–8; proper motion, 339

advance: of apse line, 486; of periastron, 486–7; of perigee, 175; of perihelion, 158–9

affine parameter, 196, 499

Algol, 412

Allen C. W., 356, 404

alt-azimuth coordinate system, 23–5, 28

altitude, 25; parallel of, 25

analogue formula, 10, 12

angle: position, 262, 469; phase, 415, 417, 419–20, 422, 437; precessional, 217–20, 234; of refraction, 84–6; in relativity, 498, 501; spherical, 3; of vertical, 97

angular diameter, 424, 430; of sun, 207, 433

angular momentum, 139, 147, 153, 159, 177; vector, 209, 227, 426

annual: equation, 174–5; proper motion, 260–2; radial velocity variation, 271; rates of precession, 209–12; variations in RA and dec., 286

annual aberration, 59, 167, 189–90, 204, 290, 316–7, 324; effect on ecliptic coordinates, 191–2; effect on RA and dec., 190

annual parallax, 59, 187–9, 204, 290, 313; effect on ecliptic coordinates, 191; effect on RA and dec., 189

annular eclipse, 441, 457–8, 483

anomalistic: month, 173; year, 241

anomaly: eccentric, 142, 144–6, 154–5, 159, 177, 470, 472, 488; mean, 142, 147–8,

anomaly: eccentric, *continued*
164, 178, 472; true, 140, 147–8, 158, 164, 178, 469–70, 472

antisolar point, 442, 444, 447

aperture synthesis, 382–7

apex, solar, 277, 346–7, 349–50, 360

aphelion, 35, 141

Apollo astronauts, 106, 175

apparent: brightness, 415–6; coordinates, 113, 166; magnitude, 335–8; orbit, 470–1; place, 122, 202–5, 290–302; sidereal time, 240; solar day, 35; solar time, 34–9

Apparent Places of Fundamental Stars, 113, 230, 304

approximate precession formulae, 222–5

argument: of periastron, 469, 477; of perigee, 164; of perihelion, 140, 158, 163, 178; of prime meridian, 426, 429

artificial satellite, 108, 123, 169, 181; geocentric place, 108–9

ascending node, 162; longitude, 162; position angle, 469, 473–4; right ascension, 164

asteroid, 160; orbit, 178–82

astrographic plate measurement, 276–7, 307, 339–42

astrolabe, 130–2, 135–6

astrometric: binary, 480; place, 325

astronomical: latitude, 97, 126; zenith, 96, 126

Astronomical Almanac, 39, 182, 224, 295, 453, 483

Astronomical Almanac Supplement, 502

astronomical constants, 503

astronomical distance scale, 342–3

astronomical unit (AU), 35, 105, 167

atmosphere: plane-parallel, 82–7, 401–3; radially symmetric, 87–93; scale height, 90–2; standard conditions, 85